LASER-INDUCED BREAKDOWN SPECTROSCOPY (LIBS)

Laser-induced breakdown spectroscopy (LIBS) is an emerging technique for determining elemental composition in real-time. With the ability to analyze and identify chemical and biological materials in solids, liquids, and gaseous forms with little or no sample preparation, it is more versatile than conventional methods and is ideal for on-site analysis.

This is the first comprehensive reference book explaining the fundamentals of the LIBS phenomenon, its history, and its fascinating applications across 18 chapters written by recognized leaders in the field. Over 300 illustrations aid understanding.

This book will be of significant interest to researchers in chemical and materials analysis within academia, government, military, and industry.

ANDRZEJ W. MIZIOLEK is a Senior Research Physicist at the US Army Research Laboratory. His work is currently concentrated on nanomaterials research and on the development of the LIBS sensor technology.

VINCENZO PALLESCHI is a researcher in the Institute for Chemical-Physical Processes at the Italian National Research Council and, in particular, the Applied Laser Spectroscopy Laboratory.

ISRAEL SCHECHTER is Professor of Chemistry in the Department of Chemistry at the Technion–Israel Institute of Technology. His main scientific interest is in new methods for fast analysis of particulate materials.

LASER-INDUCED BREAKDOWN SPECTROSCOPY (LIBS)

Fundamentals and Applications

Edited by

ANDRZEJ W. MIZIOLEK

US Army Research Laboratory

VINCENZO PALLESCHI

Instituto per i Processi Chimico-Fisici, Italy

ISRAEL SCHECHTER

Technion–Israel Institute of Technology, Haifa, Israel

CAMBRIDGE
UNIVERSITY PRESS

CAMBRIDGE UNIVERSITY PRESS
Cambridge, New York, Melbourne, Madrid, Cape Town, Singapore, São Paulo

Cambridge University Press
The Edinburgh Building, Cambridge CB2 8RU, UK

Published in the United States of America by Cambridge University Press, New York

www.cambridge.org
Information on this title: www.cambridge.org/9780521852746

First published 2006
This digitally printed version 2008

A catalogue record for this publication is available from the British Library

ISBN 978-0-521-85274-6 hardback
ISBN 978-0-521-07100-0 paperback

Contents

		page
List of contributors		x
Preface		xv
1	History and fundamentals of LIBS	1
	David A. Cremers and Leon J. Radziemski	
	1.1 Introduction	1
	1.2 Basic principles	1
	1.3 Characteristics of LIBS	5
	1.4 LIBS as an analytical technique	17
	1.5 Early LIBS instruments	27
	1.6 Components for a LIBS apparatus	30
	1.7 Conclusion	36
	1.8 References	36
2	Plasma morphology	40
	Israel Schechter and Valery Bulatov	
	2.1 Introduction	40
	2.2 Experimental imaging techniques	41
	2.3 Time-integrated morphology	59
	2.4 Time-resolved morphology: excitation by medium laser pulses (1–100 ns)	73
	2.5 Time-resolved morphology: excitation by long laser pulses (>100 ns)	110
	2.6 Time-resolved morphology: excitation by short laser pulses (fs–ps)	112
	2.7 Time-resolved morphology: excitation by double laser pulses	113
	2.8 Conclusions	118
	2.9 References	118
3	From sample to signal in laser-induced breakdown spectroscopy: a complex route to quantitative analysis	122
	E. Tognoni, V. Palleschi, M. Corsi, G. Cristoforetti, N. Omenetto, I. Gornushkin, B. W. Smith and J. D. Winefordner	
	3.1 Introduction	122
	3.2 The characteristics of laser-induced plasmas and their influence on quantitative LIBS analysis	123

3.3	Quantitative analysis	148
3.4	Conclusions	164
3.5	Appendix. Table of representative limits of detection	166
3.6	References	167
4	**Laser-induced breakdown in gases: experiments and simulation**	**171**
	Christian G. Parigger	
4.1	Introduction	171
4.2	Laser-induced ignition application	172
4.3	Focal volume irradiance distribution	173
4.4	Hydrogen Balmer series atomic spectra	176
4.5	Diatomic molecular emission spectra	177
4.6	Simulation by use of the program NEQAIR	179
4.7	Computational fluid dynamic simulations	186
4.8	Summary	189
4.9	References	191
5	**Analysis of aerosols by LIBS**	**194**
	Ulrich Panne and David Hahn	
5.1	Introduction to aerosol science	194
5.2	Laser-induced breakdown of gases	209
5.3	Analysis of aerosols by LIBS	217
5.4	Applications of aerosol analysis by LIBS	242
5.5	Future directions	245
5.6	References	245
6	**Chemical imaging of surfaces using LIBS**	**254**
	J. M. Vadillo and J. J. Laserna	
6.1	Introduction	254
6.2	LIBS chemical imaging: operational modes	255
6.3	Spatial resolution in LIBS imaging	258
6.4	Applications of LIBS imaging	262
6.5	Concluding remarks and outlook	277
6.6	References	279
7	**Biomedical applications of LIBS**	**282**
	Helmut H. Telle and Ota Samek	
7.1	Introduction	282
7.2	Investigation of calcified tissue materials	283
7.3	Investigation of "soft" tissue materials with cell structure	295
7.4	Investigation of bio-fluids	301
7.5	Investigation of microscopic bio-samples	304
7.6	Concluding remarks	309
7.7	References	309

8   LIBS for the analysis of pharmaceutical materials   314
    Simon Béchard and Yves Mouget
    8.1   Introduction   314
    8.2   Needs of the pharmaceutical industry   316
    8.3   Comparison of LIBS with the current technologies   317
    8.4   Components of a LIBS instrument for applications in the
          pharmaceutical industry   319
    8.5   Applications of LIBS to the analysis of pharmaceutical
          materials   323
    8.6   Conclusions   330
    8.7   References   331
9   Cultural heritage applications of LIBS   332
    Demetrios Anglos and John C. Miller
    9.1   Introduction   332
    9.2   Art and analytical chemistry   333
    9.3   Why LIBS in cultural heritage?   333
    9.4   Physical principles   335
    9.5   Instrumentation   336
    9.6   Analytical parameters and methodology   338
    9.7   Examples of LIBS analysis in art and archaeology   344
    9.8   LIBS in combinations with other techniques   357
    9.9   Concluding remarks   363
    9.10 References   363
10 Civilian and military environmental contamination studies using LIBS   368
    J. P. Singh, F. Y. Yueh, V. N. Rai, R. Harmon, S. Beaton,
    P. French, F. C. DeLucia, Jr., B. Peterson, K. L. McNesby and
    A. W. Miziolek
    10.1 Introduction   368
    10.2 Applications of the ADA portable LIBS unit   370
    10.3 Applications of DIAL's portable LIBS system   381
    10.4 Conclusion   396
    10.5 References   396
11 Industrial applications of LIBS   400
    Reinhard Noll, Volker Sturm, Michael Stepputat, Andrew Whitehouse,
    James Young and Philip Evans
    11.1 Introduction   400
    11.2 Metals and alloys processing   400
    11.3 Scrap material sorting and recycling   409
    11.4 Nuclear power generation and spent fuel reprocessing   417
    11.5 Miscellaneous industrial applications of LIBS   435
    11.6 References   436

12 Resonance-enhanced LIBS   440
    N. H. Cheung
    12.1 Introduction to resonance-enhanced LIBS   440
    12.2 Basic principles of spectrochemical excitation in laser-induced
          plasmas   441
    12.3 RELIPS analysis of solids   451
    12.4 Liquid samples   463
    12.5 Gaseous samples   473
    12.6 Conclusion: resonance-enhanced LIBS as an analytical tool   473
    12.7 References   474
13 Short-pulse LIBS: fundamentals and applications   477
    R. E. Russo
    13.1 Introduction   477
    13.2 Effect of pulse duration on ablation   478
    13.3 Effect of pulse duration on plasma   479
    13.4 Picosecond-induced electron plasma   480
    13.5 Femtosecond plasma   482
    13.6 Short-pulse LIBS   483
    13.7 Conclusion   487
    13.8 References   488
14 High-speed, high-resolution LIBS using diode-pumped solid-state lasers   490
    Holger Bette and Reinhard Noll
    14.1 Introduction   490
    14.2 Diode-pumped solid-state lasers   491
    14.3 State of the art   494
    14.4 Scanning LIBS   498
    14.5 Laser-induced crater geometry and spatial resolution of high-speed,
          high-resolution scanning LIBS with DPSSL   510
    14.6 References   513
15 Laser-induced breakdown spectroscopy using sequential laser pulses   516
    Jack Pender, Bill Pearman, Jon Scaffidi, Scott R. Goode and
    S. Michael Angel
    15.1 Introduction   516
    15.2 Dual-pulse LIBS   517
    15.3 Summary   532
    15.4 References   534
16 Micro LIBS technique   539
    Pascal Fichet, Jean-Luc Lacour, Denis Menut, Patrick Mauchien, Annie
    Rivoallan, Cécile Fabre, Jean Dubessy and Marie-Christine Boiron
    16.1 Introduction   539
    16.2 Experimental set-up for the micro LIBS system   543
    16.3 Results and discussion   547

16.4 Conclusion   554
16.5 References   554
17 New spectral detectors for LIBS   556
Mohamad Sabsabi and Vincent Detalle
17.1 Chapter organization   556
17.2 Introduction   556
17.3 Multidetection in LIBS   558
17.4 Evaluation of an echelle spectrometer/ICCD for LIBS applications   566
17.5 Advantages and limitations   576
17.6 Choice of an optical setup for LIBS   580
17.7 Conclusions   581
17.8 References   582
18 Spark-induced breakdown spectroscopy: a description of an electrically
     generated LIBS-like process for elemental analysis of airborne particulates
     and solid samples   585
Amy J. R. Hunter and Lawrence G. Piper
18.1 Introduction   585
18.2 Basic description of SIBS processes and hardware   586
18.3 Application-specific considerations   590
18.4 Applications and results   599
18.5 Discussion and future directions   613
18.6 References   614
Index   615

Contributors

S. Michael Angel
Department of Chemistry and
   Biochemistry,
The University of South Carolina,
Columbia, SC 29208,
USA

Demetrios Anglos
Institute of Electronic Structure
   and Laser,
Foundation for Research and
   Technology – Hellas,
PO Box 1527,
GR 71110, Heraklion, Crete,
Greece

S. Beaton
ADA Technologies, Inc.,
Littleton, CO,
USA

Simon Béchard
Pharma Laser Inc.,
75 Blvd. de Mortagne,
Boucherville, Québec,
Canada J4B 6Y4

Holger Bette
Lehrstuhl für Lasertechnik (LLT),
RWTH Aachen,
Steinbachstr. 15,
52074 Aachen,
Germany

Marie-Christine Boiron
Equipes Interactions entre Fluides et
   Minéraux,
UMR 7566 G2R - CREGU Géologie et
   Gestion des Ressources Minérales et
   Energétiques,
Université Henri Poincaré,
BP-239, 54506-Vandoeuvre-les Nancy
   Cedex,
France

Valery Bulatov
Department of Chemistry,
Technion–Israel Institute of Technology,
Haifa 32000,
Israel

N. H. Cheung
Department of Physics,
Hong Kong Baptist University,
Kowloon Tong, Hong Kong,
People's Republic of China

M. Corsi
Instituto per i Processi Chemico-Fisici
   del CNR,
Area della Ricerca di Pisa,
Via G. Moruzzi 1,
56124 Pisa,
Italy

David A. Cremers
Chemistry Division,

Los Alamos National Laboratory,
Los Alamos, NM,
USA

G. Cristoforetti
Instituto per i Processi Chemico-Fisici
   del CNR,
Area della Ricerca di Pisa,
Via G. Moruzzi 1,
56124 Pisa,
Italy

F. C. DeLucia, Jr.
US Army Research Laboratory,
AMSRL-WM-BD,
Aberdeen Proving Ground,
MD 21005–5069,
USA

Vincent Detalle
Industrial Materials Institute,
National Research Council of Canada,
75 Blvd. de Mortagne,
Boucherville, Québec,
Canada J4B 6Y4

Jean Dubessy
Equipes Interactions entre Fluides et
   Minéraux,
UMR 7566 G2R - CREGU Géologie
   et Gestion des Ressources Minérales
   et Energétiques,
Université Henri Poincaré, BP-239,
54506-Vandoeuvre-les Nancy Cedex,
France

Philip Evans
Applied Photonics Ltd,
Unit 8 Carleton Business Park,
Carleton New Road, Skipton,
North Yorkshire BD23 2DE,
UK

Cécile Fabre
Equipes Interactions entre Fluides et
   Minéraux,
UMR 7566 G2R - CREGU Géologie et
   Gestion des Ressources Minérales et
   Energétiques,
Université Henri Poincaré, BP-239,
54506-Vandoeuvre-les Nancy Cedex,
France

Pascal Fichet
CEA Saclay,
DPC/SCPA/LALES,
91191 Gif Sur Yvette,
France

P. French
ADA Technologies, Inc.,
Littleton, CO,
USA

Scott R. Goode
Department of Chemistry and
   Biochemistry,
The University of South Carolina,
Columbia, SC 29208,
USA

I. Gornushkin
Department of Chemistry,
University of Florida,
Gainesville, FL 32611,
USA

David Hahn
Department of Mechanical and Aerospace
   Engineering,
University of Florida,
Gainesville, FL 32611–6300,
USA

R. Harmon
US Army Research Laboratory,
Army Research Office,
PO Box 12211,

Research Triangle Park, NC,
USA

Amy J. R. Hunter
Physical Sciences Inc.,
20 New England Business Center,
Andover, MA 01810,
USA

Jean-Luc Lacour
CEA Saclay,
DPC/SCPA/LALES,
91191 Gif Sur Yvette,
France

J. J. Laserna
Department of Analytical
   Chemistry,
University of Málaga,
Málaga,
Spain

K. L. McNesby
US Army Research Laboratory,
AMSRL-WM-BD,
Aberdeen Proving Ground,
MD 21005–5069,
USA

Patrick Mauchien
CEA Saclay,
DPC/SCPA/LALES,
91191 Gif Sur Yvette,
France

Denis Menut
CEA Saclay,
DPC/SCPA/LALES,
91191 Gif Sur Yvette,
France

John C. Miller
Life Sciences Division,
Oak Ridge National Laboratory,
PO Box 2008,

Oak Ridge, TN 37830–6125,
USA
Present address: Chemical Sciences,
Geosciences and Biosciences Division,
Basic Energy Sciences,
Office of Science SC–14 Germantown
Building, US Department of Energy,
1000 Independence Avenue,
SW Washington,
DC 20585–1290, USA

A. W. Miziolek
US Army Research Laboratory,
AMSRL-WM-BD,
Aberdeen Proving Ground,
MD 21005–5069,
USA

Yves Mouget
Pharma Laser Inc.,
75 Blvd. de Mortagne,
Boucherville, Québec,
Canada J4B 6Y4

Reinhard Noll
Fraunhofer-Institut für Lasertechnik (ILT),
Steinbachstr. 15,
52074 Aachen,
Germany

N. Omenetto
Department of Chemistry,
University of Florida,
Gainesville, FL 32611,
USA

V. Palleschi
Instituto per i Processi Chemico-Fisici
   del CNR,
Area della Ricerca di Pisa,
Via G. Moruzzi 1,
56124 Pisa,
Italy

Ulrich Panne
Laboratory for Applied Laser
  Spectroscopy,
Institute of Hydrochemistry,
Technical University Munich,
Marchioinistrasse 17,
D-81377 Munich,
Germany

Christian G. Parigger
The University of Tennessee Space
  Institute,
Center for Laser Applications,
411 B. H. Goethert Parkway,
Tullahoma, TN 37388,
USA

Bill Pearman
Department of Chemistry and
  Biochemistry,
The University of South Carolina,
Columbia, SC 29208,
USA

Jaek Pender
Department of Chemistry and
  Biochemistry,
The University of South Carolina,
Columbia, SC 29208,
USA

B. Peterson
US Army Research Laboratory,
AMSRL-WM-BD,
Aberdeen Proving Ground,
MD 21005–5069,
USA

Lawrence G. Piper
Physical Sciences Inc.,
20 New England Business
  Center,
Andover, MA 01810,
USA

Leon J. Radziemski
Physics Department,
Washington State University,
Pullman, WA,
USA

V. N. Rai
Diagnostics Instruments and Analysis
  Laboratory (DIAL),
Mississippi State University,
205 Research Blvd.,
Starkville, MS 39759–7704,
USA

Annie Rivoallan
CEA Saclay,
DPC/SCPA/LALES,
91191 Gif Sur Yvette,
France

R. E. Russo
Lawrence Berkeley National Laboratory,
1 Cyclotron Road,
Berkeley, CA 94720,
USA

Mohamad Sabsabi
Industrial Materials Institute,
National Research Council of Canada,
75 Blvd. de Mortagne,
Boucherville, Québec,
Canada J4B 6YA

Ota Samek
Department of Physical Engineering,
Technical University of Brno,
Technicka 2, 61669 Brno,
Czech Republic

Jon Scaffidi
Department of Chemistry and
  Biochemistry,
The University of South Carolina,
Columbia, SC 29208,
USA

Israel Schechter
Department of Chemistry,
Technion–Israel Institute of Technology,
Haifa 32000,
Israel

J. P. Singh
Diagnostics Instruments and Analysis
    Laboratory (DIAL),
Mississippi State University,
205 Research Blvd.,
Starkville, MS 39759–7704,
USA

B. W. Smith
Department of Chemistry,
University of Florida,
Gainesville, FL 32611,
USA

Michael Stepputat
Fraunhofer-Institut für Lasertechnik (ILT),
Steinbachstr. 15,
52074 Aachen,
Germany

Volker Sturm
Fraunhofer-Institut für Lasertechnik (ILT),
Steinbachstr. 15,
52074 Aachen,
Germany

Helmut H. Telle
Department of Physics,
University of Wales Swansea,
Singleton Park,
Swansea SA2 8PP,
UK

Elisabetta Tognoni
Instituto per i Processi Chemico-Fisici
    del CNR,

Area della Ricerca di Pisa,
Via G. Moruzzi 1,
56124 Pisa,
Italy

J. M. Vadillo
Department of Analytical Chemistry,
University of Málaga,
Málaga,
Spain

Andrew Whitehouse
Applied Photonics Ltd,
Unit 8 Carleton Business Park,
Carleton New Road, Skipton,
North Yorkshire BD23 2DE,
UK

J. D. Winefordner
Department of Chemistry,
University of Florida,
Gainesville, FL 32611,
USA

James Young
Applied Photonics Ltd,
Unit 8 Carleton Business Park,
Carleton New Road, Skipton,
North Yorkshire BD23 2DE,
UK

F. Y. Yueh
Diagnostics Instruments and Analysis
    Laboratory (DIAL),
Mississippi State University,
205 Research Blvd.,
Starkville,
MS 39759–7704,
USA

Preface

Richard E. Russo and Andrzej W. Miziolek

LIBS (laser-induced breakdown spectroscopy) has been described as "a future super star" in a 2004 review article by Dr. James Winefordner, a world-renowned analytical spectroscopist.[1] LIBS is the only technology that can provide distinct spectral signatures characteristic of all chemical species in all environments. LIBS can be used to chemically characterize any sample: rocks, glasses, metals, sand, teeth, bones, weapons, powders, hazards, liquids, plants, biological material, polymers, etc. LIBS can be performed at atmospheric pressure, in a vacuum, at the depths of the ocean, or extraterrestrially. LIBS can respond in less than a second, indicating if a spilled white power is innocuous or hazardous, using a single laser shot. A unique attribute of LIBS is that samples do not need to fluoresce, or be Raman or infrared (IR) active. It is the simplicity of LIBS that allows this diversity of applications; simply strike any sample with a pulsed laser beam and measure a distinct optical spectrum. The laser beam initiates a tiny luminous plasma from ablated sample mass. The plasma spectrum is a signature of the chemical species in the sample; spectral data analysis provides the chemical species composition and relative abundance. Because a pulsed laser beam initiates the LIBS plasma, there is no physical contact with the sample; laboratory and open-path standoff applications are readily employed. Simply put, the LIBS phenomenon represents an efficient engine to convert the chemical information of the target material to light information that can be captured efficiently and analyzed thoroughly by modern spectroscopic instrumentation and data analysis/chemometrics software.

LIBS has been aggressively investigated for environmental, industrial, geological, planetary, art, and medical applications since the early 1980s, although initial LIBS papers appeared with the discovery of the ruby laser in 1962.[1] A comprehensive source of literature describing LIBS research and applications can be found in *Applied Optics*,[2] which dedicated a special issue to this technology, as well as an extensive review in 2004.[3] Although traditionally classified as an elemental analysis technology, the use of broadband high-resolution

[1] J. D. Winefordner, I. B. Gornushkin, T. Correll, E. Gibb, B. W. Smith, and N. Omenetto, Comparing several atomic spectrometric methods to the super stars: special issue on laser induced breakdown spectrometry, LIBS, a future super star. *J. Anal. Atom. Spectrom.*, **19** (2004), 106–108.

[2] See *Appl. Opt.*, **42** (30) (2003).

[3] W. B. Lee, J. Wu, Y-III Lee and J. Sneddon, Recent applications of laser-induced breakdown spectrometry: a review of material approaches, *Appl. Spectros. Rev.*, **39** (2004), 27–97.

spectrometers has recently extended LIBS applications to molecular species identification. The ability to detect molecular and elemental signatures with a single laser pulse offers unprecedented performance for emerging medical, biological, environmental, and security applications.

With the growth and evolution of LIBS phenomenon understanding and application areas there has been a corresponding increase in LIBS practitioners, both engineers and scientists, as well as a growth in LIBS commercial activities, in both instrument manufacturing and applications for hire. In fact, the world-wide LIBS community has established a tradition of international conferences on a two-year cycle that include LIBS 2000 (Tirrenia, Italy), LIBS 2002 (Orlando, USA), LIBS 2004 (Málaga, Spain), and LIBS 2006 (Montreal, Canada). The European LIBS community has also established the EMSLIBS (Euro-Mediterranean Symposium) series with EMSLIBS 2001 (Cairo, Egypt), EMSLIBS 2003 (Crete, Greece), and EMSLIBS 2005 (Aachen, Germany). In addition there have been a multitude of LIBS symposia associated with Optical Society of America, Pittcon, and FACSS meetings.

This book describes the history, current research in understanding fundamental processes, research to improve measurement performance, and examples of numerous applications requiring parts per million (p.p.m.) and parts per billion (p.p.b.) detection levels. Several chapters describe research efforts dedicated to improving detection capabilities. Achieving sub-p.p.b. levels would allow LIBS to compete with vacuum-based mass spectrometric measurements, without requiring a vacuum. As described throughout this book, there is a tremendous international effort to advance the LIBS technology, by addressing multiple laser pulses, short duration laser pulses, and new instrumentation. One area to increase sensitivity would be to utilize ablated mass more efficiently; current LIBS analysis detects only a fraction of the mass ablated and excited to optical emission. Focused fundamental research on laser-induced plasmas will provide advanced knowledge for efficiently generating, exciting, and detecting mass. There is a large body of supporting literature on laser ablation for other applications (micromachining, materials fabrication, nanotechnology, thin-film deposition) that is germane to LIBS; the fundamental mechanisms are the same, but the optimum parameters for application are not. Optimum parameters need to be established for analyzing diverse samples, for example organic residues compared with inorganic refractory bulk samples. Understanding plasma physics can provide new approaches for increased sensitivity by using external (for example light, radio frequency, magnetic fields) means for producing longer-lived, hotter, and denser plasmas. There have been numerous efforts to study the influence of the laser beam properties (pulse duration, wavelength, energy, and number of pulses) on LIBS analytical performance. The laser beam can deliver energy from femtoseconds to microseconds in duration. On the other hand, the LIBS plasma duration is generally several microseconds, although research needs to establish laser–plasma–property time relationships.

Most LIBS applications are based on using a laser with wavelength of 1064 nm. Wavelength contributes to plasma heating with nanosecond pulses, but research needs to establish if IR is best when using short pulsed (femtosecond and picosecond) lasers, and the role of Bremsstrahlung absorption. The use of double and triple pulses is being aggressively investigated for improving sensitivity and reducing ambient interferences. Currently, the UV–IR (ultraviolet–infrared) spectral region is interrogated for analysis, but other spectral regions, such as hyperspectral, may provide enhanced measurement capabilities. Just as the broadband spectrometer opened new vistas in LIBS applications, understanding measurement principles will advance performance specifications for existing and new LIBS applications.

Implementation of LIBS in a suite of applications requires diverse yet similar instrumentation. For example, LIBS can be used with a simple lens to focus the laser beam within a few millimeters from the laser, with an optical fiber to carry the laser beam to a remote physical location, or by using a telescope for open-path standoff applications. Improved LIBS systems for long-distance standoff measurements will benefit from advanced optical configurations. Other spectroscopic technologies (Raman, fluorescence, absorbance, light scattering) perform in open-path configurations, although they do not possess the versatility of LIBS. However, it would be easy to integrate LIBS with Raman and laser-induced fluorescence for additional measurement capabilities. An integrated system could use light scattering to identify a suspect particle based on its morphology and then Q-switch the same laser for simultaneous LIBS – all in the same system. A concern for open-path standoff laser-based analysis is eye safety. Although the FDA in the USA has established limits for pulsed exposure, these limits are for unfocused laser beams; LIBS requires a focused laser beam. As research progresses to advance LIBS sensitivity using various laser wavelengths, low-level eye-safe operation will be viable.

New applications of LIBS are expected in medical, biological, security, and nanotechnology. With the international effort to fabricate nano-devices, -structures, and -particles, new technologies will be required to ensure that these systems abide by their design criteria. LIBS can fulfill this requirement, but will need to operate on smaller spatial scales and with enhanced sensitivity. The widespread utilization of LIBS for these applications will require development of comprehensive spectral databases and data manipulation algorithms. Spectral libraries can be established for voluminous chemical species and rapidly be evaluated to determine distinct signature for classes of species. Mass spectroscopy, Raman spectroscopy, fluorescence, IR, NMR (nuclear magnetic resonance), and almost all spectral analytical technologies benefit from the use of spectral libraries – as will LIBS.

This book challenges you to benefit from the current expertise and to imagine new applications and ideas for advancing LIBS. The chapters present the current status of fundamental and applied LIBS studies, from a community excited by the numerous capabilities and possibilities. Chemical analysis is a critical component of world survivability – for understanding nature, contamination, health, climate, microelectronics, terrorism, advanced materials, and other things. We believe that LIBS will play a dominant role in every aspect of society for chemical analysis. With continued research and application, LIBS is becoming a future super star of analytical spectroscopy.

1

History and fundamentals of LIBS

David A. Cremers

Chemistry Division, Los Alamos National Laboratory

Leon J. Radziemski

Physics Department, Washington State University

1.1 Introduction

Laser-induced breakdown spectroscopy (LIBS) is a method of atomic emission spectroscopy (AES) that uses a laser-generated plasma as the hot vaporization, atomization, and excitation source. Because the plasma is formed by focused optical radiation, the method has many advantages over conventional AES techniques that use an adjacent physical device (e.g. electrodes, coils) to form the vaporization/excitation source. Foremost of these is the ability to interrogate samples *in situ* and remotely without any preparation. In its basic form, a LIBS measurement is carried out by forming a laser plasma on or in the sample and then collecting and spectrally analyzing the plasma light. Qualitative and quantitative analyses are carried out by monitoring emission line positions and intensities. Although the LIBS method has been in existence for 40 years, prior to 1980, interest in it centered mainly on the basic physics of plasma formation. Since then the analytical capabilities have become more evident. A few instruments based on LIBS have been developed but have not found widespread use. Recently, however, there has been renewed interest in the method for a wide range of applications. This has mainly been the result of significant technological developments in the components (lasers, spectrographs, detectors) used in LIBS instruments as well as emerging needs to perform measurements under conditions not feasible with conventional analytical techniques. A review of LIBS literature shows that the method has a detection sensitivity for many elements that is comparable to or exceeds that characteristic of other field-deployable methods.

1.2 Basic principles

1.2.1 Introduction

Setting . . . the surface of the Moon

The laser played across the cliff face in double waves. First a gentle scan lit up every millimeter of the sheer sedimentary surface, while widely spaced recording devices read its reflections, noting every

Laser-Induced Breakdown Spectroscopy: Fundamentals and Applications, ed. Andrzej W. Miziolek, Vincenzo Palleschi and Israel Schechter. Published by Cambridge University Press. © A. W. Miziolek, V. Palleschi and I. Schechter 2006. A. W. Miziolek's contributions are a work of the United States Government and are not protected by copyright in the United States.

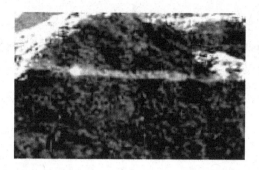

microscopic contour and color variation. Then, when that first scan was finished, the machine sent forth a much more powerful second beam, which seared away a thin layer wherever it touched. The monitors now recorded glowing spectra from these vapors, taking down elemental compositions in minute detail.

The above excerpt is from the science fiction book *Murasaki* [1] and very obviously describes the use of LIBS for the analysis of lunar terrain preceded by what is apparently reflection spectroscopy. Although at present fiction, the use of LIBS for space exploration is actively being pursued for use in the near future [2]. This potential application, along with other implemented and proposed applications of LIBS, highlights the versatility of the LIBS method. In fact a review of these applications and the LIBS literature in general leads to the conclusion that LIBS is perhaps the most versatile method yet developed for elemental analysis. In this chapter we present a brief history of LIBS, its development since the invention of the laser, and some early applications. In addition, some of the fundamental characteristics of the method are described to lay a foundation for subsequent chapters in this book. There are several reviews relating to LIBS that will serve as good background material to understanding the method [3–8].

LIBS, an analytical method born along with the invention of the laser, has had a checkered past. First, the ablation produced by the action of the laser pulse on the sample surface was exploited as a sampling method for use with the electrode-generated spark because all materials could be ablated and the finely focused laser pulse provided micro-sampling capabilities [9]. Subsequently, it was realized that the laser plasma generated during ablation could be used as an excitation source itself. However, with the development of high-performance laboratory-based elemental analysis methods (i.e. the inductively coupled plasma or ICP), the LIBS method was (temporarily) relegated to merely a scientific curiosity, with published literature devoted more to studying fundamental characteristics of the laser plasma than to its analytical capabilities. In recent years, however, there has been strong, renewed interest in LIBS as revealed by the number of published papers in refereed journals over the past several years. This is shown in Figure 1.1. Although there has always been a steady flow of publications dealing with LIBS, beginning in about 1995 the number per year has increased dramatically. The data in Figure 1.1 were compiled from a single database using LIBS as the keyword. Many more publications than listed here dealt with the phenomenology of the

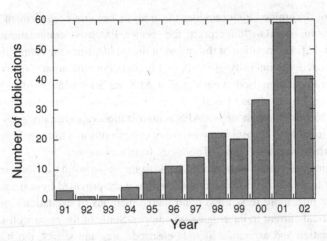

Figure 1.1. Number of LIBS-related publications during the past 11 years. At the time of this writing, not all publications for 2002 had yet been entered into the database.

laser plasma, but LIBS, which is more closely associated with analytical applications of the plasma, was the interest here.

The renewed interest in LIBS can be related to several factors. First is the need for a new method of analyzing materials under conditions not possible using current analytical methods. This is driven in part by new regulations mandating that materials and operations be monitored to ensure the health and safety of workers and the public, as well as by the need for improved industrial monitoring capabilities to increase the efficiency and reduce costs of production. Second, over the past five years there have been substantial developments in reducing the size and weight while increasing the capabilities of lasers, spectrographs, and array detectors. This makes feasible the development of compact and rugged instrumentation for use in applications outside the laboratory. The number of patents issued for LIBS-based devices also shows increased interest in the method from a technology viewpoint. Currently there are over 65 LIBS-related patents world-wide.

1.2.2 Atomic emission spectroscopy

LIBS is one method of atomic emission spectroscopy (AES). The purpose of AES is to determine the elemental composition of a sample (solid, liquid, or gas). The analysis can range from a simple identification of the atomic constituents of the sample to a more detailed determination of relative concentrations or absolute masses. Basic steps in AES are:

- atomization/vaporization of the sample to produce free atomic species (neutrals and ions),
- excitation of the atoms,
- detection of the emitted light,
- calibration of the intensity to concentration or mass relationship,
- determination of concentrations, masses, or other information.

Examination of the emitted light provides the analysis because each element has a unique emission spectrum useful to "fingerprint" the species. Extensive compilations of emission lines exist [10–12]. The position of the emission line(s) identifies the element(s) and, when properly calibrated, the intensity of the line(s) permits quantification. The specific procedures and instrumentation used in each step of AES are determined by the characteristics of the sample and by the type of analysis (i.e. identification vs. quantification). It should be noted that because the first step in AES is atomization/vaporization, AES methods are generally not suitable to determine the nature of compounds in a sample. In specific cases, however, information can be obtained about molecular origins.

The beginnings of AES can be traced back to the experiments of Bunsen and Kirchhoff (*c.* 1860) in which atomization and excitation were provided by a simple flame [13, 14]. Following this, more robust and controllable methods of excitation were developed by using electrical current to interrogate the sample. Some of the more well-known methods of vaporization and excitation include electrode arcs and sparks, the ICP, the direct coupled plasma (DCP), the microwave-induced plasma (MIP), and hollow cathode lamps [15]. These traditional sources typically require significant laboratory support facilities and some form of sample preparation prior to performing the actual analysis. In special cases, novel sampling methods have been developed for some of these sources for specific applications. Examples are an air-operated ICP providing direct analysis of particles contained in air and the introduction of particles collected on a filter into the hollow electrode of a conventional spark discharge. For various reasons, these methods saw very limited use. LIBS is an extension of the vaporization/excitation scheme to optical frequencies [16].

1.2.3 The discovery of LIBS

The production of dielectric breakdown by optical radiation, the process generating the laser plasma used by LIBS, had to wait until the development of the laser in 1960 [17]. Prior to 1960, however, the ability to produce dielectric breakdown in gases had been known for at least 100 years. These discharges can be produced fairly easily in low-pressure gas tubes with or without electrodes, at frequencies in the range of hundreds of kilohertz to a few tens of megahertz. Examination of the spectra from these sources reveals atomic emissions characteristic of the gas composition. In subsequent years, the breakdown of gases induced by frequencies on the order of gigahertz was demonstrated at reduced pressures using microwave range electromagnetic fields. Experiments were carried out at reduced pressures because the breakdown threshold shows a minimum in rarefied gases. At atmospheric pressure, the electric field required for breakdown by static and microwave fields is on the order of tens of kilovolts per centimeter. At optical frequencies the situation requires much stronger fields on the order of 10 MV/cm. Such strong fields are not attainable using conventional optical sources, thereby requiring the development of a new light source.

In 1960, laser operation was first reported in a ruby crystal. Following this in 1963 came the development of a "giant pulse" or Q-switched laser. This laser had the capability of producing high focused power densities from a single pulse of short duration sufficient to initiate breakdown and to produce an analytically useful laser plasma (also called the laser spark). This was the "birth" of the LIBS technique and in subsequent years significant milestones were made in the development of the method. Here is a list of some of the more important milestones.

- 1960 – First laser demonstrated.
- 1962 – Brech and Cross [18] demonstrate the first useful laser-induced plasma on a surface.
- 1963 – The first analytical use, involving surfaces, hence the birth of laser-induced breakdown spectroscopy.
- 1963 – First report of a laser plasma in a gas.
- 1963 – Laser micro-spectral analysis demonstrated, primarily with cross-excitation.
- 1964 – Time-resolved laser plasma spectroscopy performed.
- 1966 – Characteristics of laser-induced air sparks studied.
- 1966 – Molten metal directly analyzed with the laser spark.
- 1970 – Continuous optical discharge reported.
- 1970 – Q-switched and non-Q-switched lasers used and results compared.
- 1972 – Steel analysis carried out with a Q-switched laser.
- 1980 – LIBS developed for analysis of hazardous aerosols.
- 1980 – LIBS used for diagnostics in the nuclear power industry.
- 1984 – Analysis of liquid samples demonstrated.
- 1989 – Metals detected in soils using the laser plasma method.
- 1992 – Portable LIBS unit for monitoring surface contaminants developed.
- 1992 – Stand-off LIBS for space applications demonstrated.
- 1993 – Underwater solid analysis via dual-pulse LIBS.
- 1995 – Demonstration of LIBS using fiber optic delivery of laser pulses.
- 1997 – Use of LIBS for pigment identification in painted artworks.
- 1998 – Subsurface soil analysis by LIBS-based cone penetrometers.
- 2000 – Demonstration of LIBS on a NASA Mars rover.

1.3 Characteristics of LIBS

1.3.1 The LIBS method in brief

In LIBS, the vaporizing and exciting plasma is produced by a high-power focused laser pulse. A typical LIBS set-up is shown in Figure 1.2. Pulses from a laser are focused on the sample using a lens and the plasma light is collected using a second lens or, as shown in Figure 1.2, by a fiber optic cable. The light collected by either component is transported to a frequency dispersive or selective device and then detected. Each firing of the laser produces a single LIBS measurement. Typically, however, the signals from many laser plasmas are added or averaged to increase accuracy and precision and to average out non-uniformities in sample composition. Depending on the application, time-resolution of the spark may

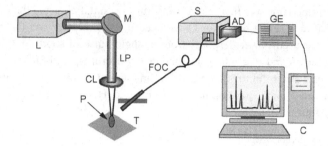

Figure 1.2. Diagram of a typical laboratory LIBS apparatus. Here: L = laser; M = mirror; LP = laser pulse; CL = lens; P = plasma; T = target; FOC = fiber optic cable; S = spectrograph; AD = array detector; GE = gating electronics; C = computer.

Figure 1.3. Left: the laser plasma formed on soil by a spherical lens is about 4–5 mm in height. Right: the long spark formed on a filter by a cylindrical lens is 7–8 mm in length.

improve the signal-to-noise ratio or discriminate against interference from continuum, line, or molecular band spectra.

Photos of laser plasmas formed on soil (by a spherical lens) and on a filter (by a cylindrical lens) are shown in Figure 1.3. To the eye, the plasma appears as a bright flash of white light emanating from the focal volume. Often the plasma formed by a spherical lens appears triangular shaped owing to formation of the initial breakdown at the focal point followed (during the laser pulse) by growth of the plasma back towards the focusing lens. Accompanying the light is a loud snapping sound owing to the shock wave generated during optical breakdown.

Because the laser plasma is a pulsed source, the resulting spectrum evolves rapidly in time. The temporal history of a laser-induced plasma is illustrated schematically in Figure 1.4a. At the earliest time, the plasma light is dominated by a "white light" continuum that has little intensity variation as a function of wavelength. This light is caused by

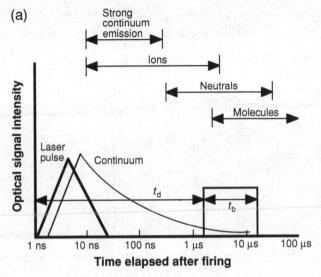

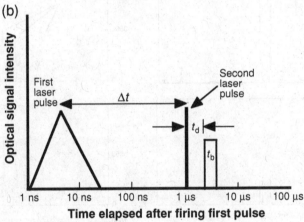

Figure 1.4. (a) The important time periods after plasma formation during which emissions from different species predominate. The box represents the time during which the plasma light is monitored using a gatable detector. Here t_d is the delay time and t_b the gate pulse width. The timing here corresponds to an RSS experiment. (b) Important timing periods for a double-pulse RSP measurement. Here Δt is the time between the closely spaced double pulses.

bremsstrahlung and recombination radiation from the plasma as free electrons and ions recombine in the cooling plasma. If the plasma light is integrated over the entire emission time of the plasma, this continuum light can seriously interfere with the detection of weaker emissions from minor and trace elements in the plasma. For this reason, LIBS measurements are usually carried out using time-resolved detection. In this way the strong white light at early times can be removed from the measurements by turning the detector on after this

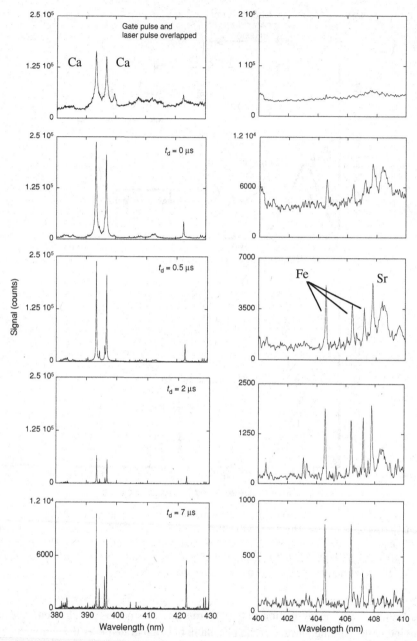

Figure 1.5. The LIBS spectrum evolves as the plasma cools. The spectra in the column on the right are expanded regions of the corresponding spectra on the left. Here, $t_b = 0.5$ μs (top eight spectra); $t_b = 2$ μs (bottom two spectra).

white light has significantly subsided in intensity but atomic emissions are still present. The important parameters for time-resolved detection are t_d, the time between plasma formation and the start of the observation of the plasma light, and t_b, the time period over which the light is recorded (Figure 1.4a).

The majority of LIBS measurements are conducted by using the RSS (repetitive single spark) in which a series of individual laser sparks are formed on the sample at the laser repetition rate (e.g. 10 Hz). In some cases, to enhance detection capabilities, the RSP (repetitive spark pair) is used. The RSP is a series of two closely spaced sparks (e.g. typically 1–10 μs separation) used to interrogate the target at the laser repetition rate. The timing arrangement in this case is shown in Figure 1.4b. Note that t_d is measured from the second laser pulse in this case. The spark pair may be formed by two separate lasers or by a single laser.

Figure 1.5 shows the evolution of the LIBS spectrum from a soil sample. Several important features should be noted. First, note the significant decrease in line widths as the delay time changes from 0 to 7 μs. This is particularly evident in the two strongest lines (once-ionized Ca) on the left side of the figure. Second, as the line widths decrease, it becomes evident at $t_d = 0.5$ μs that two additional lines (neutral Al between the Ca lines) appear that were masked by the strong Ca lines. Third, comparison of the relative intensities of the Ca and Al lines shows that these change as the plasma cools with the once-ionized Ca lines decreasing more in comparison with the neutral Al lines with increased delay time. These same features are evident in the expanded portions of the spectrum displayed on the right side of the figure showing Fe and Sr lines.

1.3.2 The physics and chemistry of the laser plasma

A life cycle schematic for a laser-induced plasma on a surface is shown in Figure 1.6. The physics of the breakdown phase was well reviewed by Weyl in 1989 [19]. Briefly, there are two steps leading to breakdown due to optical excitation [20]. The first involves having or generating a few free electrons that serve as initial receptors of energy through three-body collisions with photons and neutrals. The second is avalanche ionization in the focal region. Classically, free electrons are accelerated by the electric fields associated with the optical pulse in the period between collisions, which act to thermalize the electron energy distribution. As the electron energies grow, collisions produce ionization, other electrons, more energy absorption, and an avalanche occurs. In the photon picture, absorption occurs because of inverse bremsstrahlung. The breakdown threshold is usually specified as the minimum irradiance needed to generate a visible plasma.

Following breakdown, the plasma expands outward in all directions from the focal volume. However, the rate of expansion is greatest towards the focusing lens, because the optical energy enters the plasma from that direction. A pear- or cigar-shaped appearance results from this nonisotropic expansion. The initial rate of plasma expansion is on the order of 10^5 m/s. The loud sound that one hears is caused by the shock wave coming from the focal volume.

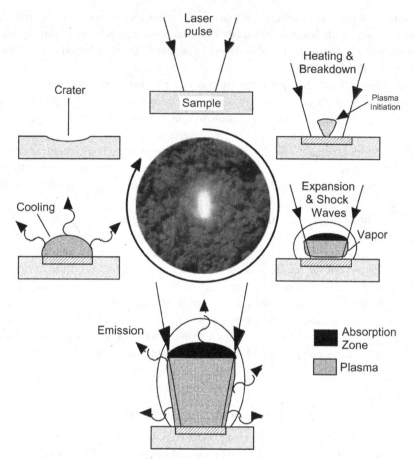

Figure 1.6. Life cycle diagram showing main events in the LIBS process.

Between its initiation and decay, the plasma evolves through several transient phases, as it grows and interacts with the surroundings. These are well described, for different irradiance regimes, by Root [21]. The three models for propagation and expansion are the laser-supported combustion (LSC), laser-supported detonation (LSD), and laser-supported radiation (LSR) waves. They differ in their predictions of the opacity and energy transfer properties of the plasma to the ambient atmosphere. At the low irradiances used in LIBS experiments, the models that most closely match experiment are LSC and LSD. In these, the plasma is at relatively low temperature and density. The plasma and the boundary with the ambient atmosphere are transmissive enough to allow the incoming laser radiation to penetrate, at least for laser wavelengths shorter than that of the CO_2 laser (10.6 μm).

Throughout the expansion phase the plasma emits useful emission signals. It cools and decays as its constituents give up their energies in a variety of ways. The ions and electrons

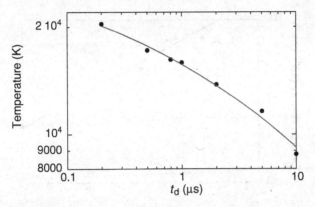

Figure 1.7. Air plasma temperature as a function of time after plasma formation. Data abstracted from reference [3] using Saha and Boltzmann data from carbon and beryllium lines.

recombine to form neutrals, and some of those recombine to form molecules. Energy escapes through radiation and conduction.

Typical plasma temperatures up to tens of thousands of degrees (several electronvolts) are achieved shortly after plasma initiation, for laser pulse energies of 10–100 mJ, and lens focal lengths of 5–20 cm, hence irradiances in the range of 10^9 to 10^{11} W/cm^2. Figure 1.7 presents composite measurements made of plasma temperatures in air as a function of time [3]. The temporal dependence leads to the experimental strategy of using a time delay for LIBS measurements. Because the early spectrum contains a bremsstrahlung and recombination continuum that decays quickly, the atomic signals (both ion and neutral) are often not sampled until after a microsecond or more into the plasma history (Figure 1.4). At that time the signal to background improves dramatically, and the atomic emission lines become much sharper.

At early times, spectral-line broadening is dominated by the Stark effect due to the high initial density of free electrons and ions. Line widths are dramatically dependent on the species, being greatest for the H$_\alpha$ line of hydrogen at 656 nm. The theory of Stark effect for atoms, including the dependence on the energy levels of the particular atom, is well developed [22]. Figure 1.8 shows measurements of the electron density as a function of time for plasmas in air at different delay times. As the plasma evolves in the post-laser-pulse regime, recombination occurs, the electron density decreases, and pressure broadening (Stark effect due to near collisions with neutrals) is often the main cause of the line width. The pressure and nature of the ambient gas influence the absolute line intensities, the line widths, and in some cases relative line intensities due to near-resonant collisions. Some experiments have shown that an argon atmosphere enhances excitation, while a helium or oxygen atmosphere suppresses excitation.

Laser wavelength may have an effect on breakdown threshold as discussed in a systematic study by Simeonsson and Miziolek [23]. They studied plasmas formed in CO and CO$_2$ gases with excitation by ArF at 193 nm and the four common Nd:YAG wavelengths: 266, 355, 532,

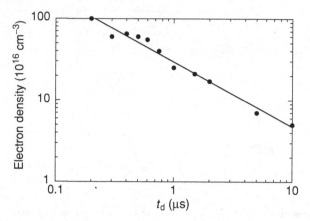

Figure 1.8. Electron density in the air plasma as a function of time after plasma formation. Data abstracted from reference [3] using Stark widths of F, Ar, N, and Cl lines.

1064 nm. The principal effect observed was the reduction of the breakdown threshold by an order of magnitude at 193 nm owing to fortuitous coincidences of two-photon excitations and dissociations. Otherwise the results showed no significant trends. Weyl [19] reviews and presents a theory that bears on the wavelength dependence. Within the past decade, there have been studies of resonance effects, where the plasma-forming laser is tuned to a strong line in an element in the sample.

As is well known, plasma temperatures can be determined in a variety of ways, including spectroscopic and probe methods. There are electron, excitation, and ionization temperatures, to name just a few. Each is determined based on different diagnostics and measurements, and they may or may not agree. The situation is complicated by the transient nature of pulsed plasmas. Generally, pulsed plasmas do not start out in equilibrium, but evolve to that state. Often electrons start out at a much higher kinetic temperature, and eventually equilibrate with the heavier atoms and ions through collisions [24]. Physics tells us that momentum transfer is small in collisions between bodies of very different masses, hence the time scale for equilibration between electrons and atoms can be quite long.

Through the 1980s much of the effort in analyzing LIBS plasmas was spent on measuring temperature and electron densities. Starting in the early 1990s researchers have attempted to model the plasma in more detail by using hydrodynamic codes and various plasma models. Outcomes include species densities as functions of time and other parameters, an understanding of excitation mechanisms in various regimes, and a clearer understanding of when LTE (local thermodynamic equilibrium) may or may not apply. Here is an example of such a model.

The interaction of 1.06 μm Nd:YAG laser radiation with metal and the resulting metal vapor at intensities of 10^6 to 10^{10} W/m^2 was studied by Mazhukin *et al.* [25]. The simulation used a collision–radiation model to describe the non-equilibrium ionization and recombination. It revealed two qualitatively different paths depending on the laser pulse intensity.

Table 1.1. *Temperature and electron density of the laser plasma for* $t_d = 1\,\mu s$
and different samples

Sample	Temperature (K)	Electron density ($\times 10^{18}$ cm^{-3})	t_b (µs)
In air [3]	17 000	0.3	1
On soil [42]	7000	—	10
On soil [41]	5000	0.069	50
In bulk liquid [28]	7950	0.9	0.25
On Ti in air [33]	7180	1.3	15
On Ti in air (RSP) [33]	7710	3.8	15
On Ti under water (RSS) [33]	n.o.	n.o.	15
On Ti under water (RSP) [33]	8880	21	15

n.o. – not observed.

If the radiation was insufficient to produce avalanche ionization, the system resided in a stationary state characterized by a temperature. At sufficiently high laser intensities optical breakdown commenced. It was a non-equilibrium transition state from a partially ionized vapor to a fully ionized plasma dominated by Coulomb collisions. In the macroscopic picture, the threshold depended on the ionization potential, the excited electronic state distribution of neutral atoms and, surprisingly, on the initial temperature of the evaporated material. Copper atoms and ions were used in this model. Details of the energy levels and ionization potentials were used. Results included electron concentrations and atom and ion temperatures as functions of time over the period of 100 ns to milliseconds. In the model, the electron and atom/ion temperatures did not equilibrate until the millisecond time regime. Other results included the calculation of neutral copper atom densities in the ground state and eight excited states as functions of time.

1.3.3 Forming the LIBS plasma in a gas, a liquid, and on solids

Initial LIBS work concentrated on the analysis of solids (e.g. metals, geological samples) but, as the unique sampling capabilities of the laser spark came to be realized, the technique was extended to a variety of other samples. Today LIBS is used to analyze gases, liquids, particles entrained in gases or liquids, and particles or coatings on solids. General characteristics of each type of sampling are described below. For comparison, Table 1.1 lists the temperature and electron density of the laser plasma for various samples. These data were collected using $t_d = 1$ µs and different values of t_b. It should be noted that the temperature and electron density depend on these parameters. For example, the temperature obtained on a soil sample with a long (50 µs) integration time was significantly lower than the temperature obtained with a five-fold shorter integration time. This is a result of the temperature being averaged over a longer time period during which the plasma temperature decays significantly.

Gases

In gases, less energy is used in the atomization process, so more energy is left for excitation. In general, the greater the irradiance, the greater is the initial ratio of ions to neutral atoms. Breakdown thresholds are slightly higher in gases than on surfaces, unless the gas contains particulate matter. The plasma volume depends on the energy per pulse and the laser wavelength. For nominal pulse energies of 200 mJ, the plasma length would be largest for high-energy CO_2 laser pulses (5–8 mm in length), and smallest for 266 nm Nd:YAG pulses (1–5 mm). Molecular gases can be completely dissociated by the plasma and the composition of neat molecular gases can be inferred. Typical plasma temperatures of 20 000 K or more are observed early in the plasma lifetime.

Liquids

Liquids can be analyzed by forming the laser plasma on the liquid surface [26] or on drops of the liquid [27]. If the liquid is transparent at the laser wavelength, a plasma can be formed in the bulk liquid below the surface [28]. Compared with LIBS analysis in air, the plasma formed in the bulk liquid decays more rapidly, emission lines appear broader, and the temperature is lower, typically starting at no more than 7000–12 000 K. Detection limits for selected elements in aqueous media can be increased by using a double-pulse RSP technique (Section 1.3.1) in which two sequential laser pulses, separated in time, typically by microseconds, interrogate the same volume of the sample. The first pulse produces a vapor cavity that is then interrogated by the second pulse, replicating an analysis that would be carried out in air. This method works for both the bulk liquid and liquid drops. Adding an absorber to the liquid may enhance plasma formation.

Particles

Particles entrained in liquids (hydrosols) or gases (aerosols) are of great interest for environmental monitoring. Two basic strategies for obtaining information on them are (1) direct monitoring in the ambient medium and (2) capture on filters with subsequent interrogation of the filter surface by the laser spark (Figure 1.3). The second option is a special case of solid sampling that will be discussed below. In the first case, some of the spark energy is used to ablate or vaporize the aerosol. Typically there is enough energy remaining to do the excitation, obtaining strong spectra and reasonable detection limits. Incomplete vaporization of the particles, a real possibility, however, can complicate quantification. There is continuing interest in determining particle compositions and loading, often for environmental applications.

Solids

Breakdown on surfaces and ablation are complex phenomena. Depending on the pressure above the surface, breakdown can be initiated by multiphoton ionization (low pressure), or inverse bremsstrahlung (high pressure), both followed by avalanche ionization. Breakdown thresholds are two to four orders of magnitude lower than in the case of gas targets. For low

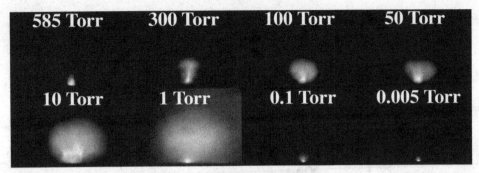

Figure 1.9. Change in visual appearance of the laser plasma formed on soil as the air pressure was reduced. Los Alamos atmospheric pressure at the time of these measurements was 585 Torr.

pressures above the surface, higher ion stages are reached for the same incident intensities. Several studies have looked at the evolution and build up of the plasma as a function of time, position, and incident laser wavelength. Regarding wavelength, long wavelengths like that from the 10.6 μm CO_2 laser have a different effect than a short wavelength such as 248 nm from a KrF laser. The long wavelength is absorbed to a much greater extent in the plasma above the surface because absorption varies as the square of the wavelength (λ^2). This effectively shields the surface from absorption of the trailing edge of the laser pulse. At short wavelengths, a higher percentage of the laser energy impacts the surface.

LIBS on surfaces depends on the ablation of material into the plasma volume. As an order of magnitude, the mass of material ablated from aluminum varies from 5 ng to 80 ng per pulse. There is an extensive literature regarding laser ablation from a variety of solids for a variety of applications. These include the analytical application of generating aerosols for ICP or other techniques, as well as applications to superconducting and other film deposition techniques. The use of laser ablation for the laser microprobe was well reviewed by Moenke-Blankenburg [5]. It was concluded that, although many applications were done in ambient air, reduced pressure or the use of other gases held advantages in some cases. Figure 1.9 shows how the visual appearance of the spark has a strong dependence on air pressure. Monitoring of element emission signals shows that the strongest signals are recorded for pressures in the range 10–100 Torr [2]. Regarding other gases, the use of argon at pressures from 10^3 to 10^6 Pa was cited as having advantages of decreasing molecular interferences, increasing spark intensities, and allowing for observations below the Shumann–Runge cutoff at 200 nm.

Reference [5] also discusses the effect of laser wavelength, pulse length, cover gas, and sample reflectivity on the amount of material ablated. In general more ablation is given by longer-pulse (i.e. 20 ns) ultraviolet (UV) lasers. In the case of LIBS, however, it is best to concentrate on optimizing the signal, not necessarily the amount of material ablated. In some cases high material content can degrade plasma properties and lead to inferior performance.

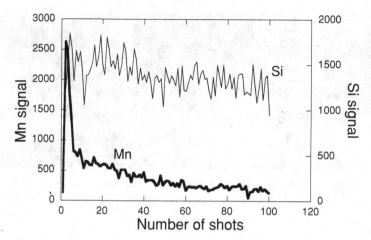

Figure 1.10. Repetitive sampling at the same spot on a weathered granite sample showing the change in Mn signal as the weathered layer is ablated. Note the uniformity of the Si signal in comparison.

Detailed fundamental studies of ablation continue. Crater dimensions influence the analytical results. The crater depth-to-diameter (aspect) ratio influences fractionation, which becomes significant when the aspect ratio is greater than 6 [29]. Laser energy coupling to a solid increases during crater formation. When the aspect ratio is more than 5, the amount of energy coupled into the surface increases to 10 times that for a flat surface [30]. More recently studies have investigated the plasma temperature and electron density at different points within the forming crater. For several years it has been known that repetitive sparks at the same location can be used for depth profiling. This has been used to determine coating thicknesses, and the age of paint layers on paintings. An example of a case where profiling revealed a variation with depth for manganese but not for silicon in weathered rock is shown in Figure 1.10.

When the laser energy is focused on solids in vacuum, higher stages of ionization are reached for the same intensity on target. The post-breakdown expansion differs from that in an ambient atmosphere because of the absence of collisions with the background atoms or molecules.

The femtosecond pulse length arena is another where ablation studies are proceeding. These pulses can lead to very high aspect ratio craters. Comparisons have been made of craters made by nanosecond and femtosecond pulses [31].

Solids under water can be interrogated using LIBS but significant emission is only observed when a double-pulse technique is used [32, 33]. If only one laser pulse interrogates a solid under water, the majority of the energy goes into formation of a vapor cavity on the surface, with little energy remaining for excitation. In the work by Pichahchy *et al.* [33], the vapor cavity was found to have a maximum diameter of about 8 mm. The vapor in that bubble was then interrogated by a second pulse forming a plasma with excitation properties similar to that of a single laser spark formed on metal in air.

1.4 LIBS as an analytical technique

1.4.1 Advantages

LIBS, like other methods of AES, has the following advantages compared with some non-AES-based methods of elemental analysis:

- ability to detect all elements,
- simultaneous multi-element detection capability.

In addition, because the laser spark uses focused optical radiation rather than a physical device such as a pair of electrodes to form the plasma, LIBS has many distinct advantages compared with conventional AES-based analytical methods. These are:

- simplicity,
- rapid or real-time analysis,
- no sample preparation,
- allows *in situ* analysis requiring only optical access to the sample,
- ability to sample gases, liquids, and solids equally well,
- good sensitivity for some elements (e.g. Cl, F) difficult to monitor with conventional AES methods,
- adaptability to a variety of different measurement scenarios,
- robust plasma that can be formed under conditions not possible with conventional plasmas.

Some of these are discussed below in detail.

Variety of measurement scenarios

Several different methods of directing the laser light on a sample to form the plasma have been implemented. Some of these have been developed for analysis at short distances in which the sample is positioned adjacent to the LIBS device and others have been devised for remote analysis in which the sample may be many meters from the instrumentation. Some of the more useful methods are discussed below.

Direct analysis. In the most basic LIBS configuration (Figure 1.2), a short focal length lens is used to focus the laser pulses onto the sample surface (solid, liquid) or into a liquid or gas to form the plasma. The plasma light can be collected using a fiber optic cable which then transports the light to a frequency selective device such as a spectrograph. Alternatively, the light can be collected by a lens which focuses the light onto a monochromator or spectrograph slit or collimates the light to pass it directly through a bandpass filter or other frequency selective device (e.g. AOTF, see below).

Fiber optic delivery. The use of fiber optics to transport light over long distances is well known. Recently, as a result of improved fiber optic materials, it is now possible to focus power densities on the order of megawatts per square centimeter onto the end of a fiber, without damage, and inject and then transport tens of millijoules of energy through the fiber (Figure 1.11a) [34, 35]. By placing a lens system at the distal end of the fiber or by merely polishing the end of the fiber to a slight curvature, the laser pulse can be focused

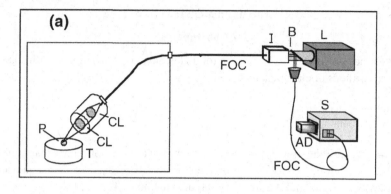

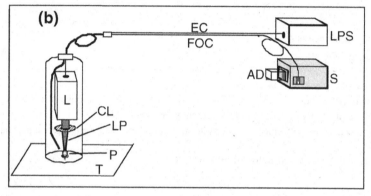

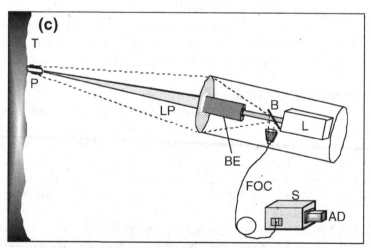

Figure 1.11. Because the laser plasma is formed by focused light, it can be adapted to different analysis scenarios as shown here: (a) fiber optic delivery; (b) compact probe; (c) stand-off analysis. Here: L = laser; B = beamsplitter; FOC = fiber optic cable; I = pulse injector for FOC; CL = lens; T = target; P = plasma; S = spectrograph; AD = array detector; LP = laser pulse; EC = electrical cables; LPS = laser power supply; BE = beam expander.

to produce a spark on a surface. Transport of laser pulses over distances up to 100 m has been demonstrated. The plasma light can be collected and transported back to the detection system using either the same fiber or a second fiber optic cable.

Compact probe. With the development of miniature solid state lasers, it has become possible to construct a small probe to use for remote LIBS measurements (Figure 1.11b) [36]. The probe contains the laser, focusing optics, and a fiber optic to collect and transport the plasma light. The laser power supply and detection system are located remotely from the probe, connected by an umbilical cable containing electrical cables for the laser and the fiber optic cable. This configuration has been used in at least one commercial LIBS unit. This method has the advantage, compared with fiber optic delivery, that spot sizes of small diameter and hence greater power density can be delivered to the target enhancing the element signal.

Stand-off analysis. Distinct from remote analysis in which some physical part of the LIBS system (Figures 1.11a and 1.11b), however small, is close to the sample, is the method of stand-off analysis. Here the laser pulse is focused onto the sample at a distance using a long focal length optical system (Figure 1.11c) [37]. The distances achievable depend on many parameters including the laser pulse energy and power, the beam divergence, spatial profile, and the optical system used to focus the pulses at a distance. Results to date are encouraging. For example, with good quality components, a laser plasma can be formed on soil at a distance of 19 m using a pulse of only 35 mJ [2]. Clearly, in this case, efficient collection of the plasma light is needed to obtain useful signals. In the other analysis scenarios described above, a bare fiber positioned a few centimeters from the plasma can be used to collect sufficient light. For stand-off analysis, however, a lens or mirror is required to increase the solid angle over which the plasma light is collected. The light-collection system can be either adjacent to or collinear with the optical axis of the system used to focus the laser pulses on the sample. The latter configuration is shown in Figure 1.11c and it avoids the problem of parallax as the sample distance varies.

No sample preparation

LIBS requires little or no sample preparation, in contrast to most conventional AES methods, because ablation and excitation occur at the same location (focal volume). For example, analysis using the ICP generally requires that the sample be chemically ashed to produce a solution that is then nebulized into the ICP plasma. This type of chemical preparation is a time-consuming and sometimes labor-intensive process that precludes rapid sample analysis. Owing to the high focused power densities required to form the plasma, on the order of 10 GW/cm^2, all materials, including non-conducting and refractory compounds, are vaporized. Focusing the laser pulses in a gas or on or below the surface of a liquid also produces an analytically useful laser plasma. In general, this ability to interrogate all materials makes LIBS a universal method of elemental analysis. It should be noted, however, that, although LIBS may require no sample preparation, sometimes ablating pulses will be useful to develop a clean surface for the subsequent, analytically useful plasmas.

In situ *analysis*

In conventional AES methods, the sample is typically brought to the instrument and intro-duced into the vaporizing/atomizing source. Because the laser plasma is generated by focused light, however, materials can be analyzed directly, often *in situ*, merely by aiming the laser pulse at the target material. In effect, in LIBS, the source is brought to the sam-ple, reversing the usual situation. Typically, the distance between the sample and the LIBS instrument is short, a few centimeters or tens of centimeters, but in some cases, as described above for stand-off analysis, it can be many meters distant.

Speed of analysis

The lack of sample preparation and the simplicity of LIBS make it amenable to what may be considered a real-time measurement. Because of the short lifetime of the plasma (a few tens of microseconds), the high speed with which a spectrum may be recorded, and the minimal time required to process the spectrum, a single LIBS measurement may be considered immediate. In practice, however, many measurements may be made to:

(1) obtain an average composition reading due to compositional inhomogeneity of the sample,
(2) ablate away an overlying surface layer having a composition that may not be representative of the underlying bulk material,
(3) average out shot-to-shot variations in the plasma characteristics.

The use of multiple pulses to overcome sampling problems associated with (1) and (2) above is obvious and is discussed in Section 1.4.2. The advantages that may be gained from using many laser shots in the analysis of even a uniform sample when plasma characteristics are changing shot-to-shot have been demonstrated [26]. Here replicate measurements, each consisting of 50, 200, and 1600 laser pulses, were performed and the relative standard deviations (%RSD) of the analyte signals computed. The sample was uranium in a 4 M nitric-acid solution. Repetitive laser plasmas were formed on the surface of the liquid in which the lens-to-sample distance and hence plasma characteristics changed on each shot owing to strong pressure waves generated that disturbed the liquid surface. The data yielded values of 13.3, 7.2, and 1.8%RSD, respectively, for the three cases. The strong dependence of precision on the number of averaged laser pulses demonstrates the advantage of repetitive measurements.

LIBS measurements made using uniform geological samples (i.e. certified reference materials in powdered form such as those available from the US National Institute of Standards and Technology), metals, and liquid samples indicate that 100 laser shots can produce measurement precision on the order of 5%–10%.

1.4.2 Considerations in the use of LIBS

Sample homogeneity

An advantage of LIBS is the ability to analyze samples with little or no preparation. How-ever, this lack of sample preparation, where preparation typically produces a homogeneous sample from a non-uniform sample, complicates the analysis of non-homogeneous samples with a point detection method such as LIBS. Well-mixed samples such as gases and liquids containing dissolved materials may be assumed to be homogeneous so every plasma interrogates a small volume having a composition representative of the bulk sample. In this case, the number of laser shots to be averaged for a measurement is determined by factors such as the method of sampling (i.e. does each laser pulse interrogate the same sample volume?) and perturbations of the laser plasma (i.e. splashing of a liquid sampled at the surface). Other samples, solids in particular, cannot always be assumed to be homogeneous and, in fact, except for some metals and plastics, etc., inhomogeneity should be assumed. Two types of inhomogeneity likely to be encountered are listed below.

Bulk non-uniformity. LIBS is a point detection method since the surface area interrogated by the laser pulse is small, typically 0.1–1 mm diameter, involving a very small mass of material. Surface features on rocks, for example, may display visual lateral irregularities in the distribution of materials that are on the order of the area sampled on each shot. These non-uniformities may be averaged out using a number of laser plasmas to repetitively interrogate different areas of a sample with the results then averaged.

Non-representative surface composition. Some samples, such as metal alloys and rocks, may have a surface layer composition that is not representative of the underlying bulk composition even though that composition may be highly uniform. For example, weathered rocks usually have a desert varnish layer ranging from 30 μm to 100 μm thick that has a composition different from the bulk rock matrix (e.g. high in Mn). On the other hand, some elements in molten metal can segregate on the surface as the metal cools, producing a surface composition not representative of the bulk material. Depending on the laser parameters and physical properties of the sample, each interrogation by the laser pulse will produce a sampling depth ranging from a few microns up to perhaps 10–20 μm. Therefore, to obtain a more representative analysis, repetitive sampling by the laser spark at the same location on the sample can be used to ablate away the outer layers revealing the true bulk composition underneath such as shown in Figure 1.10.

For samples that are significantly inhomogeneous as a result of either bulk or surface non-uniformity, it may be advisable to grind the samples and then press the resulting powder to produce a flat surface for analysis. Although this procedure eliminates the real-time and *in situ* analysis advantages of LIBS, it still preserves analysis capability without the need for chemical treatment of the sample.

Matrix effects

LIBS, like other analytical methods, displays so-called matrix effects. That is, the physical properties and composition of the sample affect the element signal such that changes in concentration of one or more of the elements forming the matrix alter an element signal even though the element concentration remains constant. For example, the signal strengths from silicon in water, in steel, and in soil appear much different even though the concentration of the element is the same in all three matrices. Even for samples that are more closely allied in matrix composition such as soils and stream sediments, significant differences in signal levels are observed for elements at identical concentrations in these materials. Because the laser spark both ablates and excites the sample, these effects can be more pronounced than in other methods that require sample preparation. Matrix effects can be divided into two kinds, physical and chemical.

Physical matrix effects depend on the physical properties of the sample and generally relate to the ablation step of LIBS. That is, differences between the specific heat, latent heat of vaporization, thermal conductivity, absorption, etc., of different matrices can change the amount of an element ablated from one matrix compared with another matrix even though the properties of the ablation laser pulse remain constant. Changes in the amount of material ablated can often be corrected for by computing the ratio of the element emission signal to some reference element known to be in the matrix at a fixed or known concentration. In this case, it is assumed that the relative ablated masses of the element and reference elements remain constant although the total mass of ablated material may change on a shot-to-shot basis. In this case, calibration is provided by using the ratio of the element signal to the signal produced by the reference element. Compensation for differences in ablated material has also been achieved through acoustic methods that monitor the sound generated by the spark [38, 39]. Other physical matrix effects have been observed, for example the dependence of detection limit on soil grain size [40]. Here it was found that the detection limit was significantly higher for fine-grain materials (e.g. clays) compared with a coarse matrix such as sand.

Chemical matrix effects occur when the presence of one element affects the emission characteristics of another element. This can complicate calibration of the technique and hence the ability to obtain quantitative results. These effects can be compensated for in the analysis if the concentration and effect of the interfering element(s) are known. Changes in the concentration of the interfering species from sample to sample, however, can be a difficult correction procedure. An example of a chemical matrix effect is the reduction in emission intensity of an ionized species (e.g. Ba(II)) upon a significant increase in the concentration of an easily ionizable species in the matrix. The easily ionizable species increases the electron density, thereby decreasing the concentration of Ba(II) [41]. In addition, there are indications that the compound form of an element (e.g. $PbNO_3$, $PbCl_2$, etc.) may result in different emission signal strengths for the same element concentrations.

Extensive work remains to be done to characterize chemical and physical matrix effects for all types of samples and to develop methods to correct for their effects and increase the quantitative ability of LIBS.

Sampling geometry

In the analysis of a solid, a plasma will be formed on the surface if the power density is sufficiently high even though the distance between the sample and the lens may be different from the focal length of the lens. These changes in the lens-to-sample distance (LTSD) can result in changes in the mass ablated as well as changes in the temperature and electron density of the plasma that, in turn, affect the element emission signals [42]. Keeping the sampling geometry constant is important to achieve the best analytical results. In the interrogation of some samples, such as soil or rocks on a conveyor belt, for example, maintaining the LTSD constant may not be possible to a high degree. This can be dealt with in several ways. These include the use of a lens of long focal length to focus the pulses on the sample so that relative changes in the LTSD are less important. Alternately, developing an active feedback system to automatically change the lens position would keep the LTSD constant. In addition, as discussed above for physical matrix effects, it is often possible to compute the ratio of the element emission to the emission of some reference element (e.g. Fe in steel or in soil) known to be in the sample at a fixed concentration. In this case, relative changes in both signals would, ideally, be the same, making the ratio less susceptible to changes in geometry.

Safety

As with other methods of AES, there are certain operational issues that must be adequately considered for the safe use of LIBS. These are: (1) the ocular hazard posed by the laser pulse, (2) the potentially lethal high voltage circuits used by the laser, (3) the explosive potential of the laser spark for certain materials, and (4) the possibility of generating toxic airborne materials. The first two hazards are discussed at length in the industry safety standard ANSI Z136.1 to which the reader is referred for further information [43]. Issues (3) and (4) can be dealt with through the preparation of a Standard Operating Procedure (SOP) or Hazard Control Plan (HCP) that identifies the hazards associated with the measurement procedure and what steps are to be taken to mitigate the hazards. Review of the operation and the SOP/HCP by properly trained industrial safety and health professionals is essential.

1.4.3 Analytical performance

Analytical figures of merit (detection limits, precision, and accuracy) for LIBS probably are more dependent on the sample properties than conventional analytical techniques because of the lack of sample preparation. In conventional laboratory analyses, sample preparation is often used to process the original sample into a standard form (e.g. a solution) removing characteristics of the original sample (e.g. non-uniformity in composition, surface irregularities) that could affect the analysis. Representative LIBS detection limits for selected elements in different matrices are presented in Table 1.2. In reviewing the table it should be noted that the different detection limits must be considered not only in light of the different matrices but also in view of the experimental parameters used (e.g. laser pulse energy, number of pulses averaged, atmosphere, etc.) in each case. For example, the higher detection limits listed for Pb in soil obtained by reference [36] compared with the values from other studies (references [41], [71], [72]) may be explained by the use of a low pulse power, compact LIBS system for the reference [36] data.

Table 1.2. *Representative detection limits for elements in selected matrices*

Element	Gas (p.p.m.)	Liquid (p.p.m.)	Surface (ng/cm^2) (on filter)	Solid (p.p.m.) (matrix)
Ag			17[60]	
Al		20[28]		54(glass,1 Torr)[61] 130(Fe ore)[63]
As	0.5 (aerosol) [62]		440[64]	
B		1200(RSS)[28] 80(RSP)[28]		30(glass,1 Torr)[61]
Ba		6.8[65]	1[60]	42(soil)[41] 265(soil)[36] 76(sand)[41] 190(glass,1 Torr)[61]
Be	0.0006 (aerosol) [62]	10[28]	1[60]	9[36]
C	36[67]			
Ca		0.8[28] 0.13[65]		85(glass,1 Torr)[61] 30(Fe ore)[66]
Cd		500[65]	300[60] 400[64]	
Cl	8[68] 0.16[69] 90[67]			
Co			100[64]	
Cr			160[60] 40[64]	40(iron)[70] 10(soil)[71] 30(soil) [72]
Cs		1[28]		
Cu			11[60] 19[64]	10(Al alloy)[73] 20(soil)[71] 30 (soil) [72]
F	38[68] 20[67]			
Fe				500(soil)[72]
Hg	0.5 (aerosol)[62] 0.005(vapor) [74]		15[60]	
K		1.2[28]		190(glass,1 Torr)[61]
Li		0.006[28] 0.013[65]		10(glass,1 Torr)[61]

(*cont.*)

Table 1.2. *(cont.)*

Element	Gas (p.p.m.)	Liquid (p.p.m.)	Surface (ng/cm²) (on filter)	Solid (p.p.m.) (matrix)
Mg		100[28]		0.5(Al alloy)[73] 130(glass,1 Torr)[61] 230(Fe ore)[63]
Mn			115[60] 30[64]	2(Al alloy)[73] 7(Al)[70] 100(soil)[72] 509(iron)[70]
Mo				
Na	0.006 (aerosol)[62]	0.0075[65] 0.014[28]		14(glass,1 Torr)[61]
Ni			185[60] 270[64]	20(soil)[71] 30(soil)[72] 64(steel)[66]
P	1.2 (aerosol)[62]			
Pb		12.5[65]	450[60] 60[64]	10(soil)[71] 50 (soil)[72] 57(soil)[41] 298(soil)[36] 17(sand)[41] 10(concrete)[76] 8000(lead paint)[36]
Rb		0.2[28]		
S	200 (aerosol)[60] 1500[67]			
Sb			280[64]	
Si				14(Al alloy)[73] 600(iron)[70] 600(Al)[70] 1500(Fe ore)[66]
Sn			50[64]	
Sr			5[60]	42[36]
Ti				410(glass,1 Torr)[61] 230(Fe ore)[63]
Tl			40[75] 100[64]	
U		100[26]		1000(soil)[60]
V			90[64]	
Zn			135[60]	160(glass,1 Torr)[61] 30(soil)[71,72]
Zr				290(glass,1 Torr)[61]

Increasing the analytical performance of LIBS is a key element in expanding its applicability to new analysis scenarios. Work is being conducted in this area. The use of acoustic methods to compensate for differences in ablated materials on a shot-to-shot basis was mentioned previously [38, 39]. In another area, theory has been combined with experiment to derive a scheme for relating relative line intensities to concentrations. CF-LIBS or calibration-free LIBS has recently been described [44, 45]. The method relies on the assumptions of thermodynamic equilibrium and an optically thin plasma. Temperatures are measured either by the Boltzmann plot method or by using the Saha equation. Then the line intensities for the analytes are used along with known f values to calculate the relative concentrations. More recently, multivariate analysis was applied to improve the quantitative outcomes [46]. Principal component regression was used to model emission dependence on multiple analytes while taking multiple matrix effects into account. The scheme was applied to the determination of nickel, iron, manganese, chromium, and copper in metal alloys.

1.4.4 LIBS combined with other techniques

Although LIBS is a stand-alone analytical method, over the years it has been combined with other analysis methods to enhance performance for selected applications. In many of these combined methods, the LIBS plasma was used solely as a vaporization/atomization source with the plasma emissions not monitored. Strictly speaking, therefore, they are not really LIBS-based techniques but, because an emissive plasma is formed, we will retain the designation. Here these methods are briefly mentioned. More detailed discussions can be found in reviews (e.g. reference [3]).

Because the plasma/plume contains free atoms, some of which may exhibit significant populations in the ground state, monitoring of species via absorption is possible. Atomic absorption spectrometry has been carried out on the laser plasma/plume in several configurations: (1) the continuum light acting as the "white light" probe source, (2) a flash-lamp or hollow cathode lamp providing the probe beam, (3) with the sample in a graphite furnace to extend the free atom lifetime, and (4) intracavity absorption with the ablated sample positioned in a laser cavity.

The free atoms in the plasma/plume may be interrogated by using laser sources tuned to excite selected fluorescing transitions. While eliminating the simultaneous multi-element detection advantage of LIBS, this LIBS-LIF (laser-induced fluorescence) technique has shown high sensitivity. Similarly, LIBS plus laser-induced fluorescence has been used in spectrochemical analysis of metals [47, 48], for analysis of metals in soils [49], and for the cleaning of limestone [50]. A good source of information about early uses of LIBS-LIF is the review article by Peslak and Piepmeier [51].

The material ablated by the laser plasma can be introduced into other types of AES sources for subsequent analysis. In this way, the desirable sampling characteristics of LIBS can be incorporated into these techniques. The laser microprobe with cross excitation, the first LIBS instrument, is an example, with the auxiliary source being the spark discharge between two electrodes. Other more recent auxiliary sources include a microwave-induced plasma and the inductively coupled plasma. Laser ablation has also been combined with non-optical detection systems such as mass spectrometry and resonance ionization spectrometry.

1.5 Early LIBS instruments

1.5.1 Introduction

Until recently, LIBS instruments were developed for very specific applications, with only one or two devices having been fabricated for a specific client. The reason for this is that LIBS applications differ significantly in system requirements. Hence, no general LIBS instrument for routine use has been developed. Here we describe a few of the early LIBS instruments. One of the earliest references to LIBS instruments is made in Adrain and Watson [52]. A survey of more recent instruments can be found in [7].

The first instrument developed for the commercial market based on the laser spark was the laser microprobe [9]. This device combined laser ablation of a solid sample with the spark produced by conventional metal electrodes. In this laboratory-based instrument, the ablating laser pulse produced an aerosol of material above the sample surface that was then interrogated by the spark produced between the electrodes. These types of instruments were made by several manufacturers and were on the market for several years. Detailed reviews of applications of these instruments can be found in the literature [9].

1.5.2 Transportable instruments

Two of the first LIBS instruments were developed in the 1980s to determine beryllium particles collected on air sampling filters. Photos of these are shown in Figure 1.12a,b. These instruments used down-sized, commercially available Nd:YAG lasers with a detection system based on a monochromator with photomultiplier tube detection. Although the monochromator could be tuned to monitor other wavelengths, these instruments were developed specifically for beryllium. The purpose of these instruments was to determine the mass of Be particles collected on air sampling filters for use in operations where Be particles were generated and where workers may be exposed. Typically, the analysis time for a filter was one minute with a detection limit of 10 ng Be on a filter (25 mm diameter). Dividing the mass so determined by the volume of air that passed through the filter gave the average mass loading of beryllium. Comparison with established exposure standards determined the risk involved.

Also included in Figure 1.12 is a more recent version (1997) of a Be monitor developed for swipe monitoring although the instrument can also analyze filters on which airborne particles have been collected. Note the considerable down-sizing of the instrument in comparison with the first two devices.

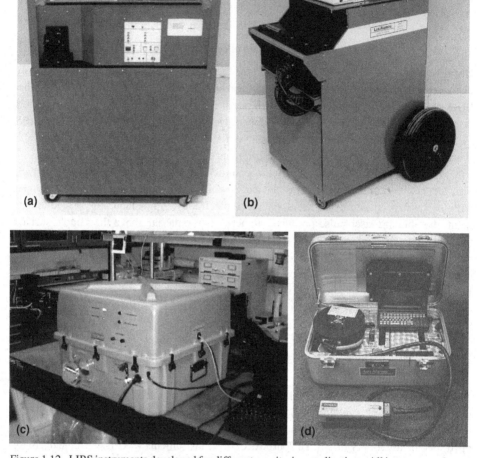

Figure 1.12. LIBS instruments developed for different monitoring applications. All instruments determine elements collected on filters or a swipe material. Because of continuing technology development, smaller and more compact instrumentation is possible. (a, b) Instruments to determine Be particle mass collected on a filter; (c) swipe monitor to determine Be particle mass on a surface swipe; (d) a person-portable surface analyzer.

A transportable LIBS analyzer for toxic elements in soils was developed in 1996 [53] as part of the US Government's Consortium for Site Characterization Technology. The instrument could be transported to the analysis site in the back of a small van and measurements carried out (1) on discrete samples inserted into the instrument or (2) by using a fiber optic probe to deliver the laser pulses to the soil surface.

1.5.3 Field-portable instrument

Because of recent developments in miniaturization of components used for LIBS, it has been possible to fabricate truly portable LIBS-based instruments [36]. An example that was developed in 1999 is shown in Figure 1.12d. This device consists of a main analysis unit connected to a hand-held probe in the configuration of Figure 1.11b. The probe contains the laser, focusing optics, and a fiber optic cable to collect the plasma light. The light is transported back to the analysis unit that houses the small spectrograph, detector, and laser power supply. The instrument is operated through the use of a micro- or laptop computer. An analysis is carried out by placing the probe on the sample and then firing the laser. The laser can be repetitively fired to average the spectra from many shots and increase measurement accuracy and precision. The resulting spectrum is then analyzed via software to determine the element signals. Quantitative analysis is possible through calibration of the instrument using calibration standards of known composition.

It is instructive to compare the four instruments shown in Figure 1.12. The first two instruments (a, b) are essentially single-element detectors (although tunable) having a volume of about 1.5 m^3 and masses in the range 90–100 kg. The swipe monitor (Figure 1.12c) shows considerable size and mass reduction while having all the capabilities of the first two units plus the capability of monitoring a spectral region via a spectrograph/CCD detection system. The volume of this 63 kg unit is about 0.15 m^3. The person-portable instrument (d) has all the capabilities of unit (c) except it incorporates a low repetition laser (1 Hz vs. 10 Hz for other units) increasing the analysis time. This unit is battery operated and has a volume of 0.02 m^3 and a mass of 13 kg.

1.5.4 Industrial instruments

Because of its advantages as an analytical tool that set it apart from conventional methods, LIBS is uniquely suited for industrial applications. One of the first reports of a LIBS-like instrument was the laser corrosion monitor (LCM) [54]. Although not used directly to monitor a spectral signal, the LCM ablated a surface exhibiting corrosion. When the corrosion was completely ablated away, an increase in reflected laser light was recorded from the shiny metal. From the number of pulses required to reach the reflective surface, and knowledge of the depth ablated per pulse, the corrosion depth was determined. The important aspect of the LCM was that it was a compact Nd:YAG laser connected to support equipment through umbilical cables to perform measurements in a hazardous environment (nuclear reactor), a precursor to more recent LIBS devices.

Around 1980, a robust, portable, remote laser microprobe device was developed to inspect reactor components [54]. The analysis head contained a compact laser that was positioned remotely against the target. The laser was fired to ablate material that was then excited by a secondary electrode spark. The spark light was collected by a fiber optic cable and relayed back to a spectrograph/vidicon arrangement to record the spectrum. Remote analysis could be carried out at a distance of about 40 m.

In the early 1990s, LIBS-based instruments were developed by Krupp for in-process quality control and assurance in an industrial plant environment [55]. The instruments were configured for each specific application. These ranged from the direct analysis of liquid steel, the determination of element distributions in a polymer matrix (tire manufacturing), and determining the composition of raw geological materials on a conveyor belt. An excimer or Nd:YAG laser was used, depending on the application.

In the early and mid 1990s there was strong interest in the development of instruments that could be used in the field to monitor toxic materials. As a result, several groups developed LIBS-based cone penetrometer systems for the subsurface analysis of soils [40, 56]. One of two configurations was used. Either a compact laser system was developed that resided in the tip of the cone penetrometer section or the laser pulses were transported to the penetrometer tip through fiber optics. In both cases, the plasma light was transported from the subsurface analysis region by a second fiber optic cable.

1.6 Components for a LIBS apparatus

1.6.1 General

A diagram of a generalized LIBS instrument is shown in Figure 1.2. The main components include a laser, a method of spectrally selecting one or many narrow regions of the spectrum to monitor emission lines, and a method of detecting the spectrally selected light. The specifications of each component of a LIBS instrument as well as the method of sampling to be used will depend on the application. Factors to consider include: (1) the elements to be monitored (number and type), (2) the characteristics of the sample (compositional complexity, homogeneity, etc.), (3) the type of analysis (e.g. a qualitative versus quantitative measurement), and (4) the state of the sample (e.g. gas, liquid, or solid).

1.6.2 Laser systems

Parameters important in the specification of the LIBS laser include: (1) pulse energy, (2) pulse repetition rate, (3) beam mode quality, (4) size/weight, and (5) cooling and electrical power requirements. The wavelength of the laser beam is not an important factor in most cases but can be a consideration if factors such as eye safety, reliability, and wavelength of scattered light are critical. On the other hand, a particular wavelength may couple more efficiently into a specific material [55].

Solid-state lasers, in particular, flashlamp-pumped pulsed and Q-switched Nd:YAG lasers having pulsewidths in the range 6–15 ns, are typically used for LIBS measurements. These lasers are a reliable and convenient source of the powerful pulses needed to generate the laser plasmas. In addition, compact Nd:YAG lasers are available for use in portable instrumentation. The fundamental wavelength of the Nd:YAG laser (1064 nm) can easily be converted to shorter wavelengths (532, 355, and 266 nm) using a crystal via passive harmonic generation techniques which may have certain advantages in terms of increased energy coupling into a particular sample. Typically, however, the 1064 nm wavelength is used because this provides the highest power density. Other types of lasers, most notably the pulsed CO_2 laser (10.6 μm wavelength) and the excimer laser (typical wavelengths of 193, 248, and 308 nm) have been used for LIBS. In comparison with solid-state lasers, however, these lasers require more maintenance (e.g. change in gases) and special optical materials because their wavelengths lie in the far infrared and ultraviolet spectral regions, respectively. For this reason these lasers are not widely used.

As noted above, the Nd:YAG laser is available commercially in a wide range of sizes. These range from laboratory-based models which can output 1 J or more of pulse energy at repetition rates between 10 and 50 Hz to small hand-held versions with a repetition rate of 1 Hz and a pulse energy of about 17 mJ. The laboratory models require 208 VAC electrical services of at least 20 A and may require external water cooling or at least a heat exchanger. The hand-held versions are air cooled, and can be operated from batteries or low-voltage direct current sources.

Femtosecond lasers are becoming more available, and LIBS experiments have been done with pulse lengths of tens to hundreds of femtoseconds. Since the surface irradiation is finished before the plasma is very large, there is no shielding effect as is common with longer pulses. This has an effect on the energy deposited and hence the ablation and crater formed. At this time, there is no conclusive evidence of how this affects the analytical results. The separation of ablation from plasma formation should cast new light on the analysis of these two parts of the technique.

1.6.3 *Methods of spectral resolution*

The basis of a LIBS measurement is the collection and analysis of an emission spectrum. The emission lines of the elements are tabulated in various sources [10–12]. Important properties of a spectrometer are: (1) the resolution, the minimum wavelength separation at which two adjacent spectral features can be observed as two separate lines, and (2) the width of spectrum that can be observed. The specifications for these depend on the particular problem at hand. Typically a wider band of observable spectrum is needed when several elements are being monitored simultaneously.

Here are some examples of different methods for the spectral component of a LIBS system.

- Narrow-bandpass (<1 nm) fixed-wavelength line filter.
- An acousto-optic tunable filter (AOTF) consisting of a crystalline material (e.g. TeO_2) to which a radio-frequency wave is applied. By adjusting the frequency of the wave, the bandpass wavelength of the AOTF can be varied continuously over a certain range.
- A monochromator is a spectrometer that is tuned to monitor a selected wavelength which is presented at the exit slit of the device for detection.
- A spectrograph is similar in basic configuration to a monochromator except it has an exit plane at which a continuous range of wavelengths is presented for detection using some type of array detector or a series of single-wavelength detectors positioned behind individual slits.
- An echelle spectrograph provides the ultimate in wavelength coverage, generally providing a span from 190 nm to 800 nm which is most useful for LIBS detection. The strongest emission lines of most elements lie in this region. Resolving power depends on the specific instrument but is in the range $\lambda / \Delta\lambda = 2500$–$10\,000$.

In the case of the narrow-bandpass filter or AOTF, only a single narrow wavelength band is passed through the wavelength selective element. Ideally, the bandpass corresponds to the emission line width. The transmitted light is then detected using a photon detector (Section 1.6.4). The advantage of the fixed-wavelength filter is very small size and low weight and cost. The AOTF, on the other hand, is somewhat larger and requires a power supply, but it can be tuned to monitor different wavelengths.

An instrument that is being used more frequently in LIBS measurements, where multiple features at widely different wavelengths need to be monitored, is the echelle spectrograph (Figure 1.13a) [57]. In the echelle, a coarse diffraction grating spectrally disperses the light as usual. A prism with its dispersion at right angles to that of the grating, stacks orders vertically over one another so as to create a two-dimensional display of wavelength vs. order. Operation at orders as high as 40 or more is typical. An echellogram image, shown in Figure 1.13b, contains the emission intensities that appear as bright dots. When using an array detector to read the echellogram, software converts the image to a spectrum (Figure 1.13c). The advantage of the echelle is that a large wavelength range may be monitored with reasonable spectral resolution. When operating over a large wavelength interval, care must be taken to calibrate the instrument response. Factors such as the grating blaze, the transmission and reflection coefficients of the optics, and the wavelength response of the detector will influence the sensitivity of the instrument in various spectral regions. There is evidence that the use of the echelle with an array detector can significantly improve analytical figures of merit for LIBS analysis [57].

The most suitable method of spectral resolution depends mainly on the analysis requirements. Factors that must be considered include: (1) the complexity of the sample (i.e. how many elements are in the sample and do these elements have many emission lines or only a few strong lines?), (2) the number of elements to be monitored, (3) whether the elements are to be monitored simultaneously or sequentially, and (4) the location of the emission lines in the spectrum.

Here is an example utilizing these factors. Emission spectra from three different samples are shown in Figure 1.14. All samples contained the element silicon having a strong emission at 288.1 nm but each sample differed in the number of other elements present as major and minor species. In general, the greater the number of elements the more complicated the spectrum. The simplest spectrum is that of Si in water (Figure 1.14a). Here the Si line appears alone without interferences from either H or O which have few emission lines making the emission spectrum particularly simple. A narrow-bandpass fixed-wavelength filter may be the method of choice in terms of simplicity and small size. Silicon in aluminum metal is readily observed in Figure 1.14b although lines due to Mg, Al, and Fe are also present. Steel represents a complex matrix because of the large number of Fe lines, and steels typically contain a large number of other elements. The Si line from a steel sample is apparent in Figure 1.14c but because it is adjacent to a Fe line it is not completely resolved in this spectrum. By using a spectral resolution method having greater resolving power, however, the Si can be separated to some extent from the adjacent Fe line as shown in Figure 1.14d.

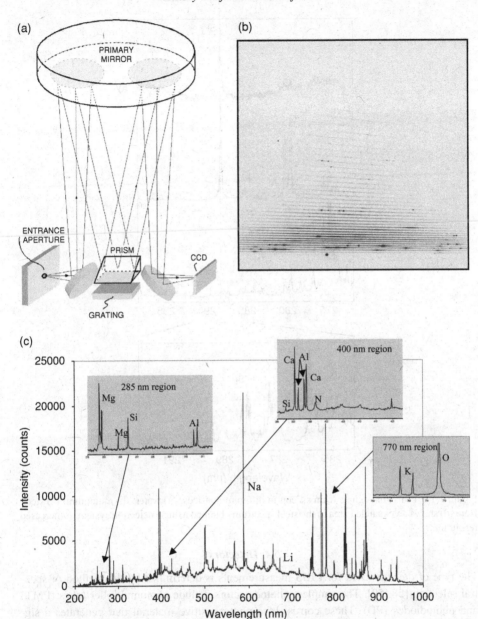

Figure 1.13. Recently developed compact echelle spectrographs permit monitoring of the entire spectral region useful for LIBS on each laser shot. A diagram of a typical echelle device is shown in (a). Here different spectral orders are stacked vertically to permit simultaneous coverage of a wide spectral region. The echellogram image (b) is converted by software into a spectrum (c). The good resolution achievable with a compact device is demonstrated by the three highlighted regions. Part (a) courtesy of Optomechanics Research, Inc.

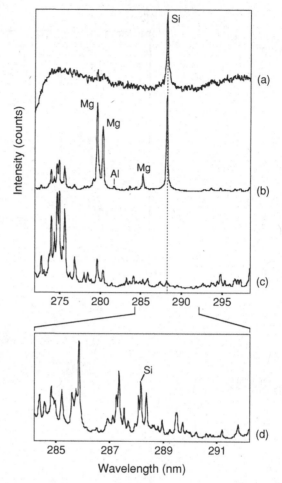

Figure 1.14. Emission signals from silicon in different matrices: (a) water; (b) aluminum; (c) steel. Here (d) shows an expanded area of the steel spectrum obtained using a detection system of increased resolution.

1.6.4 Detectors

The type of detector used for LIBS measurements is determined by the method of spectral selection [58, 59]. The simplest photodetectors include photomultiplier tubes (PMT) and photodiodes (PD). These consist of a photosensitive material that generates a signal proportional to the amount of incident light. These devices are used with spectral selection methods such as fixed filters, AOTFs, and monochromators. By placing small photosensitive elements (pixels) in either a linear or two-dimensional array, an array detector (AD) is produced that provides spatial information concerning the light pattern incident on the array. Common examples of array detectors include photodiode arrays (PDA), charge-coupled devices (CCD) and charge-injection devices (CID). Array detectors are used with spectrographs to record the continuous spectrum presented at the focal plane of the instrument. The spectra shown in Figures 1.5 and 1.14 were obtained by using a CCD detector.

There is a significant distinction that should be recognized between PMT/PD and PDA/CCD/CID devices. This is that PMTs and PDs are high-speed detectors that can provide a temporal history of the light incident on the device to subnanosecond resolution. They can be used to monitor the change in plasma light or element emission signal with time as the plasma decays if tuned to an emission feature. On the other hand, the PDA/CCD/CID devices are light-integrating devices. That is, they collect incident light for a period of time, typically microseconds, and then the charge collected on the device is read out. This difference between the two types of detectors is based on the method of readout. Because PDA/CCD/CID are array devices, the signals stored on each pixel are read out sequentially, one pixel at a time, limiting the speed with which the entire array can be read out and readied for another measurement.

As noted in Section 1.3.1, however, LIBS measurements are typically carried out by using time-resolved detection of the plasma light. It is most important to remove the spectrally broad white light that occurs at early times (0–1 μs) after plasma formation. With the PDA/CCD/CID devices this is accomplished by using a microchannel plate (MCP) in front of the array detector. This component acts as a light valve, which, when dormant, prevents light from reaching the detector behind it. When activated, however, the MCP amplifies (e.g. 1000 times) the incident light by converting it to electrons, amplifying the number of electrons and reconverting to light that is then detected by the PDA/CCD/CID. The increased light signal strikes the photosensitive array. By gating the MCP on/off at the appropriate times after plasma formation the time-resolved detection of the plasma light is accomplished. Array devices with the MCP are designated as intensified detectors (e.g. ICCD, IPDA).

Time-resolved detection using PMTs/PDs is distinctly different than that used with array detectors. The signals from PMTs/PDs consist of a current that is converted to a voltage by the recording device and so require a different procedure. The time-resolved element signal is obtained by integrating the detector current on a capacitor over the desired time after plasma formation using a sample-and-hold circuit. The accumulated charge is converted to a voltage proportional to the signal. If multiple wavelengths must be monitored simultaneously, several slit/PMT assemblies can be placed in the focal plane of a spectrograph with the slit positions aligned to the emission peaks of the elements of interest.

The spectral coverage provided by an array detector is determined by the physical size of the array and the spectral dispersion of the spectrograph. The PDA is a one-dimensional arrangement of diodes that provides spatial intensity information in one dimension. Typical spacing between individual photodiodes is 25 μm, so an array of 1024 pixels has a physical length of about 25 mm. The CCD and CID, on the other hand, are two-dimensional arrays of photodiodes that can provide intensity information along two axes. Typical pixel sizes range from 9×9 µm up to 24×24 µm and array formats range from 576×384 pixels up to 3072×2048 pixels with a large number of other formats in between these extremes.

1.7 Conclusion

Laser-induced plasma spectroscopy has evolved through several development stages in the past 40 years. It started as an optical curiosity principally investigated by plasma physicists. In the mid 1980s its advantages led to the first uses in environmental monitoring applications. Beginning in the late 1980s its potential as a robust, field-deployable instrument, led to the design of the first semi-portable pieces of LIBS apparatus. This developmental trend continued through the 1990s, with compact lasers and spectrographs now common in LIBS instruments. In the past decade, the variety of applications, as evidenced by papers presented at LIBS 2000 (Pisa), EMSLIBS 2001 (Cairo), and LIBS 2002 (Orlando) has increased considerably. The following chapters in this book will detail the leading developmental edge of this ever more-useful technique.

1.8 References

[1] R. Silverberg and M. H. Greenberg, eds., *Murasaki* (New York: Bantam Books, 1993).

[2] A. K. Knight, N. L. Scherbarth, D. A. Cremers and M. J. Ferris, *Appl. Spectrosc.*, **54** (2000), 331–340.

[3] D. A. Cremers and L. J. Radziemski, *Laser Spectroscopy and its Applications*, chapter 5 (New York: Marcel Dekker, 1987).

[4] L. J. Radziemski and D. A. Cremers, *Laser-Induced Plasmas and Applications*, chapter 7 (New York: Marcel Dekker, 1989).

[5] L. Moenke-Blankenburg, *Laser Micro Analysis* (New York: John Wiley, 1989).

[6] D. A. Rusak, B. C. Castle, B. W. Smith and J. D. Winefordner, *CRC Crit. Rev. Anal. Chem.*, **27** (1997), 257–290.

[7] K. Song, Y.-I. Lee and J. Sneddon, *J. Appl. Spectrosc. Rev.*, **37** (2002), 89–117.

[8] E. Tognoni, V. Palleschi, M. Corsi and G. Cristoforetti, *Spectrochim. Acta*, **B57** (2002), 1115–1130.

[9] H. Moenke and L. Moenke-Blankenburg, *Laser Micro-Spectrochemical Analysis* (New York: Crane, Russak, 1973).

[10] J. Reader and C. H. Corliss, *Wavelengths and Transition Probabilities for Atoms and Atomic Ions Part II. Transition Probabilities, NSRDS-NSB 68* (Washington, DC: US Government Printing Office, 1980).

[11] A. R. Striganov and N. S. Sventitskii, *Tables of Spectral Lines of Neutral and Ionized Atoms* (New York: IFI/Plenum, 1968).

[12] R. Payling and P. Larkins, *Optical Emission Lines of the Elements* (Chichester: John Wiley, 2000).

[13] G. Kirchhoff and R. Bunsen, *Chemische Analyse durch Spectralbeobachtungen* (Wien: Verl. Fabrik u. handlung, 1860).

[14] A. G. Gaydon, *The Spectroscopy of Flames* (New York: John Wiley, 1957).

[15] T. Torok, J. M. Mika and E. Gegus, *Emission Spectrochemical Analysis* (Bristol: Adam Hilger, 1978).

[16] Yu. P. Razier, *Laser-Induced Discharge Phenomena* (New York: Consultants Bureau, 1977).

[17] L. J. Radziemski, *Spectrochim. Acta*, **B57** (2002), 1109–1113.

[18] F. Brech and L. Cross, *Appl. Spectrosc.*, **16** (1962), 59.

[19] G. M. Weyl, *Laser-Induced Plasmas and Applications*, chapter 1 (New York: Marcel Dekker, 1989).

[20] T. P. Hughes, *Plasmas and Laser Light* (New York: John Wiley, 1975).

[21] R. G. Root, *Laser-Induced Plasmas and Applications*, chapter 2 (New York: Marcel Dekker, 1989).

[22] H. R. Griem, *Principles of Plasma Spectroscopy* (New York: Cambridge University Press, 1984).

[23] J. B. Simeonsson and A. W. Miziolek, *J. Appl. Phys. B*, **59** (1994), 1–9.

[24] L. J. Radziemski, D. A. Cremers and T. M. Niemczyk, *Spectrochim. Acta*, **B40** (1985), 517–525.

[25] I. V. Mazhukin, I. V. Gusev, I. Smurov and G. Flamant, *Microchem. J.*, **50** (1994), 413–433.

[26] J. R. Wachter and D. A. Cremers, *Appl. Spectrosc.*, **41** (1987), 1042–1048.

[27] H. A. Archontaki and S. R. Crouch, *Appl. Spectrosc.*, **42** (1988), 741–746.

[28] D. A. Cremers, L. J. Radziemski and T. R. Loree, *Appl. Spectrosc.*, **38** (1984), 721–729.

[29] O. V. Borisov, X. Mao and R. E. Russo, *Spectrochim. Acta*, **B55** (2000), 1693–1704.

[30] M. A. Shannon, *Appl. Surf. Sci.*, **127–129** (1998), 218–225.

[31] D. R. Alexander, *Laser Induced Plasma Spectroscopy and Applications Technical Digest*, Opt. Soc. Am. (2002), pp. 164–165.

[32] R. Nyga and W. Neu, *Opt. Lett.*, **18** (1993), 747–749.

[33] A. E. Pichahchy, D. A. Cremers and M. J. Ferris, *Spectrochim. Acta*, **B52** (1997), 25–39.

[34] C. M. Davies, H. H. Telle, D. J. Montgomery and R. E. Corbett, *Spectrochim. Acta*, **B50** (1995), 1059–1075.

[35] D. A. Cremers, J. E. Barefield and A. C. Koskelo, *Appl. Spectrosc.*, **49** (1995), 857–860.

[36] K. Y. Yamamoto, D. A. Cremers, L. E. Foster and M. J. Ferris, *Appl. Spectrosc.*, **50** (1996), 222–233.

[37] D. A. Cremers, *Appl. Spectrosc.*, **41** (1987), 572–579.

[38] H. M. Pang, D. R. Wiederin, R. S. Houk and E. S. Yeung, *Anal. Chem.*, **63** (1991), 390–394.

[39] C. Chaléard, P. Mauchien, N. André, J. Uebbing, J. L. Lacour and C. J. Geertsen, *Anal. Atom. Spectrom.*, **12** (1997), 183–188.

[40] G. A. Theriault, S. Bodensteiner and S. H. Lieberman, *Field Anal. Chem. Technol.*, **2** (1998), 117–125.

[41] A. S. Eppler, D. A. Cremers, D. D. Hickmott and A. C. Koskelo, *Appl. Spectrosc.*, **50** (1996), 1175–1181.

[42] R. A. Multari, L. E. Foster, D. A. Cremers and M. J. Ferris, *Appl. Spectrosc.*, **50** (1996), 1483–1499.

[43] *American National Standard for the Safe Use of Lasers ANSI Standard ANSI Z136.1*, American National Standards Institute (most recent issue).

[44] A. Ciucci, M. Corsi, *et al.*, *Appl. Spectrosc.*, **53** (1999), 960–964.

[45] A. Ciucci, V. Palleschi, S. Rastelli, *Las. Part. Beams*, **17** (1999), 793–797.

[46] S. R. Goode, R. Hoskins and S. Morgan, *Laser Induced Plasma Spectroscopy and Applications Technical Digest*, Opt. Soc. Am. (2002), pp. 36–38.

[47] R. M. Measures and H. S. Kwong, *Appl. Opt.*, **18** (1979), 281–286.

[48] K. Niemax and W. Sdorra, *Appl. Opt.*, **29** (1990), 5000–5006.

[49] F. Hilbk-Kortenbruck, R. Noll, P. Wintjens, H. Falk and C. Becker, *Spectrochim. Acta*, **B56** (2001), 933–945.

[50] I. Gobernado-Mitre, A. C. Prieto, V. Zafiropulos, Y. Spetsidou and C. Fotakis, *Appl. Spectrosc.*, **51** (1997), 1125–1129.

[51] W. C. Peslak and E. H. Piepmeier, *Microchem. J.*, **50** (1994), 253–280.

[52] R. S. Adrain and J. Watson, *J. Phys. D., Appl. Phys.*, **17** (1984), 1915–1940.

[53] D. A. Cremers, M. J. Ferris and M. Davies, *SPIE*, **2835** (1996), 190–200.

[54] H. Koebner (editor), *Industrial Applications of Lasers*, chapter 7 (Chichester: John Wiley, 1984).

[55] C. J. Lorenzen, C. Carlhoff, U. Hahn and M. Jogwich, *J. Anal. At. Spectrom.*, **7** (1992), 1029–1035.

[56] B. Miles and J. Cortes, *Field Anal. Chem. Techol.*, **2** (1998), 75–87.

[57] H. E. Bauer, F. Leis and K. Niemax, *Spectrochim. Acta*, **B53** (1998), 1815–1825.

[58] *Photonics Design and Applications Handbook, Book 3* (Pittsfield, MA: Laurin Publishing Company, 1997).

[59] Y. Talmi (editor), *Multichannel Image Detectors, ACS Symp. Series No. 102* (Washington, DC: ACS, 1979).

[60] D. A. Cremers and L. J. Radziemski, unpublished results.

[61] H. Kurniawan, S. Nakajima, J. E. Batubara, *et al.*, *Appl. Spectrosc.*, **49** (1995), 1067–1072.

[62] L. J. Radziemski, T. R. Loree, D. A. Cremers and N. M. Hoffman, *Anal. Chem.*, **55** (1983), 1246–1252.

[63] K. Grant, G. L. Paul and J. A. O'Neill, *Appl. Spectrosc.*, **45** (1991), 701–705.

[64] R. E. Neuhauser, U. Panne and R. Niessner, *Anal. Chim. Acta*, **392** (1999), 47–54.

[65] R. Knopp, F. J. Scherbaum and J. I. Kim, *Fres. J. Anal. Chem.*, **355** (1996), 16–20.

[66] J. A. Aguilera, C. Aragon and F. Penalba, *Appl. Surf. Sci.*, **127–129** (1998), 309–314.

[67] L. Dudragne, Ph. Adam and J. Amouroux, *Appl. Spectrosc.*, **52** (1998), 1321–1327.

[68] D. A. Cremers and L. J. Radziemski, *Anal. Chem*, **55** (1983), 1252–1256.

[69] C. Haisch, R. Niessner, O. I. Matveev, U. Panne and N. Omenetto, *Fres. J. Anal. Chem.*, **356** (1996), 21–26.

[70] L. Paksy, B. Német, A. Lengyel, L. Kozma and J. Czekkel, *Spectrochim. Acta*, **B51** (1996), 279–290.

[71] R. Wisbrun, I. Schechter, R. Niessner and K. L. Kompa, *Anal. Chem.*, **66** (1994), 2964–2975.

[72] F. Capitelli, F. Colao, M. R. Provenzano, *et al.*, *Geoderma*, **106** (2002), 45–62.

[73] M. Sabsabi and P. Cielo, *Appl. Spectrosc.*, **49** (1995), 499–507.
[74] C. Lazzari, M. De Rosa, S. Rastelli, *et al.*, *Las. Part. Beams*, **12** (1994), 525–530.
[75] S. D. Arnold and D. A. Cremers, *Am. Ind. Hyg. Assoc. J.*, **56** (1995), 1180–1186.
[76] A. V. Pakhomov, W. Nichols and J. Borysow, *Appl. Spectrosc.*, **50** (1996), 880–884.

2

Plasma morphology

Israel Schechter and Valery Bulatov

Department of Chemistry, Technion–Israel Institute of Technology, Haifa, Israel

2.1 Introduction

In this chapter we discuss plasma morphology and its relation to LIBS analysis. We limit the presentation to studies of the past decade, and only a few earlier results, which are of special importance, are mentioned. More attention is given to recent publications, of the past five years. We also limit our discussion to those studies that are relevant to chemical analysis. Many other studies, which were focused on plasma physics or on the laser ablation processes and were not related to LIBS, are not included.

The term plasma morphology refers to spatial characterization of the plasma produced by a laser pulse. The information may be either one-, two-, or three-dimensional and is usually time dependent. Moreover, the spatial data may be related to various plasma characteristics, such as its density, temperature, and compositional distribution. Spectral information is often of interest, since it indicates the presence of atomic species at various locations in the plasma and allows for calculation of temperature and pressure distributions.

The morphological information is directly related to plasma dynamics. It is generally considered that the mechanism for laser-induced breakdown involves three consecutive processes: firstly, a multiphoton absorption, which leads to the ionization and the establishment of free electrons; secondly, a nonresonant continuum absorption of the laser radiation by these free electrons and plasma charges (inverse brehmsstrahlung); and, finally, electron collisions that lead to further ionization of the gas. The last stage results in an increased electron density, heating, and expansion of the gas. Laser heating of the plasma continues, resulting in the growth of the plasma toward the laser. Therefore, the plasma evolution is interconnected with the sampling of the material and with the emission lines of the neutrals and ionic lines, which, in turn, are the basis of LIBS analysis.

We shall see in this chapter that morphological data can be applied to improving LIBS analytical figures in many ways. For example, one can allocate an area in the plasma where a certain element is best observed since, at that position, the signal-to-background ratio of its spectral emission line is the highest, at a given delay time [1, 2]. One can also apply

Laser-Induced Breakdown Spectroscopy: Fundamentals and Applications, ed. Andrzej W. Miziolek, Vincenzo Palleschi and Israel Schechter. Published by Cambridge University Press. © A. W. Miziolek, V. Palleschi and I. Schechter 2006. A. W. Miziolek's contributions are a work of the United States Government and are not protected by copyright in the United States.

morphological data for converting spatial to pseudo-temporal resolution, thus avoiding the necessity of expensive intensifiers used for gating LIBS measurements [3].

Maybe the most important contribution of plasma morphology is related to the problem of signal fluctuations in LIBS analysis. Quite often, the LIBS performance is limited by signal variations, which result from laser instability, variations in the reflectivity of the sample surface, and from the fact that the plasma formation is nonlinearly dependent on the many relevant parameters. Therefore, in many cases integration of the signal over numerous laser shots converges to a rather poor result. On the other hand, morphological information on single laser shots may be applied for characterization of each individual event. Such data can be used either for a proper re-normalization of the spectra (according to the specific conditions of the individual plasma) [4], or for selecting the plasmas of best spectral information. In these approaches, averaging over many laser shots is carried out after each signal has been compensated for its individual variations. Such an approach was utilized, for example, for improving aerosol analysis [5].

Plasma morphology can also be converted to monodimensional imaging of the sample surface. A cylindrical lens produces a long and narrow plasma. The light is projected along the spectrograph slit of an imaging spectrograph, where each ablated point in the sample generates a signal at a different height, and then acquired in a CCD (coupled-charge device) detector [6]. This way, simultaneous chemical imaging of the analyzed surface is obtained. Plasma morphology has also been applied for obtaining the best sampling conditions, by optimizing optical geometry and laser energy [1, 7]. Many other similar interesting applications of morphological data in LIBS analysis will be described here.

This chapter is organized as follows. We first describe the various experimental methods and techniques available for obtaining morphological information and discuss their advantages and limitations. Then, we discuss the morphological results, which clearly depend both on the laser used and on time on the plasma evolution scale. We initially present the results obtained in time-integrated schemes and then move to temporally resolved figures. Regarding the excitation source, we distinguish the plasmas produced by moderate (1–100 ns), long (> 100 ns), and short (fs–ps) laser pulses and discuss each condition separately. Moreover, even when using the same laser, each experimental technique reveals different morphological information. For example, some methods provide excellent temporal resolution at poor spatial definition, while others provide best spatial data with modest spectral information. Therefore, within the duration of each laser pulse, we discuss separately the information obtained by the various techniques.

2.2 Experimental imaging techniques

Several experimental techniques have been developed during the past two decades for obtaining morphological information related to laser-induced plasmas. Actually, the plasma imaging methods follow the technological progress achieved in analytical spectroscopy and in optoelectronic devices. Most of the available instrumentation has been applied to plasma imaging as soon as it has been commercialized, or even before. Understanding the morphological findings reported in the literature, their validity and limitations, requires a certain level of familiarity with the experimental setups and techniques used in these studies. Although many variants were reported, the methods used for acquiring morphological data can be grouped into several categories. These are briefly reviewed in this section.

The ideal tool for studying plasma morphology would provide time-resolved three-dimensional information with full spectrum at each pixel. Unfortunately, such a device has not been invented yet, and each of the proposed methods provides only partial information. Moreover, one can observe that in all past and current technologies there is an intrinsic trade-off, and not all plasma characteristics can be obtained simultaneously. In the following review, one should pay attention to the individual capabilities of the imaging methods, regarding the spatial dimensions, temporal gating (if any), and spectral resolution. In addition, one should pay attention to the data-acquisition mode, either from a single plasma event or averaged over many laser pulses. This point is of considerable importance, in view of the noted plasma variations.

2.2.1 High-speed photography

The simplest method of studying plasma morphology is by obtaining photographs with a high-speed framing camera. This method was widely used 30 years ago [8], although some later reports are also available [9, 10]. This technology allowed for acquisition of $4 \cdot 10^6 - 5 \cdot 10^7$ pictures per second, and was used until replaced by modern electro-optic devices. Interesting morphological information was obtained by using these cameras, such as clear photographs of the interaction of a ruby laser light with a sample and the subsequent phenomena [8].

Another method for time-resolved photography that was applied for the study of shock-wave propagation was based on stroboscopic irradiation of the plasma [11]. Pictures of the shock-wave were taken from a direction perpendicular to the laser beam generating the plasma. The flash-exposure was performed by a 1 ns dye laser and the resulting images were monitored by a video camera. This way, the pictures of the light scattered on the shock-wave were recorded. Delayed exposing pulses were created by means of a Michelson interferometer and the camera recorded stroboscopic pictures of the same plasma at two subsequent moments.

2.2.2 Slitless spectrography

As previously mentioned, a combination of spectral and spatial resolution is desired and such an instrument for two-dimensional imaging was suggested [12]. It was based on simple optical components and provided simultaneous spectral resolution. In this arrangement, the usual entrance slit of a spectrograph is removed. A demagnified image of the source is formed at the object plane of the spectrograph and effectively serves as the entrance slit. An imaging device is placed at the spectrograph exit. It may be a photographic plate, or a modern CCD camera. The inherent temporally resolved readout or time-gating can be obtained by using electronics, such as an intensifier, placed in front of the camera.

This setup works well, but suffers from a trade-off between spectral and spatial resolution. In order to obtain high spectral resolution narrow slits should be used, requiring great demagnification of the source image. However, the spatial resolution is limited by the smallest detector element.

2.2.3 Monochromatic imaging spectrometry

Another spectral imaging device was suggested and constructed using a Czerny–Turner monochromator [12, 13]. Light from the source is collimated before being fed into the monochromator, such that its image is formed on the grating. Then, the light is recollimated before passing through the exit slit. Secondary optics re-form the image of the source outside the monochromator, where an imaging detector is placed. The experimental spectral resolution is controlled by the entrance and exit slit widths, the grating angles, and the focal length of the monochromator. The spatial resolution is determined by the optics outside of the monochromator and by the detector. This setup has potential of rather high imaging performance; however, it requires handling of numerous aberrations and readjustments for each wavelength. If sufficient light is available, this setup can provide high spectral and spatial resolution simultaneously.

2.2.4 Spectrometer slit imaging

Perhaps the most widely used method for obtaining spectral information from various locations in the plasma is spectrometer slit imaging. Several such methods, which are based on imaging the laser plume onto the entrance slit of a spectrograph, have been suggested. If a non-imaging spectrometer is used (such that there is no guarantee that the light at the upper position of the entrance slit is delivered to the same position at the instrument exit), one-dimensional imaging can be performed. This is obtained by keeping a narrow slit and scanning various portions of the image on the slit [6, 8, 14–30]. Obviously, in this case single-shot plasma imaging cannot be performed, and imaging assumes plasma stability over many events. In most experiments time-resolved data were acquired; however, applications of time integrated spectrometer slit imaging to LIBS analysis were also reported [31–34].

However, when an imaging spectrometer is available, one can use imaging detectors and obtain one-dimensional imaging from single plasma events, without the need of scanning the image. The detector may be a spectrographic plate, as introduced many years ago, one or several photomultipliers [35], or a modern CCD camera. If the camera is gated, for example by using a gated intensifier placed in front of the CCD, time-resolved images are available. An example of such a setup, where the sample is placed on an X-Y-Z motorized micro-stage, is shown in Figure 2.1. An example of such slit imaging is shown in Figure 2.2 [36].

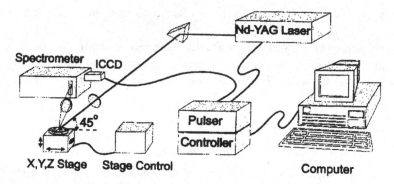

Figure 2.1. Experimental setup for time-resolved spectrometer slit imaging (Figure 1 in reference [22]).

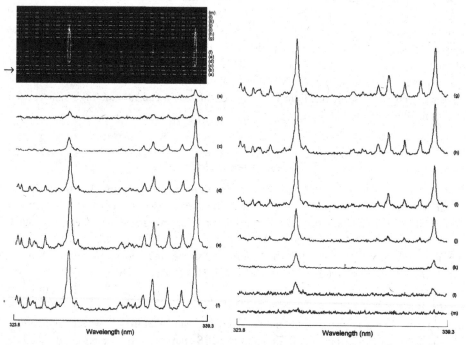

Figure 2.2. Image of a dispersed plasma of a silver–titanium interface, showing the space-resolved spectra (Figure 4 in reference [36]).

Simultaneous two-dimensional information can be achieved by using an imaging spectrometer and a CCD camera [37]. Such a setup is shown in Figure 2.3. The spectra of about 20 different points of the plasma along the z-direction were registered simultaneously, by dividing the CCD into 20 strips. In order to change the position of this recording along the x-direction, parallel to the sample surface, the imaging lens was displaced perpendicular to the optical axis.

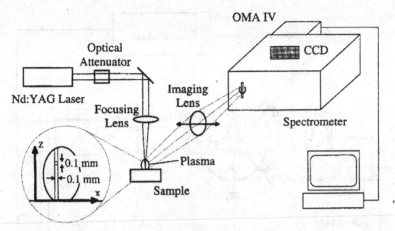

Figure 2.3. Simultaneous two-dimensional imaging (Figure 1 in reference [37]).

2.2.5 *Direct CCD imaging*

Once CCD cameras became available, they replaced photography in LIBS studies. The plasma can be directly imaged onto these two-dimensional detector arrays, thus producing digital images [24, 25, 30, 38–41]. When an intensifier is coupled to the CCD, time-resolved images can be obtained, by a proper gating of the high voltage applied to the multichannel plate. In many applications, direct CCD imaging is combined with other imaging methods, such as imaging absorption. An example of such an experimental setup is shown in Figure 2.4 [42].

Many applications of this technology have been reported [24, 25, 30, 38–41], and we provide here only a representative example. Time-resolved direct imaging of plasma produced on a silver–titanium interface, using the second harmonic of a pulsed Nd:YAG laser (532 nm, 170 mJ, 5 ns), was measured, as shown in Figure 2.5.

Obviously, if an interference filter is placed in front of the CCD detector, one can get a certain level of spectral resolution of the plasma images. Several such studies were carried out, resulting in limited spectral imaging information [22, 43, 44].

Special attention should be given to an experimental setup for recording time-dependent shape and size of the plasma, viewed simultaneously from two orthogonal directions [22]. It is shown schematically in Figure 2.6. The collecting lens was placed at its focal length from the laser-induced plasma to provide a nearly parallel beam incident on an interference filter. The focusing lens focused this transmitted parallel beam in its focal plane where the ICCD was placed to capture the spectrally filtered plasma image. The beam combiner allowed radiation from the plasma as viewed from the front via one of the plane mirrors and a side view via the other plane mirror to reach the detector simultaneously. By slight rotation of the mirrors, the frontal and the side-on image could be laterally offset from one another.

Although CCD detectors provide two-dimensional information, their dynamic range is rather limited. This is a considerable drawback when LIBS signals are concerned, because

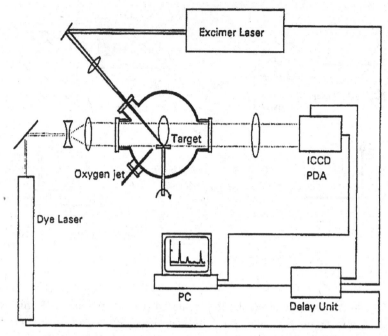

Figure 2.4. Direct CCD imaging setup (Figure 1 in reference [42]).

the plasma light intensity spans several orders of magnitude. Therefore, in some applications it is preferable to use photodiode arrays (PDA). These detectors accept higher light intensities and are less expensive; however, they provide only one-dimensional information [45]. Here, temporal resolution can be achieved by gating an intensifier in front of the PDA, similarly to the CCDs. Wavelength-integrated two-dimensional images were obtained by imaging the plasma onto the PDA (x-direction) and then repeating the acquisition at a series of y-positions. This results in three-dimensional array $I(x, t, y)$, which can be reorganized and presented as two-dimensional plots of $I_t(x, y)$.

An alternative method for ultrafast imaging using picosecond (ps) laser pulses and a video camera has also been suggested [46, 47]. This method allowed for a temporal resolution of 100 ps and was applied mainly for studying the laser ablation processes and the associated blast-wave generation.

2.2.6 Time-resolved imaging through tunable filters

We have seen that two-dimensional images can be obtained by imaging the plasma onto a regular intensified CCD detector, where temporal resolution is reached by gating the multichannel plate intensifier. Now, if a tunable filter is placed between the plasma and the camera, a certain level of spectral imaging may be obtained. The tunable filter may be liquid crystal type or acousto-optic type. The advantage of the tunable filters over the regular ones is that they allow for easy spectral acquisition.

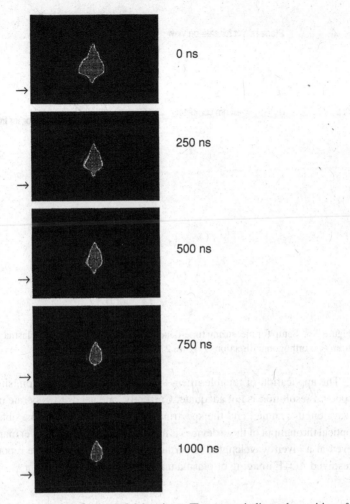

Figure 2.5. Direct CCD plasma imaging, for several delay times. The arrows indicate the position of the sample surface (Figure 2 in reference [36]).

The spectral resolution achieved by these devices depends on several parameters. A tunable liquid crystal (LC) filter provides spectral resolution of about ±5 nm, which is adequate in simple cases for resolving species in the plasma [48]. Better spectral results were obtained when the plasma was imaged using an acousto-optic tunable filter (AOTF) [32]. This device permitted continuous tuning of a narrow spectral bandpass (1.24 nm at 543 nm) from 360 nm to 560 nm (Figure 2.7). Unfortunately, the angle at which the selected wavelengths emerge from the AOTF is wavelength dependent; therefore it is necessary to adjust the position of the ICCD horizontally when changing the bandpass wavelength by more than *c*. 50 nm.

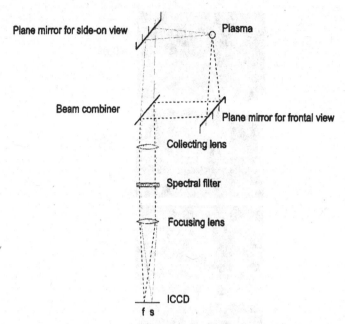

Figure 2.6. Setup for measuring time-dependent shape and size of the plasma, viewed simultaneously from two orthogonal directions (Figure 2 in reference [22]).

The application of tunable filters to LIBS is not very common, since in most cases the spectral resolution is not adequate. Currently, one needs to generate new plasmas for each wavelength scanned and the spectrum of single shots cannot be obtained. Moreover, the optical throughput of these devices is not high, thus integration over many pulses is necessary even at a given wavelength. Nevertheless, several studies were reported, including time-resolved AOTF imaging of plasma under water [49].

2.2.7 Single fiber scanning

A simple setup for obtaining plasma morphologic information was based on the viewing solid angle of a single optical fiber [16]. The image of the plasma was collected by an optical fiber, as shown in Figure 2.8, and sent into the entrance slit of a monochromator. The fiber was placed on a two-dimensional stage with the x- and y-axes parallel and perpendicular to the laser beam direction, respectively. Spectra were acquired with a photodiode array detector, operated in the time-integrated mode. When temporal resolution was desired, an intensified CCD detector was applied [32, 50].

2.2.8 Time-resolved fiber-assisted simultaneous spectroscopy

Although the previously mentioned setup provides temporal resolution, it cannot be applied for morphologic study of single shots. This limitation is caused by the necessity of scanning

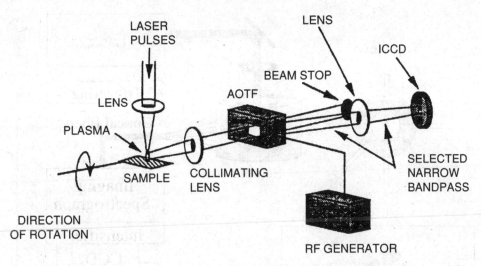

Figure 2.7. Setup for time-resolved plasma imaging using an acousto-optic tunable filter (AOTF) (Figure 1a in reference [32]).

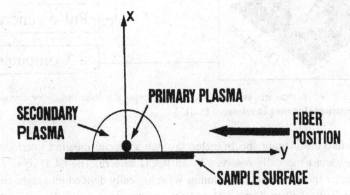

Figure 2.8. Schematic setup for single fiber scanning (Figure 1 in reference [16]).

the fiber location. Therefore, a new device that overcomes this drawback was suggested [51]. This experimental setup was designed for simultaneous spectra acquisition at various locations along the plasma plume, with excellent wavelength and temporal resolutions. The main advantage of this setup is that it obtains time-resolved spectra along the laser beam axis for single shots. The experimental arrangement is shown in Figure 2.9. Plasma excitation was carried out with a Q-switched Nd:YAG laser at single-shot mode. The light from the plasma was simultaneously collected by an eight-ended optical fiber bundle. All collection ends of the bundle were placed along the axis of the laser plasma plume, 1.5 mm apart. These fibers were evenly spaced around the the plasma and adjacent fibers had a vertical displacement from each other. The optical geometry used provided a small collection solid angle, which resulted in a spatial resolution of about 1 mm. The other end of the bundle

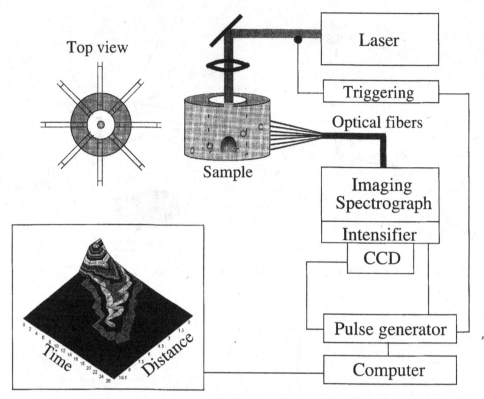

Figure 2.9. Setup for spectral analysis with spatial and temporal resolutions. Information on single shots can be obtained (Figure 1 in reference [51]).

was coupled into an imaging spectrometer, through a special imaging interface. Spectra were recorded and temporally resolved by an ICCD camera. The CCD rows were used for recording the spectra, while the columns were logically divided into eight strips, each devoted to an optical fiber. The CCD and the spectrometer were aligned such that no cross-talk between the fibers was observed [52, 53].

The above experimental concept has been repeated using a much larger optical bundle [54]. The proximal end of the fiber array incorporated a 17×32 matrix of square close-packed fibers that were ordered into a 544 linear array at the distal end. The latter was coupled to a spectrometer and an ICCD detector. Plasma images at a specific lead line could be resolved this way. In this particular setup, both spatial and spectral resolutions were not adequate for LIBS accurate analysis owing to the limited number of pixels in the ICCD chip.

2.2.9 Streak cameras

Streak cameras were frequently applied to plasma diagnostics, mainly because of their exceptional temporal resolution. Several experimental setups have been suggested for study

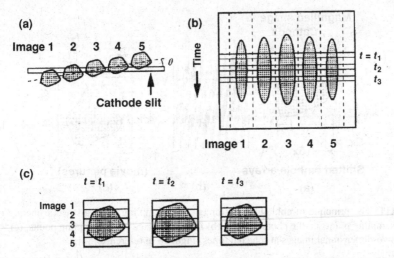

Figure 2.10. The principle of obtaining the two-dimensional imaging with streak cameras: (a) multiple image array aligned on the photocathode slit with a tilting angle; (b) time-resolved images are obtained by streaking the camera; (c) time-resolved two-dimensional images are reconstructed by rearranging the images (Figure 1 in reference [56]).

of plasma dynamics and morphology. The streak camera technology has changed during the past decade and the effective temporal resolution has been improved, from the nanosecond to the picosecond range. Traditionally, this technology was applied to understanding of the plasma dynamics in its first stage of evolution. Generally, the streak cameras were used for three different modes of measurements: (a) space (z–y) mode, (b) time–space (t–z) mode, and (c) wavelength–space (λ–z) mode [55].

Ultrafast two-dimensional X-ray imaging with steak cameras has been obtained by two instrumental methods, which will be briefly described. These are: (a) the multi-imaging camera, which provides temporal resolution of 12 ps and spatial resolution of 15 µm; and (b) the two-dimensional sampling camera, which improves the spatial resolution to about 5 µm [56].

In the multi-imaging camera method, the X-ray streak camera has a slit on the cathode to observe a one-dimensional image that is dispersed with time in the direction perpendicular to the slit, thus obtaining a time-resolved one-dimensional spatial image. However, a two-dimensional image can be converted into a set of one-dimensional images by sampling the image in another direction. Then the set of one-dimensional images converted from the original two-dimensional image can be recorded on a streak camera by placing each one-dimensional image element in order on the cathode slit. After streaking the whole image, the original two-dimensional image can be generated again by selecting a set of one-dimensional images at a certain time and reconstructing them in order.

Figure 2.10 shows the principle of obtaining the two-dimensional imaging. A pinhole camera with an array of N pinholes aligned on a line with a constant separation length is

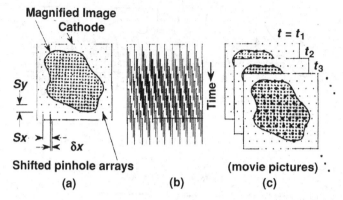

Figure 2.11. The principle of obtaining two-dimensional sampling using a streak camera. (a) Distributed sampling points on the photocathode. (b) Streaked image of the sampling points. (c) Reconstructed two-dimensional images at selected times (Figure 3 in reference [56]).

used to form N images on the horizontal photocathode slit of the streak camera. The image array is slightly tilted from the cathode slit. Thus, sampled images at different positions of the source, separated by a constant vertical distance, are selected by the cathode slit as a set of divided image elements. The set of image elements is streaked and recorded with the streak camera. Streaked image data are sliced and each divided image element rearranged to reconstruct the original two-dimensional configuration of the image. The reconstructed two-dimensional image is then a snap-shot at the selected time.

Better spatial resolution can be obtained with two-dimensional sampling X-ray streak camera technique. Here, a two-dimensional image is sampled with a set of sampling points distributed regularly over the whole image. The distributed sampling points on the cathode of the streak camera act as a set of detector arrays with time-resolving capability. This setup is shown in Figure 2.11. A mask with arrayed pinholes is set on a wide cathode of an X-ray streak camera. The array is shifted by every line, to avoid overlapping of the streaked sample images. The streaked image consists of temporal traces of each sampling point and time-resolved two-dimensional images can be reconstructed by taking the intensity data of each point at the corresponding time.

Two-dimensional space-resolved plasma imaging was obtained by a combination of an imaging optical bundle of fibers and a streak camera [57]. This setup, shown in Figure 2.12, allowed for time resolution better than 10 ps. The plasma image is related by an optical telescope to the end of the bundle fibers. The two-dimensional image sampled by the bundle is transported to the other end of the fibers, which are arranged on a line. The line-arranged fibers are set on the streak slit to obtain a time resolution with a streak camera. The time-resolved two-dimensional images can be reconstructed by rearranging the streak data at the same time position taking into account the bundle fiber arrangement. Other similar arrangements were also recently reported [58, 59].

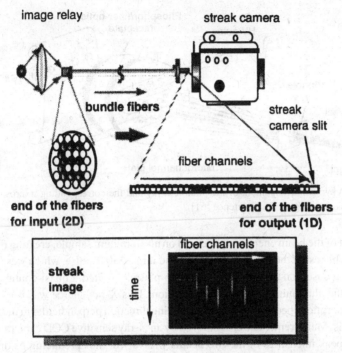

Figure 2.12. Setup of high-speed sampling camera for obtaining two-dimensional spatial and temporal resolutions. It is composed of optical bundle fibers coupled to a streak camera (Figure 1 in reference [57]).

2.2.10 High-speed gated X-ray cameras

Compact multiframe X-ray imaging systems, with both temporal and spatial resolution, were developed [59–62]. In these systems, each impinging X-ray photon is converted into an electron by a multichannel plate, which is intensified and directed toward a second multichannel plate. Finally, electrons hit a photocathode and emit visible photons that are recorded on a CCD camera. The recorded X-rays were in the 1–2 keV range.

The high-speed gating is obtained by a combination of a pinhole camera and a microstrip that propagates the gating voltage on the multichannel plate intensifier. A typical arrangement of the pinhole array and the corresponding microstrip is shown in Figure 2.13. In this particular setup, the pinhole array projects 14 separate X-ray images of the target onto the multichannel plate. Each image is gated on and off in turn as the voltage pulse propagates along the microstrip. Each image is recorded with a temporal resolution of about 80 ps. The apparent spatial resolution was 25 μm.

Another interesting setup for point-projection X-ray absorption spectroscopy with picosecond time resolution was recently described [63]. It provided spatio-temporal mapping of ion distribution within the plasma produced by a nanosecond pulsed laser. The experimental setup included a laser that provided coupled nanosecond and sub-picosecond

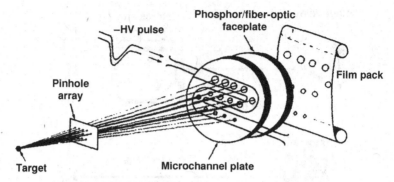

Figure 2.13. A typical arrangement of the pinhole array and the corresponding microstrip used for multiframe imaging (Figure 2 in reference [61]).

pulses. A part of the beam energy was focused on an aluminum sample, creating the expanding plasma. The rest of the beam was compressed into a 350 fs pulse, which was frequency-doubled and focused on an Sm target. The Sm plasma emitted a quasi-continuum X-ray spectrum in the aluminum K-shell spectral region. This X-ray source was used for point-projection absorption spectroscopy along a viewing direction perpendicular to the aluminum expansion axis. Measurements were achieved by an X-ray sensitive CCD camera. The delay between the peak intensities of the two beams was varied, thus producing ps time resolution. Space resolution was limited to 60 μm by the X-ray source dimensions. This setup allowed for mapping the expansion of the aluminum plasma from the surface of the target as a function of time.

2.2.11 Near-resonant photographic absorption/shadow imaging

Another powerful tool for studying plasma morphology was developed, based on photographic absorption/shadow imaging [64]. Although very detailed information is provided this way, interpretation of the data depends on proper modeling of the plasma [59, 62, 65, 66]. In this method, the plasma plume is created by a laser and absorption measurements are obtained through photographic recording of the transmitted light of a secondary pulsed dye laser (Figure 2.14). The dye laser beam was enlarged and spatially filtered by a telescope to illuminate the laser spark as homogeneously as possible over its whole extent. Both direct (absorbed) and reflected rays were refocused to produce a sharp image on a film plate. A small hole was positioned in the focal plane of the imaging lens to reject the plasma light. This technique results in very good spatial resolution and the use of a tunable laser allows both resonant and near-resonant absorption by selected excited ionic species.

The plasma images obtained by absorption photography show two distinct features: (a) a white, dark-fringed line external to the laser spark body, which is produced by the refraction of the dye laser light at the shock-wave propagating in the medium; and (b) dark areas which are produced by absorption of laser light by the ionized gas.

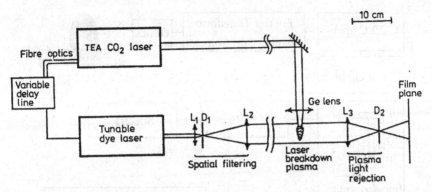

Figure 2.14. Experimental setup for near-resonant photographic absorption measurements (Figure 1 in reference [64]).

Clearly, the transmission of the tunable laser light through the plasma depends on its optical depth. Therefore, a proper interpretation requires information on the density at which plasma becomes optically thick, which is a function of plasma dimension and electron density and temperature. Such data are provided by various models and are very useful when choosing the wavelength at which a given electron density level appears in the pictures. Therefore, application of this technique, combined with time resolution, produces well-defined space- and density-resolved images of plasma inhomogeneity and ionization dynamics. At the end of the breakdown event, where electron densities are greatly reduced, resonant absorption is preferable. However, at electron densities greater than 2×10^{17} per cm^3, the optical depth is lower at resonance than if detuned by 0.2 nm, owing to Stark line shifts.

In a similar setup, absorption imaging was combined with emission imaging, using a fast gated ICCD detector, which provided temporal resolution of about 5 ns [42]. Other modified methods were also suggested. The shadowgraph technique consists of passing a light beam through the test section and letting it fall directly, or via an imaging lens, onto a recording device such as a photographic plate or a CCD matrix [49]. Simple time-resolved plasma shadowgraphs were obtained by transferring a collimated green He–Ne laser light through the focal volume after formation of the plume. A dye laser was also used for this purpose [40]. Time-resolved resonance shadow imaging was applied for investigation of laser-produced lead and tin plasmas [67]. The excitation was carried out by an excimer laser (308 nm) at power density of 10^9 W/cm^2. A dye laser, tuned in resonance with a strong atomic transition of the target material, was directed through the plasma plume for resonance shadowing. The shadow images obtained in this way were created on a fluorescent screen placed behind the target and recorded with a TV camera.

2.2.12 *Fourier transform imaging spectroscopy*

An electro-optical setup for plasma imaging with Fourier transform spectroscopy is illustrated in Figure 2.15 [1]. It consists of imaging optics (zoom lenses) coupled to a step-scan Fourier transform spectrometer. This instrument is based on a mechanically driven

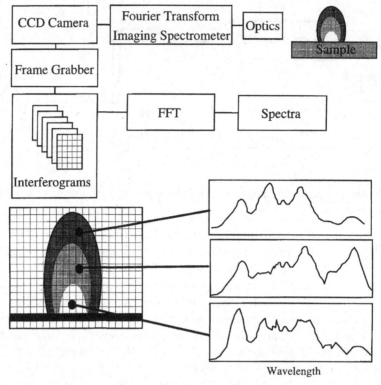

Figure 2.15. Schematic electro-optical setup for plasma imaging with Fourier transform spectroscopy. Spectrum at each pixel is obtained (Figure 1 in reference [1]).

interferometer, coordinated with a CCD camera. In this way an image is collected for each step of the interferometer. The interferograms are imaged onto a thermoelectrically cooled frame-transfer CCD camera. The interferometer steps have been fitted to the maximum data acquisition rate of the CCD (in its frame-transfer mode of operation). The imaging spectra were obtained from these interferograms by a common FFT algorithm. This system provided spectral intensities with 12-bit accuracy, for *c.* 100 wavelengths in the range 400–800 nm. This spectral information was provided for each of the 100 × 100 CCD pixels.

In this method, each analysis consists of *c.* 10 laser pulses at each interferometer step. Thus, the interferograms are averaged over accidental fluctuations. The CCD camera collects the emitted light at a repetition rate that is matched to the laser firing rate. (Actually, the laser was driven at the pre-defined rate of the camera.) In this operational mode, a full 100 × 100 imaging spectrum can be collected in a few minutes

2.2.13 *Double-pulse holography*

Double-pulse holography technique allows for reconstruction of the neutral and electron densities behind the shock-wave front [68]. The plasma is induced by one laser and a second laser, suitably delayed with respect to the first pulse, is applied for following the time expansion of the breakdown event. The holographic laser beam is transmitted through the region of interest, while the reference beam directly lights up the holo-plate. The holograms obtained this way are diffuse, infinite fringe type. Diffusion of the holographic beam allows for obtaining three-dimensional images of the object. The fringe pattern produced by the relative phase change between the two exposures is caused by the neutral and electron densities' perturbation following the explosion.

2.2.14 Resonance laser-induced fluorescence

Laser-induced fluorescence can be used for measurements of relative number densities of atoms and ions at various locations in the plasma. The measurements can be carried out as functions of time, on the plasma time scale. In a particular application of these principles, the plasma was generated on a metal surface using an Nd:YAG laser, while a second dye laser was used for atomic fluorescence excitation of Mg [69]. The fluorescence radiation was detected by a spectrograph, coupled with a CCD detector. Time resolution was achieved by the intensifier placed in front of the camera. The emission light of the atoms in the initial plasma was strong and had to be subtracted. In order to measure the LIF in dependence on the height above the sample surface, the surface was moved by micrometer positioners. The lateral distribution of the Mg atoms and ions was measured by varying the lateral position of a lens, as illustrated in Figure 2.16. By moving the focus laterally through the plasma, the laser-induced fluorescence from different areas could be measured.

The scanning LIF method can be applied for quantitative three-dimensional mapping of elements in the plasma, especially when an intense probe laser is used. In this case optical saturation of the transition is approached, where the fluorescence radiance becomes independent of fluctuations in the excitation source intensity. Yet the fluorescence signal remains proportional to the analyte concentration. Also under these conditions prefilter effects are minimized owing to optical bleaching, and fluorescence radiances may be much larger than thermal excited emission.

Similar LIF setups, combined with optical absorption spectroscopy, were also suggested and applied to LIBS analysis [66, 70]. A related Raman method was also reported [71].

2.2.15 Nonresonance laser-induced fluorescence

The experimental setup for scanning nonresonance laser-induced fluorescence is basically the same as for the resonant LIF measurements; however, this method may have several advantages. When plasma analysis is concerned, the high spectral irradiance of the probing laser can result in a significant amount of scatter from particles in the plume; this degrades the precision and detection limits of the resonance LIF experiments. This problem is severe in saturated fluorescence measurements since the scatter signal remains proportional to the excitation irradiance while the fluorescence signal levels off and becomes independent

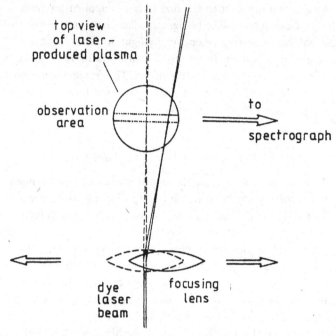

Figure 2.16. Experimental setup for measuring the lateral distribution of analytes in the plasma, by varying the lateral position of a lens (Figure 1 in reference [69]).

of irradiance. Therefore, the nonresonance LIF measurements may eliminate scattering problems [72].

There are four groups of nonresonant atomic fluorescence methods: (a) direct-line fluorescence, (b) stepwise-line fluorescence, (c) sensitized fluorescence, and (d) multiphoton fluorescence. This variety of excitation schemes gives nonresonance fluorescence detection greater freedom from spectral interferences. Another major advantage of nonresonance detection is that the linearity of analytical curves may be extended because postfilter absorption is often not as severe as when resonance fluorescence is used.

2.2.16 Acousto-optical scanning of laser-enhanced ionization

Imaging of the vaporized material can be achieved by scanning a well-focused probe beam across the plume region [73]. Scanning is performed with commercially available acousto-optical beam deflectors, which provide over 10^3 resolution elements in each dimension and scan rates of <100 ns per spot. By using two beam deflectors in tandem, a two-dimensional scan is produced. The spatial resolution of the map depends on the size of the probe beam at the plume and 50 μm resolution can be easily achieved. Therefore, this technique can be used for acquiring detailed time- and space-resolved information about species in the plasma. Particle sizes in the plume can also be measured in this way.

In principle, this scanning technique could be coupled with several detection methods, such as absorption and fluorescence. However, useful information on the vaporization processes could be obtained when coupling this scanning method with laser-enhanced ionization. The ionization is monitored with two electrodes. The spatial and temporal information is determined by the excitation pulse, even though the produced ions take additional time to migrate to the electrodes after they are formed.

2.3 Time-integrated morphology

In the previous section we described the most common experimental techniques designed for the investigation of plasma morphology. Now we focus on the main results obtained by these methods, in relation to LIBS analysis. We have seen that the ultimate device, which provides both spatial and temporal resolution, together with excellent spectral information for single shots, has not been invented yet. Therefore, a certain compromise has to be considered in all morphological studies. We divide all results according to the principal criterion of temporal resolution. We first describe the morphological results obtained under time-integrated schemes (the current chapter) and then move to those obtained with temporal resolution.

Naturally, each family of experimental methods provides different sorts of data and can illuminate different aspects of the complicated issue of plasma morphology. Therefore, we divide the discussion into subsections, each devoted to results obtained using a specific experimental method.

2.3.1 Spectrometer slit imaging

Time-integrated plasma morphology was investigated by recording the spectra at various heights above the sample surface. In the early 1970s, such data were acquired by imaging the plume created on an aluminum disk onto the entrance slit of a spectrograph [8]. The main conclusion was that significant improvement in the slopes of the analytical curves can be obtained by spatially selecting an area of the plume for analysis. This observation was the basis of further improvements of LIBS performance using morphological data. The effects due to several experimental parameters were studied as well.

It was found that the actual results depend on the *laser power* used, and the background increases with laser power. At higher laser power, the spectral lines are considerably more broadened than at lower power. *Self-absorption*, on the other hand, was found to be more in evidence in the low-power originating plumes. (This was especially noticeable in the case of the Al (I) 308.2 and 309.2 nm lines.) Three power densities of a ruby laser were applied in that study, and the results are briefly presented for each of them. However, it should be noted that the results are a function of the temporal profile of the laser pulse, and some of them were obtained using old types of laser.

Low-power regime (1 MW/cm^2; 850 μs pulse): spectra photographed at 1 mm above the sample surface were different from those at 10 mm above the surface. The spectra at the upper region of the plume were relatively weak compared with the lower region and the former contained molecular bands, identified as those of aluminum oxide. In the lower region of the plume these bands were not visible, because of the background. The spectral background at 1 mm above the sample surface was more intense than that of the upper region and was localized in the core of the plasma column.

In the lower region aluminum (III) lines, of excitation potential of 17.8 eV, were detected. These high-temperature lines were confined to the core of the plasma column, while aluminum (II) lines (such as 11.8 eV) radiated strongly and uniformly across the sampled cross section.

Medium-power regime (10 MW/cm^2; 30 μs pulse): aluminum oxide molecular bands were most intense at heights of 1–2 mm and 6–7 mm, and less intense at 4 mm. The intensity of the background continuum was high near the sample surface and became lower for increasing height above the surface. Although the background decreased with increasing height, atomic line emission remained fairly constant up to a height of approximately 4 mm, and then decreased. The highest line-to-background ratio for lines of low excitation potential (*c.* 3 eV) was found at a height of 3–4 mm above the sample surface.

Also at this energy density, the aluminum (III) lines were confined to the core of the plasma and were detected up to a height of 3 mm. Aluminum (II) lines were found up to a height of 5 mm and were not as localized as the aluminum (III) lines.

Self-reversal and extreme broadening of the resonant lines were characteristic of these spectra. Line broadening was at a maximum in the lower region near the sample surface in the plasma core.

High-power regime (4000 MW cm^{-2}; complex temporal profile): at such laser power densities, no intense aluminum oxide molecular bands were present in the spectra. Only one weak such band was detected in the lower region of the plasma.

The background radiation was more intense than that of the plasma produced at lower laser power densities. Background intensity was very high throughout the lower region. A very narrow band of continuum was seen in the core of the plasma. This corresponds to the path of the laser pulse through the plasma, caused by the absorption of the tail of the laser pulse in the plasma. High-temperature lines of aluminum (III) were only faintly seen in the spectra. These lines were broadened as a result of the high pressure associated with the plasma.

Ultra-high-power regime: an interesting observation was the effect of the laser power density, studied in the range 80–900 GW/cm^2 [37]. For metallic samples, a decrease in the total spatially and time-integrated emission intensity was found at intermediate laser power densities, accompanied by changes in the spatial profiles along the laser direction. Beyond 700 GW/cm^2, a sharp increase of the line intensities was observed. It was suggested that these effects are related to shielding by the air plasma, where a part of the laser energy is absorbed by the air plasma during its expansion. This results in a reduction of the power density at the sample surface. Figure 2.17 shows the emission profiles along the z-direction of Cu(I) and N(II) lines for several power densities. At low and high power densities, the N(II) emission is produced nearer to the sample than the Cu(I) emission. However, at intermediate power density the situation is reversed. The shielding effect of the air plasma

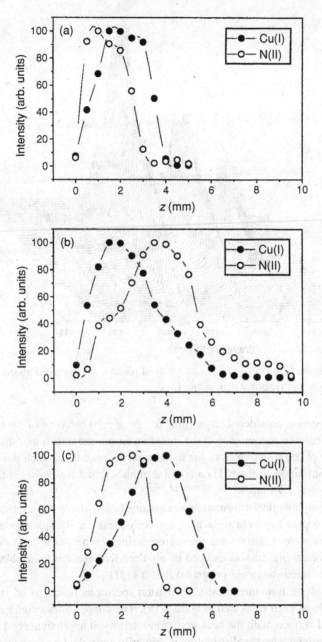

Figure 2.17. Spatial profiles along the z-direction of the plasma, obtained for different power densities: (a) 80 GW/cm², (b) 300 GW/cm², (c) 900 GW/cm² (Figure 9 in reference [37]).

Israel Schechter and Valery Bulatov

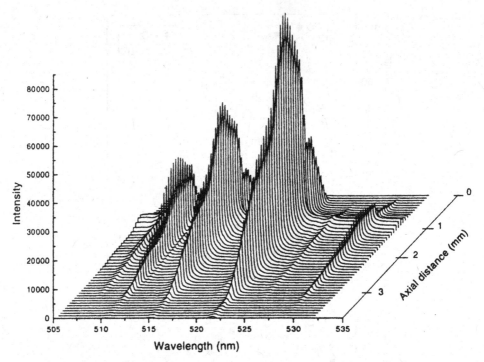

Figure 2.18. Integrated Cu emission intensity vs. axial position in the plasma generated by power density of 4.93 GW/cm^2 (Figure 2 in reference [31]).

(nitrogen emission) is considered responsible for the special behavior of the emission. At higher power densities, the absence of the shielding air plasma results in a strong increase in the intensity of the emission from the metal samples plasma. At such power densities the relative contribution of the N(I) emission diminishes, and the emission from surface components becomes dominant.

Additional morphological information was obtained by studying the laser-induced plasmas from solid copper targets in argon at atmospheric pressure. The spatial profiles of their emission spectra were measured and the corresponding temperatures were calculated. In these measurements, plasma was induced by a pulsed KrF excimer laser (248 nm, 30 ns) and the emission intensities were integrated for 10 s [31].

Spatially resolved, time-integrated Cu emission spectra as functions of axial position in the plasma have been measured (Figure 2.18). The copper lines extend about 3 mm above the metal surface, with the peak emission intensity at approximately 1.5 mm from the surface. The axial intensity distribution is primarily dependent on the power density of the laser, as for the previous case (Figure 2.19). The spatial extent of the plasma diminishes when the power density decreases. The peak emission intensity is closer to the target surface at lower power densities. For low laser power densities, there is only one peak in the Cu emission intensity spatial profile, in comparison with extensive shoulders at the higher laser

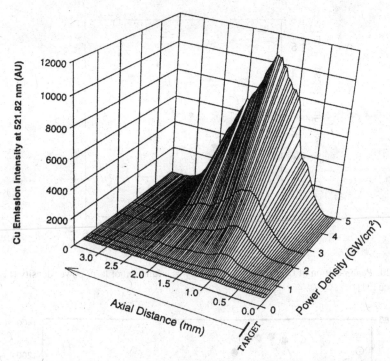

Figure 2.19. Spatial distribution of Cu(I) emission line as a function of laser power density (Figure 3 in reference [31]).

power densities. The position of peak emission intensity occurs farther from the surface as the power density is increased (Figure 2.20).

The spatial profile of the *plasma temperature* was also calculated. The excitation temperature was a weighted average of the temperature at different times. Apparently, the maximum in the temperature spatial profile does not overlap with the maximum plasma emission intensity. The maximum temperature is always closest to the target surface before the emission intensity has peaked. This general finding was validated by other studies as well.

The effects of the *ambient atmosphere* and *irradiation wavelength* on the copper plasma emission characteristics were studied with time-integrated spatially resolved spectroscopy [21]. The spatial distribution of the emission intensities of the neutral atomic line, Cu(I) at 521.82 nm, and the ionic line, Cu(II) at 508.83 nm, was measured, as shown in Figure 2.21. The maximum intensity of the Cu(I) line was observed at 3.2 mm from the target surface and gradually decreased at further distance from the surface. The copper ion lines were observed near the surface, owing to the thermal ionization of the hot target vapor. Its intensity rapidly decreased within a short distance from the surface.

The surrounding atmosphere had a considerable influence on line-broadening and self-absorption phenomena. Generally, the *line broadening* was large near the surface,

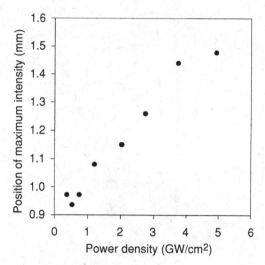

Figure 2.20. Peak position of Cu(I) emission intensity as a function of laser power density (Figure 5 in reference [31]).

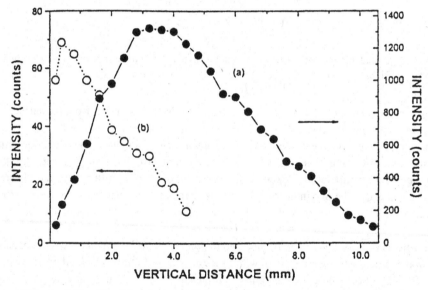

Figure 2.21. Spatial profile of the plasma emission intensities of (a) the neutral atomic line Cu(II) at 521.82 nm and (b) the ionic line Cu(II) at 508.83 nm (Figure 2 in reference [21]).

mainly as a result of the Stark effect and then decreased in further regions of the plasma.

Self-absorption was observed only near the target surface. The self-absorption under helium atmosphere was observed even in 760 Torr, and a little appeared in neon. In the argon atmosphere, no self-absorption appeared under 769 Torr, since sufficient atomic species did not exist in the plasma because of the lower efficiency in laser ablation. When the argon pressure was decreased below 200 Torr, the self-absorption occurred near the target surface. These results indicate that the *ambient gas* acts not only as a damping material to prevent the free expansion of the propelled atoms but also as a cooling material for the hot vapor plasma. The differences in the cooling effect were attributed to the differences in thermal conductivity of the ambient gases.

One-dimensional imaging, providing a combination of morphological and chemical information, was studied by focusing a pulsed excimer laser (193 nm, 100 mJ, 10 ns) on a metal sample [15]. The plasma emission spectra were spatially resolved by moving the position of the sample in the slit imaging method. It was found that copper, which has high thermal conductivity and high boiling point, produced a confined plasma and high excitation temperature. Lead, which has low thermal conductivity and low boiling point, generated an extended plasma and relatively low excitation temperature. The spectral information showed that laser ablation of metals leads to the generation of different chemical species, which depend on the position in the plasma. Investigation of the optimum position for spectrochemical analysis showed that the best observation point in the plasma depends on the metal itself.

Two-dimensional imaging was applied later for morphological investigation of the plasma produced by a Nd:YAG laser (1064 nm, 4.5 ns) on metal surfaces, under ambient atmospheric pressure [37]. This method allowed for acquiring spectra as functions of distance from the surface (z-direction), and at various perpendicular locations (x-direction). Figure 2.22 shows a typical plot of the plasma emission spectra as functions of z, for two x-positions: emission from the central z-position and from the border of the plasma. Clearly, more-intense lines are emitted at the center of the plasma, with a higher extension in the z-direction. Also, the emission lines are broadened by the Stark effect, and an intense continuum is emitted at the plasma center, where a high electron density and temperature exist in the early stages of the plasma evolution.

Chemical maps, namely the spatial distribution of Cr(I), Fe(I), and Fe(II) spectral lines, are presented in Figure 2.23, as contour lines of constant intensity. It was found that, under these experimental conditions (900 GW/cm^2), the spatial distribution of neutral atom emission is quite similar. However, the ion emission is confined to a narrower region in the x-direction than the neutral atom emission, corresponding to a higher temperature at the center during the expansion process of the plasma.

The time-integrated spatial distribution of atomic lines obtained in vacuum is of interest, since the damping and cooling effects are not present under these conditions. Such measurements were carried out, when a Nd:YAG laser (1064 nm, 8 ns) was focused on a brass sample placed in a vacuum chamber and the spectral line of Zn(I) at 481 nm was measured [28]. Typical results are shown in Figure 2.24. It is seen that the emission intensity experiences a step climb from 2 mm to 3 mm, and declines rapidly between 3 mm and 5 mm. Differently, the background intensity exhibits a practically flat featureless profile, which is slightly higher close to the surface.

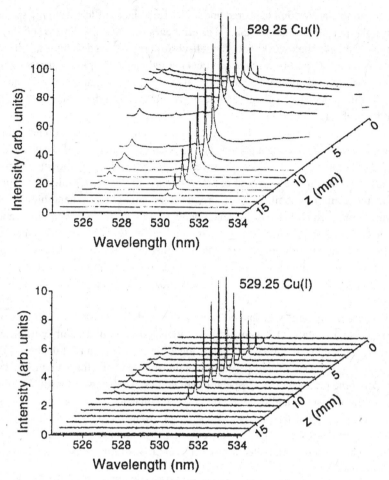

Figure 2.22. Spectra of the time-integrated plasma emission for different distances to the sample (z). Upper part: emission from the central position in the plasma. Lower part: emission from the border of the plasma (Figure 2 in reference [37]).

2.3.2 *Monochromatic imaging spectrometry*

Monochromatic imaging spectrometry was applied to morphological characterization of the plasma generated by a Q-switched Nd:YAG laser (150mJ, 10 ns) focused on aluminum metal sample, under controlled buffer gas pressure of Ar, He, and air [13]. The setup was designed for spectral resolution of 1 nm and spatial resolution of 0.2 mm. The effects of the gas pressure and composition were studied by using plasma imaging and it was found that, in general, the plasma expanded with reducing the ambient pressure. The degree of expansion was more emphasized in He than in Ar or air. Images of Al(I) 396.15 nm line emission of laser-induced plasmas obtained in different atmospheres are shown in

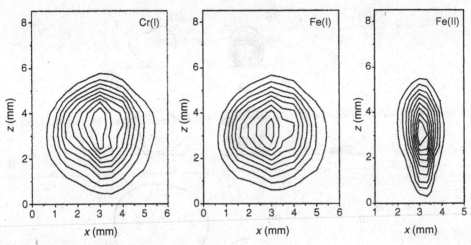

Figure 2.23. Spatial distribution of the time-integrated emission intensity at selected spectral lines in steel plasma (Figure 4 in reference [37]).

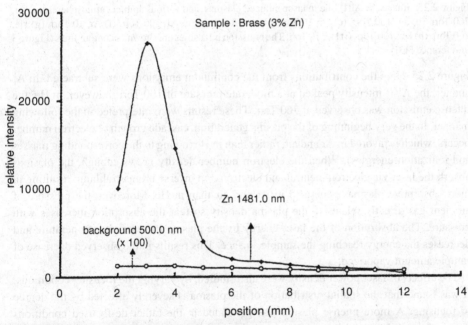

Figure 2.24. Time-integrated spatial distribution of Zn(I) line and background. The laser energy of 26 mJ was focused onto a brass sample (Figure 6 in reference [28]).

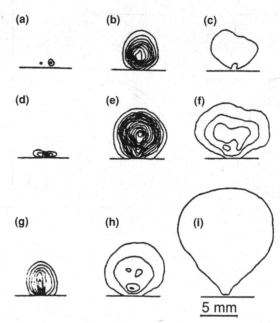

Figure 2.25. Images of Al(I) line in laser-induced plasmas, and various ambient atmospheres: (a) air, 760 Torr, (b) air, 100 Torr, (c) air, 10 Torr, (d) Ar, 760 Torr, (e) Ar, 100 Torr, (f) Ar, 10 Torr, (g) He, 760 Torr, (h) He, 100 Torr, (i) He, 10 Torr. The emission intensities are shown as contour lines (Figure 3 in reference [13]).

Figure 2.25. Here the contributions from the continuum emissions were subtracted. In Ar and air, the Al(I) intensity peaked at a moderate pressure of 100 Torr; however, in He, the intense emission was observed at 760 Torr. These results were interpreted in the following manner. In the very beginning of the plasma generation, cascade growth of electron number occurs, which is favored in Ar and air, rather than in He (owing to the corresponding masses and ionization energies). When the electron number density grows enough, the plasmas absorb the laser via electron–neutral and electron–ion inverse bremsstrahlung, resulting in more-absorptive plasmas generated in Ar and air than in He. Moreover, the pressure of ambient gas directly relates to the plasma density so that the absorption increases with pressure. The absorption of the laser energy by the plasma increases its temperature and decreases the energy reaching the sample surface. This results in the observed decrease of sample amount vaporized.

The effects of laser power density were also studied, by varying the focusing conditions. It was found that the spatial distribution of the plasma is severely affected by the degree of focusing. A more intense plasma was generated in the rather de-focused conditions than in the tightly focused cases. But in the more de-focused conditions, the intensity and the emissive region of the plasma was much decreased. It was concluded that a proper adjustment of the focusing conditions may improve the signal-to-background ratio of LIBS measurements.

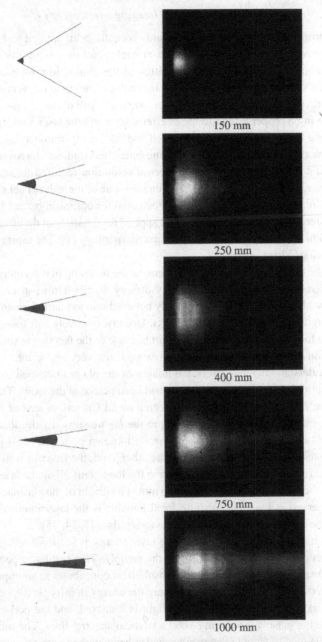

Figure 2.26. The effect of the focusing lens on the morphology of the induced plasma (Figure 2 in reference [1]).

2.3.3 *Fourier transform imaging spectroscopy*

Fourier transform imaging spectroscopy has only recently been applied to LIBS studies. This method provides full spectral resolution at each pixel of the plasma; therefore, it may result in mapping the elemental composition of the plume. In principle, one cannot obtain results on single shots with the current technology, owing to the very nature of the Fourier transform spectral acquisition. However, one could still obtain temporal resolution, simply by applying a proper timing to the interferometer and the laser. Unfortunately, this timing procedure is experimentally complicated and such an instrument has not yet been constructed. The current reports are only for time-integrated studies. Moreover, the current instrumentation does not provide the best spectral resolution required for accurate LIBS analysis; however, it may be improved in the future. In spite of these drawbacks, this method provides useful morphological data, which are essential for optimization of LIBS analysis.

Fourier transform imaging spectroscopy was applied for the study of the effects of several optical and geometrical factors upon laser plasma morphology [1]. The most relevant ones are briefly mentioned here.

Lens effect: the effect of the laser focusing lens on the resulting plasma morphology was studied. Each lens produces a different energy density distribution in space, resulting in different plasma characteristics. It is commonly believed that too hard focusing conditions (small focal length) do not provide good results. Usually, relatively soft focusing (using a lens with a long focal length) is preferred, in part because of the fact that in such conditions the exact location of the sample along the laser beam is not very important.

In order to understand the lens effect, the images of the plasma obtained by a variety of lenses were acquired. In each case the sample had been placed at the focus. The results are shown in Figure 2.26, where the target has been a metal Cu foil. A lens of $f = 150$ mm produced a very small and hot plasma, owing to the high energy density localized at the focal point. This is probably the reason for the well-known poor analytical performance obtained under such focusing conditions. On the other hand, the lens of $f = 1000$ mm was also not suitable for analytical purposes, owing to the long "tail" along the laser beam. The best conditions were obtained at about $f = 400$ mm. The origin of the plasma "tail," which was clearly observed at long and medium focal lengths, is the laser interaction with the aerosol above the sample, generated by the previous pulses [5, 74, 75].

At $f = 1000$ mm a considerable part of the laser energy is scattered and only a small portion of the energy reaches the surface of the sample. This results in poor analytical performance. On the other hand, the same aerosol effect contributes to an improvement of the results at medium focusing conditions, where the energy density (at the same distance from the surface) is lower. In this case, less light is scattered, and the persistent aerosol is excited in the hot plasma, producing good analytical spectral lines. The morphological effects due to the lens-to-sample distance have also been studied with this technique, and the results were similar to those obtained with AOTF imaging [32].

Similarity maps: the morphological structure of the plasma has been examined by using similarity maps. The construction of these maps required identifying regions that have

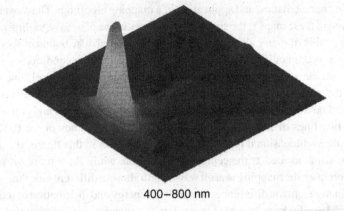

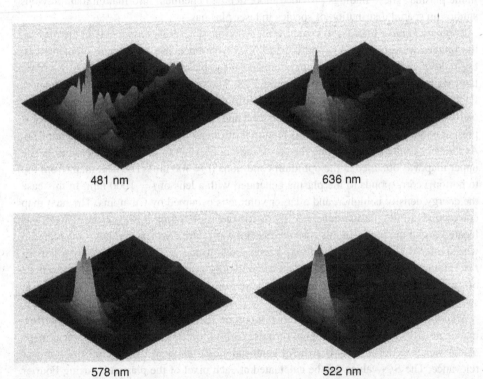

Figure 2.27. A three-dimensional presentation of the integrated intensities of the full spectral image of the plasma (top) and of the images obtained by application of logical filters at several emission wavelengths (four bottom images) (Figure 5 in reference [1]).

similar spectral characteristics, and application of a mapping algorithm. The most important feature observed in these maps is the spherical symmetry of the plasma regarding its spectral characteristics. This symmetry allows for calculations of radial distribution of spectra, since Abel transformation can only be performed under some symmetry conditions.

Spectral maps: since this method provides full spectrum at each pixel, one can draw the image of the plasma at any desired wavelength. Figure 2.27 shows the image of the plasma (induced on a brass sample) at the spectral rage of 400–800 nm and at four selected wavelengths: two lines of Zn (481 nm and 636 nm), and two lines of Cu (522 nm and 578 nm). The three-dimensional representation is also shown in this figure. It can be seen that, generally, Cu is located at the center of the plasma, while Zn is more present in the outer shell. Moreover, the mapping at each wavelength shows a different morphology, which probably originates from the differences in transition energy and distribution of temperature in the plasma. These findings provided more detailed chemical information than previous studies on spatial distribution of species in laser plasmas.

Elemental maps: in order to visualize the location of several components in the plasma, the images were classified according to known reference spectra of pure elements. An example of such an elemental map, obtained for the plasma created by a pulsed Nd:YAG laser on brass sample, is shown in Figure 2.28. The presence of Zn and Cu is indicated. Each map in this figure was obtained under different focusing conditions. Generally, these maps show that Zn is present at an *outer* shell and in the "tail" of the plasma, while Cu is present mainly in an *inner* shell. This feature is attributed to the lower melting point of Zn. At the very center of the plasma the spectra are noisy (due to the high temperature). The upper map was obtained at the optimum conditions ($f = 400$ nm). The next map (from top to bottom) corresponds to the plasma generated with a lens of $f = 1000$ mm. In this case the energy density is higher, and a larger volume is occupied by Cu atoms. The next map corresponds to the plasma generated by a lens of $f = 400$ mm; however, the sample was located just at the focus. In this case the plasma was relatively localized but some material was sputtered away. The bottom map corresponds to the plasma generated by a lens of $f = 150$ mm. The plasma was small and hot, and only a thin shell of Cu and some residuals of Zn were observed. This was probably the reason of the poor analytical performance at -such conditions.

Signal-to-noise maps: one of the most important factors affecting the analytical performance in LIBS is the spectral signal-to-noise (S/N) ratio. Therefore, morphological maps of the signal-to-noise ratio, drawn for each individual element (or spectral line), are of relevance. The S/N values can be calculated at each pixel of the plasma by using Fourier transform imaging spectroscopy, resulting in three-dimensional maps. Usually, the best S/N was obtained along the laser beam axis, not too close to the center of the plasma. Several local maxima were found along this coordinate, and the S/N values decayed at long distances. A maximum was also observed when moving along the perpendicular coordinate; however, it was much lower. The clear global maximum observed in such S/N maps indicates that analysis can be improved by a proper optical setup that is designed to collect the light emitted from the best location.

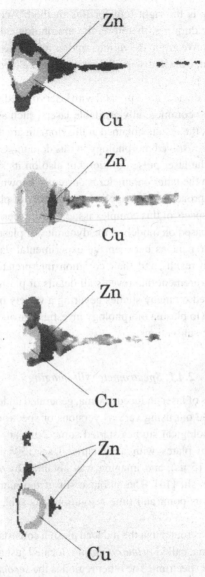

Figure 2.28. A result of the classification algorithm performed on the spectral imaging data. Each image was obtained at different focusing conditions, using a lens of $f = 400$, 385, 360 and 250 mm (from top to bottom) (Figure 6 in reference [1]).

2.4 Time-resolved morphology: excitation by medium laser pulses (1–100 ns)

The most important morphological information has been obtained in the time-resolved operational mode, since it allows for understanding the plasma evolution and optimization of the analytical performance. Since accurate LIBS analysis usually requires temporal gating, time-resolved morphology is the right tool for this method. Actually, plasma dynamics depends on the laser pulse duration; therefore, the morphological data must be classified according to this parameter. We group the various studies into those related to long, medium, and short laser pulses. Then, we will discuss the morphology related to the plasma produced be a sequence of two short pulses.

Most LIBS experimental devices are equipped with lasers of medium pulse duration (commonly, around 10 ns). Most commercially available lasers, such as Nd:YAG and excimer, operate in this range. Thus, the results obtained at this domain are the most relevant and we start the discussion on time-resolved morphology in this domain. Clearly, plasma morphology depends not only on the laser pulse duration but also on its spatial profile. However, only a little information on the latter parameter is currently known.

Several experimental approaches were applied to the study of plasma morphology, each one illuminates different aspects of this complex issue. Most experiments support a general theoretical interpretation, based on modeling the dynamics of plasma expansion; however, there still exist some discrepancies between the experimental data and the models. We will now present the main results and their common interpretation. It should be noted that there is no single measurement that reveals all details of plasma morphology, and the conclusions drawn are based on many studies covering a variety of parameters. The most relevant techniques applied to plasma morphology investigation and their results are briefly mentioned in the following subsections.

2.4.1 Spectrometer slit imaging

Most morphological studies of laser-induced plasma, generated under conditions of medium pulse duration, were carried out using various versions of spectrometer slit imaging. The early time-resolved morphological studies started some 20 years ago, when plasma was produced on brass and Zn plates with an excimer laser (308 nm, 6 mJ, 10 ns) and dye laser (585 nm, 4 mJ, 15 ns), and imaging was obtained by varying the observation point of the spectrometer's slit [14]. The entrance slit of the monochromator was set at 3 mm distant from the focus point and time resolution was achieved by a fast rise time photomultiplier.

In these experiments it was found that the induced plasma consists of two distinct regions. One is a small region plasma, called *primary plasma*, located just above the surface of the sample. It emits continuous spectrum. The other region is the *secondary plasma*, spreading around the primary plasma. Its edge is clear and it has a hemispherical shape. The diameter of the hemisphere increases with decreasing pressure. The temperature of the secondary plasma was in the range of 6000–10 000 K, depending on the pressure. The time behavior of spectral lines indicates that, immediately after the cessation of the emission of the primary plasma, atoms gush from the target, producing the secondary plasma. The primary plasma plays a role as a source of blast wave expanding around it with a spherical wave front. It was considered that the luminous, high-temperature secondary plasma is formed because of the heating effect due to the blast wave.

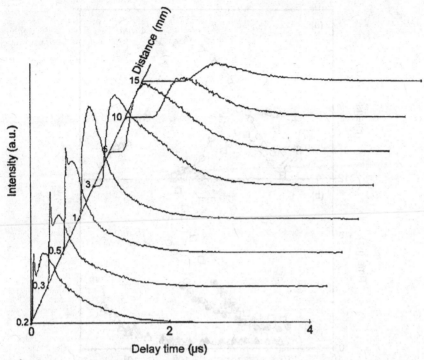

Figure 2.29. Temporal profiles of Cu(I) emission at different distances from the target (Figure 4 in reference [35]).

As previously mentioned, time-integrated morphological studies indicated that laser ablation of metals leads to the generation of different chemical species, which depend on the position in the plasma and the optimum position for spectrochemical analysis depends on the metal itself. Similar conclusions were drawn for time-dependent morphology as well; however, much more dynamical information was available in this case.

The *time-dependent morphology* of laser-induced plasma is of special interest, since it provides insight in to plasma dynamics. The dynamics of copper plasma generated by a XeCl excimer laser (308 nm, 15 ns) was investigated by time- and spatially resolved measurements, under reduced pressure environment [35]. The temporal profiles were acquired by a fast photomultiplier. It was found that there are three types of plume emissions according to their time evolution and space distribution: the Cu(I) emission at 427.5 and 465.1 nm (group-a), the Cu(I) emission at 324.8, 327.4, 510.6, 515.3, and 521.8 nm (group-b) and the copper ion Cu(II) at 283.7 nm (group-c). An example of the temporal profiles observed at different distances from the copper target is shown in Figure 2.29, for the 515.3 nm Cu(I) line. One can see the continuum appearing as the first peak of the profile in the vicinity of the target. Far away from the target (4–15 mm), the time profile of the signal does not change drastically. At a distance of about 1 mm, the peak intensity of the emission has its maximum value.

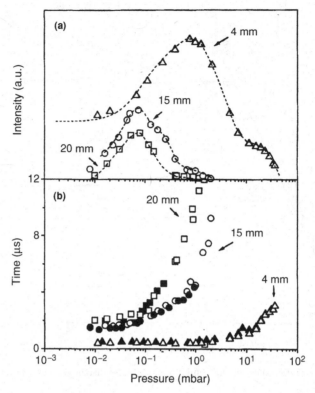

Figure 2.30. Maximum emission intensity (a) and its delay with respect to the laser pulse (b), as functions of the oxygen pressure applied. The results were recorded at various distances from the target surface, as indicated (Figure 3 in reference [19]).

However, it is interesting that the evolution of group-a lines differs significantly from that of group-b lines. Results indicate that group-a and group-c (corresponding to high-energy excited Cu(I) species and copper ions, respectively) are very similar in their behavior: they emerge promptly after the laser pulse and decay rapidly with time and space. In contrast, the group-b emissions emerge at a later time. In addition, they extend to a larger region and last a longer time.

The above behavior of the plume emission led to the conclusion that the plasma undergoes several stages: (a) it forms as the result of the laser ablation, (b) it evolves from the optical and collision-induced breakdown of the ablation-created vapor (the stage when a significant fraction of the ablated material appears as group-c and group-a species), and (c) it moves to the electron-collisional expansion dynamics (the stage when the plasma consists mainly of low-energy excited Cu(I) emissions).

Obviously, plasma morphology depends on the *surrounding gases*. Generally speaking, the plasma generated under vacuum conditions has been studied in relation to vacuum deposition technologies, while the ambient conditions were usually related to analytical

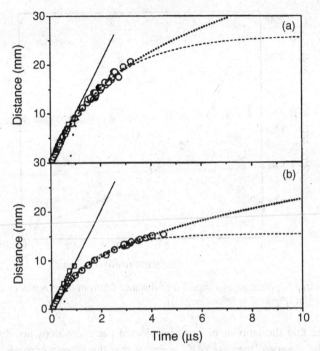

Figure 2.31. Distance at which the maximum emission intensity is observed as a function of its delay with respect to the laser pulse. Results were obtained for Ba ion (493.4 nm), in 0.1 mbar (a) and in 1 mbar (b) of oxygen. The results obtained in vacuum are denoted by squares and the full line shows the free-expansion regime (Figure 5 in reference [19]).

applications. For example, the expansion dynamics of the plasma generated by laser ablation of a $BaTiO_3$ target in the presence of two gases over a wide pressure range was studied by a similar experimental method [19]. The excitation laser was ArF excimer (193 nm, 12 ns, 2 J/cm^2). It was found that the plasma emission intensity, observed at a given distance from surface, is time dependent. It reaches a maximum, I_m, at a time t_m, and then decays slowly. However, these values are pressure dependent, as shown in Figure 2.30 for measurements performed at three distances above the sample surface. Both the pressure at which the maximum of I_m is observed and that at which I_m becomes zero depend on the distance from the target. The higher the distance, the lower the pressure. Note that t_m remains constant up to a pressure threshold that also depends on the distance to the target. For higher pressures t_m increases sharply, being consistent with the reduction in the ejected species velocity in the presence of the gas environment.

The measured dependence of the distance from the target at which the maximum emission intensity (I_m) occurs on the delay time of this maximum (t_m) is shown in Figure 2.31. The results obtained in vacuum are also presented, indicating a linear dependence over the whole measured range. When ambient gas pressure is applied, the emission can be detected at greater distances and higher delays than in vacuum.

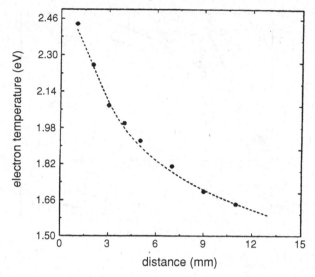

Figure 2.32. Electron temperature as a function of distance (z) from target surface. The dotted line represents the $z^{-0.1}$ fit (Figure 2 in reference [33]).

If we assume that the amount of material ejected per pulse does not depend on the gas pressure, the observed increase of t_m suggests that the plasma expands more slowly at high pressures. The density of the plasma is, thus, increased in the region close to the target, promoting the confinement of the plasma. Therefore, the emission observed at longer distances from the target would tend to disappear. The sharp increase of t_m above a threshold of pressure or distance is a clear indication that the plasma expansion dynamics in the presence of a gas change with respect to the free-expansion dynamics followed in vacuum.

Another morphologic parameter of interest is the distribution of *electron density* in the plasma. The electron density and *temperature* of the plasma generated on graphite sample were studied using a Q-switched Nd:YAG laser (1064 nm, 275 mJ) [33]. These values were calculated from spatially resolved spectroscopic data, by using Stark broadening and atomic line ratio. The calculated time-integrated spatial dependence of electron temperature and electron density of the carbon plasma are shown in Figures 2.32 and 2.33. They both show a decreasing behavior with distance from sample surface. With an increase in *laser irradiance*, both electron temperature and density increase and saturate at higher irradiance levels. The saturation at high irradiance levels was attributed to plasma shielding, as was supported by other similar investigations [34, 76].

The time-dependent *ionic temperature* in silicon plasma, produced by a Nd:YAG laser (532 nm, 5 ns), was studied as a part of a general investigation of spatial distribution [30]. Figure 2.34 shows the ionic temperature at variable axial positions in the plasma, for three delay times. The temperature decreases with the increase in distance from the sample surface with a stepped decrease at 0.25 mm.

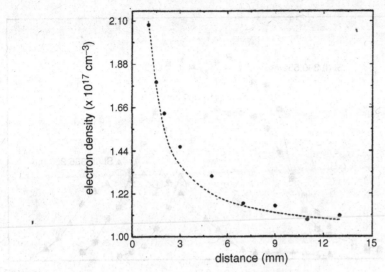

Figure 2.33. Electron density of graphite plasma as a function of distance from the target surface. The dotted line represents the $1/z$ curve (Figure 3 in reference [33]).

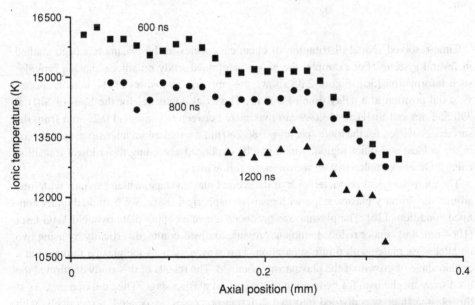

Figure 2.34. Ionic temperature for silicon plasma in the axial direction calculated at different positions from the surface at variable delay times, as indicated (Figure 8 in reference [30]).

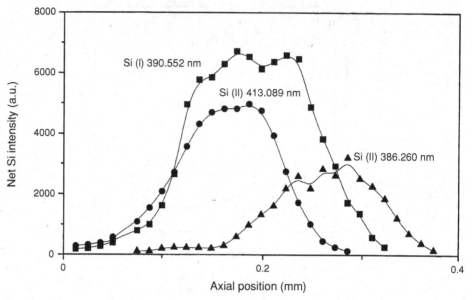

Figure 2.35. Spatial distribution of silicon species vs. axial position from the target. Results were obtained at various spectral lines, as indicated and at a delay time of 600 ns (Figure 9 in reference [30]).

Time-resolved spatial distribution of chemical species in the plasma has been studied in several systems. For example, the above-mentioned study on silicon plasma provides such information [30]. Figure 2.35 shows the emission intensities of the silicon species vs. axial position at a delay time of 600 ns. The axial intensities for the lines of Si(I) at 390.552 nm and Si(II) at 413.089 nm maximize between 0.1 mm and 0.25 mm from the surface. However, for the ionic species at 386.260 nm the maximum intensity was observed in the coldest and outer region, which can be explained according to its lower transition energy. Other examples will be shown in the following.

The morphological characteristics of the second plasma stage, which expands with time around the primary plasma stage with near-hemispherical shape, were studied under confined conditions [26]. The plasma was produced above a copper plate by a Nd:YAG laser (1064 nm, 8 ns) under reduced ambient pressure and was confined vertically by using two parallel glass plates with 6 mm separation. This way, a view of the plasma radiation at a certain stage of growth of the plasma was obtained. The results of this study confirmed the blast wave mechanism for generation of the second plasma stage. The emission process at this plasma stage was divided into two different processes, associated, respectively, with the "shock formation" and the "plasma cooling." During the former, the atoms gushing out from the target are adiabatically compressed against the surrounding gas. During the latter stage the temperature of the plasma decreases gradually.

The effect of the wavelength and fluence on the laser ablation of graphite was studied by observing the spatially resolved emission spectra of the carbon plume [29]. A Nd:YAG laser (266 nm, 5 ns and 1064 nm, 20 ns) and KrF excimer laser (248 nm, 18 ns) were applied. The temporal variations of the emission intensity of C^+ (426.7 nm) were measured at various distances from the target. It was found that at a distance of 8 mm from the target, the emission intensity was less than 10% of that near the target. The C^+ velocity is highly dependent upon the laser wavelength and slightly dependent on the fluence. The C^+ velocity decreased with increasing laser wavelength at the same fluence and increased slightly with increasing laser fluence.

It should be mentioned that the morphological data obtained by the spectrometer slit imaging method are integrated along the axis connecting the emission source (plasma) and the slit. Resolution along this axis is possible, in principle, using the Abel transformation; however, this treatment is only justified under certain assumptions and conditions, as discussed in the following. The validity of the Abel transformation in the study of plasma morphology was experimentally proven by using Fourier transform imaging technique [1].

The Abel-inverted spatially and temporally resolved distributions of plasma temperature and electron densities were measured, when a Nd:YAG laser pulse (532 nm, 10–13 ns) was focused in ambient air [25]. The optical intensities obtained by scanning the plasma image with the spectrometer slit represent line-of-sight measurements, or path-integrated intensities. If the plasma is optically thin and cylindrically symmetric, Abel inversion may be used to determine radially resolved emission intensities. For this purpose, measurements must be performed at many distances from the axis of symmetry. Such data were acquired with an imaging spectrometer and a CCD camera, by orienting the spectrometer slit perpendicular to the plasma axis. Then, each camera image provides wavelength in one direction and radial position in the other. Abel inversion was performed using the convolution backprojection algorithm and applied to analysis of the effect of ambient gas conditions upon plasma morphology. The main finding was that variations in laser energy, gas composition, particulate levels, and humidity levels produce very little variation in plasma temperature and electron density distributions. The small variations could be explained by the laser-supported radiation wave model.

The above experimental measurements should be compared with theoretical calculations carried out using the blast-wave theory [77]. The time-dependent spatial distributions of the number densities of different species in the plasma, formed on a carbon surface, were calculated by using the blast-wave theory with ionization described by the Saha equations and the charge conservation equations. Good agreement was found between the theoretical and experimental time-profiles. The calculations lead to the conclusion that the plasma produced by a focused laser beam on a solid surface evolves in two phases. During the first phase, the plasma is initiated at the surface and in the surrounding gas. The plasma grows rapidly, generating a detonation wave, which absorbs the incident laser radiation. The second phase of the plasma evolution occurs after the laser pulse is over, during which the plasma expands and cools.

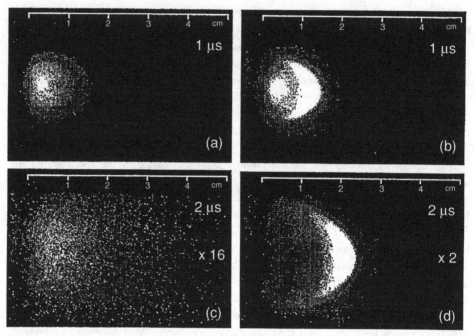

Figure 2.36. ICCD images of Ba(II) emission at 1 μs and 2 μs, recorded using 20 ns integration time. The plasma was generated on YBCO target in vacuum, (a) and (c), and in 180 mTorr oxygen, (b) and (d) (Figure 2 in reference [42]).

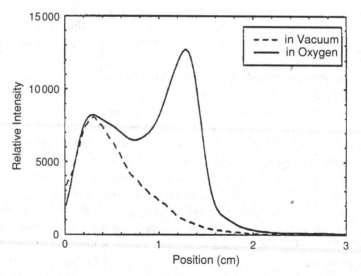

Figure 2.37. Plots of Ba(II) emission intensity in vacuum and oxygen at 1 μs delay as functions of distance from the target along the normal at the laser ablation spot (Figure 3 in reference [42]).

2.4.2 Direct CCD imaging

Recently, CCD detectors have become very popular in analytical applications, especially since their cost has dropped and their operation has become simplified. At the beginning, these devices just replaced the photographic plates, but when coupled to gated intensifiers they provided time-resolved imaging, with temporal resolution in the nanosecond range. Thus, the intensified CCD cameras allowed for time-resolved morphological information on laser-induced plasma. They are perfectly suited for most applications, and almost the only drawback comes from their restricted dynamic range. This is caused by the limited capacitance of the individual detectors in this two-dimensional array. We now discuss the various results related to plasma morphology obtained by the method of direct CCD imaging.

Time-resolved *spectral morphology* of an individual emission line has been studied by placing a narrow-band filter (10 nm) in front of the CCD camera. In this study, time-resolved direct CCD imaging was applied to plasma produced by an excimer laser (248 nm, 2.6 J/cm) on $YBa_2Cu_3O_7$ solid samples [42]. The filter was selected to record the Ba(II) ($6^2P^0_{1/2} \rightarrow 6^2S_{1/2}$) 493.41 nm emission. Measurements were carried out at two delay times at vacuum and at 180 mTorr oxygen. The resulting images are shown in Figure 2.36. Slices through the 1 μs data in the direction normal to the target are shown in Figure 2.37. Clearly, the plume in oxygen is more confined. An enhancement of all ion emissions at the expanding front of the plume in ambient oxygen is observed.

A more sophisticated setup was applied for *simultaneous imaging of a spectral line observed from two orthogonal directions* [22]. Registration of two perpendicular views of the plasma, observed through interference filters, was carried out simultaneously. The plasma was generated by a Nd:YAG laser (1064 nm, 9 ns) over a metal sample. One filter was used to observe mainly the Pb(II) ionic line, while the other filtered some Pb(I) atomic lines. An example of the time-resolved images is shown in Figure 2.38. Non-normal incidence of the laser beam onto the target material leads to nonsymmetrical (cylindrical) plasma formation. Inspection of the images obtained by the two filters indicates that the lead ions are confined to a smaller volume and that they do not extend out into the plasma tail. The ionic emission is also shorter lived. From these measurements it is obvious that the elemental spectral line emission in the plasma is a complicated three-dimensional spatial and temporal function that is different for ionic and atomic lines. This observation indicated that the common assumption of local thermodynamic equilibrium might not be valid.

Direct time-resolved CCD imaging was applied for investigation of laser-induced metal and polymer plasmas and provided evidence of a certain *plasma stratification* [9, 39, 48]. In such an experiment a Nd:YAG laser (1064 nm, 8 ns) generated the plasmas above the solid samples under ambient gases (Ar, air, He, and O_2), at pressure regime from 10^{-3} to 10^{-1} Torr [39]. Typical ICCD images of the expanding plume, generated on aluminum sample at 10^{-3} Torr of He, at different time delays are shown in Figure 2.39. The intensity profiles of the photographs are shown as well.

The intensity profiles of both aluminum and Teflon indicate the splitting of the plume into two components, which is called plasma stratification. The intensity profiles show two

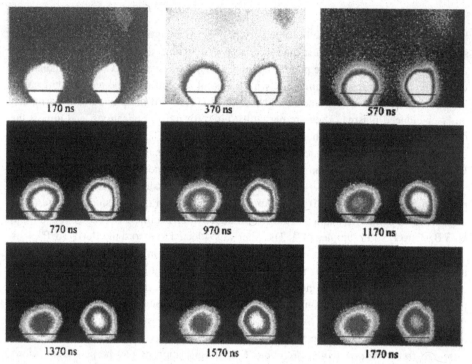

Figure 2.38. Images showing the temporal and spatial development of the atomic emission from the laser-induced breakdown plasma transmitted by a 280 nm filter (Figure 6 in reference [22]).

peaks at an earlier time that merge into one at a later time. This stratification is assumed to be a result of the velocity difference of the two stratified components. A simple stratification mechanism was suggested for the metal plasma: the earlier part of the laser pulse removes a small amount of material from the surface, which is further heated by absorption of the incoming laser radiation. It makes the material thermally ionized and opaque to the incident radiation, therefore, absorbing most of the laser energy. Near the end of the laser pulse this material becomes so hot that it begins to re-radiate thermally. Some of this radiation may reach the target surface causing further vaporization. Double vaporization of the material from the target surface during a single laser pulse results in the stratification of the plasma.

An attempt at obtaining the *radial emission distribution* in the plasma was carried out using direct time-resolved imaging [40]. The plasma was generated by a Nd:YAG laser (8 ns) and the corresponding isointensity contour lines were studied. Clearly, within this experimental setup, the measured two-dimensional images, $I_{(x,y)}$, are actually integrated over the three-dimensional distribution of emitters in the plume

$$I_{(x,y)} = \int_{-z/2}^{+z/2} \int_{\text{line}} \varepsilon_\nu(r) \mathrm{d}z \mathrm{d}\nu$$

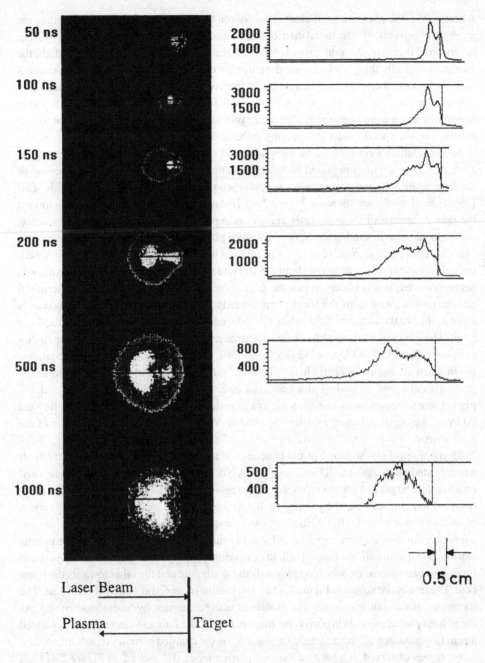

Figure 2.39. ICCD images of expanding aluminum plasma at 1 mTorr of He pressure at various time delays. The intensity profiles are also shown (Figure 2 in reference [39]).

where $\varepsilon_\nu(r)$ is the emission coefficient at the photon frequency ν, and z the total length of the emitting plasma chord. The radial distribution of the emitters, $N(r)$, can be obtained by Abel inversion of the above integral, provided that: (a) the observed light rays are paraxial, (b) the plasma is optically thin, in which case the emission coefficient $\varepsilon_\nu(r)$ is linearly proportional to $N(r)$, and (c) there is radial symmetry in the (y,z) plane. From the experimental data it is evident that condition (c) seems readily satisfied; however, the validity of the other conditions is not always ensured. This study provided complementary information to that obtained by the spectrometer slit imaging technique.

Several studies were focused on morphological observations related to the laser power. Direct CCD spectrally integrated imaging was applied for measuring the lengthening of the laser spark with increasing laser energy, when the pulses are focused into ambient air [25]. Typical results are shown in Figure 2.40. It was found that the spark lengthens toward the laser beam with increasing laser energy, as expected by the laser-supported radiation wave model. Also in another study, time-resolved plasma morphology was obtained when a Nd:YAG laser pulse (5.6 ns, 1064 nm) was focused in ambient air [44]. Plasma images were obtained at a spatial resolution of about 8.5 μm and the absolute location of the plasma was also monitored. It was observed that the higher the incident energy, the farther the initial plasma moves away from the focal point towards the laser source. If the laser power is close to the breakdown threshold value, the induced plasma appears very close to the focal point. The effect of the laser fluence upon plasma morphology was also studied for silicon samples, using Nd:YAG laser (532 nm, 5 ns) [30]. It was found that, at low laser energy, the shape of the plasma is nearly hemispherical. With an increase in laser energy, the shape gradually deformed, extending along an axial direction from the surface. As expected, the plasma size increases, presumably owing to a growth in the amount of ablated mass. Beyond 90 J/cm^2 the deposited energy causes the plasma expansion to grow in the direction of the laser source.

Morphological investigation of the plasma produced under *non-uniform magnetic fields* was also studied by direct CCD imaging [24]. A Nd:YAG laser (1064 nm, 8 ns) was focused on a graphite target in a vacuum chamber at a pressure of 0.1 mTorr. In addition to the time-resolved imaging (obtained by using an ICCD camera), the emission spectra of carbon plasma were recorded at different distances away and parallel to the target surface by moving a monochromator in a plane perpendicular to the surface. In these experiments, the plasma expanded in a non-uniform magnetic field of maximum 3.5 kG. The plume was observed to break into two symmetric lobes, one towards the north pole and the other towards the south pole. The results are shown in Figure 2.41 for the plasma observed at a delay time of 1 μs. The formation of the multiple peaks was explained using a magnetohydrodynamic model. As the plasma plume expands in transverse magnetic field, the ions and electrons get separated due to Lorentz force till the magnetic force acting on the charged particles equals the electric force. It was observed that each of the two plasma lobes (L1 and L2 in Figure 2.41) has three components: the fast, the intermediate, and the slow one. These are also depicted in this figure and are denoted by F, I, and S, respectively. Each of these components for both lobes has been observed in the temporal profiles of C(II) transition at 426.7 nm and C(IV)

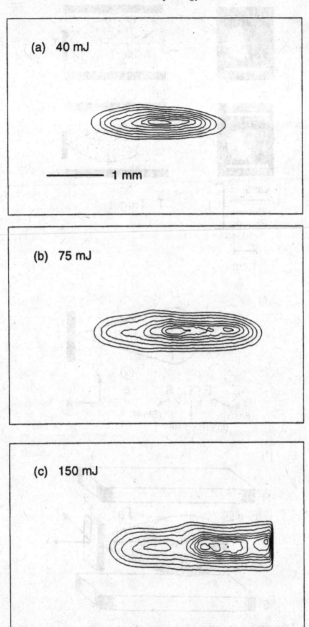

Figure 2.40. Two-dimensional images of laser-induced plasmas in air, obtained at various laser energies taken just after the laser pulse (Figure 11 in reference [25]).

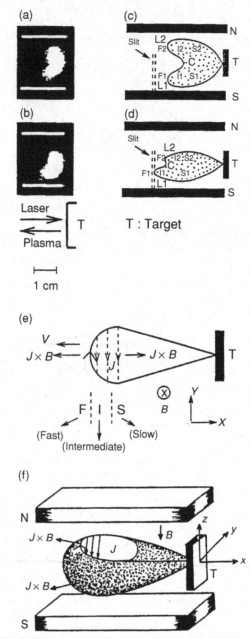

Figure 2.41. (a) ICCD images of the plume breaking into two symmetric lobes (at 1000 ns) when the focal spot is exactly between the two poles. (b) The plume breaking into two asymmetric lobes when focal point in nearer to one of the poles. (c) Pictorial presentation of (a). (d) Pictorial presentation of (b). (e) Schematic diagram of the direction of current density J, giving rise to fast, intermediate, and slow components. (f) Schematic diagram showing the cross-sectional area of the plume. It shows that the $J \times B$ term at the plume boundary is more towards the poles, resulting in two lobes, which move towards the poles (Figure 6 in reference [24]).

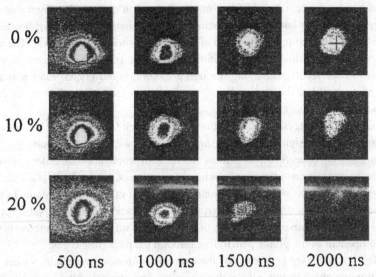

Figure 2.42. A series of ICCD images of plasmas formed on targets of varying weight percentage of water (Figure 4 in reference [38]).

transition at 580.1 nm from 6 mm onwards (from target), and C(III) transition at 465.0 and C(I) transition at 399.7 nm from 16 mm onwards. Instabilities are observed in the temporal profile of these ions at the edge of the plume, attributed to edge instability in magnetic field.

Another interesting application of direct CCD imaging together with independent spectral information was carried out for studying the *matrix effect* related to the water content of the sample [38]. The CCD images of the plasma generated on targets of different water content, obtained at a series of delay times, were recorded (Figure 2.42). Measurements indicate that, as water content increases, the intensity of the plasma decreases. The addition of water to a target can directly affect the density and the thermal diffusivity of the sample, thereby affecting the energy density required for breakdown. Water addition decreases the electron number density in the plasma, while the excitation temperature remains roughly constant. This explanation could be responsible to the previous measurements of such matrix effects [75].

2.4.3 Time-resolved imaging through tunable filters

Tunable filters coupled to two-dimensional time-resolved detectors are suitable for providing interesting morphological information. Such a setup was applied for studying the *chemical composition* of the plasma with spatial resolution, in relation to its dynamics [48]. A KrF excimer laser (25 ns) was focused on several solid samples in argon ambient, in order to study time- and spatially resolved emission from the plasma, and time-resolved imaging was obtained through tunable liquid crystal (LC) filter.

This detection method was applied to several surfaces and it was found that, in each case, the plasma image contained, in addition to the obvious bright shock region, two additional regions of luminescence: (a) a shielded, unaffected region close to the target that appears nearly equal in intensity and shape to that in vacuum expansion, and (b) a region extending to long distances (fast component) ahead of the bright, slowing shock region. The target of this study was to determine the composition of these two plume components propagating into the gas. Time-resolved imaging through a tunable liquid crystal filter resulted in a series of photographs of the plasma at various delay times, corresponding to the emission of known species.

The imaging data suggested that a portion of the original target material is transmitted to long distances without significant scattering interactions with the background gas, thereby maintaining velocities nearly equal to those for vacuum propagation. In order to investigate the chemical composition of the plume during the splitting of the ion flux, optical absorption spectra were obtained as well. It was found that the fast component contains fast plume ions originating from the target (and not from the gas). The combined data (absorption and emission) indicated relative depopulation of neutrals (and an associated increase in ions) in the fast component of the plume penetrating through the background gas.

The effects of *sampling geometry* on characteristics of plasma morphology were studied by using time-resolved imaging through an acousto-optic tunable filter (AOTF) [32]. The data were acquired for the plasma induced by focusing Q-switched Nd:YAG laser pulse (1064 nm, 10 ns) on various solid surfaces. Two main geometrical effects were varied: the lens-to-sample distance (LTSD) and the angle of incidence. The main findings are discussed in the following.

The lens-to-sample distance (LTSD) has been known to affect LIBS results. Time-resolved images of the plasma formed by focusing the laser with a spherical lens that had a focal length of 150 mm on an Al sample were acquired by using the AOTF-ICCD system. For a short delay time (0.11 µs), no major differences in the shapes or intensities of the plasma images were recorded with the AOTF tuned to Al(I) at 394.40 nm, Al(II) at 466.68 nm, or off-line (392.50 nm). This indicates that the emissions observed at early times are mainly caused by the spectrally broad continuum light from the plasma. However, for longer delay time (0.69 µs), tuning the AOTF on and off spectral features indicated that the background continuum light had decayed sufficiently to ensure that observed emission is mainly the result of the selected atoms. Comparison of the images obtained at this delay time shows that, generally, the Al(I) emissions were more laterally dispersed across the sample than the Al(II) emissions, whereas the Al(II) emissions were observed at a some-what greater vertical height above the sample than the A(I) emissions. These images can be seen in Figure 2.43. At longer delay time (1.81 µs) the only observable difference between the Al(I) and Al(II) emissions, other than the maximum intensities, was that Al(I) emitted over a larger lateral area.

Similar images recorded from plasma produced on Ti metal imply that the maximum horizontal extent of the atomic cloud increased as the LTSD decreased. Comparison of the images taken at early and late delay times indicates that the atomic cloud expanded significantly at late times beyond the area of the sample originally interrogated by the laser pulse. The emitting atomic species expand outward significantly beyond the plasma width indicated by the background continuum present at early times.

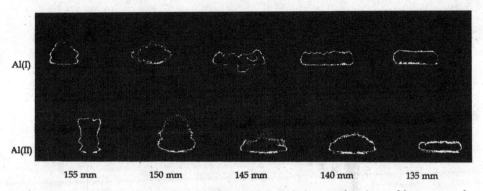

Figure 2.43. Images of Al(I) (top) and Al(II) (bottom) emissions as functions of lens to sample distance, at delay time of 0.69 μs (Figure 2 in reference [32]).

Vertical and horizontal expansions of atomic emissions as functions of time for three LTSD values are shown in Figure 2.44. The maximum height decreased as the LTDS decreased. However, the widths observed were approximately the same for all LTSD values.

Effects of the LTDS upon the morphology of analyte emissions were also studied. For spherical lenses, the emission intensities of all analytes were at maximum at LTDS for which the focal point was 4–12 mm below the sample surface. These positions correspond to LTDS values at which the higher plasma temperatures were recorded. However, for cylindrical focusing lenses, the emission intensities were maximum when the LTSD was 1.5–3 mm greater than the back focal length of the lens, as observed for the central maxima of the plasma temperature. The results show that the relative change in the analyte emission intensity with a change in the LTDS was about four times greater for the cylindrical lenses than for the spherical lenses.

Another interesting effect studied by time-resolved AOTF imaging is the relation of the *angle of incidence* with plasma morphology. Multiple images of the plasma were recorded for various angles of incidence. At early times (< 0.3 μs), the background continuum light from the plasma was seen along the vertical direction of the incident laser pulse for all rotations, indicating that the most intense background light from the plasma occurs along the path of the laser. After *c*. 0.3 μs, the continuum emission decreases and the atomic emissions can be observed. The central axis of these atomic emissions appeared perpendicular to the surface for all sample orientations. Images of Cr(I) emissions observed at delay times of 4.81 μs and 49.81 μs, at three incident angles, are shown in Figure 2.45. This figure clearly demonstrates that the atomic emissions are always perpendicular to the sample surface.

A detailed inspection of the plasma images as a function of time, at a given angle of incidence (50°) shows that at 0.11 μs delay time, the laser plasma is formed on the sample and expands upward along the path of the incident laser pulse [7]. (Figure 2.46). This result is independent of the filter bandpath wavelength, owing to the strong spectrally broad continuum emission from the plasma at this initial stage. At later times, <0.5 μs, the continuum emission decays and the atomic lines are observed. However, the morphological

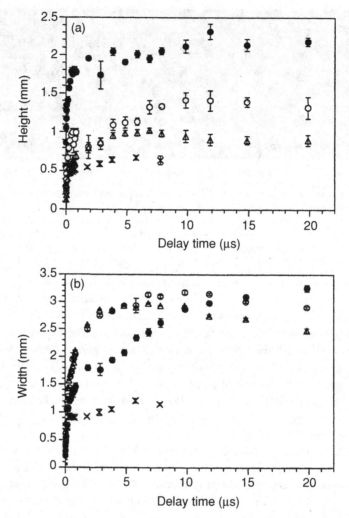

Figure 2.44. The vertical (a) and horizontal (b) expansions of the atomic emissions as functions of time for three lens-to-sample distances: 150 mm (solid circles), 142.5 mm (hollow circles), and 125 mm (triangles). Results for cylindrical lens at 150 mm (crosses) (Figure 4 in reference [32]).

shape of the atomic emission appears symmetric about the normal to the surface rather than along the axis of the incident laser pulse. The observation that atoms emerge normal to the surface independent of the angle of incidence of the laser pulse can be attributed to the formation of high-pressure vapors in the focal volume at the surface that subsequently expand away from the sample, normal to the surface.

A quantitative description of the normalized spectral peaks of several atomic species, as functions of the incident angle, at fixed delay time of 0.81 μs, is shown in Figure 2.47.

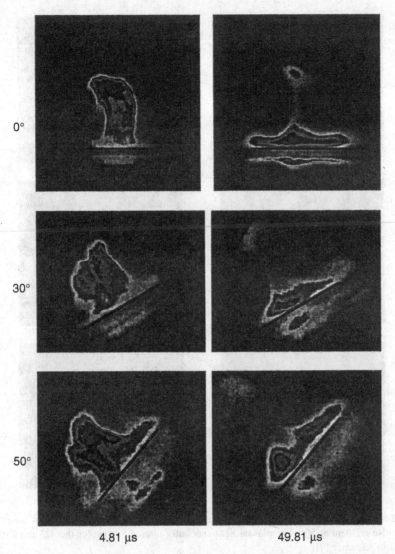

Figure 2.45. Images of Cr(I) plasma emissions at two delay times and three angles of incidence (Figure 11 in reference [22]).

The maximum peak intensity was observed at normal incidence for all species. A minimum was reached at about 50°. These findings could be explained by a simple geometric model, considering only the observation that the atoms are ejected normal to the sample surface and that the average power density on the sample surface decreases with increased angle of incidence. As the sample is rotated, the overlap of the ejected atomic material with the initial plasma volume, defined by the conical shape of the laser beam focused onto the surface,

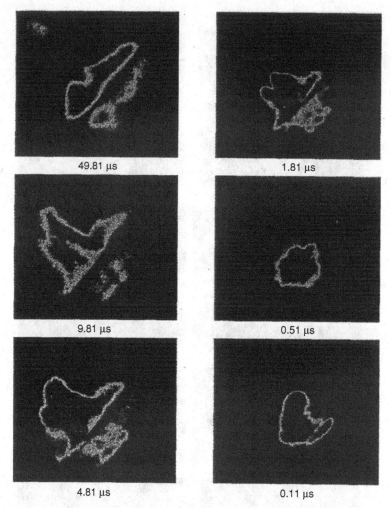

Figure 2.46. Temporal evolution of the laser plasma formed on a Cr-coated aluminum sample with a laser pulse incident at 50°. The plasma light spectrally filtered to monitor the 425 nm Cr(I) line (Figure 1 in reference [7]).

decreases, leading to a decrease in the emission intensities. The decrease is especially noted for ionized species that require greater excitation. Beyond approximately 50°, the overlap of the ejected material with the incident plasma volume increases, producing an increase in the atomic emission intensities.

2.4.4 Single-fiber scanning

Single-fiber scanning was applied to imaging of plasma produced by an excimer laser (308 nm, 50 mJ, 20 ns), hitting glass samples under various gas pressures [16]. It was

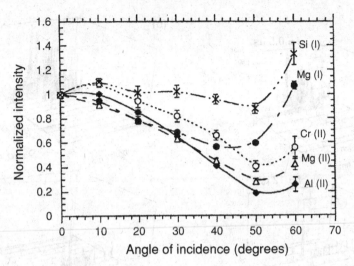

Figure 2.47. Normalized peak areas of atomic emission signals as functions of the angle of incidence of the laser pulse (Figure 2 in reference [7]).

observed that, in contrast to the results obtained with a longer laser pulse (100 ns), the excimer laser always generated a hemispherical plasma on both low melting point and high melting point glasses.

2.4.5 Time-resolved fiber-assisted simultaneous spectroscopy

A study of plasma morphology was applied for understanding some of the matrix effects involved in LIBS analysis of sand and soil samples [51, 52]. A Nd:YAG laser (1064 nm, 7 ns) was applied to plasma generation above the soils. The fiber-assisted simultaneous spectroscopy method allowed for both spatial and temporal resolution of the plume.

Figure 2.48a shows the LIBS spectra as functions of distance from the sample surface, composed of 60%/40% sand/soil matrix. After a short delay of 0.5 μs, most intensities are observed close to the surface (0.3 mm). After a longer delay (say, 12 μs), the plasma has already propagated and the maximum intensity is observed at a distance of 2.9 mm. Since the plasma continuously cools down, the intensities are already lower; nevertheless, the signal-to-noise ratios become higher. Observation of the plasma at an even longer time (22.5 μs), under the same conditions, reveals that the plasma has cooled down (very low spectral intensities) and the maximum is reached at distance of 5.5 mm. Some non-negligible intensities are observed close to the surface as well, supporting the existence of a back reflected shock-wave.

A different presentation is obtained by observing the spectra as functions of time at a constant distance from the sample. Typical results are shown in Figure 2.48b. Again, the plasma dynamics is clearly observed: being close to the surface, the plasma event is intensive and decays in about 4 μs. The measured decrease is a result of a convolution of emission lifetime, quenching and plasma propagation. A second intensity increase is sometimes observed at

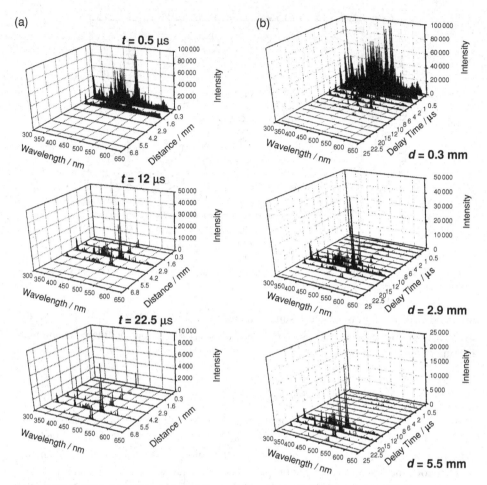

Figure 2.48. (a) Spectra as functions of distance from soil sample (60% sand) measured at three delay times. (b) Spectra as functions of delay time measured at three observation distances from the soil sample (Figure 2 in reference [51]).

longer times, owing to the backscattered shock-wave. At a large distance (say, 5.5 mm), the emission starts after 12 μs. Relatively weak signals are observed, decaying within a few more microseconds. Such measurements were repeated for a series of matrix compositions, in order to understand the nature of the observed matrix effects upon LIBS analysis.

These matrix effects are related to complicated phenomena involved in plasma formation and sample ablation. Many of the associated processes are highly nonlinear and are not fully understood. In the following discussion, they are represented by the case of Pb in soils.

Plots of Pb signals as functions of Pb concentration and matrix composition are shown in Figure 2.49. Clearly, these surfaces form the calibration plots for analysis of Pb contamination. It is well known that the signals from sand are higher than those from soil, therefore

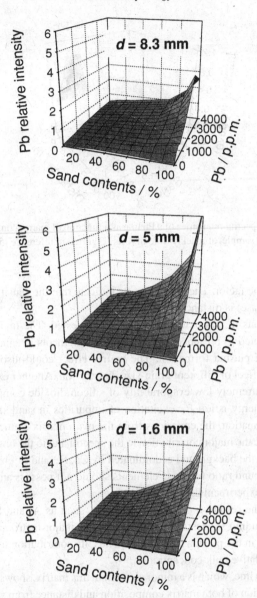

Figure 2.49. Pb relative spectral intensities as functions of matrix composition and concentration, measured at three distances from the surface. The delay time was 12 µs (Figure 3 in reference [51]).

it is not surprising that the slope of the calibration plots increases with the sand content in the examined matrix. However, the distance dependence of this effect was revealed from the morphological information. Clearly at a given matrix, there is an optimum distance for obtaining the highest calibration slope. Actually, the absolute intensity of Pb line increases more than three times in sand compared with soil, contrary to the background level, which

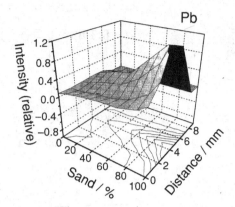

Figure 2.50. Relative spectral intensities of a Pb line as a function of matrix composition and distance from the contaminated sample, at a delay time of 4 μs (Figure 6 in reference [51]).

decreases by the same factor. The above results indicate that the sensitivity to a particular element (Pb in this case) is distance dependent.

Several explanations to this matrix effect were suggested. The higher emissions from sand could be attributed to presence of easily ionized elements in sand, which contribute to the first stages of plasma formation. The matrix effect could also be the result of a simple geometrical effect of different grain-size distribution. Another explanation could be that, owing to the extremely low evaporability of silicon dioxide compared with calcium compounds, more energy is left for excitation of impurities in sand samples. In order to find the correct explanation, the emissions from the main matrix elements were analyzed. These elements make the major contribution to the spectra and to the observed background. It was found out that the background is dominated by Ca, Fe, and Sr emissions. Therefore, the signal-to-background ratio for these elements should not possess any matrix effects, as indeed was verified experimentally.

On the contrary, the results for other components, such as Al and Si, indicate a clear matrix effect. This finding suggests that although Si is a major matrix constituent, it is not properly represented in the plasma (unlike Ca). Thus, its contribution to the background is much less than its relative bulk concentration.

Focusing on the Pb line, which is a minor element in this matrix, shows that the relative Pb intensities are a function of both matrix composition and distance from surface, as shown in Figure 2.50. This plot shows an important feature, namely, that at a given Pb concentration, there is a unique optimum distance valid for all matrixes (about 4.5 mm, in this case).

The contribution of the morphological information to the analytical performance was pointed out by measuring the Pb calibration plots as a function of plasma location. Typical results are shown in Figure 2.51. It is of both practical and theoretical interest that the calibration curves for Pb actually depend on the observation point within the plasma. In this particular case, the best sensitivity is obtained at a distance of 4.5 mm from the surface.

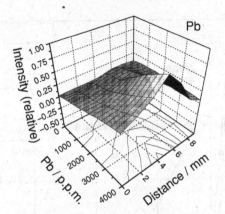

Figure 2.51. Pb relative intensity as a function of concentration and distance from 30% sand sample, measured at a delay time of 4 μs (Figure 8 in reference [51]).

The sensitivity slowly decays at smaller distances and rapidly decays at larger distance. It is clear from this plot that an averaged calibration plot (over the whole plasma) is much worse than any measurement taken in the vicinity of the optimum distance of 4.5 mm.

The origin of analytical matrix effects in LIBS was investigated by plasma morphology. Matrix effects in LIBS analysis were characterized by the parameters of the laser-induced shock-wave propagation [53]. In particular, matrix effects in sand/soil mixtures were addressed. An explanation for the increase in the spectral response of trace elements (at constant concentration) with sand percentage was suggested. It was found that the energy coupled in the plasma, and which can be calculated from the propagation of the laser-induced shock-wave, indeed characterizes the matrix. Results indicate that the main matrix effects are attributed to the depth of the laser-induced crater, which was correlated to a portion of the laser energy that penetrates into sand particulates and does not cause direct ablation. This explanation holds when no other effects are present (e.g. grain-size distribution). The hypothesis was validated by experimental data.

Typical results of the plasma emission as functions of time are shown in Figure 2.52. In this particular case, the sample contained 10% sand and 90% soil. The propagation of plasma emission front along the axis of the laser beam is clearly indicated. Assuming that the shock-wave front propagates according to the Sedov equation

$$R = \xi_0 \left(\frac{E}{\rho_0} \right)^{1/5} t^{2/5}$$

one can use such data to find out the energy involved in the formation of the plasma. Here, R is a distance from the sample, t is the shock-wave transient time, ρ_0 is the normal air density, and ξ_0 is a dimensionless parameter determined from the condition of energy conservation that depends on the specific heat ratio γ ($\xi_0 = 1.033$ for $\gamma = 1.4$, for air). It is well known that this simple theory of strong explosion explains well the propagation of the front emission up to 15–20 μs after the laser pulse.

Israel Schechter and Valery Bulatov

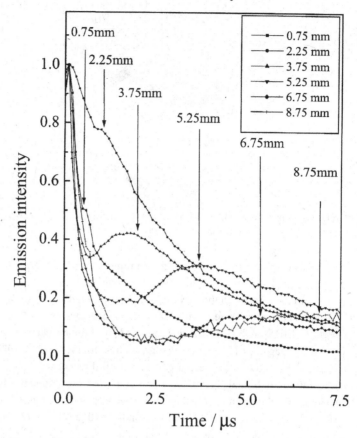

Figure 2.52. Plasma emission intensity measurements as functions of time for a soil matrix containing 10% sand. The corresponding observation points, located 1.5 mm apart, are indicated (Figure 3 in reference [53]).

The (normalized) plasma energies calculated in this way, obtained for a series of sand/soil matrices, are shown in Figure 2.53. It can be seen that the amount of energy involved in the shock-wave increases with the sand concentration up to a certain point (of about 20% sand) and then it linearly decreases as the amount of sand goes up. The LIBS spectral intensities of some elements are also shown in Figure 2.53. These include Ca, which is the main matrix component, as well as Fe and Sr, which are minor matrix elements. The relative intensities of the plasma emission, calculated according to the Stefan–Boltzmann law of black-body radiation, $\propto T^4$, are also presented. A good correlation between the calculated intensities and the measured data has been observed for the sand/soil matrix. However, only partial correlation is observed between the energy measurements and the integral spectral intensities of the matrix elements. The data in Figure 2.53 indicate that (a) the plasma energy decays linearly with the sand percentage, except for the first point (no sand at all) and, maybe,

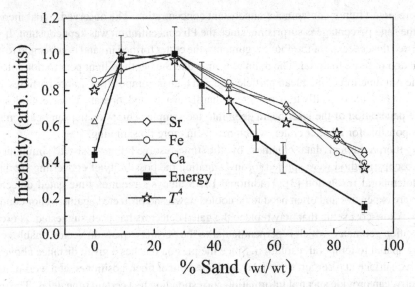

Figure 2.53. Plasma energies as functions of matrix composition for sand/soil matrices, calculated from shock-wave propagation. The corresponding spectral intensities of some matrix components are also shown, as well as the corresponding T^4 normalized values (Figure 6 in reference [53]).

the second point (10% sand); and that (b) generally, the plasma energy curve follows the trend of the spectral intensities of the major and minor constituents of soil. However, the slope of the energy decay is higher that that of the spectral intensities.

These findings imply that the sand matrix results in a lower plasma energy, which means that the traditional explanations of the matrix effects upon Pb signals actually fail. However, it is clear that measurements of the plasma energy may provide a correct characterization of the matrix (at least in the simple case of sand/soil mixtures). The suggested explanation of the plasma energy decay was that the laser light that hits a sand particulate is partly scattered and partly refracted into the particle. The latter energy portion is dissipated in the sample. It cannot ablate the hard sand particulate, but it goes to a deeper layer and contributes to the formation of a larger crater. This explanation was based on the hypothesis that the energy of the laser pulse is split among the matrix components. The laser beam that encounters the sand or quartz particles is partly reflected (internally and externally) and then passes through the particle with some losses. These losses are basically responsible for the matrix effect. The out-coming beam encounters other solid material and may take part in plasma formation. The energy thus captured inside the quartz or silica particles (which are hardly evaporated) is lost as regards plasma formation. This model may explain the finding that the measured plasma energy decreases linearly with sand concentration.

The explanation of the Pb matrix effects was based on the same grounds. Unlike the soil elements, whose spectral intensities are really expected to change with sand percentage,

Pb represents a minor contaminant at constant concentration. The observed signal increase with the sand percentage is surprising, since the Pb concentration was kept constant. It was suggested that, because of the above arguments, the crater formed in sand is larger and results in ablation of more material. The main reason is that the depth of light penetration into the sample is a function of the mean particle size and matrix composition. Large particles such as sand, which are optically transparent and hardly evaporated, operate as wave-guides that enable penetration of the laser beam deep into the sample. This way, the sand particulates are responsible for a larger crater, which results in more Pb sampling.

The morphological data obtained by the time-resolved fiber-assisted simultaneous spectroscopy method were applied for investigating the possibility of converting spatial to pseudotemporal resolution [3]. Traditional LIBS analysis requires time-gated detectors, which are expensive and often need to be cooled down and protected against vapor condensation. A low-cost setup that may replace the gated detectors has been suggested. A proper observation geometry, which is perpendicular to the plasma expansion vector, enables converting spatial to temporal resolution. Since the plasma reaches a given distance above the analyzed surface at a certain time delay, a single optical fiber, positioned at a well-defined geometry, can provide spectral information corresponding to a certain time delay. The multifiber imaging spectrometer setup provides information corresponding to a series of delay times, which is adequate for analysis of a variety of matrices. It was found that the performance of the non-gated detector observing a narrow solid angle is similar to that of a gated one observing the whole plasma. An example of the signal-to-background obtained for both the gated and the non-gated experimental arrangements is presented in Figure 2.54. Generally, the non-gated results obtained at a specific location in the plasma provide calibration plots that are comparable to those obtained in the gated experiments, over a wide concentration range (however, not over the full range). It was concluded that sensitive LIBS analyses can be carried out by less expensive (non-gated) detectors, when acquisition of morphological data is possible.

2.4.6 Streak camera

Streak cameras allow for morphological data obtained at fine temporal resolution, much better than that reached by using ICCD or photomultiplier detectors. For instance, a streak camera provided temporal resolution of 0.1 ns, from the first 100 ns life of the breakdown event [55]. Plasma was induced on a high T_c superconductor surface using an excimer laser ($\lambda = 248$ nm, 10 ns). Interesting information was obtained when the light emission from the plume was focused on the entrance slit of the camera. This way, the camera observed one line, in the center of the plume, as a function of time. The results are shown in Figure 2.55 for three laser energies. This figure shows the presence of two distinct light-emitting plasma components: a slow-moving one and a fast-moving one. The average expansion velocity of the slow component was 6×10^5 cm/s, and 5×10^6 cm/s for the fast one. At lower energy densities, only the fast component pattern can be seen. At medium energy densities the slow component appears and becomes dominant at higher energies.

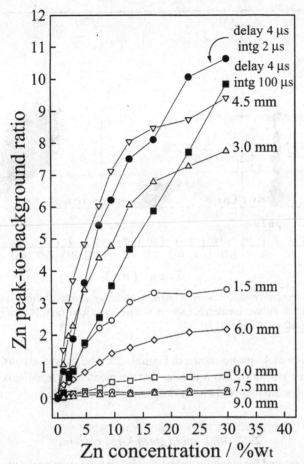

Figure 2.54. Calibration plots for Zn in brass, obtained in gated (solid symbols) and in non-gated (open symbols) experiments, at various distances from surface. Note that, despite the nonlinear behavior at high concentrations, best sensitivity is obtained with the non-gated setup, at a distance of 4.5 mm (Figure 8 in reference [3]).

When repeating the measurements for off-center portions of the plasma only the fast component was observed. Therefore, it was concluded that the slow component of the ablated particles expands with more forward direction in space than the fast component. Analysis of space/time-resolved data showed that particles ejected from the surface absorb laser energy after their ablation but exhibit a delay in their light emission from the peak of laser pulse. Mass analysis indicated that the majority of particles ablated initially under these conditions are clusters below 100 nm. They are further excited by the later part of the laser pulse.

Two-dimensional space-resolved images of the plasma produced by 2.2 ns laser pulses on Al surface were recorded using a streak camera [57]. Typical results obtained at a fine

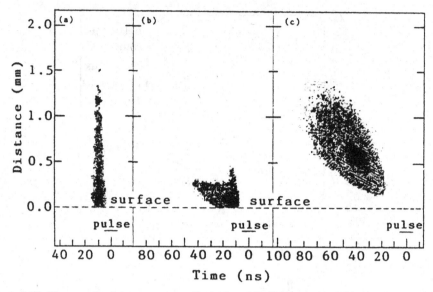

Figure 2.55. Time-resolved measurements of intensity of the light emission from laser-ablated particles along the direction of laser incidence. Laser power densities were (a) 0.14 J/cm^2, 0.4 J/cm^2, and (c) 3.6 J/cm^2 (Figure 2 in reference [55]).

temporal resolution of 47 ps are shown in Figure 2.56. The plasma expansion is sampled at a very high time resolution and it is clear that the picosecond framing is not needed in LIBS, since changes are observed on a nanosecond time scale only.

2.4.7 High-speed gated X-ray cameras

This method was applied for obtaining morphological information on the plasma generated by very high-power lasers. Such a laser pulse (8.1 kJ, 1.9 ns) was focused for creating high density implosion on a plastic target [60]. In the imploding phase (first 500 ps) the shell was found to have nearly spherical symmetry. During the expanding phase, deviation from sphericity was observed. In another study a high-power laser pulse (0.9 kJ, 1 ns) was focused on aluminum microdots (200 μm × 200 μm, and 1 μm thick) coated onto a 6 μm plastic foil [62]. In this case it was reported that the plasma as a whole is laterally expanding. It was suggested that while no lateral expansion of the Al plasma exists, the plastic plasma is cooled rapidly by the expansion process, and so the Al plasma cools by lateral heat transport.

2.4.8 Near-resonant photographic absorption/resonance shadow imaging

This powerful tool was applied for several morphological investigations of laser-induced plasmas and the related analytical/physical characteristics. Several aspects were studied, as we will now briefly describe.

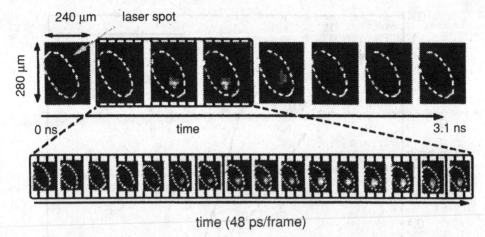

Figure 2.56. Typical framing images of Al plane target showing the shock breakthrough obtained with a high-speed sampling camera. The temporal resolution was 47 ps and the spatial resolution was 45 μs (Figure 4 in reference [57]).

Morphological data providing information on the expansion of fragments in the plasma were obtained by focusing a KrF excimer laser (258 nm, 25 ns) on a high-T_c superconductor ($YBa_2Cu_3O_{7-\delta}$) target [65]. The expansion of the produced fragments was imaged with a second dye laser (1 ns), using the near-resonant photographic absorption method. The temporal evolution of the fragments and the resulting shock-wave above the target in oxygen ambient were imaged. There was no evidence of any fragments leaving the surface for about 10–15 ns from the start of the ablation pulse. The expelled fragments were clearly observed for longer delay periods. They remain near the target surface. At the end of the laser pulse, the density of fragments above the target decreases slowly as the fragments expand away, and they are no longer visible after about 500 ns. The shock-wave had an initial velocity of about 1.5×10^5 cm/s and seemed to expand with apparent hemispherical symmetry and with decreasing velocity as a function of time. It should be noted that a similar study of the spatial and temporal distribution of species formed from surfaces by laser vaporization was carried out using acousto-optical scanning of laser-enhanced ionization [73].

Information on the velocity components of the various species in the plasma, as functions of gas pressure, could also be obtained by the same technique. Time-resolved absorption imaging, using an ICCD camera, was applied to plasma produced by an excimer laser (248 nm, 2.6 J/cm) on $YBa_2Cu_3O_7$ solid samples [42]. Also here, the plasma was back-lit by the expanded beam from a short-pulse dye laser. Absorption profiles of the excited Ba(II) $6^2P^0_{1/2}$ population at 1 μs in vacuum and 180 mTorr oxygen were recorded by the expanding plume with the dye laser tuned to 452.49 nm. The absorption as a function of position in the plume is shown in Figure 2.57. In vacuum a substantial proportion of the Ba(II) ions have high z-component velocities of 4×10^6 cm/s which corresponds to ion kinetic energies in excess of 500 eV. It is evident that the Ba(II) ground-state absorption

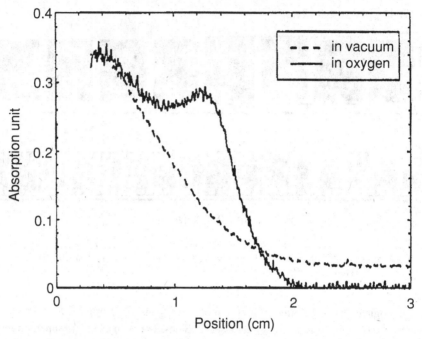

Figure 2.57. Absorption profiles of excited Ba(II) population at 1 μs in vacuum and oxygen (Figure 4 in reference [42]).

shows the presence of an initial very fast component which is strongly forward-directed along the z-axis and followed by a slower component with x- and z-velocity components of comparable magnitudes. This is also evident in the ICCD images recorded at lower oxygen pressures; however in 180 mTorr oxygen this fast z-directed ion component has been completely suppressed. In ambient oxygen the fast ion component is progressively damped and the propagating plume develops a sharp contact front with the ambient atmosphere. Ions are the dominant plasma species in this expanding front of the plume in both oxygen and vacuum with ion/neutral ratios >8 observed for both Ba and Y species at delays <2 μs.

Spatial profiles of the line-integrated Ba(II) ground-state *number densities* were measured in vacuum and oxygen, at 1 μs delay, using absorbance values on the 455.4 transition. The results indicate that oxygen leads to increased local ion number densities. Typically at 1.5 cm from target, the Ba(II) ground-state number densities are increased by 1.7 in the oxygen case.

Resonance shadow imaging was applied to laser-produced lead and tin plasmas, generated by an excimer laser (308 nm) under various ambient pressures [67]. This study was aimed at the *time-resolved plume morphology*. It was found that at the early stage of plume development (1.5 μs) the shadow images consist of three distinct parts: (a) a compact dark spot attached to the target surface, which corresponds to the dense cloud of neutral atoms; (b) a thin circle around the spot corresponding to the shock-wave detached from the plume front and propagated in the argon medium; and (c) a dark cone above the target, which reflects the shape of the focused ablating beam. In this method, the shock-wave was visualized by deflection of laser light by the density gradient in a region of maximum compression of the medium. The cone above the target was probably formed due to photodecomposition of target aerosol particles within the laser light cone.

As the delay time increases, the size of the plume becomes larger owing to its expansion and the cone becomes weaker. After about 75 μs, the cone disappears and, after a delay of 200 μs, almost no absorption of laser light can be observed for all pressures studied.

The nature of the *ablating cone* above the target was studied by wavelength-integrated images obtained with an ICCD camera. It was concluded that, under the experimental conditions of this study (pressure of 50–1000 nbar and delay times of 1.5–200 μs), dimer formation is not efficient, whereas the condensation of lead into larger particulates is more likely. The condensation was efficient beyond about 75 μs and about 20 μs with respect to the ablation pulses for lead and tin, correspondingly.

The target material was also found to affect plasma morphology. Analysis of the shadow images obtained from lead and tin targets at 100 mbar argon pressure showed a different character of the light–surface interaction for the two targets. The lead plume is uniform and grows toward the laser source, whereas the tin plume consists of two regions of different neutral atom density and has almost spherical symmetry. In addition, the tin absorbing plume, along with the photodecomposition cone, disappears much faster than that of lead.

The larger neutral atom density found in the lead plume was attributed to the larger amount of material ablated from lead, compared with tin, because of the lower vaporization temperature and lower thermal conductivity of lead, resulting in more vaporized material with less heat dissipation. Additionally, the amount of laser energy coupled to the lead surface is larger than that coupled to the tin surface, owing to the larger reflectivity of the latter.

The differences in plume shape and densities were related to memory effect from the early stage of the plasma development. Because of the high initial density of free electrons, the plasma may become opaque owing to inverse bremsstrahlung at some instant during the laser ablation pulse, blocking the further penetration of laser light to the surface. This results in additional heating of the plasma, which grows toward the laser beam.

Shadowgraphic and interferometric methods were applied for morphological investigation of the plasma plume produced by a Nd:YAG laser focused on Si_3N_4 target [78]. Morphological features were compared for three excitation wavelengths (1064, 532, and 355 nm). The differences between the plume morphologies were interpreted in terms of dominant absorption regions within the plasmas. Whereas, for ultraviolet (UV) radiation, absorption in a region close to the target leads to spherical expansion, a laser-supported detonation wave has been found for infrared (IR) laser light with an absorption region at the very tip of the shock-wave. A transition between these two absorption modes was observed for the green laser excitation. Information obtained for electron densities indicates that inverse bremsstrahlung absorption dominates at 1064 nm, while photoionization plays the major role at 532 nm.

2.4.9 Double-pulse holography

Double-pulse holography was applied for investigation of plasma expansion [68]. Plasma was generated by a Nd laser of 20 ns pulse. Three-dimensional images were obtained and the data were analyzed in terms of time-resolved plasma expansion in various gases and pressures. Almost spherical shock-waves were measured under these conditions. The experimental results on shock-wave radius and velocity, reported as functions of time during the expansion of the shock front itself, fitted well the strong explosion theory for shock-wave Mach number greater than 2. The experimental fringe patterns were analyzed to obtain the electron density distribution by Abel inversion from the center to the plasma boundary.

2.4.10 Resonance laser-induced fluorescence

The *time-resolved spatial distribution* of neutral and ionic species in plasma was measured by using the LIF method. The effect of the buffer gas pressure was investigated when plasma was generated on copper surface containing magnesium traces by a Nd:YAG laser (8 ns, 9 mJ), in argon gas [69]. The distribution of neutral and ionic magnesium was measured at various argon pressures. It was found that the buffer gas pressure has two major effects upon the plasma: at very low gas pressure there is an insufficient atomization of the laser-ablated matrix, because of the small energy deposition in the gas. At higher pressures, the plasma stores more energy and becomes denser and hotter, but the penetration of particles is more difficult. These opposing effects produce a maximum of analyte concentration at about 140 hPa, which is located in the middle of the plasma.

The *spatial distribution* of neutrals and ions at constant delay (55 μs) was also measured. The relative lateral distribution of magnesium ground-state ions becomes independent of the buffer gas pressure for higher pressures. However, for low pressures a wider ionic distribution was measured than for high pressures. In contrast to the ions, the spatial distribution of magnesium neutrals was found to be strongly pressure dependent. The relative distribution of neutral magnesium atoms becomes narrower with increasing gas pressure, while the relative ions distribution is unchanged in this range. This behavior reflects the fact that the spatial confinement of the plasma becomes stronger with increasing gas pressure.

Regarding the *temperature evolution*, it was found that the laser-produced plasma has a hot kernel with an almost unchanged relative temperature distribution in the time up to about 100 μs. At 140 hPa and delay times larger than about 50 μs the ion temperature of the plasma was found to increase vertically in the kernel. At larger delay times, where the ratio of ions to neutral atoms was almost constant, the degree of ionization was still about 30%, and larger in the center of the plasma.

Investigation of neutral and ion number densities in the plasmas produced in vacuum and ambient environments was also carried out using resonance laser-induced fluorescence [66]. Resonance LIF images were recorded from plasmas created by a KrF excimer laser

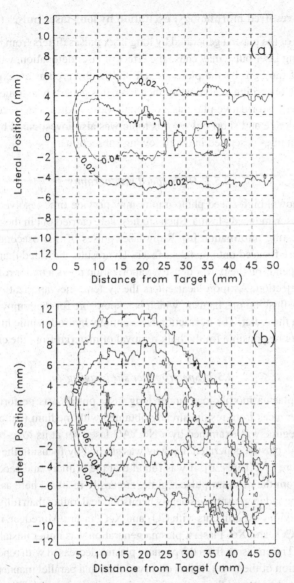

Figure 2.58. Resonance LIF images observed for Ti(II) at 3 μs delay in vacuum (a) and in 25 mTorr nitrogen (b) (Figure 6 in reference [66]).

(248 nm, 30 ns) on titanium samples, as shown in Figure 2.58. This figure indicates that the plasma in the vacuum case undergoes essentially collision-free self-similar expansion. For N_2 (25 mTorr), the LIF image shows strong suppression of the expansion along the ablating laser direction and spatial widening along the perpendicular fluorescence excitation direction.

2.5 Time-resolved morphology: excitation by long laser pulses (>100 ns)

The morphology of the plasma generated by long laser pulses differs from that produced by short shots. A long laser pulse may cause the effect of laser detonation, which results from the absorption of the light by the plasma itself and is observed as an expansion towards the laser source. Concerning analytical applications, it is generally assumed that long laser pulses produce less informative plasmas. Although most studies were focused on plasmas produced by moderate and short pulses, this topic was also investigated by using some of the imaging techniques described in the following subsections.

2.5.1 High-speed photography

As previously shown, high-speed photography may provide time-resolved morphological information on laser-induced plasma. Such studies were carried out in the early 1970s, and resulted in interesting information [8]. Ruby laser pulses of microsecond to millisecond time duration were focused onto metal surfaces. The results indicated that, after a heating period, which depends on the power density, vaporization plays a major role, accompanied by fine-particle ejection. As more heat enters the system, the vaporization process gives way to a melting–flushing mechanism, which is responsible for the removal of the bulk of the material from the crater. The interaction is essentially hydrodynamic in nature and pure vaporization is not responsible for the bulk removal of material from the crater.

2.5.2 Spectrometer slit imaging

Investigation of plasma morphology under long laser pulses was performed in order to find out any differences from results already obtained with medium (nanosecond) pulses. The long-pulse regime was examined by a Nd:YAG laser, set at its long-pulse mode. The morphology of the plasma generated this way was studied by focusing the laser on a brass target, at a pressure of 1 Torr [20]. It was found that the primary and secondary plasmas, already known from the medium-pulse regime, are easily resolved here as well. However, the shape of the secondary plasma was nearly hemispherical, rather than fully hemispherical, as observed in previous experiments with CO_2 and XeCl lasers. Moreover, in contrast to the case of the TEA CO_2 and XeCl lasers, plasma generation was rather unstable for the long-pulse YAG laser. The emission from the primary plasma decreased with repeated irradiation on the same position of the target. It was observed that, in a parallel manner, the size of the secondary plasma shrinks and the total emission intensity is reduced. The collected data in this case were found to be consistent with the shock-wave being responsible for the generation of the secondary plasma. Similar conclusions were drawn in other studies as well [23].

2.5.3 Single-fiber scanning

Imaging of plasma produced by a TEA CO_2 laser (10.6 μm, 300 mJ, 100 ns) hitting glass samples under various gas pressures was achieved by the method of single-fiber scanning [16]. It was observed that, in contrast to the results obtained with a medium-pulse laser

(20 ns), the CO_2 laser did not generate a hemispherical plasma on low-melting-point glasses. However, hemispherical plasmas were observed when hitting high-melting-point glasses, for both long-pulse and medium-pulse lasers.

An explanation of these findings has been suggested. Because the CO_2 laser pulse is long, the relatively long exposure of glasses with low melting points to the laser radiation generally results in softening or even melting of the glass surface. As a consequence, expulsion of atoms by the surface is weakened because softened glass absorbs recoil energy, and atoms gushing out from the primary plasma do not acquire enough speed to form a shock-wave. In contrast to the case of glass with a low melting point, higher-melting-point glasses do not suffer from this problem of premature softening and melting of the surface, and they thereby allow the formation of nearly hemispherical secondary plasma. This explanation was supported by measurements of the amount of evaporated material from the various glasses.

2.5.4 Near-resonant photographic absorption

Near-resonant absorption photography was applied to laser induced plasma in helium gas at 2 atm. In these studies a CO_2 laser produced a 100 ns (FWHM) pulse of 1 J, and a nitrogen-pumped tunable dye laser (3 ns) was used for absorption sampling [64]. Several wavefronts were observed at 510 ns after the laser pulse: (a) a shock-wave propagating in the neutral gas surrounding the spark; (b) a longitudinal ionization wave propagating backward; (c) a radial breakdown wave; and (d) a late ($t > 500$ ns) turbulent, forward wave originating near the focal plane, known as the secondary breakdown. This absorption photography allowed for visualization of the evolution of a high-temperature region behind the ionization front, propagating forward, which was associated with the re-entry of the excitation laser in the plasma.

In all cases, the plasma was surrounded by a hydrodynamic shock-wave. The space between the shock-front and the sharp plasma boundary was optically thin even at resonance. The region behind the shock-wave is not responsible for atomic line emission. It was observed that after 200 ns the laser shock-front clearly separates from the spark body. This indicates the negligible role of the shock-wave in ionization dynamics. The high-temperature internal front due to laser beam re-entry and post-heating of the plasma was also shown by these measurements.

2.5.5 Nonresonance laser-induced fluorescence

Laser-induced direct-line nonresonant atomic fluorescence was used to study the spatial distribution of titanium, zirconium, and hafnium in laser plumes generated by long laser pulses [72]. The plasma was generated by a pulsed dye laser (560–630 nm, 110 mJ, 1 μs). The fluorescence excitation laser was a tunable pulsed dye laser, emitting in the range of 427–720 nm (and in the UV from 265 nm to 385 nm, using doubling crystals). In this particular setup, nonresonance fluorescence and absorption were monitored simultaneously. The spatial resolution, obtained by moving the sample with a stage, was 5 μm.

It was found that iron fluorescence and absorption profiles obtained during a series of spatial scans along the emission axis at various points on the excitation axis and at various times, indicate that severe pre- and postfilter effects occur up to at least 53 μs after vaporization. The plume appears to be smaller and more concentrated at 98 μs than at 33 μs after vaporization.

Measurements made during spatial scans along the emission axis at various heights above the sample surface show that the iron concentration is greatest near the sample surface and decreases at greater heights until the iron is no longer detectable at about 8 mm above the sample.

A series of spatial scans along the vaporization axis in helium and argon atmospheres at various delay times after vaporization show that the plume is contiguous to the sample surface at short delay times, then separates from the surface and eventually disperses into the atmosphere.

2.6 Time-resolved morphology: excitation by short laser pulses (fs–ps)

Application of femtosecond and picosecond laser pulses to LIBS analysis is of considerable importance, since the spectral lines obtained under such excitation are expected to be better resolved and less affected by the background. At short laser pulses the light is absorbed by the sample itself and there is no "tail" to heat the plasma already produced by the pulse front. This should clearly affect the plasma morphology. Although several recent results indicate the actual improvements due to the short laser excitation, very little is known on the morphology of the plasma generated under such conditions. However, the ablation process and the morphology of the induced craters have been studied in this excitation regime [79–81].

The light-absorption mechanism depends both on the laser spatial and temporal profiles, and on its power. These characteristics change when applying femtosecond lasers. It was found that the fraction of energy absorbed by the plasma and the resulting electron temperatures depend on the scale length of the plasma at the surface. Two-dimensional effects considerably increase the amount of absorption into hot electrons over the amount predicted using one-dimensional models [82]. The time-dependent electron density and temperature conditions in the plasma produced by 500 fs pulses were assessed from spectroscopic data [83]. Densities and temperatures in the range of 10^{23} per cm^3 and 300 eV, respectively, were measured 2–4 ps after the laser pulse. The plasma conditions reached a plateau near these values for 5 ps after that.

An analytical study of LIBS with femtosecond laser pulses, under time-resolved space-integrated data acquisition, was recently carried out [84]. The results show a faster decay of continuum and spectral lines and a shorter plasma lifetime than in the case of longer laser pulses. This results in a significant reduction of the continuum emission in the early phase of the plasma formation [84, 85]. It was concluded that the plasma generated by ultra-short pulses exhibits a faster thermalization, compared with nanosecond excitation conditions.

As previously mentioned, very few *morphological* reports are currently available. Numerical simulations of ultra-short laser pulse ablation and plasma expansion in ambient air were performed [86, 87]. According to these calculations, the axial plasma temperature and density are strongly inhomogeneous and the maximum radiation emission occurs in the front of the plasma.

The temporal evolution of copper plasma generated by Ti–Al$_2$O$_3$ laser, of pulse duration in the range 70 fs–10 ps, was measured by a direct CCD imaging technique [41]. The longitudinal and transverse plasma dimensions could be derived from the plasma images. In these experiments, both dimensions exhibited identical temporal evolution. During the first nanoseconds, the ablated matter was observed to escape from the surface with a fast expansion in the direction normal to the target. The initial rates of plasma plume expansion were not found to depend significantly on the laser pulse duration in the studied range. In all cases, after *c*. 25 ns there was no further evolution of the plasma volume.

2.7 Time-resolved morphology: excitation by double laser pulses

Several studies have dealt with the plasma generated by a sequence of two laser pulses. Such a pulse pair may assist LIBS analysis under certain experimental conditions. For example, the first pulse may clean the surface while the second pulse samples it. In other cases, the first pulse may ablate material, while the second may excite the atoms. When the sample is placed under water, the first pulse may produce a bubble, into which the second one may generate the plasma plume.

A pulse pair can be accomplished by triggering the Q-switch Pockels Cell using separate Marx bank generators [49]. The time between the triggering of the Marx banks determines, the time between the laser pulses. The energies of the two pulses depend on the time delay between the triggering of the flash-lamp and the first Marx bank to generate the first pulse and the time separation between the two pulses. Typical separations between the pulses obtained in this way are 30–180 μs. The relevant timing parameters in such a setup are defined in Figure 2.59. An alternative setup, using two Q-switched Nd:YAG lasers, was also applied [85].

The various LIBS parameters were characterized for double-pulse excitation, and here we present only the experiments dealing with the effects upon plasma morphology.

2.7.1 Time-resolved imaging through tunable filters

The morphology of the plasma formed by the double-pulse technique was studied by using AOTF imaging [49]. Images of the plasmas produced on a metal under water are shown in Figure 2.60, where the filter was tuned to an Al(I) line. The data were recorded for several separation times. It is seen that as the separation increases, the size of the bubble increases. Therefore, it was concluded that the use of larger separation times may increase the intensity of the analyte emission.

Results on plasma size and temperature were recently obtained under dual-pulse conditions [79]. It was found that although the plasma temperature increases for dual-pulse excitation, the signal enhancement is attributed to increased sample ablation. Plasma

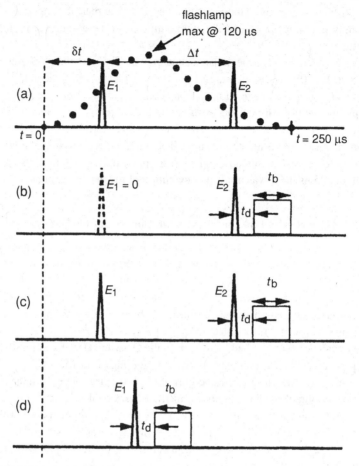

Figure 2.59. Definition of timing parameters used. (a) Relative timing between the first and the second laser pulses. (b, d) Detector gating used to obtain the data from repetitive single sparks. (c) Detector gating used to obtain the data from repetitive spark pair (Figure 2 in reference [49]).

imaging indicates that the magnitude of the enhancement can be affected by the collection optics and geometry.

2.7.2 *Time-resolved plasma shadowgraphs*

Time-resolved plasma shadowgraphs, obtained by transferring a green HeNe laser through the plume, were applied to studying laser-induced plasmas formed on a metal under water [49]. Generally, it was found that the image of the plume obtained from the second pulse of the repetitive spark pair was much brighter than that recorded from repetitive single pulses. The process responsible for the strong excitation produced by the second pulse has been attributed to the formation of a bubble by the first pulse. This bubble produces an interface

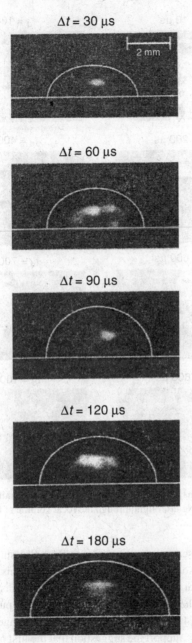

$\Delta t = 30$ µs

$\Delta t = 60$ µs

$\Delta t = 90$ µs

$\Delta t = 120$ µs

$\Delta t = 180$ µs

Figure 2.60. Images of the distribution of atomic emission from Al(I) at 394.4 nm from the repetitive spark pair timing, formed on aluminum for different values of Δt. The plasma light was filtered by AOTF to monitor Al(I) emissions (Figure 9 in reference [49]).

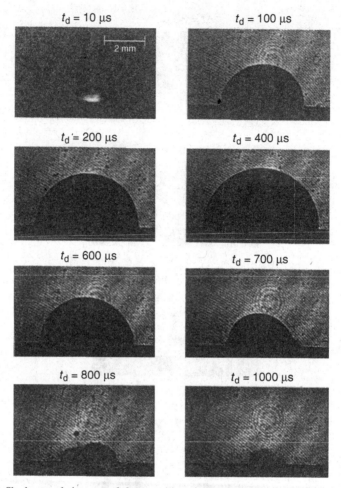

Figure 2.61. Shadowgraph images of the repetitive single spark timing at different delay times, obtained by passing green HeNe laser light through the focal volume of the first laser pulse (Figure 5 in reference [49]).

that permits the second laser pulse to interrogate the solid surface in a way similar to the analysis in air. Time-resolved shadowgraphs obtained in this way are shown in Figure 2.61. These images show that an opaque hemispherical-shaped disturbance expands radially outward from the area of the metal interrogated by the laser pulse. Expansion reaches a maximum at approximately 400 μs after the laser pulse. The images of the bubble expansion and its collapse at later times support the suggestion that the first pulse forms a vapor cavity, in which the second pulse can be generated.

The effect of the time delay between the two pulses was investigated by monitoring the plasma image formed by the second pulse. Such images obtained at various time intervals are shown in Figure 2.62. It is clear from these images that as the pulse-separation time

Δt = 30 µs

Δt = 60 µs

Δt = 120 µs

Δt = 180 µs

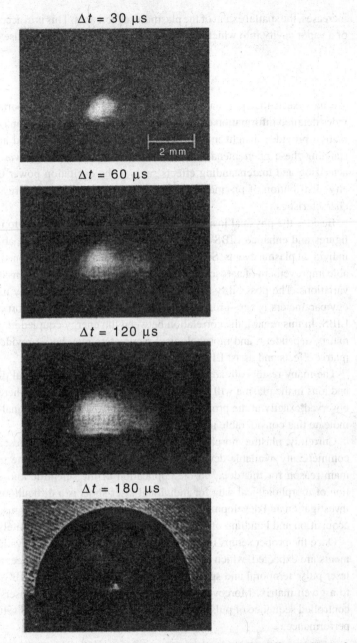

Figure 2.62. Images recorded after the second pulse of the repetitive spark pair for different values of Δt, obtained by passing green HeNe laser light through the focal volume of the laser pulses (Figure 8 in reference [49]).

increases, the spatial extent of the plasma also increases. This is in accordance with formation of a vapor cavity into which the plasma formed by the second pulse expands.

2.8 Conclusions

We have seen that plasma morphology is of considerable importance in LIBS. It provides detailed information on the dynamics of plume generation and expansion. This information provides insight into the complicated processes involved and is useful in understanding these phenomena. For instance, plasma morphology was instrumental in characterizing and understanding effects such as the excitation power density, optical geometry, distribution of plasma temperature, gas composition and pressure, and spectral line characteristics.

Besides the physical insight, plasma morphology was proved to improve the analytical figures and enhance LIBS performance. This was related to the ability of characterizing individual plasma events. Several achievements should be mentioned, such as the considerable improvement of aerosol analysis, which was based on compensation for pulse-to-pulse variations. The possibility to characterize the analyzed matrix by using plasma morphology parameters is very promising, since it may result in a new sort of matrix-independent LIBS. In this regard, the correlation between the energy coupled into the plasma, which is matrix-dependent, and morphological parameters may both provide explanation of some matrix effects and assist LIBS analysis.

The many results already obtained on the spatial and temporal distribution of neutrals and ions in the plasma will probably lead to new LIBS designs where the spectral lines are observed exactly at the proper time and location. Maps of the signal-to-noise data already indicate that considerable improvement is expected in this way.

Currently, plasma morphology analysis is mainly limited to research laboratories, and commercially available devices do not take full advantage of the potential benefits. The main reason for this delay in the application of the scientific know-how is that acquisition of morphological data for individual laser shots is a difficult task. Therefore, further investigation and developments on novel LIBS experimental designs, capable of automatic acquisition and handling of morphological information, are definitely necessary.

Once the proper setups for morphological LIBS analysis are available, further improvements are expected, which have not been studied yet. For instance, a proper design of the laser pulse temporal and spatial profile may be optimized for LIBS and may be designed to a given matrix. Moreover, the availability of femtosecond lasers that may generate a controlled sequence of pulses could be integrated in such a LIBS setup and result in better performance.

2.9 References

[1] V. Bulatov, l. Xu and I. Schechter, *Anal. Chem.*, **68** (1996), 2966.
[2] I. Schechter, *Rev. Anal. Chem.*, **16** (3) (1997), 173.

[3] V. Bulatov, R. Krasniker and I. Schechter, *Anal. Chem.*, **72** (2000), 2987.

[4] L. Xu, V. Bulatov, V. V. Gridin and I. Schechter, *Anal. Chem.*, **69** (1997), 2103.

[5] I. Schechter, *Anal. Sci. Technol.*, **8** (1995), 779.

[6] M. P. Mateo, S. Palanco, J. M. Vadillo and J. J. Laserna, *Appl. Spectrosc.*, **10** (2000), 1429.

[7] R. A. Multari and D. A. Cremers, *IEEE Trans. Plasma Sci.*, **24** (1996), 39.

[8] R. H. Scott and A. Strasheim, *Spectrochim. Acta.*, **25B** (1970), 311.

[9] O. B. Anan'in, Y. A. Bykovskii, Y. V. Eremin, *et al.*, *Kvantov. Elektron.* **18** (1991), 869.

[10] O. B. Anan'in. Y. A. Bykovskii, Y. V. Eremin, *et al.*, *Kvantov. Elektron.* **18** (1991), 1483.

[11] Zs. Márton, P. Heszler, A. Mechler, *et al.*, *Appl. Phys. A.*, **69** (1999), S133.

[12] J. W. Olesik and G. M. Hieftje, *Anal. Chem.*, **57** (1985), 2049.

[13] Y. Iida, H. Morikawa, A. Tsuge, Y. Uwamino and T. Ishizuka, *Anal. Sci.*, **7** (1991), 61.

[14] K. Kagawa, S. Yokoi and S. Nakajima, *Opt. Commun.*, **45** (1983), 261.

[15] Y. Lee, S. P. Sawan, T. L. Thiem. Y. Teng and J. Sneddon, *Appl. Spectrosc.*, **46** (1992), 436.

[16] H. Kurniawan, S. Nakajima, J. E. Batubara, *et al.*, *Appl. Spectrosc.*, **49** (1995), 1067.

[17] S. S. Harilal, R. C. Issac, C. V. Bindhu, *et al.*, *Pramana*, **46** (1996), 145.

[18] Z. Fu, E. Fang, Y. Gu, *et al.*, *Huaxue Wuli Xuebao*, **9** (1996), 125.

[19] J. Gonzalo, C. N. Afonso and I. Madariaga, *J. Appl. Phys.*, **81** (1997), 951.

[20] H. Kurniawan and K. Kagawa, *Appl. Spectrosc.*, **51** (1997), 304.

[21] Y. Lee, K. Song, H. Cha, *et al.*, *Appl. Spectrosc.*, **51** (1997), 959.

[22] B. C. Castle, K. Visser, B. W. Smith and J. D. Winefordner, *Appl. Spectrosc.*, **51** (1997), 1017.

[23] K. Kagawa and H. Kurniawan, *Trends Appl. Spectrosc.*, **2** (1998), 1.

[24] A. Neogi and R. K. Thareja, *J. Appl. Phys.*, **85** (1999), 1131.

[25] S. Yalçin. D. R. Crosley, G. P. Smith and G. W. Faris, *Appl. Phys. B*, **68** (1999), 121.

[26] W. S. Budi, H. Suyanto, H. Kurniawan, M. O. Tjia and K. Kagawa, *Appl. Spectrosc.*, **53** (1999), 719.

[27] E. J. Iglesias, H. R. Griem, R. C. Elton and H. Scott, *Aip Conf. Proc.*, **467** (1999), 143.

[28] H. Kurniawan, T. J. Lie, K. Kagawa and M. O. Tjia, *Spectrochim. Acta*, **55B** (2000), 839.

[29] T. Shinozaki, T. Ooie, T. Yano and M. Yoneda, *Jap. J. Appl. Phys.*, **39** (2000), 6272.

[30] M. Milan and J. J. Laserna, *Spectrochim. Acta*, **56B** (2001), 275.

[31] X. L. Mao, M. A. Shannon, A. J. Fernandez and R. E. Russo, *Appl. Spectrosc.*, **49** (1997), 1054.

[32] R. A. Multrari, L. E. Foster, D. A. Cremers and M. J. Ferris, *Appl. Spectrosc.*, **50** (1996), 1483.

[33] S. S. Harilal, C. V. Bindhu, R. C. Issac, V. P. N. Nampoori and P. G. Vallabhan, *J. Appl. Phys.*, **82** (1997), 2140.

[34] S. S. Harilal, C. V. Bindhu, V. P. N. Nampoori and C. P. G. Vallabhan, *Appl. Spectrosc.*, **52** (1998), 449.

[35] J. D. Wu, Q. Pan and S. C. Chen, *Appl. Spectrosc.*, **51** (1997), 883.

[36] J. M. Vadillo, M. Milán and J. J. Laserna, *Fres. J. Anal. Chem.*, **355** (1996), 10.

[37] C. Aragon and J. A. Aguiler, *Appl. Spectrosc.*, **51** (1997), 1632.

[38] D. A. Rusak, M. Clara, E. E. Austin, *et al.*, *Appl. Spectrosc.*, **51** (1997), 1628.

[39] R. K. Thareja, A. Misra and S. R. Franklin, *Spectrochim. Acta,* **53B** (1998), 1919.
[40] W. Whitty and J. Mosnier, *Appl. Surf. Sci.,* **127–129** (1998), 1035.
[41] B. Salle, O. Gobert, P. Meynadier, *et al., Appl. Phys. A,* **69** (1999), s381.
[42] R. A. Al-Wazzan, J. M. Hendron and T. Morrow, *Appl. Surf. Sci.,* **96–98** (1996), 170.
[43] S. Beatrice, M. N. Libenson, P. Manchien, *et al., Proc. SPIE – Int. Soc. Opt. Eng.,* **3822** (1999), 56.
[44] Y.-L. Chen, J. W. L. Lewis and C. Parigger, *J. Quant. Spectrosc. Radiat. Transfer,* **67** (2000), 91.
[45] A. Borghese and S. S. Merola, *Appl. Opt.,* **37** (1998), 3977.
[46] T. Zyung, H. Kim, J. C. Postlewaite and D. D. Dlott, *J. Appl. Phys.,* **12** (1989), 4548.
[47] J. P. Fischer, T. Juhasz and J. F. Bille, *Appl. Phys. A,* **64** (1997), 181.
[48] D. B. Geohegan and A. A. Puretzky, *Appl. Surf. Sci.,* **96–98** (1996), 131.
[49] A. E. Pichahchy, D. A. Cremers and M. J. Ferris, *Spectrochim. Acta.,* **52** (1997), 25.
[50] A. D. Giacomo, V. A. Shakhatov and O. D. Pascale, *Spectrochim. Acta,* **56B** (2001), 753.
[51] V. Bulatov, R. Krasniker and I. Schechter, *Anal. Chem.,* **70** (1998), 5302.
[52] V. Bulatov and I. Schechter, *Int. J. Chem.,* **1** (1998), 8.
[53] R. Krasniker, V. Bulatov and I. Schechter, *Spectrochim. Acta,* **56B** (2001), 609.
[54] M. P. Nelson and M. L. Myrick, *Appl. Spectrosc.,* **53** (1999), 751.
[55] O. Eryu. K. Murakami, K. Masuda, A. K. Asuya and Y. Nishina, *Appl. Phys. Lett.,* **54** (1989), 2716.
[56] H. Shiraga, N. Miyanaga, M. Heya, *et al., Rev. Sci. Instrum.,* **68** (1997), 745.
[57] R. Kodama, K. Okada and Y. Kato, *Rev. Sci. Instrum.,* **70** (1999), 625.
[58] M. E. Sherrill, R. C. Mancini, J. E. Bailey, *et al., Proc. SPIE – Int. Soc. Opt. Eng.,* **3935** (2000), 14.
[59] A. E. Burgov, I. N. Burdonsky, V. V. Gavrilov, *et al., Rev. Sci. Instrum.,* **72** (2001), 652.
[60] M. Katayama, M. Nakai, T. Yamanaka, Y. Izawa and S. Nakai, *Rev. Sci. Instrum.,* **62** (1990), 124.
[61] D. K. Bradaley, P. M. Bell, J. D. Kilkenny, *et al., Rev. Sci. Instrum.,* **10** (1992), 4813.
[62] D. M. Chambers, S. H. Glenzer, J. Hawreliak, *et al., J. Quant. Spectrosc. Radiat. Transfer,* **71** (2001), 237.
[63] M. Fajardo, P. Audebert, H. Yashiro, *et al., J. Quant. Spectrosc. Radiat. Transfer,* **71** (2001), 317.
[64] J. P. Geindre, J. C. Gauthier and N. Grandjouan, *J. Phys. D: Appl. Phys.,* **13** (1970), 1235.
[65] A. Gupta, B. Braren, K. G. Casey, B. W. Hussey and R. Kelly, *Appl. Phys. Lett.,* **59** (1991), 1302.
[66] T. P. Williamson, G. W. Martin, A. H. El-Astal, *et al., Appl. Phys. A,* **69** (1999), s859.
[67] I. B. Gornushkin, M. Clara, B. W. Smith, *et al., Spectrochim. Acta,* **52** (1997), 1617.
[68] M. Gatti, V. Palleschi, A. Salvetti, D. P. Singh and M. Vaselli, *Opt. Commun.,* **69** (1988), 141.
[69] W. Sdorra and K. Niemax, *Spectrochim. Acta.,* **45B** (1990), 917.
[70] V. S. Burakov, N. V. Tarasenko and N. A. Savastensko, *Pub. Astron. Ops. Beogradu,* **68** (2000), 27.
[71] D. A. Akimov, A. M. Zheltikov, N. I. Koroteev, *et al., Kvantov. Elektron.,* **25** (1998), 1105.
[72] W. C. Pesklak and E. H. Piepmeier, *Microchem. J.,* **50** (1994), 253.
[73] E. S. Yeung, *Anal. Sci.,* **7** (1991), 1447.

[74] H. Schröder, I. Schechter, R. Wisbrun and R. Niessner. In *Excimer Lasers*, ed. L. D. Laude (Dordrecht, Netherlands: Kluwer Academic Publishers (1994), p. 269.

[75] R. Wisbrun, I. Schechter, R. Niessner, H. Schröder and K. L. Kompa, *Anal. Chem.*, **66** (1994), 2964.

[76] S. Tulapurkar, A. G. Bidve, S. S. Patil and S. Itagi, *Indian J. Phys.*, **72B** (1998), 515.

[77] P. Shan, R. L. Armstrong and L. J. Radziemski, *J. Appl. Phys.*, **65** (1989), 2946.

[78] D. Breitling, H. Schittenhelm, P. Berger, F. Dausinger and H. Hügel, *Appl. Phys. A*, **69** (1999), s505.

[79] H. Kumagai, K. Midorikawa, K. Toyoda *et al.*, *Appl. Phys. Lett.*, **65** (1994), 1850.

[80] B. K. F. Young, B. G. Wilson, G. B. Zimmerman, D. F. Price and R. E. Stewart, *J. Quant. Radiat. Transfer*, **58** (1997), 991.

[81] D. von der Linde and K. Sokolowski-Tinten, *Appl. Surf. Sci.*, **154–155** (2000), 1.

[82] S. C. Wilks and W. L. Kruer, *IEEE. J. Quant. Electron.*, **33** (1997), 1954.

[83] K. B. Fournier, B. K. F. Young, S. J. Moon *et al.*, *J. Quant. Spectrosc. Radiat. Transfer*, **71** (2001), 339.

[84] B. Le Drogoff, J. Margot, M. Chaker *et al.*, *Spectrochim. Acta B*, **56** (2001), 987.

[85] S. M. Angel, D. N. Stratis, K. L. Eland *et al.*, *Fres. J. Anal. Chem.*, **369** (2001), 320.

[86] F. Vidal, S. Laville, T. W. Johnston *et al.*, *Spectrochim. Acta B*, **56** (2001), 973.

[87] D. Stratis, K. Eland and S. M. Angel, *Appl. Spectrosc.*, **55** (2001), 1297.

3

From sample to signal in laser-induced breakdown spectroscopy: a complex route to quantitative analysis

E. Tognoni, V. Palleschi, M. Corsi and G. Cristoforetti

IPCF, Italy

N. Omenetto, I. Gornushkin, B. W. Smith and J. D. Winefordner

Department of Chemistry, University of Florida

3.1 Introduction

The aim of this chapter is to provide basic information on the use of the technique of laser-induced breakdown spectroscopy (LIBS) for quantitative analysis. It begins with a discussion of the theoretical assumptions on the state of the plasma that must be made in order to ensure reliability of the analysis. A review is then presented of some of the methods developed to extract quantitative information from experimental LIBS data.

In 1997, Castle *et al.* stated that at the time only a limited number of studies had reported on the use of LIBS as a quantitative technique [1]. This paucity of results was attributed to the inadequate level of the analytical figures of merit (accuracy, precision and detection limits) attainable by this technique in comparison with other well-established techniques. Since then, however, many papers have appeared in the literature reporting on the use of the LIBS technique for quantitative analysis. In fact, owing to the peculiar advantages of LIBS, including short measurement times, the ability to use samples without any pre-treatment and the capability for simultaneous multi-element detection, many researchers have focused their efforts on developing new methods for reliable LIBS-based quantitative analysis. Undoubtedly, in some particular situations (screening, *in situ* measurement, process monitoring, hostile environments, etc.) LIBS may be the technique of choice. Thus, the main research efforts have been aimed at exploiting the technique's potential and minimizing its drawbacks.

Most of the drawbacks of LIBS are, however, side effects of its intrinsic advantages. For example, because of the small size of the focused laser beam and the small sample mass vaporized by the spark, the accuracy is heavily dependent on the homogeneity of the sample. In addition, the lack of sample preparation means that small amounts of surface contaminants may affect the analysis and, at the same time, reproducibility of the laser spark may be reduced by changes in surface composition. Also, the pulsed operation of the spark yields a lower integrated emission signal and less reproducible sample excitation

Laser-Induced Breakdown Spectroscopy: Fundamentals and Applications, ed. Andrzej W. Miziolek, Vincenzo Palleschi and Israel Schechter. Published by Cambridge University Press. © A. W. Miziolek, V. Palleschi and I. Schechter 2006. A. W. Miziolek's contributions are a work of the United States Government and are not protected by copyright in the United States.

than a continuous excitation source such as an inductively coupled plasma [2]. However, most of these problems can be alleviated through proper choice of the experimental conditions.

In this chapter, we review the approaches used for extracting quantitative data from LIBS. Before presenting a brief discussion of the methods, however, it is worthwhile to examine the fundamental spectrochemical hypotheses underlying the laser-induced plasma state. Since plasma-state theory is dealt with systematically elsewhere in this book, no attempt will be made to provide a complete description. Only those issues strictly related to achieving precise quantitative analysis via LIBS will be covered. The chapter concludes with a table of the detection limits for several elements, compiled from the literature.

In keeping with the literature, most of the discussion refers to the analysis of solid samples. However, some interesting results achieved in the field of liquid or aerosol LIBS analysis have been included as well.

3.2 The characteristics of laser-induced plasma and their influence on quantitative LIBS analysis

Laser-induced plasma emission consists of atomic and ionic spectral lines characteristic of the constituent species, superimposed on a broad-band continuum that is the result of electron–ion recombination and free–free interactions. Identification of the spectral lines and measurement of their intensities provide qualitative and quantitative information, respectively. Quantitative analysis is not a trivial task: the spectral emission intensity in the plasma is determined not only by the concentration of the element in the sample, but also by the properties of the plasma itself, which in turn depend on factors such as the characteristics of the excitation source (energy, power density, wavelength), the sample and the surrounding gas. Furthermore, the laser ablation process (a term which includes the processes of evaporation, ejection of atoms, ions, molecular species and fragments; hydrodynamic expulsion; shock waves; plasma initiation and expansion; plasma–solid interactions; etc.) influences the amount and composition of the ablated mass and must be understood and controlled in order to achieve accurate and sensitive quantitative analysis [3].

The complexity of the phenomena involved can be appreciated by resorting to a simple derivation of the dependence of the LIBS signal upon the various processes leading from the (solid) sample to the measured signal photons emitted from the (gas phase) atoms and ions excited in the plasma volume. The fundamental parameters governing the overall process can then be explicitly factored out. The signal, S (counts), due to emission of a particular atomic or ionic line of an element is given by the product of the excited state number density, n_u (cm^{-3}), the spontaneous transition probability of the transition chosen, A_{ul} (photons s^{-1}), and the *detection function*, f_{det} (cm^3 counts photon^{-1} s). This function can be defined here as the product of the excitation volume V_{exc} (cm^3), seen by the detector, and of an overall detection efficiency, η_{det} (counts photon^{-1} s), which includes such "trivial" parameters as optical transmission, detector gain (counts photon^{-1}) and integration time (s), and a calibration function, f_{cal} (no units), which describes the plasma characteristics in terms of optical depth, i.e. the probability that photons emitted in the plasma center escape through the remaining plasma volume and reach the detector. Therefore, f_{cal} includes self-absorption (and self-reversal) effects. In symbols,

$$S = n_u A_{ul} f_{det} = n_u A_{ul} V_{exc} f_{cal} \eta_{det}. \tag{3.1}$$

The total number of excited atoms (ions) in the emitting state ($\equiv n_u V_{exc}$) must be related to the total number of atoms (ions) created by the laser and present in the gas phase in the plasma volume, $(N_T)_g$, multiplied by an *excitation/ionization function*, f_{exc} (no units), which gives the probability of occupation of that particular atom/ion emitting level among all other possible levels. Therefore,

$$S = (N_T)_g f_{exc} A_{ul} f_{cal} \eta_{det}. \tag{3.2}$$

The total number of atoms (ions) in the excitation volume, $(N_T)_g$, must be related to the total number of atoms in the sample through an *ablation/vaporization function*, f_{abl}, which describes the mechanism by which a certain fraction of the solid material is ablated and carried in the vapor phase by the developing plasma plume. Accordingly,

$$S = (N_T)_s f_{abl} f_{exc} A_{ul} f_{cal} \eta_{det}. \tag{3.3}$$

The product $(N_T)_s f_{abl}$ can also be factored out and represented by the product of three factors: (i) the ratio $(m_s N_A / M_s)$, where m_s(g) is the ablated sample mass, N_A is the Avogadro number (atoms/mole) and M_s is the atomic weight (g/mole) of the element; (ii) the weight fraction of the element in the sample, χ_s; and (iii) a stoichiometric factor, f_{st}, which can be defined as the deviation of the elemental composition of the sample in the gas phase as compared with that in the solid phase. In other words, if only two elements (*i* and *j*) are considered, f_{st} can be defined by

$$f_{st} \equiv \frac{(N_i/N_j)_g}{(N_i/N_j)_s}. \tag{3.4}$$

Combining all the above expressions (3.1–3.4) we obtain

$$S = A_{ul}(m_s(N_A/M_s)\chi_s f_{st}) f_{exc}(f_{cal}\eta_{det}) \tag{3.5}$$

or

$$S = A_{ul} f_{int} f_{exc} f_{det}. \tag{3.6}$$

It can then be concluded that the signal is influenced by three interrelated functions, describing the initial interaction between the sample and the laser, f_{int} (leading to ablation/vaporization of solid material), the excitation/ionization mechanism leading to atomic (ionic) emission, f_{exc}, and the characterization of the radiation environment, f_{det} (thin or thick plasmas). These functions can now be discussed separately and their relevance to analytical LIBS assessed. In this way, it should be possible to point out the problems and pitfalls of the technique, to review the attention and solutions given to each of these functions in the literature and to indicate the areas where further work is necessary.

In an attempt to simplify a very complex phenomenon, most of the methods developed for quantitative LIBS analysis assume, either implicitly or explicitly, that:

- the composition of the plasma volume under observation is representative of the sample composition (stoichiometric ablation),
- the plasma volume under observation is in Local Thermodynamic Equilibrium (LTE), and
- the spectral lines measured are optically thin.

In the following, it will become clear that even when such hypotheses seem to be superfluous (for example, when quantitative LIBS analysis is performed using the calibration curves obtained from known reference samples), the reproducibility of the quantitative results is, in most cases, assured only when these basic assumptions are satisfied.

3.2.1 Stoichiometric ablation: the interaction function

Understandably, the problems of ablation efficiency and stoichiometric material removal have been addressed by numerous studies and authors. Only a few pertinent references will be included in this overview. The hypothesis of stoichiometric ablation forms the very basis of the LIBS method (as well as other techniques which rely on laser ablation for the sampling stage: LA-ICP, LA-MS, etc.). Indeed, in 1991 Chan and Russo [4] demonstrated that laser ablation is stoichiometric when the power density on the target exceeds 10^9 W cm^{-2}, a value that is commonly reached in LIBS measurements. In several subsequent papers, Russo and co-workers [3–5] provided further insight into the phenomenon as well as a more detailed explanation of the processes which lead to the establishment of stoichiometric ablation. According to their studies, laser–material interactions can be described by using two different models: vaporization or ablation. A vaporization process is generally involved at power densities $\leq 10^6$ W cm^{-2}, typically corresponding to microsecond or longer laser pulses. Phonon relaxation rates are on the order of 0.1 ps; therefore, the absorbed optical energy is rapidly converted to heat. Heat dissipation and vaporization are rapid in comparison with the laser pulse duration. Differential vaporization is possible in such cases because the higher vapor-pressure elements will be enriched in the vapor phase with respect to the original solid sample. At higher power densities, $\geq 10^9$ W cm^{-2}, corresponding to nanosecond and shorter laser pulses, an explosion occurs. The vaporization temperature of the surface is exceeded within a fraction of the laser pulse duration. However, before the surface layer can vaporize, the underlying material reaches its vaporization temperature, causing the surface to explode. The rapidly heated material has the same composition as the solid, and the process results in stoichiometric ablation. Recently, Russo *et al.* [6] discussed five distinct regimes in the laser/solid interaction process with four transition points or thresholds. Nano-, pico- and femtosecond lasers were considered. The processes described by Russo and co-workers offer an important guideline to the researcher preparing a LIBS experiment. In any event, it should be kept in mind that this model cannot cover all possible experimental situations, so that in practical LIBS measurement, the occurrence of stoichiometric ablation should be checked a posteriori for the specific class of materials under analysis.

For example, a case of non-stoichiometric ablation has been reported by Mao *et al.* [7], in which laser ablation of a brass sample was performed using different lasers with various wavelengths and pulse durations. For a 30 ns pulse-duration excimer laser at lower power density, the interaction was dominated by thermal vaporization, and the vapor was enriched in zinc (the latent heat of vaporization for zinc is lower than that for copper by a factor of about 3). Using a picosecond Nd:YAG, on the contrary, the resulting vapor was enriched in copper at lower power density. In order to explain such findings, a non-thermal mechanism was proposed, which involves an interaction between space charges and ionized species at the sample surface. The fast photoelectrons generated by a picosecond pulse can, in fact, produce space charges. In this case, at lower power density the ionization of copper (ionization potential = 7.72 eV) could be favored over that of zinc (ionization potential = 9.29 eV). However, for both types of laser, increasing the laser power density eventually leads to stoichiometric ablation. Nouvellon et al. [8], on the other hand, using a KrF laser at 248 nm, could not verify stoichiometric ablation in the case of copper in brass samples.

Another condition affecting stoichiometric ablation is the formation of a deep crater by repetitive irradiation of the same surface position. Borisov et al. [9] investigated the dependence of the Pb/U line emission intensity ratio on the number of laser pulses focused on the same spot in a NIST glass standard. This particular element pair was chosen for the large difference between the melting points of their corresponding oxides. For power density values above a certain threshold (previously determined by the same group), the ablation was stoichiometric as long as the crater was shallow. However, when the crater aspect ratio (depth/diameter) exceeded a value of about 6, the composition of the ablated material diverged from stoichiometric, becoming enriched in lead. Two interpretations of this behavior are possible: the first attributes the deviation from stoichiometry to the effect of reduction of the actual power density within the crater, owing to geometric factors, to values below the threshold for stoichiometric ablation; the second, instead, points to the shielding and thermal effects of the plasma generated inside the crater. In any event, the authors conclude that good stoichiometric ablation can be achieved when experimental conditions are carefully selected.

Investigation of an issue analogous to stoichiometric ablation in solids has been undertaken by Dudragne et al. [10] for the detection of fluorine, chlorine and carbon compounds in air. They found that the slope of the calibration curves of the carbon-line intensity versus the concentration of the compounds in air (expressed as volume/volume ratio) is proportional to the number of carbon atoms in the molecule. The same has been observed for chlorine and fluorine compounds, demonstrating that the gaseous molecules are completely dissociated in the plasma (similar conclusions have been reached by Cremers and Radziemski [11]). This finding enabled the authors to calculate a calibration curve normalized to one atom per molecule for each element considered and to determine the stoichiometric ratio between fluorine, chlorine and carbon for unknown compounds. Similarly, Winefordner and co-workers [12] have demonstrated a stoichiometric relationship for the determination of C:H:O:N ratios in solid organic compounds. As a final remark, it should be noted that, because only a small amount of material is sampled and analyzed in LIBS, the accuracy and precision of the measurement are heavily dependent on the homogeneity of the sample. A particular case is represented by the analysis of aerosols, in which deviation from stoichiometry may be the result of incomplete vaporization of particles, because some elements are segregated on the grain surface or interior. For instance, in a study based on Be particles, Cremers and Radziemski [13] suggested the existence of an upper limit to the mass that can be vaporized by single laser pulses and observed that under the specific experimental conditions used, particles greater than 10 μm in diameter were undersampled. This value has been often quoted in the literature as the upper limit for complete particle vaporization in LIBS but it is certainly very dependent upon laser fluence and particle composition and may vary widely. Recent measurements by Carranza and Hahn [14] on SiO_2 particles have yielded an upper size limit of about 3 μm.

In general, the use of shorter laser pulses should be beneficial with respect to fractionation problems [6]. Also, the ablation efficiency (defined here as the ratio of the ablated matter volume to the laser pulse energy) was found to be better when a femtosecond laser was used [15].

3.2.2 The excitation/ionization function

Definition of thermodynamic equilibrium in laser-induced plasmas

In laser-induced plasmas, the duration of the plume emission is long compared with both the radiative lifetimes of the emitting species and the laser pulse length. Therefore, plasma emission is not a direct consequence of the photo-excitation mechanism. Rather, a longer-lived secondary process, such as impact excitation by thermal electrons, has been invoked to explain the phenomenon [16]. Owing to the nature of the particles making up the plasma, we would expect the kinetic, excitation, ionization, and radiative energies to contribute to the description of the system state. The distributions corresponding to the above-mentioned energy forms are described respectively by the Maxwell, Boltzmann, Saha and Planck functions. The equilibrium distribution of energy among the different states of the assembly of particles is determined by the temperature, T, defined for each particular form of energy. It may happen that an equilibrium distribution exists for one of these forms of energy, but not for another. Complete thermodynamic equilibrium would exist when all forms of energy distribution are described by the same temperature. Under such conditions, the principle of detailed balance must hold. In practice, this situation cannot be fully realized, and some approximations must be adopted to describe the plasma state. The form of energy that is most often decoupled from the others is radiation energy, since radiative equilibrium requires the plasma to be optically thick at all frequencies. However, typical LIBS plasmas, in which electron collision is the rate-determining mechanism, can be described by a state known as Local Thermodynamic Equilibrium (LTE). In this state, the collision processes must be much more important than the radiative ones, so that the non-equilibrium of radiative energy can be neglected, while for every point it is still possible to find a temperature parameter that satisfies the Boltzmann, Saha and Maxwell distributions. Thus, the plasma electronic excitation temperature, T, and the electron density, n_e, which can be derived from the plasma emission data, can be used to describe the plasma characteristics. Among the many books on this topic, the reader is referred to Thorne [17] for a succinct overview.

In referring to the plasma constituents, we need to distinguish between chemical elements (whose concentration we wish to measure in the sample) and species corresponding to different ionization stages of the same element present in the plasma. By convention, spectroscopic notation indicates neutral and single ionized species of the element Pb, for instance, as Pb(I) and Pb(II), respectively. It is widely recognized that, in typical LIB plasma and within the typical measurement time window, only neutral atoms and singly charged ions are present to a significant degree. Therefore, in the following, only neutral and singly ionized particles will be considered. In any case, all the relations can, if necessary, be easily generalized to include higher ionization states. Under LTE conditions, the population of the excited levels for each species follows a Boltzmann distribution (see the f_{exc} in equation (3.2)):

$$n_i^s = \frac{g_i}{U^s(T)} n^s e^{-E_i/kT},$$ (3.7)

where n_i^s indicates the population density of the excited level i of species s, g_i and E_i are the statistical weight and the excitation energy of the level, respectively, n^s is the total number density of the species s in the plasma, k is the Boltzmann constant and $U^s(T)$ is the internal partition function of the species at temperature T:

$$U^s(T) = \sum_i g_i e^{-E_i/kT}.$$ (3.8)

Here, and in what follows, the ground state of the atom or ion corresponds to zero energy. Owing to the crowding of energy levels toward the ionization limit, calculation of the partition function should, in principle, include infinite terms, especially at high plasma temperatures, causing the sum to diverge. However, the sum extends to infinity only if the possible principal quantum number and, hence, the atomic radius extend to infinity. In the plasma environment, however, because of screening by the other charged particles, the electron is attracted by the nucleus until it is within a finite distance (corresponding to the radius of the Debye sphere). This is equivalent to reducing the effective ionization potential E_{ion} for each species in the plasma by a factor ΔE_{ion}. This same factor defines a cut-off limit to the sum of the partition function, thereby removing the problem of divergence [17].

   The condition that atomic and ionic states should be populated and depopulated predominantly by electron collisions, rather than by radiation, requires an electron density which is sufficient to ensure a high collision rate. The corresponding lower limit of electron density n_e is given (cm^{-3}) by the McWhirter criterion:

$$n_e \geq 1.6 \times 10^{12} T^{1/2} (\Delta E)^3$$ (3.9)

where ΔE (eV) is the highest energy transition for which the condition holds, and T (K) is the plasma temperature. This criterion is a necessary, though insufficient, condition for LTE, and is typically fulfilled during the first stages of plasma lifetime. It is, however, difficult to satisfy for the low-lying states, where ΔE is large. However, for any n_e, it is possible to find high excitation levels where the states are close enough for equation (3.9) to hold. In this case, the plasma is said to be in partial LTE [17].

As already mentioned, LTE plasmas can be characterized by a single temperature that describes the distribution of species in energy levels, the population of ionization stages or the kinetic energy of electrons and heavier particles. Consequently, the excitation temperature which controls the population of the atomic (and ionic) energy levels should be the same as the ionization temperature, which determines the distribution of atoms of the same element in the different ionization stages. This latter distribution is described by the Saha equation, which in the case of the neutral and singly ionized species of the same element can be written as

$$n_e \frac{n^{II}}{n^I} = \frac{(2\pi m_e kT)^{3/2}}{h^3} \frac{2U^{II}(T)}{U^I(T)} e^{-\frac{E_{ion}}{kT}}, \qquad (3.10)$$

where n_e is the plasma electron density, n^I and n^{II} are the number densities of the neutral atomic species and the single ionized species, respectively, E_{ion} is the ionization potential of the neutral species in its ground state, m_e is the electron mass, and h is Planck's constant. It is worth mentioning that in accurate calculations, the ionization potential lowering factor ΔE_{ion} should be taken into account (the typical value being on the order of 0.1 eV).

It should be noted that, by definition, LTE might also result in spatial decoupling, possibly leading to different plasma temperatures at different spatial positions. This should be considered especially when comparing results obtained in different observation geometries [18].

Measurement of plasma temperature

Many methods have been described for determining the plasma temperature based on the absolute or relative line intensity (line pair ratio or Boltzmann plot), the ratio of line to continuum intensity, etc. Depending on the experimental conditions, one of these methods may be more suitable than others. For the diagnosis of early phase plasma, for instance, Liu *et al.* [19] used the line-to-continuum intensity ratio, because line and continuum intensities are typically comparable at the start of plasma evolution. As our main interest here is in the analytical applications of LIBS (i.e. involving observations of the plasma at later times), this method will not be discussed further.

Provided that the LTE hypothesis described above is fulfilled, the plasma temperature can be calculated from the intensity ratio of a pair of spectral lines originating in different upper levels of the same element and ionization stage. In fact, assuming that the level population obeys a Boltzmann distribution (equation (3.7)), the total spectrally integrated radiant emissivity ($W\ m^{-3}\ sr^{-1}$) corresponding to the transition between the upper level i and the lower level j is given by

$$e_{ij} = \left(\frac{hc}{4\pi}\right) \frac{A_{ij} g_i}{\lambda_{ij} U^s(T)} n^s e^{-E_i/kT}, \qquad (3.11)$$

where λ_{ij}, A_{ij} and g_i are the wavelength, the transition probability and the statistical weight for the upper level, respectively; c is the speed of light, and the other symbols have already been defined. With the detectors typically used in LIBS measurements, an alternative formula in terms of the integrated line intensity (number of transitions per unit volume per unit time) is preferred, i.e.

$$I_{ij} = n_i^s A_{ij} = \frac{A_{ij} g_i}{U^s(T)} n^s e^{-E_i/kT}. \tag{3.12}$$

Now, by considering two lines, λ_{ij} and λ_{mn}, of the same species, characterized by different values of the upper level energy ($E_i \neq E_m$), the relative intensity ratio can be used to calculate the plasma temperature

$$T = \frac{E_i - E_m}{k \ln\left(\frac{I_{mn} g_i A_{ji}}{I_{ij} g_m A_{mn}}\right)} \tag{3.13}$$

When selecting a line pair, it is advisable to choose two lines as close as possible in wavelength and as far apart as possible in excitation energy. This is to limit the effect of varying spectral response of the apparatus, as well as to minimize the sensitivity to small fluctuations in emission intensity. Assuming that the intensity values are the only factors affected by the experimental error, the uncertainty in the temperature determination based on equation (3.13) can be given as

$$\frac{\Delta T}{T} = \frac{kT}{\Delta E} \frac{\Delta R}{R}, \tag{3.14}$$

where $\Delta E = E_i - E_m$ is the difference in energy of the two states observed, $R = I_{ij}/I_{mn}$ is the measured ratio of emission intensities, and ΔR is the uncertainty associated with the ratio. As is clear from equation (3.14), large values of ΔE will minimize the effect of the uncertainty in R on the uncertainty in T [20].

As we have seen, the emitted spectral line intensity is a measure of the population of the corresponding energy level of a certain species in the plasma. Under the assumptions that the plasma is both in LTE and optically thin, if we have information on the intensity emitted from several excited levels, we can then determine the temperature which is responsible for the observed population distribution. Once again, we use the Boltzmann equation (equation (3.7)) to relate the population of an excited level i to the total number density n^s of the species s in the plasma, and equation (3.12) to represent the intensity of the transition starting with level i. After linearization of expression (3.12), the familiar form of the Boltzmann plot equation is obtained

$$\ln \frac{I_{ij}}{g_i A_{ij}} = \ln\left(\frac{n^s}{U^s(T)}\right) - \frac{E_i}{kT}. \tag{3.15}$$

Measurement of the intensities of a series of lines from different excitation states of the same species allows evaluation of the plasma temperature, provided that the transition probabilities and statistical weights are known. A plot of the left-hand side of equation (3.15) vs. E_i has a slope of $-1/kT$. Therefore, the plasma temperature can be obtained via linear regression, without knowing n^s or $U^s(T)$.

The use of several different lines instead of just one pair leads to greater precision of the plasma temperature determination. In fact, though the precision of the intensity values can be improved by increasing the signal intensity, the transition probability values reported in

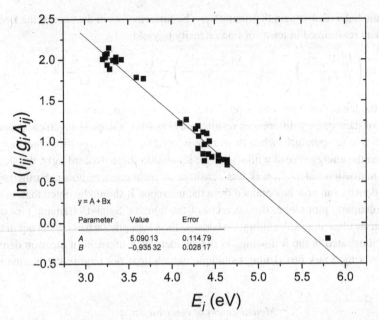

Figure 3.1. Boltzmann plot of the neutral iron lines observed in the LIBS spectrum of an aluminum alloy sample. The line intensity values have been determined as the integral area of the best fitting analytical function. The values have been corrected for the wavelength response of the system. The Δy error bars (on the order of $\pm$ 0.1) have been omitted for the sake of readability. The resulting excitation temperature is 1.24×10^4 K $\pm$ 3%.

the literature exhibit significant degrees of uncertainty (from 5% to 50%). The use of many lines in some sense "averages out" these uncertainties. An example of the application of this method for calculating the temperature is shown in Figure 3.1, where selected lines of neutral iron have been plotted in the Boltzmann plane. The sample is an aluminum alloy with 0.2% iron content; under these conditions the plasma can be considered to be optically thin for the iron radiation. Because of the experimental difficulties associated with absolute intensity calibration, the information provided by the intercept in the Boltzmann plot is rarely used [21].

Because emission lines from different ionization stages are usually present in a laser-induced plasma, a combination of the Saha ionization and Boltzmann excitation distributions can be used to measure the electron temperature. The most common form of the coupled Saha–Boltzmann relation takes the form of the ionic/atomic emission radiance ratio

$$\frac{e_{ij}^{\mathrm{II}}}{e_{mn}^{\mathrm{I}}} = \left(\frac{A_{ij}^{\mathrm{II}} g_i^{\mathrm{II}} \lambda_{mn}^{\mathrm{I}}}{A_{mn}^{\mathrm{I}} g_m^{\mathrm{I}} \lambda_{ij}^{\mathrm{II}}}\right) \left(\frac{2(2\pi m_e k T)^{3/2}}{n_e h^3}\right) \mathrm{e}^{-\left(\frac{E_{\mathrm{ion}} - \Delta E_{\mathrm{ion}} + E_i^{\mathrm{II}} - E_m^{\mathrm{I}}}{kT}\right)}. \tag{3.16}$$

The superscripts I and II denote atomic and ionic parameters, respectively. Here, E_{ion} is the first ionization potential and ΔE_{ion} is the lowering correction parameter. The coupled form of the Saha–Boltzmann distribution can be linearized as in the case of the Boltzmann relation and rearranged in terms of line intensity to yield

$$\ln\left(\frac{I_{ij}^{\text{II}} A_{mn}^{\text{I}} g_m^{\text{I}}}{I_{mn}^{\text{I}} A_{ij}^{\text{II}} g_i^{\text{II}}}\right) = \ln\left(\frac{2(2\pi m_e kT)^{3/2}}{n_e h^3}\right) - \frac{\left(E_{\text{ion}} - \Delta E_{\text{ion}} + E_i^{\text{II}} - E_m^{\text{I}}\right)}{kT}. \quad (3.17)$$

Plotting the logarithmic ratio of several ionic and atomic emission line combinations as a function of their energy differences results in a line whose slope is inversely proportional to the electron temperature (when the source is in LTE). The energy difference is typically larger than the energy spread within a single ionization stage. Accordingly, the slope from a linear regression calculation is less sensitive to measurement noise. Furthermore, the electron density can now be obtained from the intercept. It should be noted that, in contrast to the Boltzmann plot alone, the intercept of the coupled Saha–Boltzmann plot does not require an absolute intensity calibration because the geometric factors cancel out in the ratio [21]. As illustrated in the following, an independent measurement of electron density can be made via the Stark broadening technique, which does not require the plasma to be in LTE.

Measurement of electron density

Two spectroscopic methods are commonly used to measure the plasma electron density, n_e. The first requires measuring the Stark broadening of plasma lines; the second requires measuring the population ratio of two successive ionization states of the same element.

Stark effect   The profile of a line is the result of many effects, but under typical LIBS conditions the main contribution to the line width comes from the Stark effect (see Gornushkin *et al.* [22], for a discussion of the different broadening effects influencing the spectral line shape in LIBS). In fact, the electric field generated by electrons in the plasma perturbs the energy levels of the individual ions, thereby broadening the emission lines from these excited levels. The Stark broadening of a well-isolated line is thus a useful tool for estimating the electron density, provided that the Stark broadening coefficient is known (by measurement or by calculation). The Stark broadening of a line, expressed as the FWHM (full width at half maximum) in nanometers, is given by

$$\Delta\lambda_{\text{Stark}} = 2w\left(\frac{n_e}{10^{16}}\right) + 3.5A\left(\frac{n_e}{10^{16}}\right)^{1/4}[1 - BN_{\text{D}}^{-1/3}]w\left(\frac{n_e}{10^{16}}\right). \quad (3.18)$$

In the above equation, B is a coefficient equal to 1.2 or 0.75 for ionic or neutral lines, respectively, w is the electron impact parameter (or half-width), and A is the ion broadening parameter. The first term on the right side comes from the electron interaction, while the second one is generated by the ion interaction; n_e is the electron density (cm^{-3}) and N_{D} is the number of particles in the Debye sphere

$$N_{\text{D}} = 1.72 \times 10^9 \frac{T^{3/2}}{n_e^{1/2}}, \quad (3.19)$$

where T is the temperature $(K/1.16 \times 10^4)$. For typical LIBS conditions, the contribution from ion broadening is negligible, and equation (3.18) becomes

$$\Delta\lambda_{\text{Stark}} = 2w\left(\frac{n_e}{10^{16}}\right) \qquad (3.20)$$

from which, assuming that other sources of broadening (natural, Doppler, etc.) are negligible (i.e. $\Delta\lambda_{\text{line}} \cong \Delta\lambda_{\text{Stark}}$), the value of n_e can be derived. Values of w, the electron impact half-width (which is temperature dependent), can be found in the literature, for example, in the extensive tables given by Griem [23]. Determination of n_e by this method is independent of any assumptions regarding LTE conditions.

Saha–Boltzmann method   When the plasma is sufficiently close to LTE conditions, the electron density can be derived from the intensity ratio of two lines corresponding to different ionization stages of the same element. In fact, while the formulation of the Saha equation reported in equation (3.10) refers to the ratio of the total number densities of two ionization stages of the same element, a similar expression holds for the population ratio of two excited levels i and m of different ionization stages of the same element (singly ionized and neutral, respectively)

$$n_e \frac{n_i^{\text{II}}}{n_m^{\text{I}}} = \frac{2(2\pi m_e kT)^{3/2}}{h^3} \frac{g_i^{\text{II}}}{g_m^{\text{I}}} e^{-\frac{E_{\text{ion}}+E_i^{\text{II}}-E_m^{\text{I}}}{kT}}. \qquad (3.21)$$

For the sake of simplicity, the ionization potential lowering factor has been omitted here and in the following equations. However, it should be included for accurate measurements. By rearranging this equation, we obtain the explicit relation between n_e and the population ratio

$$n_e = \frac{2(2\pi m_e kT)^{3/2}}{h^3} \frac{n_m^{\text{I}} g_i^{\text{II}}}{n_i^{\text{II}} g_m^{\text{I}}} e^{-\frac{E_{\text{ion}}+E_i^{\text{II}}-E_m^{\text{I}}}{kT}}. \qquad (3.22)$$

From equation (3.12), for the population of level i and the intensity emitted at the transition ij, one obtains the final expression

$$n_e = \frac{2(2\pi m_e kT)^{3/2}}{h^3} \frac{I_{mn}^{\text{I}} A_{ij} g_i^{\text{II}}}{I_{ij}^{\text{II}} A_{mn} g_m^{\text{I}}} e^{-\frac{E_{\text{ion}}+E_i^{\text{II}}-E_m^{\text{I}}}{kT}}. \qquad (3.23)$$

Evaluation of thermodynamic equilibrium conditions in laser-induced plasmas

The problem of the existence of thermodynamic equilibrium in laser-induced plasmas has been investigated by many researchers from both the experimental and theoretical points of view. After building an appropriate model of plasma evolution, computer simulation has been used extensively to predict the physical characteristics of the plasma, especially the patterns of electron density and temperature. A number of experimental characterizations of laser-induced plasmas in terms of the T and n_e values have also been reported: typically, at a laser fluence of the order of a few joules per square centimeter, T ranges from a fraction of an electronvolt to a few electronvolts, and n_e ranges from 10^{16} to 10^{19} cm^{-3}.

From an experimental perspective, the occurrence of LTE in LIBS plasmas has been assessed by comparing either the actual population of atomic levels with the theoretical Boltzmann distribution or the temperature values derived from the excited level populations with those derived from the ionization stage populations and the electron kinetic energy.

Iida [24] calculated the temperature of a plasma induced by a Nd:YAG laser on an aluminum target at different delays and in different ambient gases. He reported the time-resolved Boltzmann plot for a set of iron neutral lines at a pressure of 100 Torr, showing the atomic level population to be compatible with the assumption of thermal equilibrium already during the first microseconds of plasma evolution. Castle *et al.* [1] arrived at the opposite conclusion in a very detailed paper, where the evolution over time of the population of five excited levels of neutral lead was described. This work reports a discrepancy between the ratios of level populations, which the authors conclude could only be explained by the occurrence of a selective population mechanism.

The use of the Boltzmann plot alone for the assessment of LTE conditions can lead to erroneous conclusions, as demonstrated by the work of Quintero *et al.* [25], who demonstrated that an apparently straight line can be obtained in the Boltzmann plane even if the level population is quite different from its equilibrium value. The apparently straight line turned out to be a consequence of the experimental constraints, which limited the number of lines measured and hence the level energy spread represented in the Boltzmann plot. Simeonsson and Miziolek [26] measured the time evolution of excitation and ionization temperatures and electron density in the plasma produced by an ArF laser in different gases. The excitation temperature was evaluated by the intensity ratio of two lines of the same element at the same ionization stage, while the ionization temperature was calculated, via the Saha equation, from two lines of the same element corresponding to different ionization stages. The electron density was calculated through the Stark broadening coefficient. Some differences were observed between the temperature values measured by the two methods. After discussion of the uncertainties associated with the calculated values in both cases, the authors conclude that the ionization temperature must be considered a better estimate of the real plasma conditions. Le Drogoff *et al.* [27] obtained the same values, within the experimental margin of error, for the excitation and ionization temperatures with delays of greater than 1 μs after the laser shot. The authors attributed the differences observed at smaller delay values to the methods used for the optical collection of the LIBS signal, which led to a spatial average of the spectroscopic quantities.

Rusak *et al.* [28] calculated the excitation, vibrational and rotational temperatures for a laser-induced plasma formed on graphite. They concluded that, while a complete thermodynamic equilibrium (with the same excitation, vibrational, rotational and kinetic temperature values) could not be established in the plasma during the first microseconds after the laser pulse, irradiances below 1 GW cm^{-2} produced plasmas in which at least the excitation and vibrational temperatures were similar. Hermann *et al.* [29] studied the temporal evolution of laser plasmas by comparing the electron temperature (measured from the spectral line profile) and the vibrational temperature. The two values were in agreement for delays greater than 1 μs after the laser shot, indicating thermalization between electrons and heavy particles. In another work, Hermann *et al.* [30] studied the characteristics of a laser-induced plasma during the first 200 ns after the laser pulse. They measured the electron temperature and density using the Boltzmann plot and Stark broadening, respectively, then refined the temperature value by comparing the measured spectral profiles with a theoretical simulation that accounted for self-absorption of the lines. The experimental profiles were in satisfactory agreement with the model of an LTE cylindrical plasma, although even closer agreement was achieved by modeling the plasma as an internal denser core with a less dense peripheral region, both in LTE and with the same temperature. In plasmas formed on a silicon target, Milan and Laserna [31] compared (with temporal and spatial resolution) the excitation temperature obtained via the Boltzmann plot with the ionization temperature calculated using the coupled Saha–Boltzmann equation. They found that, for their experimental conditions, the two values correspond at times greater than 2 μs. As for the spatial characterization, the ionization temperature was revealed to decrease with increasing distance from the target. The slight experimental differences between excitation and ionization temperatures, which are more pronounced for short delays, could also be attributed to spatial inhomogeneity, rather than deviation from LTE conditions. Temperature calculations are generally based on a space-averaging method; so the contribution of the cooler outer regions of the plume may decrease the averaged excitation temperature, while not affecting the ionization temperature obtained via the Saha equation, which represents the hotter central part [27].

Ng *et al.* [16] emphasized the three criteria to be fulfilled in order to guarantee LTE. The first, regarding electron density, has been already introduced in equation (3.9), and the authors claim it is commonly fulfilled in laser-induced plasmas. Similarly, Riley *et al.* [32] calculated the level population ratios resulting from simulations with different values of electron density and obtained an equilibrium distribution for $n_e > 10^{16}$ cm^{-3} in a plasma with a temperature in the range of 1–3 eV. The second criterion deals with the transient nature of the plasma and requires that the temporal variations in T and n_e be small during the time τ required to establish the excited state populations distribution. In practice, τ may range from sub-picoseconds to tenths of nanoseconds, i.e. much shorter than the tens of nanoseconds characteristic of variations in the plasma plume. This issue has also been investigated by Riley *et al.* [32] who agree on the validity of the steady-state assumption during the early stages of plasma lifetime. It should be noted that, in their study, the plasma was made to expand in a vacuum, thus leading to conditions different from LIBS measurements at atmospheric pressure. The third criterion requires T and n_e variations to be negligible over the distance d, through which the species diffuse during the time interval τ. The quantity d depends heavily on T: for $T > 1$ eV, the diffusion length is less than 1 μm, which is of course shorter than the spatial dimensions of the plasma plume, but for $T < 1$ eV, as the necessary excitation time is longer, the lighter elements may diffuse over tens of micrometers.

On the other hand, Capitelli *et al.* [33] have expressed doubts about the actual occurrence of LTE in LIBS plasmas. According to their computer simulations, the time scales involved in the processes of ionization, relaxation and plume expansion are of the same order of magnitude. Another important issue raised by these authors addresses the problem associated, even in equilibrium plasmas, with the calculation of the partition function, which stems from the difficulty of determining the correct value of ΔE_{ion}, the lowering factor of the ionization potential. Uncertainty in the value of ΔE_{ion} will affect the proper cut-off setting in the partition function sum. Simeonsson and Miziolek [26] have also calculated the values of other characteristic plasma time-scale parameters. In particular, the time needed to establish a kinetic equilibrium between electrons and heavier ions and neutrals was estimated at about 10 ns. On the other hand, the time needed for the excited states to reach a Boltzmann-like distribution was less than 50 ns for the species investigated (H, C, O and Cl). The authors conclude that fulfilment of the aforementioned theoretical criteria represents a strong indication of the effective existence of equilibrium. The same argument has been upheld by Amoruso *et al.* [34], who described the mechanism of plasma energy redistribution. During the laser pulse, the primary absorption mechanisms in the plasma involve electrons, which, through collisions, transfer their energy to ions and neutral particles. In laser-ablation plasmas, the equipartition time for energy transfer from electrons to ions (10^{-10}–10^{-11}s, according to Ready [35]) is generally much shorter than the laser pulse duration, τ_p. Thus, electrons can effectively transfer absorbed energy to ions during the laser pulse, and the electron and ion temperatures can be assumed to be nearly the same.

In conclusion, experimental and simulation results confirm that in typical LIBS measurements the LTE state is a good approximation for plasma modeling, at least at delays longer than 1–2 μs. However, it is also clear that the reliability of this hypothesis depends heavily on parameters such as the laser energy, pulse duration, ambient gas, acquisition gate, etc., that are used in the actual experiment. It is therefore advisable to report all of these parameters as accurately as possible, so that one can check on the validity of the LTE approximation and reproduce the same experimental conditions.

3.2.3 The calibration function

In order to effectively use line intensity values to calculate a sample's composition, one must make certain that the plasma is optically thin for the evaluated lines. This is especially true for the calculation of plasma temperature. This assumption means that the radiation is not reabsorbed along the optical path between the emission volume and the detector. Atoms in the lower energy levels can easily reabsorb the radiation emitted by other atoms of the same element in the plasma. A measure of the self-absorption effect can be obtained from the intensity ratio of two emission lines from a species, s, having the same upper energy level. In the absence of self-absorption, the intensity ratio should be the same as the ratio of the corresponding transition probabilities. Self-absorption is less troublesome in small plasmas and for trace elements, i.e. situations in which path lengths are short or only a small concentration of the analyte is present in the plasma [18]. In some cases one may also detect marked self-absorption of some lines emitted by the central hot plasma and reabsorbed in the colder region near the plasma boundaries, leading to a typical dip at the center of the emission profile (self-reversal) [36, 37].

Optically dense plasmas are considerably more complex to interpret than those within the optically thin limit. Riley *et al.* [32] report on a simulation of the effect of radiation trapping on the calculation of line intensity ratios. The results obtained under the assumption of optically thin plasma were completely unreliable when relatively strong lines were chosen for ratioing (i.e. transition probabilities of the order of $(1–3) \times 10^7$ s^{-1}). Reliable results were obtained when weak transitions were used (transition probabilities $<1 \times 10^7$ s^{-1}). The self-absorption effect can be modeled and compensated for, but in most cases the simplest approach is to just avoid using self-absorbed lines for the purposes of quantitative analysis. Figure 3.2 shows the importance of the self-absorption effect for the major component lines in a metallic alloy. When all the measured aluminum lines are used for calculating the temperature, the result is a rather scattered Boltzmann plot leading to unreliable plasma temperature values. On the other hand, if the strongly self-absorbed resonance lines ($\lambda = 308.2, 309.3, 394.4, 396.1$ nm) are removed from the data set used for the calculation, the accuracy and precision of the measurement are substantially improved.

Understandably, working with emission lines characterized by small transition probabilities in order to overcome the effect of high number density and reduce the optical thickness of the plasma is a viable approach to minimize self-absorption [38]. The lines chosen, however, should still provide an adequate signal/noise ratio so that the precision of the measurement is not degraded.

The analytical calibration function

From the preceding discussion, it is clear that a calibration function exists between the spectral line intensities and the number densities of the emitting species present in the plasma. The most general approach to derive this function is the so-called curve of growth (COG) method. This method was applied to a plasma induced by a Nd:YAG laser on the surface of steel samples (containing 0.007%–1.3% Cr) by Gornushkin *et al.* [39] According to classical theory [20, 40, 41], the integrated spectral line irradiance (in units of W m^{-2}), often indicated as intensity, is given by

$$I = \alpha \frac{8\pi hc}{\lambda^3} \frac{n_1}{n_0} \frac{g_1}{g_0} \int\limits_{\nu_0-\infty}^{\nu_0+\infty} (1 - e^{-k(\nu)l})d\nu \tag{3.24}$$

where α is a constant factor which accounts for the response of the measuring instrument, h is the Planck constant (J s), c is the speed of light (m s^{-1}), λ is the transition wavelength (m), n_1, n_0, g_1, and g_0 are the number densities (m^{-3}) and the degeneracies (dimensionless) of the upper and the ground atomic levels (for a resonance transition), respectively, $k(\nu)$ is the frequency dependent absorption coefficient (m^{-1}), and l is the absorption length (m).

The frequency dependence of the absorption coefficient $k(\nu)$ is given by

$$k(\nu) = k_0 \frac{a}{\pi} \int\limits_{-\infty}^{\infty} \frac{e^{-t^2}dt}{(t-x)^2 + a^2} \tag{3.25}$$

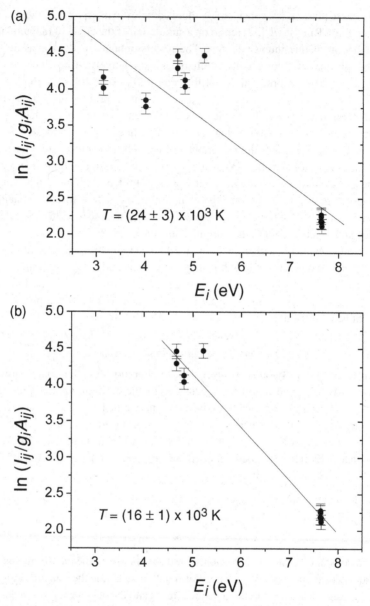

Figure 3.2. (a) Boltzmann plot of the aluminum lines emitted by an aluminum alloy (Al content about 98%). The line intensity values have been determined as the integral area of the best fitting analytical function. The values have been corrected for the wavelength response of the system. The intensity of the two most intense doublets ($\lambda = 308.2$, 309.3, 394.4, 396.1 nm) is clearly lower than expected, indicating the occurrence of self-absorption effects. (b) Omitting the self-absorbed lines from the temperature calculation provides more reliable results.

where k_0, a, and x are defined as follows:

$$k_0 = 2\pi^{3/2}\frac{e^2}{m_e c}\frac{n_0 f}{b} \tag{3.26}$$

$$b = \frac{\pi \Delta \nu_D}{\sqrt{\ln 2}} \tag{3.27}$$

$$a = \frac{(\Delta \nu_N + \Delta \nu_L)}{\Delta \nu_D}\sqrt{\ln 2} \approx \frac{\Delta \nu_L}{\Delta \nu_D}\sqrt{\ln 2} \tag{3.28}$$

$$x = \frac{2(\nu - \nu_0)}{\Delta \nu_D}\sqrt{\ln 2}. \tag{3.29}$$

Here e is the elementary charge (C), m_e is the electron mass (kg), f is the transition oscillator strength (dimensionless), ν_0 is the frequency of the center of the line (Hz), and $\Delta \nu_N$, $\Delta \nu_L$, and $\Delta \nu_D$ are the natural, Lorentzian, and Doppler line half-widths (Hz), respectively.

The spectral line intensity introduced in equation (3.24) can also be rewritten in terms of *total absorption*, A_t, which is defined as

$$A_t = 2\pi \frac{\Delta \nu_D}{\sqrt{\ln 2}}\int_0^\infty \left[1 - \exp\left(-\int_0^\ell k(x)dx\right)\right]d\nu \tag{3.30}$$

where $\int_0^\ell k(x)dx$ can be replaced by kl for a homogeneous plasma.

The double-logarithmic plot of $A_t/2b$ vs. $n_0 fl/b$ represents the theoretical curve of growth. This curve has two asymptotes with slopes 1 and 1/2. Obviously, the asymptotes for the emission COG have the same form as for the absorption COG but both include a constant proportionality factor. The experimental COG is usually constructed by measuring the integrated line intensity in arbitrary units as a function of analyte concentration. A linear relationship between the atom number density in the plasma and the analyte concentration in the original sample is presumed so that the calibration function can be directly related to the theoretical COG.

The curve of growth contains valuable information on several fundamental plasma parameters.

(1) The a-parameter can be measured by comparing the experimental curve with a set of theoretical COGs. This yields information about collisional vs. Doppler line broadening.
(2) If the a-parameter is known and values for $\Delta \nu_L$ or $\Delta \nu_D$ are found, the plasma temperature or collisional cross section can be obtained.
(3) The $n_0 fl$ value can be determined by matching the experimental and theoretical COGs. If the oscillator strength and the absorption path length are known, the number density of ground state atoms n_0 can be found. Conversely, if n_0 and l are known, this is a good means to determine the oscillator strength of the transition.

(4) The intersection of the initial and final asymptotes marks the point where transition from low to high plasma optical density occurs. The abscissa of this point yields the upper limit of the linear calibration range, providing a useful criterion for the choice of appropriate spectral lines for quantitative analysis.

To determine a range within which the theoretical COG would give the best match with the experimental one, Gornushkin *et al.* [39] made a preliminary estimate of the *a*-value. The measurement of plasma temperature by a Boltzmann plot method provided an estimate for Doppler and Lorentzian line half-widths. For the case of collisional broadening, it was assumed that the main collision partner for Cr atoms was nitrogen (N_2) with a collisional cross section estimated as 50–100 Å^2. The value for the *a*-parameter was then estimated to fall between 0.12 and 0.24. The prediction accuracy was then confirmed by the direct measurement of the line shape. In the experiment, a frequency doubled narrow band (80 MHz) titanium sapphire laser was scanned across the Cr 425.4 nm transition. The resulting line shapes are shown in Figure 3.3. The top spectrum represents a purely Doppler-broadened line profile from a low temperature line source (Cr hollow cathode lamp), measured using the opto-galvanic technique. The other line profiles in the laser plasma were obtained from samples with different concentrations of Cr. Only one spectrum in this series (0.025% Cr in steel, second row) was obtained under optically thin plasma conditions; all other spectra were broadened by self-absorption. Doppler broadening was found to be dominant with respect to Lorentzian broadening. The relative contributions of the Doppler and Lorentz half-widths to the total line half-width ($\Delta\nu_D/\Delta\nu_T$ and $\Delta\nu_L/\Delta\nu_T$) were also estimated from well-known values for the Voigt function and were 0.86 and 0.26 for the Doppler and Lorentzian fractions, respectively. This provided a value of the *a*-parameter of 0.2, in close agreement with the first estimate (0.12–0.24).

The theoretical COGs were calculated and matched to the experimental points. The fit of the theoretical COG with $a = 0.2$ to the experimental points is shown in Figure 3.4, obtained for ablation of Cr-doped stainless steel at the 5 μs delay time and 20 mJ of the laser energy. The fit accuracy was within the values for the *a*-parameter of 0.20 ± 0.05. As follows from equation (3.30), the two asymptotes intercept at the ordinate value of $2a$, corresponding to an $n_0 fl/b$ value of 5. The absolute number density of neutral Cr atoms n_0 was then calculated by using a measurement of the absorption path length obtained by optical imaging of the plasma. It was found that the number density of neutral Cr atoms corresponding to the transition from low to high optical density was $6.5 \cong 10^{12}$ cm^{-3}. This corresponded to a concentration of Cr in the stainless steel matrix between 0.05 wt.% and 0.1 wt.%. With the known *a*-value, a collisional cross-section, σ, of (66 ± 16) Å^2 was also obtained, close to the one observed in flame spectroscopy.

Overall, the COG method provides important practical information concerning the linear dynamic range of the calibration and assists in the choice of appropriate lines for a particular range of concentrations. It is, therefore, a valuable tool for obtaining both fundamental and analytical spectroscopic information. Other, albeit less general, approaches have been discussed in the literature that aim to account for the effect of self-absorption on the calibration curve.

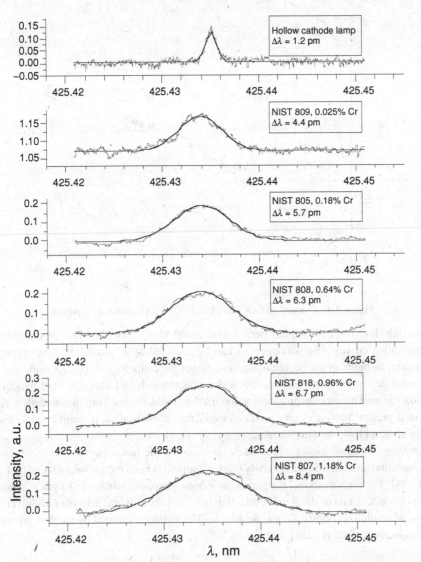

Figure 3.3. The Cr 425.435 nm line width in the 800-series NIST stainless steel standards. Ablation energy is 20 mJ, delay time is 5 μs.

Without resorting to the general expression for line intensity (equation (3.24)), Aragon *et al.* [42] proposed the following simplified empirical function:

$$I \cong a + bc_0(1 - e^{-x/c_0}),\qquad(3.31)$$

where x represents the concentration ratio and I is the intensity ratio of the analyte line to an internal standard line. This expression, which at low concentrations is approximated by

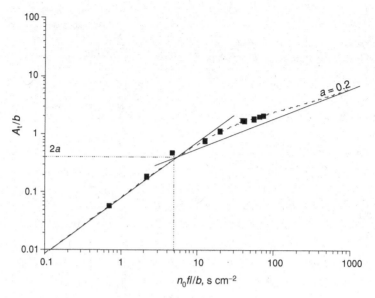

Figure 3.4. Theoretical COG matched to the experimental data points.

the straight line $I = a + bx$, has been found useful to describe the onset of nonlinearity in calibration curves. The parameter c_0 can be regarded as a "saturation concentration," furnishing an index of the degree of nonlinearity of the calibration curve. Its value is given by the concentration at which the curve's slope decreases by a factor $1/e$ from its value at $x = 0$. The maximum concentration value used to build the calibration curve should not be much greater than c_0, otherwise the slope of the calibration curve would be too low to insure an acceptable measurement accuracy.

Another simplified model which takes into account the nonlinear relationship between concentration and intensity in optically thick plasmas has been recently developed by Lazic *et al.* [43]. The authors model the plasma as a homogeneous cylinder of length L, centered along the optical axis of the signal-collection lens, and divided into a number of thin layers, δL, within which self-absorption is negligible. The radiation emitted by the first layer next to the sample surface is then given by

$$\left(I_{ki}^{s}\right)^{(1)} = F \frac{C^{s}}{L/\delta L} \frac{f_{ki}^{s}(T)}{U^{s}(T)}, \tag{3.32}$$

where $f_{ki}^{s}(T)$ is a function incorporating all the spectroscopic parameters for the specific ki transition. The second layer emits the same amount of radiation as the first, while at the same time attenuating the radiation coming from the underlying layer, with a mean absorption coefficient β_{ik}. The radiation escaping from the second layer and directed towards the collecting optics is given by

$$\left(I_{ki}^{s}\right)^{(2)} = F \frac{C^{s}}{L/\delta L} \frac{f_{ki}^{s}(T)}{U^{s}(T)} + \left(I_{ki}^{s}\right)^{(1)} e^{-\beta_{ik} c_{i}^{s} \delta L}, \tag{3.33}$$

where c_i^s indicates the species concentration at the lower transition level i. After summing the contributions from all the layers, the overall line intensity emerging from the plasma is given by

$$I_{ki}^s = \sum_n \left(I_{ki}^s\right)^{(n)} = F \frac{C^s}{L/\delta L} \frac{f_{ki}^s(T)}{U^s(T)} \times \sum_n e^{-\beta_{ik} c_i^s \delta L(n-1)}$$

$$= FC^s \frac{f_{ki}^s(T)}{U^s(T)} \frac{1 - e^{-\beta_{ik} c_i^s L}}{L/\delta L \left(1 - e^{-\beta_{ik} c_i^s \delta L}\right)}. \tag{3.34}$$

This result has been obtained considering a homogeneous LTE plasma. Because in actual laboratory practice the species concentration and plasma temperature are not uniform over the whole plasma volume sampled by the detection system, some further assumptions are needed. In order to account for the different behavior of the less dense, outer plasma, the authors assume that self-absorption can be neglected in this region, i.e. the outer plasma is optically thin at the wavelength considered. Therefore, the contribution of the outer plasma to line intensity is modeled as an additional linear term

$$I_{ki}^s = F_1 C^s \frac{f_{ki}^s(T)}{U^s(T)} \frac{1 - e^{-\beta_{ik} c_i^s L}}{L/\delta L \left(1 - e^{-\beta_{ik} c_i^s \delta L}\right)} + F_2 C^s \frac{f_{ki}^s(T)}{U^s(T)}. \tag{3.35}$$

The constants F_1 and F_2 correspond to optically thick and thin plasmas, respectively, and depend heavily on the experimental geometry. This model has been used by the authors to correct for self-absorption in measured intensity values, extending the range of linearity of their calibration curves. The model, in fact, enables one to relate the measured intensity value to the intensity effectively emitted, and therefore the "measured" concentration C_m to the effective species concentration, C_E

$$C_m = \frac{a_1}{a_2} \left(1 - e^{-a_2 C_E}\right) + a_3 C_E. \tag{3.36}$$

Depending on the specific line, either the first (thick plasma) or second (thin plasma) term will predominate. The authors have successfully applied their method to several species present in the plasma at concentrations high enough to result in a nonlinear calibration curve.

A realistic approach to modeling LIB plasmas must always take into account the fact that such plasmas *may be optically thick and inhomogeneous*. These problems have already been considered in detail in the literature [44–51]. Tondello *et al.* [47] observed spatially resolved, time integrated spectra from Be(IV) in the soft X-ray region. The plasma was initiated on a beryllium target under vacuum by a 10 J, 10 ns ruby laser. Line profiles with asymmetrical shift reversal were calculated under the assumption of local thermodynamic equilibrium (LTE) and uniform temperature distribution across the plasma. General agreement was reached between the computed and experimental profiles. Malvezzi *et al.* [48] proposed a model which took into account Stark and Doppler broadening, Doppler shift and optical opacity. The plasma was induced on a polyethylene target in vacuum and analyzed in the XUV spectral range (2–30 nm). The observed profiles of the Lyman C(VI) lines were reproduced numerically and their build-up within the plasma was explained. Tallents [49] developed a method of determining laser-produced plasma conditions from the shape of a spectral line for an optically thick plasma. Model parameters, such as the plasma emissivity, absorption coefficient and ion velocity, were adjusted until a close fit between the computed and experimental profiles was obtained. The experimental data from Malvezzi *et al.* [48] were used to verify the model. Hermann *et al.* [30] performed time- and space-resolved diagnostics of the early stage plasma ($\leq$200 ns) induced by an excimer laser on a Ti target in low pressure nitrogen. It was shown that self-absorption has an influence on spectral line profiles even for transitions between highly excited levels. A model of a non-uniform plasma divided into two uniform zones of different densities and temperatures was applied to describe self-absorbed and self-reversed line profiles. The authors also concluded that measurements of electron temperature by the Boltzmann plot method may predict temperature values that are too high.

In a recent paper by Gornushkin *et al.* [36], a semi-empirical model of an optically thick inhomogeneous plasma was developed based on a standard thermodynamic description. The model described plasma emission in the vicinity of a strong self-reversed non-resonance line. The Si(I) 288.16 nm line was chosen; this has a lower level only 6269 cm^{-1} (0.78 eV) above the ground state. This line is strongly self-reversed in the laser-induced breakdown plasma in air, indicating that the plasma is heterogeneous and optically thick with a strong temperature gradient. For many years, these plasmas have been of great interest in astrophysics, where information about celestial objects could only be extracted from spectroscopic information. The appropriate spectroscopic methods and models were thus developed to obtain important plasma parameters like temperature, temperature distribution, and electron density. These same spectroscopic methods and models can also adequately describe the dense LIB plasma.

The model establishes the relation between the observed spectral radiance I_ν (Js^{-1} m^{-2} sr^{-1} Hz^{-1}) and the distribution of volume emission coefficient, $\varepsilon_\nu(\nu, x, T)$ (Js^{-1} m^{-3} sr^{-1} Hz^{-1}), and absorption coefficient, $\kappa(\nu, x, T)$ (m^{-1}), within the optically thick inhomogeneous reabsorbing plasma. The general equation of radiation transfer, given by equation (3.24), can be rewritten as

$$I_\nu = \int_{-x_0}^{x_0} \varepsilon_\nu(\nu, x, T) e^{-\tau(\nu, x, T)} dx, \qquad (3.37)$$

where $\tau(\nu, x, T)$ is the optical thickness of the plasma, $\tau(\nu, x, T) = \int_{-x_0}^{x_0} \kappa(\nu, x, T) dx$ and x_0 and $-x_0$ denote the edge-to-edge plasma size along the line of sight. The values for the absorption, $\kappa_\nu(\nu, x, T)$, and the volume emission, $\varepsilon_\nu(\nu, x, T)$, coefficients in a heterogeneous plasma will vary depending on position along this line of sight. Most models of the LIB plasma simplify equation (3.37) either by assuming homogeneous distribution of plasma species or by assuming optically thin conditions. This model does neither. Bartels [45] and, later, Zwicker [46] have shown that in the absence of stimulated transitions, equation (3.37) can be written as the product of three terms

$$I_\nu = A(T_m) \cdot M(E_l, E_u) \cdot Y[\tau(\nu, x, T), p], \qquad (3.38)$$

where the model parameters are explicitly given in the parentheses. Here, T_m is the maximum temperature in the plasma center, E_l and E_u are the lower and upper atomic transition levels, respectively, and p is a formal parameter which will be defined below. The first term, $A(T_m)$, can be considered as a source function depending only on T_m:

$$A = \frac{2h\nu^3}{c^2} \exp\left(-\frac{h\nu}{kT_m}\right). \tag{3.39}$$

This term is the exact expression for Wien's law. Omenetto *et al.* [52] have clearly shown that a source function in the form of Wien's law can be used for plasmas in which stimulated transitions are insignificant and the condition $h\nu/kT_m \gg 1$ is satisfied, whereas Planck's law is used when stimulated transitions are considered.

The second term in equation (3.38) accounts partially for the effect of plasma heterogeneity on emission. $M(E_l, E_u)$ is the factor by which the emission along the line of sight of the *inhomogeneous* plasma column is smaller than the emission of a *homogeneous* column with the same optical thickness. For naturally broadened and van der Waals broadened lines,

$$M = \sqrt{\frac{E_l}{E_u}} \quad \text{if} \quad \frac{kT_m}{E_l} \ll 1. \tag{3.40}$$

For spectral lines of singly ionized atoms and for Stark broadened neutral lines,

$$M = \sqrt{\frac{E_l + 0.5\chi_0}{E_u + 0.5\chi_0}} \quad \text{if} \quad \frac{kT_m}{E_l + 0.5\chi_0} \ll 1. \tag{3.41}$$

Here E_l and E_u are the excitation energies of the lower and upper levels and χ_0 is the ionization energy. The conditional statements in the above expressions are not fulfilled for resonance lines of neutral atoms; however, they are reasonable assumptions for ionic emission and for non-resonance emission lines of neutral atoms.

The third term in equation (3.38) accounts for the effect of optical thickness as well as heterogeneous mixing. The function $Y[\tau(\nu, x, T), p]$ can be well approximated by

$$Y[\tau(\nu, x, T), p] = e^{-\frac{\tau(\nu, x, T)}{2}} \left[\frac{\tau(\nu, x, T)}{2}(1 - p) + p \sinh\left(\frac{\tau(\nu, x, T)}{2}\right) \right.$$
$$\left. + \frac{1}{\sqrt{p}} \sinh\left(\frac{\tau(\nu, x, T)}{2}\sqrt{p}\right) \right], \tag{3.42}$$

where the parameter p is calculated from

$$p = \frac{6}{\pi}\arctan\frac{M^2}{\sqrt{1 + M^2}}. \tag{3.43}$$

Substitution of the values calculated via equations (3.42) and (3.43) into equation (3.38) allows an estimation of the spectral emission intensity, I_ν, for an optically thick, heterogeneously mixed plasma. The superposition of line and continuous radiation can also be incorporated into the model by assuming that the optical thickness consists of two separable components: $\tau = \tau_L + \tau_C$.

The following simplifying assumptions were applied in the model [36]: the plasma was assumed to be spherically symmetrical; the number of elements in the plasma did not exceed 2; the plasma composition did not change with time; ionization stages higher than 1 were neglected; and the distribution of number densities was not affected by trapping of radiation and was determined entirely by collisions. The principal inputs for the model were the total number of atoms in the plasma plume, the temperature function, the plasma expansion function, and the atomic stoichiometry of plasma species. All these parameters can be estimated from experiment. The number of atoms can be found from direct or indirect pressure measurements (for example, from measuring the speed of a shock wave front). The plasma expansion function can be evaluated from plasma shadowgraphy, whereas the atomic stoichiometry can be estimated from intensity ratio measurements of optically thin lines.

For model calculations, ablation of silicon in nitrogen was considered. Plasma emission was modeled in a spectral window of a few nanometers around the strong 288.16 nm Si(I) line. The time window was set in arbitrary units, from 1 to 10, which covered a range of approximately 2 μs starting from about 20–40 ns delay with respect to plasma ignition. During this time, the plasma core was assumed to cool down from 40 000 K to 8000 K. Figure 3.5 shows the calculated emission profiles for an initial number of atoms of 10^{16} and different Si/N atom ratios taken at several discrete delay times (from 1 to 4 a.u.). The profiles are very instructive in terms of understanding processes taking place inside the inhomogeneous plasma and the dependence of these processes upon the plasma constituents. One can see from Figure 3.5 that if the total number of atoms, which is directly related to number densities of all species in the plasma, is not too high (10^{16}), the plasma continuum (curves at t_1, t_2) monotonically decreases as the plasma progresses in time. However, at higher number densities, a significant amount of early plasma continuum is absorbed within the plasma, which results in lower emission intensity at time t_1 than at times t_2 and t_3. This causes the plasma to approach a black body emitter at early times. It is also evident from Figure 3.5 how the Si/N ratio affects the plasma emission. Because Si has a much lower ionization potential than N (8.15 eV vs. 14.53 eV), a relative increase in silicon number density leads to a significant increase in electron number density which, in turn, results in a stronger continuum and a larger Stark width and shift of the atomic line. On the other hand, an absolute increase in total Si and N number densities results not only in a depression of the plasma continuum, as discussed above, but also in an earlier and deeper self-reversal on both the continuum and line spectra.

The model thus predicts the spatial and temporal distributions of atom, ion, and electron number densities, evolution of an atomic line profile and optical thickness, and the resulting absolute intensity of plasma emission in the vicinity of a strong non-resonance atomic transition. For spectrochemical analysis, the model can be of interest as it predicts the behavior of a spectral line in terms of its usefulness for analysis; it shows if the line is self-absorbed or self-reversed and to what extent. For example, even without further refinements, the model can be used to validate and extend to real plasmas the "standardless" analytical procedure described by Ciucci *et al.* [53].

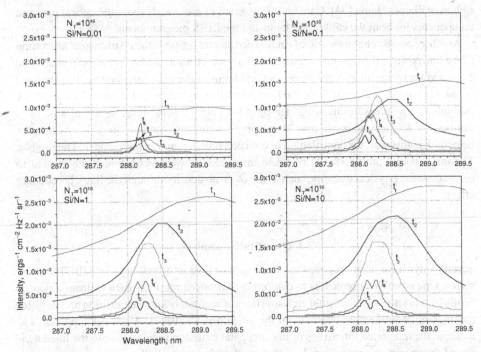

Figure 3.5. Calculated emission profiles for early LIB plasma consisting of 10^{+16} Si and N atoms in different proportions: from Si/N = 0.01 to Si/N = 10. Times t_1–t_5 correspond to the values of 1; 1.5; 2, 3, and 4 in relative units.

3.2.4 The detection efficiency function

Each experimental apparatus exhibits a unique spectral response. The collecting optics, lenses and fiber optics, the spectrometer grating, the detector sensor and intensifier, all have a characteristic wavelength dependent efficiency. Thus, the data from measurements are a convolution of the actual physical data and the system response. It is generally unnecessary to normalize intensity data to account for system response because one routinely compares the intensity of a single spectral line in the standard material with the same line emitted by an unknown. This is, by far, the most common approach to quantitative spectrochemical measurements. However, in many situations it is necessary to account for the spectral response of the measurement system. Several of the plasma diagnostic methods discussed above require accurate relative line intensities, as do methods which make use of large numbers of spectral lines, such as the calibration-free approach proposed by Ciucci *et al.* [53]. In such cases, the wavelength sensitivity of the apparatus can be derived either by direct calibration with a radiation standard, or by determining the optical properties of all the components of the detection system. The former method calls for using a well-characterized continuum radiation source close to a black body radiator, such as, for example, a tungsten ribbon or filament lamp [54]. Care must be exerted to use exactly the same optical path and components for both the calibration lamp and the LIBS measurements.

Another specific characteristic of each experimental set-up is the instrumental aberration of the line profile. The possible occurrence of asymmetries in the line profile can usually be corrected for by optimizing the alignment and focus of the collecting system. Instrumental broadening, on the other hand, should be measured using a narrow-line spectral source (for example, a low-pressure discharge lamp). Alternatively, some long-lasting plasma lines can be used to measure instrument broadening by choosing suitably long delay times for detection; in this way, the contribution of collision broadening can be considered negligible. The instrumental profile should be taken into account whenever a line width is used to calculate some physical quantity, such as the electron density through the Stark broadening effect.

3.3 Quantitative analysis

Several ways have been proposed to obtain quantitative information from LIBS measurements. One possible approach is to determine the concentration of each element independently by working with elemental emission lines of accurately known transition probabilities and making absolute measurements of the integrated intensities. A second approach is to measure the integrated intensities of the elemental emission lines relative to the line intensity of the specimen's most abundant element. In this case, only the ratios of the transition probabilities of the emission lines need to be known. Both of these methods rely upon a calibrated spectroscopic measurement system. Moreover, the quantitative data obtained reflect the composition of the laser-induced plasma, and not necessarily of the actual sample. The most common approach is to measure the LIBS intensities in relation to known calibration standards [38]. This approach is generally the most practical for extracting quantitative information on sample composition, despite the fact that the laser material interaction is highly matrix dependent and therefore variations in the matrix between the unknown sample and the standard must be minimal.

Most current LIBS applications aim to determine various elements in an approximately constant matrix. At present, no generally accepted approach exists for LIBS calibration when the aim is to determine the main components as well as trace elements in different matrices. The variation in composition of such samples (e.g. soil, sludge, or steel, cement, and glass), which leads to variability in their surface absorption, reflection, and thermal conductivity, affects the interaction function in an unpredictable, non-reproducible manner. Thus, the analytical determination of the elements may be affected, as well as the characteristics of the plasma [18]. The sensitivity for each element is in fact influenced by the plasma parameters, which are in turn strongly influenced by the sample matrix.

The following sections are devoted to a description of the most common approaches to quantitative LIBS analysis, beginning with the methods based on internal standardization with a major matrix constituent.

3.3.1 Methods based on internal standards

Such methods rely upon knowledge of the concentration of an internal standard, either already present (for example, nitrogen in atmospheric air) or purposely introduced into the sample. Quantitative determination of the concentration of the element in question can be made by measuring the intensity of one of the lines emitted, which is proportional to the population of the corresponding energy level, via the transition probability between the upper and lower levels of the transition (see equation (3.12)). In actual measurements, the efficiency of the collecting system acts as a scaling factor affecting the measured line intensity; so we can write

$$\overline{I_{ij}} = F C^s A_{ij} \frac{g_i e^{-\frac{E_i}{kT}}}{U^s(T)}, \tag{3.44}$$

where $\overline{I_{ij}}$ represents the measured integral line intensity, C^s is the concentration of the emitting species in the sample, and F is an experimental parameter that takes into account the optical efficiency of the collection system as well as the plasma density and volume. The plasma temperature can be calculated, for instance, by using the line-intensity ratio method (see equation (3.13)). If the concentration of the internal standard is known, measuring the intensity of at least one of its emission lines allows the F parameter to be calculated from equation (3.44). The unknown concentration of another element in the sample is then obtained by substituting this parameter into the same equation corresponding to the measured emission line [55]. Although this method appears to be quick and simple, one cannot necessarily identify the concentration of an element as present in the analytical sample with its concentration as a species in the LIB plasma. Equation (3.44) applies to all the different species present in the plasma, while the internal standard method mistakenly suggests using it to calculate the elemental concentration in the original sample. The known concentration of the internal standard is limited to its elemental concentration in the sample while the concentrations of the various species resulting from the excitation/ionization characteristics of the plasma cannot be known without recourse to the Saha equation. In 1989, Kim [38] stressed the importance of the degree of ionization of all the elemental species present in the plasma. When the degree of ionization of each species is not negligibly small (i.e. the number density of singly ionized atoms of a given elemental species is not negligible compared with that of the neutrals), the number density n^s appearing in the expression for line intensity (equation (3.12)) underestimates the total number density of the given element. On the other hand, a sufficiently high degree of ionization is necessary in order to fulfil the requirements of thermodynamic equilibrium. Moreover, within a LIBS plasma, the distribution of the atomic population among the neutral and ionized states differs widely from element to element, depending on the atomic ionization potential. This introduces further errors into the elemental composition analysis. In conclusion, the internal standard method is limited in its applicability by the availability in the sample of an internal standard of known concentration (which is by itself a strong constraint), and one should be cautious in using it because of the misleading results that can, in principle, be obtained.

3.3.2 Calibration curves

As in most analytical methods, quantitation in LIBS relies upon the use of calibration curves. Starting with equation (3.44), it can be deduced that the concentration values for each element can be obtained by comparing a selected line intensity from an unknown sample with the corresponding one from a certified sample, as all factors in the equation are common except the concentration and the line intensity. Provided that several reference samples are available, whose elemental compositions cover the range of interest for the elements to be measured, this method allows the establishment of a calibration curve (linear in the most favorable cases) which relates the specific line intensity and its corresponding element concentration. By comparing the line intensity measured for an unknown sample, the corresponding unknown elemental concentration can be directly inferred from this curve. Unfortunately, the same issues already raised for the internal standard method also apply to the calibration curve method, since equation (3.44) strictly applies to the plasma species concentration. One must then assume that there is a consistent relationship between elemental concentration in the sample and species concentration in the plasma among all the standard and unknown materials. This approximation can be considered satisfactory only if all parameters affecting the plasma characteristics (including the sample morphology and composition) are constant during calibration and measurement. In this case, we can assume that the distribution of the elements among the different excitation and ionization states remains the same. When the matrix of the certified samples differs appreciably from the unknown sample, this assumption can fail. The problems related to this matrix effect and the analytical methods formulated to overcome them are described below.

Both the terms "intensity" and "concentration" can be defined in different ways, which often reflect the actual experimental conditions. Several authors define the line intensity as the area of the analytic function used to fit the line profile (usually a Lorentzian or Voigt function). In this case, the background is automatically subtracted. Other researchers take the line intensity to be the peak height i.e. the maximum of the line profile minus the baseline, which is evaluated in a spectral region free from significant line emission. Concentrations are usually expressed as the mass of the analyte per unit mass of the sample (w/w). Another possibility is to indicate the concentration as the number of analyte atoms per 100 atoms in the sample (or per 10^6 atoms in the sample). A different notation is used for gaseous samples, for which the ratio of analyte volume to total volume is commonly reported (v/v). It is important to keep in mind that, in the ideal case, the analyte spectral line intensity is strictly proportional to the number of atoms of that particular element removed from the sample. The calibration function is therefore correctly formulated in terms of the moles of analyte present in the sample, rather than the mass [56].

In order to increase the reliability of the calibration method, the working curve is usually the result of averaging repeated measurements, possibly including multiple laser probings for each measurement. The uncertainty associated with any individual point is then determined by assuming a Gaussian distribution of the experimental results and a confidence level, for instance, of 95%, corresponding to 2σ [57]. The set of experimental points obtained from the reference samples is then fitted with a linear or polynomial function, depending on the data. In principle, the calibration function obtained should pass through the origin, i.e. zero concentration of the analyte should correspond to no LIBS signal from the sample. As discussed above, a linear relationship is expected at low analyte concentrations, limited by the onset of self-absorption at some well-defined upper limit. The quality of the calibration function can differ considerably for the various emission lines of a given element. One can try to predict the quality of the calibration for a given line from known spectroscopic data, such as the excitation potential, line strength, and population of the lower state. Nevertheless, experimental results indicate that such considerations should be used merely as general guidelines; detailed measurements are always necessary [58]. In any event, the choice of the analysis line should be dictated by the concentration range involved and the potential occurrence of spectral interferences from matrix components. Resonance lines, which are typically among the strongest spectral lines of certain elements, should be chosen in the case of low concentration. Weaker lines are a better choice for high relative concentrations, in order to avoid self-absorption effects [38, 59].

Correction methods for shot-to-shot fluctuations

In general, excellent linear calibration curves spanning several orders of magnitude in concentration can be obtained and are typical of LIBS analysis. However, along with these favorable cases, many researchers have drawn attention to problems of poor reproducibility, wide data scatter, nonlinearity of the calibration curves and resulting poor accuracy. Excluding problems associated with inadequate standards, shot-to-shot variations in plasma conditions, which are in turn caused by the stochastic nature of the laser–material coupling, seem to be the major cause of these effects. A number of correction methods have therefore been investigated.

The simplest way to reduce the effects of shot-to-shot variations is to normalize the analyte line intensity to the intensity of a matrix reference line. In practice, the intensity ratio is usually plotted against the concentration ratio of the analyte to the matrix reference element. Pakhomov *et al.* [60] defined the calibration function to be fitted to the experimental data as

$$R_a = SC^n \qquad (3.45)$$

where R_a is the ratio of the integrated intensity of the analyte line to the reference line, C is the analyte concentration and S is the sensitivity. The exponent n, in the absence of self-absorption, is unity. In this work, the authors plotted the calibration curve for Pb in concrete using the 405.78 nm line and the O(II) line at 407.59 nm as the reference. They also evaluated the analyte line intensity as a function of the laser pulse energy from the plasma threshold to the maximum laser output. Both the analyte line and the reference intensities increased. The resulting calibration curve for the ratio R_a was found to be independent of laser energy.

An alternative method has been adopted by Fichet *et al.* [57], who calculated the line intensity value as the maximum of the line profile divided by the background. This particular method was chosen because of the high spectral resolution of their measurements, which necessitated the use of a narrow spectral window. This made it impossible to find a proper matrix line for normalization of the analyte line intensity in the same spectral window. The authors found, however, that normalization by the background improved the reproducibility of their results.

Single shot analysis

A different approach, based on analysis of single shot spectra, has been proposed by Xu *et al.* [61] in order to overcome the problems related to signal instabilities. The authors developed this method to overcome the particular difficulties encountered when working with aerosols, in which the intensity fluctuations of the LIBS signal are especially pronounced. This is because optical breakdown may occur at different positions along the laser beam path, leading to unstable plasma location. In contrast to common internal calibration procedures based on the ratio of spectral lines, which are applicable only when the concentration of one component is constant, this approach requires no constant constituent, and should provide the absolute concentrations of the elements without the need to normalize to a known reference element concentration. The basic idea is that shot-to-shot signal variations can be described as a multiplicative effect, for both the spectral peaks and a component of the baseline. As a consequence, in order to perform quantitative analysis, a single spectrum corresponding to an individual breakdown event must be processed. Furthermore, those spectra whose signals are particularly weak (corresponding to poor laser shots), which would only contribute to increasing the noise, are eliminated from the data set. The authors assume that the baseline value, obtained with shot i, can be expressed as

$$B_i = b_0 + b_i = b_0 + k_1 f_i, \qquad (3.46)$$

where b_0 is a (sample-dependent) constant, and b_i is the contribution characterizing shot i, which is related to the fluctuation f_i through the proportionality factor k_1. The peak height measured for shot i, on the other hand, is modeled as

$$P_i = B_i + Ck_2 f_i, \qquad (3.47)$$

where C is the element concentration, and k_2 is a proportionality factor that correlates the fluctuation with the peak intensity. After proper substitution, the authors obtain

$$P_i = B_i(1 + kC) - kCb_0, \qquad (3.48)$$

where $k = k_2/k_1$. Thus, a plot of the peak intensity versus the baseline intensity should yield a straight line with a slope $\alpha = 1 + kC$. A plot of $\alpha - 1$ versus the concentration C would therefore result in a linear calibration curve passing through the origin and free of fluctuation effects. Such a plot can be used in place of a traditional calibration curve. The routine previously described requires performing, apart from the measurements needed to construct the calibration curve, a sequence of single shot determinations for the analysis of the unknown sample. The authors propose an approximate formula to be used when one single spectrum of the unknown sample is available, although they note that it provides less accurate results. A theoretical rationale for the procedure was also attempted, identifying the electron density as the main parameter affecting the fluctuations in the spectral intensity. They report that more precise experimental results and improved LOD values were obtained with their new approach, involving single spectrum analysis, than with the traditional technique, in which all the spectra (of the same data set) were averaged together. It should be noted, however, that although the method does not require an internal standard element of known concentration, standard samples are still needed to build the calibration curves. Finally, although the authors claim greater precision of the results obtained with the proposed method, they caution that its accuracy has yet to be verified.

The above method has also been studied by Gornushkin *et al.* [62], who obtained contradictory results. For a set of 1000 laser shots, the peak intensity of an aluminum line did in fact show a linear dependence on the baseline value, but with pronounced scatter of the experimental data, which led to imprecise evaluation of constant α. As a result, the calibration curve of the $\alpha - 1$ parameter versus the aluminum concentration in the sample was not linear, indicating a possible lack of generality in the applicability of the model proposed by Xu and co-workers. Gornushkin and colleagues [62] tried to find a physical explanation for the results obtained in their laboratory. Considering the analytical expression for line intensity and continuum in the plasma, they observed that the ratio of these two quantities is not a simple linear function of the element concentration, but also depends on the plasma temperature. Thus, the method proposed by Xu *et al.* [61] seems able to provide correct results only in the case that the shot sequence always generates the same temperature in the plasma, even for the different samples needed to construct the calibration curve.

3.3.3 Detection limits

Detection limits are usually determined by means of data extracted from calibration curves (see, for example, reference [63] for a general discussion of detection limits in plasma emission spectroscopy). According to the 3σ-IUPAC definition, the LIBS limit of detection (LOD) for a given element is the concentration producing a net line intensity equal to three times the standard deviation of the background, s_B:

$$\text{LOD} = 3\frac{s_B}{S} \tag{3.49}$$

where S is the slope of the calibration curve for the specific atomic emission, also called sensitivity, at the lowest measured concentration. Some authors (see, for instance, Dudragne *et al.* [10] and Berman and Wolf [64]) use a multiplicative factor of 2, instead of 3, in equation (3.49). Alternatively, Eppler *et al.* [65] define the detection limit in the same formal way, but take the value of s_B to be the standard deviation of different measurements on the least concentrated sample. Regardless of the definition adopted, when reporting limits of detection, it is important to state the basis for their determination and the measurement bandwidth or time constant. In the case of LIBS, the latter might be stated as the number of laser probings averaged and the repetition rate of the laser.

An appendix (see Section 3.5) presents a table of LOD values compiled from recent reports in the literature. The relative concentration limits are listed for a series of elements (in weight/weight ratio, unless otherwise specified). For several elements, multiple detection limits are reported, depending upon the different matrix in which the element was detected. In fact, several matrices permit only high LODs for some elements because of the presence of matrix interference lines, poor reproducibility or pronounced heterogeneity of the material. However, because of the possibility of obtaining results in field measurements, even LOD values considerably higher than those attained by other laboratory techniques are often considered satisfactory [66]. Regarding toxic elements, the LODs are often lower than those imposed by regulatory limits (see, for example, Marquardt *et al.* [67] for Pb in paint, and Lazzari *et al.* [68] for Hg in air). On the other hand, Dudragne *et al.* [10] reported an LOD for sulfur in air that was unable to satisfy the regulatory requirement.

3.3.4 *Multivariate analysis*

Under certain conditions, both the detection limit and the precision of a measurement can be improved by applying multivariate data analysis, e.g. by accounting for as many spectral lines as possible in the analysis [69]. As with all atomic emission methods, LIBS has an inherently high information content. There is a large redundancy in spectral information, which is seldom used efficiently in conventional, single line calibration methods. Modern array detectors have made it practical to take advantage of the full spectral content of LIBS emission.

An automated analytical procedure based on multivariate analysis has been described by Wisbrun *et al.* [58]. Information on many lines for each element of interest was extracted from the spectral data collected during the measurement. Two different calibration techniques were applied and assessed in these trials: the generalized linear model (GLM) and principal component regression (PCR). The GLM technique calls for fitting each of the calibration plots corresponding to a different spectroscopic line with a multi-parameter function

$$y = f_\lambda(\alpha, \beta, \gamma, \ldots, x), \tag{3.50}$$

where y is the signal collected at wavelength λ, corresponding to a concentration x. The fitting parameters $(\alpha, \beta, \gamma, \ldots)$, as well as the quality of fit, q_λ^2, are stored for future use in the inverse process. Each fitted line intensity is transformed using the transformation A_λ that linearizes the fitting function f_λ with unit slope and zero intercept

$$A_\lambda f_\lambda(\alpha, \beta, \gamma, \ldots, x) = x. \tag{3.51}$$

The transformed points are plotted against the concentration and given a weight corresponding to the quality of the original fitting procedure, q_λ^2. The new linear model contains information obtained at all wavelengths, with a relative contribution that depends on the quality of the individual calibration plots. Afterwards, the same calibration plot is used for the analysis by applying the same known set of parameters (α, β, γ, ...), weights, q_λ^2, and inverse transformation, A_λ^{-1}, to the experimental intensities.

The technique of PCR provides a calibration model rather than a single plot, and it can compensate for mutual influences of the elemental lines. This model, which is applied to simultaneous multi-elemental analysis, is based on the inspection of many spectral lines. The so-called loading plots and leverage plots provide information on the relative contribution of the various spectral lines to the overall calibration model. The input to the PCR calibration method is the peak information obtained from analysis of the spectral data. The training set consists of the peak data for all objects (samples) and variables (wavelengths) in the model. Influence plots (residual variance versus leverage value) are used to estimate the importance of the variable in the overall model and to identify and remove non-influential observations. The leverage value shows the influence of each spectral peak on the calibration model (zero leverage means no influence at all). The residual variance, instead, shows how well the model describes the particular variable (zero residual variance means that the variable fits the model perfectly). An evaluation of the calibration model is provided by the plot of predicted versus measured concentrations, using a cross-validation technique.

3.3.5 Conditional analysis

Hahn *et al.* [70] developed a LIBS-based technique devoted to monitoring the composition (particularly the metal content) of discrete aerosol particles originating in waste incineration processes. The main goal of their work was to detect sub-micrometer particles escaping from collection by the air-pollution control systems of waste treatment plants. The authors evaluated the mean number of particles involved in a single plasma shot, which in a realistic case is less than unity. This finding suggested to the authors that the problem could be treated as a statistical sampling problem in order to enhance the sensitivity of the analysis. Instead of a mere average of the data, a random LIBS sampling was performed, followed by a conditional analysis of the resulting data. Such an approach is particularly suited to the analysis of very dilute discrete aerosols: only the LIBS spectra obtained by vaporizing a particle, which thus contain information on metal loading, are retained for analysis, while the spectra corresponding to pure gaseous background are discarded after increasing the shot counter. On the basis of information on the plasma volume, effluent gas flow and laser repetition rate, the concentration of metal in the effluent can be calculated. Accurate concentration values were obtained at levels lower by a factor of 30 than the results obtained by simple time averaging of the signal.

3.3.6 Matrix effects

In recent years, the availability of improved instrumentation, a better understanding of the relevant physical processes and the strong need for rapid analytical sampling methods have all stimulated growing interest in sampling techniques based on laser ablation. Many research groups are currently working in this field, and an increasing number of publications are devoted to the subject each year. Nevertheless, techniques based on laser ablation are not yet recognized as reliable in analytical chemistry, mostly because they are known to be highly matrix dependent [69]. The dependence of the mass ablation rate on the nature of the matrix forces matrix-matched reference standards to be used for calibration. If matrix-matched standards are available, laser ablation can provide accurate quantitative analysis even if fractionation occurs, since the phenomenon affects both the known standard and the unknown sample [71]. On the other hand, methods based on the use of calibration curves are unreliable when applied to a matrix different from that used in constructing the curve.

The ablated mass is not the only parameter on which the matrix effect has an impact. The variables of plasma temperature and electron density both depend on the sample, as well as on the laser properties. As in the internal standard method, the calibration curve is supposed to correlate the emission intensity of a single plasma species to the total element concentration in the sample. Therefore, if the atomic distribution of the same element among the different excitation/ionization states in the plasma differs from sample to sample (a likely outcome in view of the differences in plasma temperature and electron density), the results furnished by the calibration curve will be erroneous. In conclusion, when matrix-matched standards are not available, some normalization procedure is advisable.

The procedures developed for correcting statistical fluctuations based on normalization with respect to a reference line provide higher accuracy and reproducibility of the analytical results than methods using simple line intensity. However, such corrections are effective only for small variations in the experimental parameters. A number of cases have been reported in the literature for which the calibration method proved to be inapplicable because of the occurrence of physical effects different from statistical fluctuations. One example was shown by Eppler *et al.* [65], who investigated the effects of chemical speciation on analyte emission intensity in sand and soil samples. The authors prepared a set of samples doped with a known amount of a monitor element (in this case, Ba and Pb) in different chemical forms (oxide, carbonate, sulfate, chloride, nitrate). The emission intensity for the same quantity of monitor element in the sample was found to depend on the chemical speciation. Furthermore, the calibration curves built for the different compounds had different slopes. Consequently, the authors caution that, because of the difficulty in preparing an appropriate calibration curve, the accuracy of measurements performed following this technique may be reduced for samples whose exact speciation is unknown.

A classic example of nonlinear calibration curves was reported by Nouvellon *et al.* [8], Borisov *et al.* [72], and by Gagean and Mermet [73] in the LIBS analysis of Cu–Zn binary alloys. Although the specific papers refer to laser ablation ICP-MS [72], and ICP-AES [73], the results are interesting since they illustrate the fundamental mechanisms of laser ablation. Borisov *et al.* [72] obtained nonlinear calibration curves, regardless of the laser used for ablation (nanosecond or picosecond pulse, UV or visible wavelength). In particular, the copper signal exhibited an almost flat dependence on copper concentration, which made it impossible to use the curve for concentration determinations. The resulting nonlinearity was shown to depend on the amount of ablated mass: alloys containing more Zn were easier to ablate, resulting in a higher emission intensity for both Zn and Cu. Linear calibration curves could, however, be obtained by normalizing the Zn intensity to an internal standard (in this case, the Cu line intensity). Since an internal standard is not always available, especially for unknown samples, the authors suggest the possibility of normalizing the experimental intensity to an external standard, such as the total mass removed from the sample surface, which can be determined, for example, by measuring the crater volume with an interferometric microscope. Margetic *et al.* [74, 75] showed that the observed differences in the ablated mass could be related to changes in the phase composition of the material. In their study, sampling was performed with a femtosecond laser.

Currently, work in the field of LIBS matrix effects is progressing along two parallel tracks. Some researchers have undertaken a systematic study of the elements responsible while several other groups have developed models of the plasma state in an attempt to obtain a general description that would make it possible to account for any kind of sample matrix and which would therefore be independent of the specific type of matrix to be analyzed. An example of the first approach is the work published by Russo and co-workers, who deal with the matrix effect in ICP-AES measurements. In a recent study [76], the authors investigated the role of 31 elements in generating matrix effects in ICP-AES by monitoring the plasma excitation conditions (temperature and electron density). They found that the matrix effect is strongly correlated with the presence of elements having low second ionization potentials. Since a strong matrix effect was observed using argon as the carrier gas, and disappeared with helium, a hypothesis was advanced involving an interaction between the matrix active elements and some specific levels of the argon atoms. Once again, although the study was specifically devoted to the ICP source, the conclusions made should also be of value for laser-induced plasmas, where the ambient gas is an important parameter to be taken into account.

Over the past few years, several models have been developed to compensate for matrix effects by parameterizing the plasma state. Some of these methods are described in the following section.

Methods developed to correct for matrix effects on calibration curves

A straightforward method to minimize the effect of plasma state variations on quantitative measurements has been proposed by Aragon *et al.* [42]. According to these authors, careful selection of the reference lines can improve analytical results. In a quantitative analysis of the minor components in steel samples, they used reference iron lines having the same degree of ionization and an upper-energy value close to the analytical lines. Lines with close upper-energy levels should be used when possible, because their intensity ratio is proportional to the ratio of Boltzmann factors, and can therefore be expected to be relatively insensitive to small variations in the plasma temperature. Moreover, as shown by the Saha equation, the ratio of ion densities to neutral atom densities will also be susceptible to variations depending on the plasma temperature, which will result in fluctuations in the intensity ratio. In the work cited above, the authors compared the precision for the intensity of a Si(I) line and the ratios of Si(I)/Fe(I) and Si(I)/Fe(II) lines, obtained in 10 consecutive measurements, each corresponding to 100 accumulated pulses. Better precision was obtained by using the ratio of lines emitted by species of the same ionization state. Calibration curves with very high correlation coefficients have been obtained by using this method.

The approach described by Panne *et al.* [18] stems from analogous considerations on the pulse-to-pulse variability of the plasma electron temperature and density, and proposes a method for normalizing the line intensity ratio. The normalization procedure is based on a model which includes the equilibrium Saha and Boltzmann relationships. The authors use the most general expression for the specific intensity e of an atomic or ionic line of the element A:

$$^q e_{ik}^A = \frac{g_i^A e^{-\left(E_i^A/kT\right)}}{^q U^A(T)} \frac{hc}{4\lambda_{ik}\pi} A_{ik}^{Aq} W^A(T, n_e) n^A, \qquad (3.52)$$

where q ($=$ I or II) is the ionization state, i and k are the respective indices of the higher and lower energy quantum states of the same ionization state, g is the statistical weight, E_i is the energy of the level measured with respect to the ground state in the same ionization state, $^q U^A$ is the partition function of the quantum states of the particular ionization state, λ_{ik} is the emission wavelength, A_{ik} is the transition probability for spontaneous emission, $^q W^A$ is the probability of occurrence of the ionization state q, and n^A is the total number density of atoms from analyte A.

In order to derive the probability $^q W^A$, it should be noted that only neutral and singly ionized species are usually observed in the typical LIBS plasma. The ratio between the ion and the atom number density of element A in a multi-component plasma can be derived using the Saha equation

$$\frac{^{II}n^A}{^I n^A} = 2\frac{^{II}U^A(T)}{^I U^A(T)} \frac{1}{n_e} \left(\frac{2\pi m_e kT}{h^2}\right)^{3/2} e^{-\frac{E_{ion}^A - \Delta E_{ion}^A}{kT}}, \qquad (3.53)$$

where E_{ion} is the ionization energy of element A, and ΔE_{ion} is the lowering in ionization energy of the element in the plasma. With the substitutions

$$\xi = \frac{^{II}n^A}{^I n^A} \quad \text{and} \quad n^A = {}^{II}n^A + {}^I n^A \qquad (3.54)$$

the probability of observing the ionization state q is given by

$$^{II}W^A = \frac{\xi}{1+\xi} \quad \text{and} \quad {}^I W^A = \frac{1}{1+\xi} = 1 - {}^{II}W^A. \qquad (3.55)$$

This equation completes the set of expressions that allow evaluation of the line emission intensity. The goal of the authors was to determine the concentration ratios of the main components in glass samples from the intensity ratios. The equations derived above were therefore used to calculate the intensity ratio of two different lines from elements A and B, in their corresponding ionization states q and q', respectively.

$$\frac{^q n^A}{^{q'} n^B} = \frac{A_{mn}^B g_m^B}{A_{ik}^A g_i^A} \left(\frac{\lambda_{ik}^A}{\lambda_{mn}^B}\right) e^{-(E_i^A - E_m^B/kT)} \frac{^q U^A(T)}{^{q'} U^B(T)} \frac{^{q'} W^B}{^q W^A} \frac{^q I_{ik}^A}{^{q'} I_{mn}^B}. \qquad (3.56)$$

While the first factor of the expression can be derived from known atomic properties, the authors suggest determining the electron temperature from the ratio of two lines from an element or from the corresponding Boltzmann plot. On the other hand, in order to calculate the ratio $^{q'}W^B/^qW^A$, the electron density n_e must be determined via measurement of the Stark broadening or, alternatively, by using the Saha equation. Calculation of a concentration ratio of two elements, A and B, following this procedure then requires a set of at least three other lines: two lines from element C in different ionization states, to calculate the electron density n_e, and one additional line for calculating the temperature, T (in principle, C can be the same as A or B). As a further simplification, the authors suggest neglecting the temperature dependence of the partition function, which thus becomes a constant in the equation. As the reduction in the ionization energy, ΔE_{ion}, is on the order of 0.1 eV, they suggest neglecting this as well. Furthermore, since the procedure described above is intended for normalization of the intensity ratio used in the construction of the calibration curves, the partition function ratio in equation (3.56) can be considered as part of the calibration curve slope, and does not need to be determined explicitly. As a result of the normalization, the pulse-to-pulse variation in the line ratios was reduced, and linear calibration of LIBS intensity ratios versus the concentration ratios was achieved.

A different analytical procedure has been proposed in order to correct for the matrix effect when normalization of the analyte signal by a reference signal is not possible due to the lack of a reference element. This frequently happens in "on-line" analysis, where the matrix may change over time as the process evolves, as well as in microanalysis, because the matrix composition may vary from one location to the next on the solid. Chaléard *et al.* [77] developed an analytical model that describes the line emission intensity as a function of the vaporized mass and the plasma excitation temperature, two parameters which, under the hypothesis of stoichiometric ablation, contribute to the matrix effects observed when the sample composition changes. This model can be represented in mathematical terms by defining the emission intensity of a line of a given element as

$$I = KCM_{pl}e^{-E/kT} \tag{3.57}$$

where K is a constant that accounts for collection efficiency and the spectroscopic parameters of the line being considered, M_{pl} is the total mass vaporized in the plasma plume, C is the concentration of the given element in the plasma (equivalent to the concentration in the solid phase) and the exponential term accounts for the excitation temperature, assuming the plasma is in LTE. According to equation (3.57), the matrix effects observed in laser ablation experiments would depend only on the sample-to-sample variations in the vaporized mass and the plasma excitation temperature, while the variations in electron density would play a less important role. Therefore, a correction for the matrix effects could be achieved if both the vaporized mass and the plasma temperature are measured simultaneously with the analytical signal. In the experimental conditions chosen, it was demonstrated that the acoustic signal emitted by the plasma is proportional to the vaporized mass (this was verified for aluminum, nickel, steel and glass) and that the excitation temperature can be measured by applying the line ratio method, after choosing a given element as the "temperature sensor."

The authors note that the slope of the acoustic signal vs. vaporized mass is quite low. However, the resulting limit to the precision is compensated for by the good reproducibility of the acoustic measurements. After calculating the plasma temperature, the normalized intensity can be written as

$$\frac{I}{A_s e^{-E/kT}} = \frac{1}{\alpha} KC, \qquad (3.58)$$

where $A_s = \alpha M_{pl}$ is the acoustic signal intensity. Because of the uncertainties in the spectroscopic parameters, it was suggested that the same element should be chosen as the "temperature sensor," both for building the calibration curves and for analysis of the unknown sample. The experimental results reported in the paper show that the differences in ablation rate and excitation temperature are independent and could occur separately for different matrices. Different calibration curves were then constructed for different matrices. The authors distinguish a first-order matrix effect, occurring between completely different materials, from a second-order matrix effect, occurring between samples with the same major element and different percentages of minor constituents. The second-order effect was found to be negligible when analyzing Mn in steel samples and Cu in Al samples. A pronounced first-order matrix effect was observed, for example, in the analysis of Cu in Al, steel and brass. Such normalization of the Cu emission intensity to the acoustic signal and excitation temperature provided efficient correction for the matrix effect and provided a linear multi-matrix calibration curve [77].

The calibration-free technique

In 1998, Ciucci et al. [53] proposed a new method for standardless LIBS analysis, called calibration-free LIBS (CF-LIBS). The basic idea behind the method is to compensate for the matrix effect by applying the basic equations derived from the LTE assumption to the spectral data in order to make each single measurement self-consistent and avoid the need for any comparison with calibration curves or reference samples. The method rests on the assumptions of stoichiometric ablation, local thermal equilibrium and, in its initial formulation, optically thin plasma. From a practical point of view, the method requires measurement of at least two emission lines from the same species, in order to calculate the plasma electron temperature, as well as at least one line for each element in the plasma, with known spectroscopic parameters. The parameterization of the measured line intensity given by equation (3.44) is used to construct a Boltzmann plot that contains the points corresponding to all the lines observed in the experimental spectrum

$$\overline{I_{ij}} = FC^s A_{ij} \frac{g_i e^{-\frac{E_i}{kT}}}{U^s(T)} \qquad (3.59)$$

where $\overline{I_{ij}}$ represents the measured integral line intensity (counts), and C^s is the concentration of the emitting species in the plasma. The experimental parameter F takes into account the optical efficiency of the collection system, as well as the total plasma number density and its volume. Moreover, the spectral response of the collecting optics, spectrometer and detector, which is a characteristic of the experimental apparatus, should be measured radiometrically once for all determinations. The spectral intensity values are then normalized with this efficiency curve. It is important to note that only the relative response of the system must be determined, because the information on the absolute intensity of the lines is included in the constant parameter F. Now, the line integral intensity $\overline{I_{ij}}$ in equation (3.44) is the result of the measurement, the spectroscopic parameters A_{ij}, g_i and E_i can be obtained from spectral databases, and the F, C^s and T values must be determined from the experimental data. The partition function of each species, in turn, can be calculated from the known spectroscopic data once the plasma temperature has been determined.

By defining the quantities

$$y = \ln \frac{\overline{I_{ij}}}{g_i A_{ij}}, \quad x = E_i, \quad m = -\frac{1}{k_B T}, \quad q^s = \ln \frac{C^s F}{U^s(T)} \quad (3.60)$$

and taking the logarithms of both sides, equation (3.44) can be rewritten in the linear form

$$y = mx + q^s. \quad (3.61)$$

A similar relation can be written for each species in the plasma: using the definitions in equation (3.60), we can represent the intensity value of each experimental LIBS line as a point in the Boltzmann plane. Thus, the entire data set of peaks in the spectrum can be graphically represented as a Boltzmann plot, where the different points lie on several straight parallel lines, with slope m and intercept q^s, each of which corresponds to a different atomic species. By way of example, the Boltzmann plot obtained from the analysis of an aluminum sample is shown in Figure 3.6.

According to the definitions in equation (3.60), the slope m of the plots is related to the plasma temperature, while the intercept q_s is proportional to the logarithm of the species concentration times the experimental factor F, which takes account of the efficiency of the detection system, the plasma volume, the experimental geometry, etc. Provided that experimental data are available for each species present in the plasma, after calculating the electron temperature, which according to the LTE assumption is the same for all species, one can calculate the concentration of the species and consequently the concentration of the corresponding element (times the experimental factor F) as the sum of the concentrations of the neutral and single ionized species.

In some particular cases, the LIBS spectrum exhibits lines corresponding to only one species of a certain element. In this case, the concentration of the other ionization states can be derived via the Saha equation (3.10), which relates the species concentrations of successive ionization states of the same element. The plasma electron density, n_e, which must be known if the Saha equation is to be applied, can be directly measured through determination of the Stark broadening of the plasma lines. Alternatively, the electron density can be calculated by means of the Saha equation itself, provided the concentration ratio of two successive ionization states has been determined for at least one element. Note that in this case the information corresponding to just one element is sufficient for the calculation,

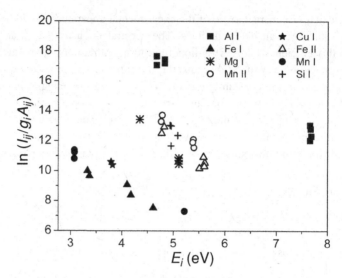

Figure 3.6. Detail of the Boltzmann plot obtained from analysis of an aluminum alloy sample. The data sets corresponding to the different species lie on parallel lines. Visual inspection of the plot provides a qualitative estimate of the corresponding concentration ratios: the higher the intercept value, the higher the relative concentration. Quantitative determination of the actual concentration value must be performed through calculation of the partition functions.

because the plasma electron density, like the temperature, is a global plasma parameter and is the same for all the elements. In order to remove the unknown experimental factor F, one can either take advantage of the known concentration of an internal standard, if this information is available, or otherwise use the normalization relation

$$\sum_e C^e = 1 \qquad (3.62)$$

since the sum of the relative concentrations of all the elements must equal unity.

In principle, assuming that the uncertainties in the measurement and the spectroscopic parameters are negligible, the minimum error associated with the results is on the same order of magnitude as the concentration of the most abundant undetected element in the sample. As LIBS is generally able to detect most elements with p.p.m. sensitivity, CF-LIBS measurements can theoretically achieve very high accuracies. However, it should be borne in mind that LIBS measurements are often characterised by reproducibility no better than 3%–5%, and that transition probability values are often known only to within 10%–20%. Therefore, the uncertainties associated with this method are usually greater than the theoretical minimum.

In principle, after determination of the plasma temperature, the concentration of a species can be obtained by measuring just one spectral line with known spectroscopic parameters. However, more precise results can be obtained by using as many spectral lines as possible, in order to average out the possible effects of the uncertainties on the transition probability values. The calibration-free procedure requires all the atomic components of a sample to be detected by measuring at least one of their characteristic spectral lines. Therefore, an unknown sample should be probed extensively, collecting as wide a spectrum as possible (typically from 200 to 800 nm). Clearly, broad-band detection systems are best suited to exploiting the potentialities of the CF-LIBS procedure, because they enable acquisition of a wide spectral region in a single shot.

Recently, the CF-LIBS procedure has been updated in order to account for self-absorption [78, 79]. In its original formulation, the CF-LIBS procedure [53] skirted the issue of self-absorption by simply excluding self-absorbed resonant lines of the major elements from the analysis. In the new formulation, a self-absorption correction step has been implemented in the CF-LIBS procedure via a recursive algorithm. The procedure calls for carrying out an initial evaluation of sample composition by application of the CF-LIBS procedure to all measurable lines, without any correction. This allows first estimates to be obtained for the temperature, from the Boltzmann plot, and the electron density, from the Saha equation. The next step is to calculate the number densities n^s of all the different species from these first estimates and the relative abundances of all species, C^s, together with the conditions of plasma neutrality. In fact, as each electron in the plasma practically corresponds to a single charged ion, we can write the following expression:

$$\sum_s n^s(\text{II}) = n_e \tag{3.63}$$

where $n^s(\text{II})$ represents the number density of the ion of element s in the plume. By introducing the known relative abundance of each species as $C^s(\text{II}) = n^s(\text{II})/n_{\text{TOT}}$, equation (3.63) becomes:

$$n_{\text{TOT}} \sum_s C^s(\text{II}) = n_e \tag{3.64}$$

from which it is a simple matter to calculate the absolute density n_{TOT}. In order to simulate the COGs of all lines in the Boltzmann plot, their Gaussian and Lorentzian broadening must also be known. Gaussian broadening, $\delta\nu_D$, is produced by thermal effects and is easily determined from the plasma temperature by using the familiar relationship

$$\delta\nu_D = \frac{2\nu_0}{c} \sqrt{\left(\frac{2RT \ln 2}{M} \right)} \tag{3.65}$$

where M is the atomic mass number and R is the universal gas constant. The often present instrument contribution to Gaussian broadening can be easily measured as well. Evaluating the Lorentzian broadening is more difficult. Unfortunately, the parameters to be found in the literature (collisional cross section, etc.) do not cover all the needed transitions. Moreover, the experimental width found by fitting the spectral line is not immediately applicable to determination of the self-absorption coefficient, since the saturation effect produces distortion of the line profile, which therefore yields an apparent experimental width much greater than the real one. Using the experimental Lorentzian width furnished by the fit would lead to an overestimation of $\delta\nu_L$ and consequent underestimation of the self-absorption effect. However, it has been demonstrated [80] that the growth of the total observed line broadening in the presence of self-absorption varies as:

$$\Delta\lambda_{obs} = \frac{\Delta\lambda_{orig}}{\sqrt{SA}} \tag{3.66}$$

where SA is the ratio of the experimental peak height to the intensity that the line would have if growth were linear. Assuming that the self-absorption coefficient, SA, is known, the total original width and the contribution of $\delta\nu_L$ can be calculated (since $\delta\nu_D$ is known). However, as SA is itself derived from the COG, which is a function of $\delta\nu_L$, the procedure must be iterated to find the proper values for SA and $\delta\nu_L$.

Finally, the self-absorption correction for each line is calculated in order to determine the intensity that the line would have in the absence of self-absorption. When these corrections are applied to each point in the Boltzmann plot, the spread of points from each species around the best fit line is reduced, demonstrating that the scatter is mainly the result of self-absorption rather than of experimental error, fitting error or uncertainties in A_{ki}. Changes in the Boltzmann plot also affect calculation of the temperature and electron density, thus making it necessary to iterate the algorithm until convergence. The results obtained with this improved CF-LIBS procedure are more reliable than the original formulation, first because they are obtained by using more lines, and also because the quality of the regression fit in the Boltzmann plane is improved. For the same reasons, the determinations of plasma temperature, electron density and ablated mass, the most critical plasma parameters affecting quantitative LIBS measurements, are also improved. An example of the application of the SAC algorithm is shown in Figure 3.7.

3.4 Conclusions

The plasma generated in LIBS measurements is very complex. Many approximate models have been developed in order to improve the usefulness of quantitative data and continuing research in the field is bearing the fruit of greater understanding. All methods for quantitative LIBS analysis are based on simplifying hypotheses. These typically include stoichiometric ablation, local thermodynamic equilibrium and optical thinness. Many critical evaluations of the validity of such hypotheses have been performed, as testified to by the considerable literature devoted to the subject. The experimental conditions required to fulfil these basic hypotheses have been determined, aiding in the design of LIBS measurements. For those cases in which the experimental conditions cannot be freely adjusted, some methods have been proposed in order to apply corrections to the results (for example, modeling of optically thick plasmas). At present, the different methods of quantitative LIBS analysis produce accuracies of the order of a few percent in the best cases. In practice, the choice of quantitative analysis method should therefore be made after careful consideration of the specific aims of the experiment. For laboratory measurements, where only one or a few elements are to be determined in approximately constant matrices, the classical calibration curve technique is probably still the method of choice, albeit with some reservations. With the aid of the

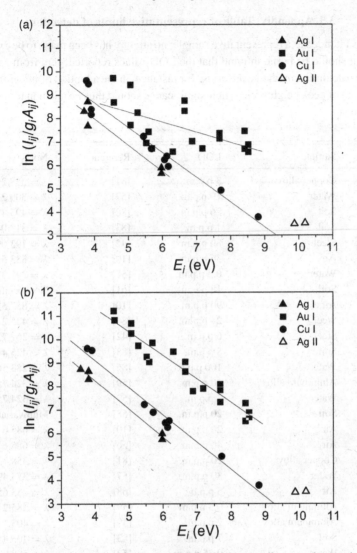

Figure 3.7. (a) Boltzmann plot obtained from analysis of a ternary alloy sample containing Au, Cu and Ag (the Au content was 91.7%). The reliability of the plasma temperature evaluation is limited by the scattering of the data sets (including all the lines observed in the spectrum). (b) Application of the SAC algorithm corrects the absorbed-line intensity values, thus reducing the scatter and yielding better precision and accuracy of the plasma temperature determination.

corrective methods available to account for variations in plasma conditions, this method can generally afford good reliability. On the other hand, when characterization of an entire sample is needed or when the matrix is completely unknown or likely to change substantially from sample to sample, the self-consistent measurements provided by the CF-LIBS method should have attractive advantages.

3.5 Appendix. Table of representative limits of detection

This table is intended as a representative sample; no attempt has been made to be exhaustive. Moreover, it should be borne in mind that the LOD values reported stem from application of different calculation methods, differing, for instance, in the adoption of integral intensity values instead of peak heights, etc. Their significance should thus be taken in terms of order of magnitude only.

Element	Sample	LOD	Reference	Notes
Ag	Copper alloy	3 p.p.m.	[81]	$\lambda = 338.29$ nm
Al	Water	10 p.p.m.	[57]	$\lambda = 309.27$ nm
Ba	Soil	26 p.p.m.	[82]	$\lambda = 493.41$ nm
Be	Soil	1 p.p.m.	[83]	$\lambda = 313.04$ nm
C	Steel	80 p.p.m.	[42]	$\lambda = 193.09$ nm
C	Air	36 p.p.m.	[10]	$\lambda = 833.51$ nm
Ca	Water	0.3 p.p.m.	[57]	$\lambda = 393.37$ nm
Cd	Soil	18 p.p.m.	[61]	Not specified
Cl	Air	90 p.p.m.	[10]	$\lambda = 837.59$ nm
Cr	Steel	24 p.p.m.	[84]	$\lambda = 425.2$ nm
Cr	Steel	6 p.p.m.	[42]	$\lambda = 267.72$ nm
Cr	Soil	8 p.p.m.	[83]	$\lambda = 425.44$ nm
Cr	Water	100 p.p.b.	[85]	$\lambda = 283.56$ nm
Cu	Aluminum alloy	10 p.p.m.	[59]	$\lambda = 324.75$ nm
Cu	Water	7 p.p.m.	[57]	$\lambda = 324.75$ nm
Cu	Soil	20 p.p.m.	[58]	Multivariate analysis
F	Air	20 p.p.m.	[10]	$\lambda = 685.6$ nm
F	Air	40 p.p.m.	[66]	$\lambda = 685.6$ nm
Fe	Copper alloy	20 p.p.m.	[81]	$\lambda = 358.12$ nm
Fe	Water	30 p.p.m.	[57]	$\lambda = 373.49$ nm
Hg	Air	5 p.p.b	[68]	$\lambda = 253.65$ nm
Mg	Aluminum alloy	0.5 p.p.m	[59]	$\lambda = 285.2$ nm
Mn	Aluminum alloy	2 p.p.m.	[59]	$\lambda = 403.1$ nm
Mn	Soil	7 p.p.m.	[83]	$\lambda = 403.45$ nm
Na	Water	0.55 p.p.m.	[57]	$\lambda = 588.99$ nm
Ni	Copper alloy	10 p.p.m.	[81]	$\lambda = 349.4$ nm
				$\lambda = 341.5$ nm
Ni	Steel	50 p.p.m.	[42]	$\lambda = 231.60$ nm
Ni	Water	18 p.p.m.	[64]	$\lambda = 341.5$ nm
				$\lambda = 352.4$ nm
				$\lambda = 361.9$ nm
Ni	Soil	20 p.p.m.	[58]	Multivariate analysis
Pb	Concrete	10 p.p.m.	[60]	$\lambda = 405.78$ nm
Pb	Soil	5 p.p.m.	[83]	$\lambda = 405.78$ nm

Element	Sample	LOD	Reference	Notes
S	Steel	70 p.p.m.	[86]	$\lambda = 182.03$ nm
S	Air	1500 p.p.m.	[10]	$\lambda = 921.29$ nm
Sn	Water	100 p.p.m.	[57]	$\lambda = 283.99$ nm
Si	Steel	30 p.p.m.	[84]	$\lambda = 288.2$ nm
Si	Aluminum alloy	14 p.p.m.	[59]	$\lambda = 251.6$ nm
Sr	Soil	3 p.p.m.	[83]	$\lambda = 407.77$ nm
Zn	Water	120 p.p.m.	[57]	$\lambda = 334.50$ nm
Zn	Soil	30 p.p.m.	[58]	Multivariate analysis

3.6 References

[1] B. C. Castle, K. Visser, B. W. Smith and J. D. Winefordner, *Spectrochim. Acta B*, **52** (1997), 1995–2009.

[2] D. A. Cremers, *Appl. Spectrosc.*, **41** (1987), 572–579.

[3] R. E. Russo, *Appl. Spectrosc.*, **49** (1995), 14A.

[4] W. T. Chan and R. E. Russo, *Spectrochim. Acta Part B.*, **46** (1991), 1471–1486.

[5] X. Mao, W. T. Chan, M. Caetano, M. A. Shannon and R. E. Russo, *Appl. Surf. Sci.*, **96–98** (1996), 126–130.

[6] R. E. Russo, X. Mao and S. S. Mao, *Anal. Chem.*, **74** (2002), 70A–77A.

[7] X. L. Mao, A. C. Ciocan, O. V. Borisov and R. E. Russo, *Appl. Surf. Sci.*, **127–129** (1998), 262–268.

[8] C. Nouvellon, C. Chaleard, J. L. Lacour and P. Mauchien, *Appl. Surf. Sci.*, **138–139** (1999), 306–310.

[9] O. V. Borisov, X. L. Mao and R. E. Russo, *Spectrochim. Acta Part B,* **55** (2000), 1693–1704.

[10] L. Dudragne, P. Adam and J. Amouroux, *Appl. Spectrosc.*, **52** (1998), 1321–1327.

[11] D. A. Cremers and L. J. Radziemski, *Anal. Chem.*, **55** (1983), 1252.

[12] M. Tran, Q. Sun, B. W. Smith and J. D. Winefordner *J. Anal. Atom. Spectrom.*, **16** (2001), 628–632.

[13] D. A. Cremers and L. J. Radziemski, *Appl. Spectrosc.* **39** (1985), 57.

[14] J. E. Carranza and D. W., Hahn, *Anal. Chem.* **74** (2002), 5450–5454.

[15] A. Semerok, C. Chaleard, V. Detalle, *et al.*, *Appl. Surf. Sci.*, **138–139** (1999), 311–314.

[16] C. W. Ng, W. F. Ho and N. H. Cheung, *Appl. Spectrosc.*, **51** (1997), 976–983.

[17] A. P. Thorne, *Spectrophysics* (London: Chapman and Hall, 1974).

[18] U. Panne, C. Haisch, M. Clara and R. Niessner, *Spectrochim. Acta Part B*, **53** (1998), 1957–1968.

[19] H. C. Liu, X. L. Mao, J. H. Yoo and R. E. Russo, *Spectrochim. Acta Part B*, **54** (1999), 1607–1624.

[20] C. Th. J. Alkemade, Tj. Hollander, W. Snelleman and P. J. Th. Zeegers, *Metal Vapours in Flames* (Oxford: Pergamon Press, 1982).

[21] C. A. Bye and A. Scheeline, *Appl. Spectrosc.*, **47** (1993), 2022–2030.

[22] I. B. Gornushkin, L. A. King, B. W. Smith, N. Omenetto and J. D. Winefordner, *Spectrochim. Acta Part B*, **54** (1999), 1207–1217.

[23] H. R. Griem, *Spectral Line Broadening by Plasmas* (New York: Academic Press, 1974).

[24] Y. Iida, *Spectrochim. Acta Part B*, **45B** (1990), 1353–1367.

[25] M. C. Quintero, A. Rodero, M. C. Garcia and A. Sola, *Appl. Spectrosc.*, **51** (1997), 778–784.

[26] J. B. Simeonsson and A. W. Miziolek, *Appl. Opt.*, **32** (1993), 939–947.

[27] B. Le Drogoff, J. Margot, M. Chaker, *et al.*, *Spectrochim. Acta Part B*, **56** (2001), 987–1002.

[28] D. A. Rusak, B. C. Castle, B. W. Smith and J. D. Winefordner, *Spectrochim. Acta Part B*, **52** (1997), 1929–1935.

[29] J. Hermann, C. Vivien, A. P. Carricato and C. BoulmerLeborgne, *Appl. Surf. Sci.* **129** (1998), 645–649.

[30] J. Hermann, C. BoulmerLeborgne and D. Hong, *J. Appl. Phys.*, **83** (1998), 691–696.

[31] M. Milan and J. J. Laserna, *Spectrochim. Acta Part B*, **56** (2001), 275–288.

[32] D. Riley, I. Weaver, T. Morrow, *et al.*, *Plasma Sources Sci. Technol.*, **9** (2000), 270–278.

[33] M. Capitelli, F. Capitelli and A. Eletskii, *Spectrochim. Acta Part B*, **55** (2000), 559–574.

[34] S. Amoruso, M. Armenante, V. Berardi, R. Bruzzese and N. Spinelli, *Appl. Phys. A*, **65** (1997), 265–271.

[35] J. F. Ready, *Effects of High Power Laser Radiation* (New York: Academic Press, 1971).

[36] I. B. Gornushkin, C. L. Stevenson, B. W. Smith, N. Omenetto and J. D. Winefordner, *Spectrochim. Acta Part B*, **56** (2001), 1769–1785.

[37] A. E. Pichahchy, D. A. Cremers and M. J. Ferris, *Spectrochim. Acta Part B*, **52** (1997), 25–39.

[38] Y. W. Kim, *Laser-Induced Plasmas and Applications* (New York: Marcel Dekker, 1989).

[39] I. B. Gornushkin, J. M. Anzano, L. A. King, *et al.*, *Spectrochim. Acta Part B*, **54** (1999), 491–503.

[40] R. Mavrodineanu and H. Boiteux, *Flame Spectroscopy* (New York: Wiley, 1965).

[41] A. C. G. Mitchell and M. W. Zemanski, *Resonance Radiation and Excited Atoms* (New York: Cambridge University Press, 1961).

[42] C. Aragon, J. A. Aguilera and F. Penalba, *Appl. Spectrosc.*, **53** (1999), 1259–1267.

[43] V. Lazic, R. Barbini, F. Colao, R. Fantoni and A. Palucci, *Spectrochim. Acta Part B*, **56** (2001), 807–820.

[44] R. D. Cowan and G. H. Dieke, *Rev. Mod. Phys.*, **20** (1948), 418–455.

[45] H. Bartels, *Z. Phys.*, **125** (1949), 597.

[46] H. Zwicker, Evaluation of plasma parameters in optically thick plasmas. In W. Lochte-Holtgreven (Editor), *Plasma Diagnostics* (New York: John Wiley and Sons, 1968).

[47] G. Tondello, E. Jannitti and A. M. Malvezzi, *Phys. Rev. A*, **16** (1979), 1705–1714.

[48] A. M. Malvezzi, L. Garifo, E. Jannitti, P. Nicolosi and G. Tondello, *J. Phys. B*, **12** (1979), 1437–1447.

[49] G. J. Tallents, *J. Phys. B*, **13** (1980), 3057–3072.

[50] P. Pianarosa, J. H. Gagne, G. Larin and J. P. Dizier, *J. Opt. Soc. Amer.*, **72** (1982), 392–394.

[51] C. Aragon, J. Bengoechea, J. A. Aguilera, *Spectrochim. Acta Part B*, **56** (2001), 619–628.

[52] N. Omenetto, J. D. Winefordner and C. Th. J. Alkemade, *Spectrochim. Acta Part B*, **30** (1975), 335–341.

[53] A. Ciucci, M. Corsi, V. Palleschi, *et al.*, *Appl. Spectrosc.*, **53** (1999), 960–964.

[54] J. Metzdorf, A. Sperling, S. Winter, K. H. Rastz and W. Moller, *Metrol.*, **35** (1998), 423–426.

[55] B. J. Goddard, *Trans. Inst. Meas. Cont.*, **13** (1991), 128–139.

[56] R. Q. Aucelio, B. C. Castle, B. W. Smith and J. D. Winefordner, *Appl. Spectrosc.*, **54** (2000), 832–837.

[57] P. Fichet, P. Mauchien, J. F. Wagner and C. Moulin, *Anal. Chim. Acta*, **429** (2001), 269–278.

[58] R. Wisbrun, I. Schechter, R. Niessner, H. Schroder and K. Kompa, *Anal. Chem.*, **66** (1994), 2964–2975.

[59] M. Sabsabi and P. Cielo, *Appl. Spectrosc.*, **49** (1995), 499–507.

[60] A. V. Pakhomov, W. Nichols and J. Borysow, *Appl. Spectrosc.*, **50** (1996), 880–884.

[61] L. Xu, V. Bulatov, V. V. Gridin and I. Schechter, *Anal. Chem.*, **69** (1997), 2103–2108.

[62] I. B. Gornushkin, B. W. Smith, G. E. Potts, N. Omenetto and J. D. Winefordner, *Anal. Chem.*, **71** (1999), 5447–5449.

[63] P. W. J. M. Boumans, *Spectrochim. Acta Part B*, **46** (1991), 917–939.

[64] L. M. Berman and P. J. Wolf, *Appl. Spectrosc.*, **52** (1998), 438–443.

[65] A. S. Eppler, D. A. Cremers, D. D. Hickmott, M. J. Ferris and A. C. Koskelo, *Appl. Spectrosc.*, **50** (1996), 1 175–1181.

[66] C. K. Williamson, R. G. Daniel, K. L. McNesby and A. W. Miziolek, *Anal. Chem.*, **70** (1998), 1186–1191.

[67] B. J. Marquardt, S. R. Goode and S. M. Angel, *Anal. Chem.*, **68** (1996), 977–981.

[68] C. Lazzari, M. De Rosa, S. Rastelli, *et al.*, *Laser Part. Beams*, **12** (1994), 525–530.

[69] H. E. Bauer, F. Leis and K. Niemax, *Spectrochim. Acta Part B*, **53** (1998), 1815–1825.

[70] D. W. Hahn, W. L. Flower and K. R. Hencken, *Appl. Spectrosc.*, **51** (1997), 1836–1844.

[71] R. E. Russo, X. L. Mao, O. V. Borisov and H. C. Liu, *J. Anal. Atom. Spectrom.*, **15** (2000), 1115–1120.

[72] O. V. Borisov, X. L. Mao, A. Fernandez, M. Caetano and R. E. Russo, *Spectrochim. Acta Part B*, **54** (1999), 1351–1365.

[73] M. Gagean and J. M. Mermet, *Spectrochim. Acta Part B*, **53** (1998), 581–591.

[74] V. Margetic, A. Pakulev, A. Stockaus, *et al.*, *Spectrochim. Acta Part B*, **55** (2000), 1771–1785.

[75] V. Margetic, K. Niemax and R. Hergenroder, *Spectrochim. Acta Part B*, **56** (2001), 1003–1010.

[76] G. C. Y. Chan, W. T. Chan, X. L. Mao and R. E. Russo, *Spectrochim. Acta Part B*, **56** (2001), 77–92.

[77] C. Chaléard, P. Mauchien, N. André, *et al.*, *J. Anal. Atom. Spectrom.*, **12** (1997), 183–188.

[78] M. Corsi, G. Cristoforetti, V. Palleschi, A. Salvetti and E. Tognoni, *STS Press* (2001), 807–812.

[79] D. Bulajic, M. Corsi, G. Cristoforetti, *et al.*, *Spectrochim. Acta Part B*, **57** (2002), 339–353.

[80] W. Demtröder, *Laser Spectroscopy* (Berlin: Springer-Verlag, 1988).

[81] M. Sabsabi and P. Cielo, *J. Anal. Atom. Spectrom.*, **10** (1995), 643–647.

[82] D. A. Cremers, J. E. Barefield and A. C. Koskelo, *Appl. Spectrosc.*, **49** (1995), 857–870.

[83] R. A. Multari, L. E. Foster, D. A. Cremers and M. J. Ferris, *Appl. Spectrosc.*, **50** (1996), 1483–1499.

[84] F. Leis, W. Sdorra, J. B. Ko and K. Niemax, *Microchim. Acta*, **2** (1985), 185–199.

[85] G. Arca, A. Ciucci, V. Palleschi, S. Rastelli and E. Tognoni, *Appl. Spectrosc.*, **51** (1997), 1102–1105.

[86] A. Gonzalez, M. Ortiz and J. Campos, *Appl. Spectrosc.*, **49** (1995), 1632–1635.

4

Laser-induced breakdown in gases: experiments and simulation

Christian G. Parigger

The University of Tennessee Space Institute

4.1 Introduction

Measurements and simulations of laser-induced optical breakdown spectra are presented in this chapter. Of special interest is the determination of spectroscopic temperature and species concentrations following nominal 1–10 ns laser-induced optical breakdown. Early in the micro-plasma decay, the use of Stark-broadened atomic lines for determination of electron number densities is indicated, and the use of Boltzmann plots is indicated for determination of excitation temperatures. Later in the plasma decay, molecular recombination spectra can be used to characterize the micro-plasma. Spectroscopic temperature and species number densities are inferred by use of the non-equilibrium air radiation (NEQAIR) code. For selected diatomic species that dominate the recombination emission spectrum in the plasma decay, accurate diatomic line strengths are used to find temperature information. The species number densities are estimated from calculated plasma emission spectra for the experimental conditions. Computational fluid dynamics modeling is used to characterize post-breakdown fluid physics phenomena. Laser spark decay phenomena are studied by the use of a two-dimensional, axially symmetric, time-accurate model. The initial laser spark temperature distribution is generated to simulate a post-breakdown profile that is consistent with focal volume investigations. In the developed computational model, plasma equilibrium kinetics are implemented in ionized regions, and non-equilibrium, multi-step, finite-rate reactions are implemented in non-ionized regions.

This chapter addresses simulations of laser focusing, analysis of atomic and molecular spectra, and the computational model to describe laser-induced post-breakdown phenomena. The research discussed here is primarily work performed during the past 12 years at The University of Tennessee Space Institute, Center for Laser Applications.

Laser-induced optical breakdown of gases and subsequent inverse Bremsstrahlung-laser heating of gases are known to generate high-temperature plasma. Such plasma shows rich emission spectra. At early times following plasma formation, the spectroscopic signature consists of atomic and ionic products of the parent gas. Subsequently, recombination

Laser-Induced Breakdown Spectroscopy: Fundamentals and Applications, ed. Andrzej W. Miziolek, Vincenzo Palleschi and Israel Schechter. Published by Cambridge University Press. © A. W. Miziolek, V. Palleschi and I. Schechter 2006. A. W. Miziolek's contributions are a work of the United States Government and are not protected by copyright in the United States.

molecular spectra occur [1]. Several years after the first reported observation of optical breakdown [2], the spectroscopy study of laser-induced optical breakdown is now better known as laser-induced breakdown spectroscopy (LIBS).

In this chapter it is of interest to investigate laser-induced optical breakdown that is achieved by the use of laser radiation of nominally a few nanoseconds in duration. Specifically the emphasis is on the characterization of the so-called post-breakdown micro-plasma decay. Early in the plasma decay, electron number densities can be inferred by measuring the width of Stark-broadened atomic lines. The electron temperatures of the laser-induced plasma can be inferred from Boltzmann plots. At later times in the plasma decay, measurement of recombination spectra can be used to characterize the temperature field. Analysis of sufficiently isolated molecular spectra can yield temperature information. Frequently one has the formidable task of investigating overlap-spectra of different atomic and molecular species. Appropriate simulation of these emission spectra allows one to infer both number density and temperature of the plasma. The micro-plasma decay also includes typical fluid dynamics phenomena. These fluid dynamics phenomena are driven by the specific deposition of laser energy during the laser pulse. This chapter first addresses laser-beam focusing, followed by analysis methods of atomic and molecular spectra. Subsequently, non-equilibrium air radiation (NEQAIR) is used to infer plasma temperature and species number densities. Further, the laser spark phenomena are studied in laboratory air by use of a two-dimensional, axially symmetric, time-accurate computational fluid dynamic model.

For the experimental investigations discussed here, optical breakdown plasma is generated by focused Nd:YAG infrared laser radiation of 1–100 TW cm^{-2} irradiance. Typically, rich emission spectra are observed. Time-resolved spectroscopy techniques are applied to characterize the decaying laser-induced plasma. Initial spectroscopic signatures show contributions from free-electron radiation and atomic species emissions. By measuring atomic line broadening and line shifts, the initial plasma state of laser-generated plasma in hydrogen gas is characterized using Balmer series emissions [3–6]. Measurements of spectroscopic features of the plasma decay show primarily recombination molecular spectra. Diatomic molecular spectra of, for example, CN, C_2, N_2^+, OH, and NO, are recorded following optical breakdown in selected gas mixtures and laboratory air [7–13]. From these measurements, temperature, and species density information are inferred. In addition, shadow graphs of optical breakdown are recorded to show development of the shock wave and shock wave propagation, including vortex formation [14], and these phenomena are simulated using computational fluid dynamic modeling [15,16].

4.2 Laser-induced ignition application

Applications of laser-induced breakdown include laser spark ignition of combustible gases [17, 18]. Following the initial pulsed ruby studies, laser ignition was extended into the ultraviolet and infrared and, as a result, new laser ignition mechanisms were discovered. These mechanisms can, in general, be grouped into four categories [19, 20]: (1) thermal ignition, where the combustible gas is heated by the presence of the laser beam to the auto-ignition temperature (this type of ignition usually involves infrared wavelengths and c/w lasers); (2) photochemical ignition, in which the absorption of photons initiates chemical reactions that lead to the formation of free radicals, which drive the combustion processes; (3) non-resonant spark ignition, in which a laser beam (or pulse) is focused to sufficient irradiance to cause dielectric breakdown of the gas; and (4) resonant spark ignition, in which a single- or multi-photon resonance is exploited to aid in the initial processes of optical breakdown. A thorough review of these four mechanisms is given in the article by Ronney [20]. Laser spark ignition has been extensively investigated at The University of Tennessee Space Institute, Center for Laser Applications [21–26]. In this chapter, however, discussions concentrate on non-combustible gas mixtures.

4.3 Focal volume irradiance distribution

Detailed characterization of laser beam focusing is instrumental for the study of laser-induced optical breakdown. Specifically, knowledge of the focal volume irradiance distribution is required. Spectroscopic investigations depend on the computed and experimentally verified spatial irradiance distribution. Fluid dynamics modeling is critically dependent on the initial conditions derived from the peak intensity irradiance distribution in the focal volume.

Typically, high peak power is needed to achieve optical breakdown. To increase the irradiance of a focused laser beam, expansion of the source beam is usually indicated. However, since spherical aberrations of a single lens typically increase the focal spot size relative to the diffraction-limited value, a decrease in the focused irradiance can be expected with continued increase in the diameter of the focus lens. Consequently, the highest irradiance will occur for an optimum beam diameter and focal length of the lens [27].

Figure 4.1 illustrates the effects of spherical aberration in and near focus of a single lens for the fundamental Nd:YAG wavelength of 1.064 μm, and it shows the results for the irradiance map in focus of an ideal lens.

The details of the theoretical model used for computing the focal volume distribution are summarized below. Results for Gaussian beam propagation [28] for an ideal (aberration-free) lens show a focal spot diameter of $d_0 \approx 2f^{\#}\lambda$, where the f-number is defined by $f^{\#} \equiv f/D$, with f the focal length of the lens and D the diameter of the laser beam. The confocal parameter or depth of focus ($2z_R$) is given by $2z_R = 6.28(f^{\#})^2\lambda$, and the focal volume (area × confocal parameter) is $V_{focus} = 19.2(f^{\#})^4\lambda^3$. The peak irradiance at the center of the focused spot is $I_0 \approx P_0/2(f^{\#}\lambda)^2$, where P_0 is the total power of the beam.

For the experimental work discussed here, typically f-numbers in the range 4–10 are realized. For example, 4× beam expanding a nominal 6 mm diameter Nd:YAG beam to 20 mm, and using a singlet with focal length 100 mm, gives a value of 4.2 for the f-number. For the presented axially symmetric irradiance distributions (Figure. 4.1) data for a plano-convex lens (Newport model KPX016) of focal length $f = 19$ mm are used for $f^{\#} = 3.2$ focusing of Nd:YAG infrared radiation. Spherical aberration effects are diminished for f-number 10 focusing, yet typically the peak on-axis irradiance is a factor of 2 smaller than that obtained from the aberration-free result.

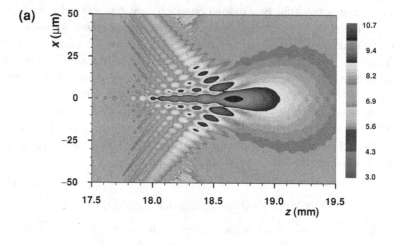

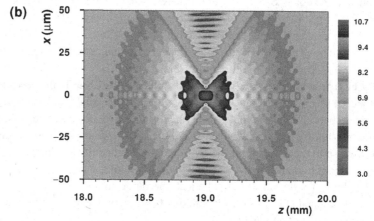

Figure 4.1. Calculated focal irradiance distributions for a plano-convex lens of focal length $f = 19$ mm for $f^{\#} = 3.2$ focusing in (a), including spherical aberrations. For comparison, (b) shows the results for Gaussian beam focusing of the ideal (aberration-free) lens. The aspect ratio is 1:20 for the logarithmic maps.

The theoretical model is developed from the scalar Kirchhoff theory of diffraction. The three-dimensional (Fresnel) distribution near focus of waves emanating from a circular aperture of radius a can be evaluated from the Debye integral yielding the usual Fresnel–Kirchhoff integral [29]. At the distance z from the circular aperture and at the distance r from the optical axis, the cylindrically symmetric amplitude distribution can be written as (omitting constant phase factors)

$$U(r, z) = \frac{k}{z} \int_{0}^{a} A(r') \exp\left\{-ik\frac{r'^2}{2z}\right\} J_0\left(\frac{krr'}{z}\right) r'\mathrm{d}r', \tag{4.1}$$

where $k = 2\pi/\lambda$ is the wavenumber for light of wavelength λ, and J_0 is the Bessel function of zero order. The amplitude distribution $A(r')$ at the aperture can be written as

$$A(r') = A(r')_{\text{Beam}} A(r')_{\text{Lens}}, \tag{4.2}$$

where the input beam distribution is $A(r')_{\text{Beam}}$, and $A(r')_{\text{Lens}}$ is the phase retardation of the lens.

The collimated Gaussian input irradiance profile [28] with beam waist w at the lens and on-axis irradiance I_0 (for a large radius-of-curvature spherical wave) is described by

$$A(r')_{\text{Beam}} = \sqrt{I_0} \exp\left\{-\left(\frac{r'}{w}\right)^2\right\}. \tag{4.3}$$

The phase retardation of a thin lens of refractive index n_{Lens} is calculated from the varying thickness of the lens from axis to margin [30]. In terms of the radii of curvature, R_1 and R_2, the lens factor is

$$A(r')_{\text{Lens}} = \exp\left\{-ik(n_{\text{Lens}} - 1)\left(R_1\left[1 - \sqrt{1 - \frac{r'^2}{R_1^2}}\right] - R_2\left[1 - \sqrt{1 - \frac{r'^2}{R_2^2}}\right]\right)\right\}. \tag{4.4}$$

The irradiance profiles are obtained by numerically solving the Fresnel–Kirchhoff integral (equation (4.1)) using equations (4.2)–(4.4). Effects of spherical aberrations from the thin lens are included. Invoking the standard (paraxial) approximation by expanding the square root expressions in equation (4.4) in a Taylor series and neglecting higher-order terms yields the standard, aberration-free lens term

$$\overline{A}(r')_{\text{Lens}} = \exp\left\{-ik\frac{r'^2}{2f}\right\}, \tag{4.5}$$

where the focal length f of the thin lens in terms of the radii of curvature is

$$\frac{1}{f} = (n_{\text{Lens}} - 1)\left[\frac{1}{R_1} - \frac{1}{R_2}\right]. \tag{4.6}$$

An alternative method of calculating the effects of spherical aberrations of a thin lens consists of developing the lens contribution (equation (4.4)) in terms of the circle polynomials of Zernike and evaluating the various aberrations. The preferred method was to numerically solve the Fresnel–Kirchhoff integral (equation (4.1)) without decomposing the lens aberrations. The diffraction integral was evaluated point by point by direct integration. The results from this integral-solver procedure are consistent with single lens results that are obtained applying convolution-integral and fast Fourier transform methods [31].

Note the apparent asymmetric (about the focal plane) irradiance distribution that is caused by the spherical aberrations (see Figure 4.1a). The on-axis peak irradiance is lower by a factor of around 6 and is more spatially distributed than is found for the ideal (aberration-free) lens (see Figure 4.1b). Also, multi-mode output of a Q-switched Nd:YAG laser may increase the spot size by the so-called M-factor (e.g. $M \approx 1.4$ for the YG680S-10 Continuum Nd:YAG laser) and further decrease the peak irradiance levels. The modeling discussed here primarily addresses the spatial characteristics of a pulsed laser beam; however, the temporal pulse shape can further affect generation of optical breakdown. Specifically, micro-plasma may be generated near one of the peaks that are a result of spherical aberrations. Consequently, the beginning of the laser-induced plasma process is spatially and temporally distributed when using low *f*-number focusing and irradiance levels well above threshold for optical breakdown to occur. For computational fluid dynamic modeling, summarized in Section 4.7, an asymmetric laser energy deposition profile is used to generate the initial conditions for the numerical studies.

4.4 Hydrogen Balmer series atomic spectra

For laser-induced, optical breakdown plasmas in nominal high-pressure gases, the time-varying electron–ion recombination region is investigated by measurement of the electron density and electronic excitation temperature. The relatively large Stark-broadening of the H-atom spectra can be used to determine the electron number density. The excitation temperature can be inferred by use of line-to-continuum ratio measurements and by use of Balmer series Boltzmann plots [3–6].

In the first few microseconds following laser breakdown, electron number densities are found in the range of 10^{19}–10^{16} cm^{-3}. The corresponding excitation temperatures found are in the range of 100 000–6600 K. The simulation of individual atomic lines is based on the use of theoretical profiles [32], and is based on the use of Boltzmann plots except at early times in the plasma decay [3]. The tabulated profiles are weakly temperature dependent for a given density.

Figure 4.2 shows experimental and simulated results for a delay of 5 ns after optical breakdown. Optical breakdown was generated by focusing 1064 nm Nd:YAG laser radiation of 150 mJ energy per 10 ns pulse width into a cell containing research grade hydrogen gas at a pressure of 810 ± 25 Torr.

The full-width at half-maximum (FWHM) $\Delta\lambda_{1/2}$ of the hydrogen α (H$_\alpha$) line is 25.4 nm, and the line shift $\Delta\lambda$ amounts to 2.7 nm. For the fitting procedure, the relation $\Delta\lambda_{1/2} \approx n_e^{2/3}$ is used to interpolate and extrapolate the tabulated theoretical profiles [32] to find the electron number density n_e for the measurement series. For the results illustrated in Figure 4.2, $n_e = 1.1 \times 10^{19}$ cm^{-3} was initially inferred [3–5].

An investigation of the post-breakdown gas-dynamic expansion showed that the electron number density is related to temperature by $n_e \cong T^{1/(\gamma-1)}$ and, assuming frozen chemistry and internal energy exchange, a slope of $\gamma - 1 = 0.667$ was found from a plot of log T vs. log n_e [3]. It should be noted that temporally and spatially resolved hydrogen spectra that show regions of large n_e concentrations do not necessarily correspond to regions of large, excited state concentrations of hydrogen atoms [4].

For the measured, broad hydrogen Balmer series lines, use of Oks' generalized theory of Stark broadening [33] is, however, recommended. Application of the generalized theory for LIBS has been of recent interest in the analysis of measured Balmer series H$_\beta$ line profiles

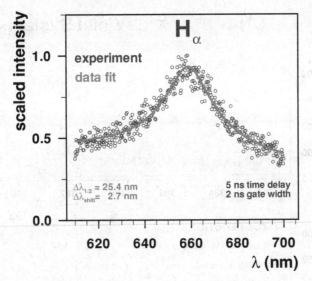

Figure 4.2. Measured (dotted) and fitted (solid) hydrogen α line early in the micro-plasma decay following laser-induced optical breakdown.

following optical breakdown [6]. The H_β line is approximately a factor of four wider than the H_α line and, therefore, is spectroscopically too broad to be used for electron number density diagnostics for number densities larger than approximately 10^{18} cm^{-3}, or for the optical breakdown generated above (150 mJ in 10 ns focused into a cell containing hydrogen gas at a pressure of 810 Torr) for delays smaller than 0.1 μs. The detailed analysis of H_β profiles, and comparison with results from the measured H_α profiles [6], show that deduced values for the electron density are consistent with each other early in the plasma decay. For the electron number very early in the plasma decay (see Figure 4.2), n_e is found to be in the range of $(0.7-1.0) \times 10^{19}$ cm^{-3} from the H_α line width, and $(0.4-0.8) \times 10^{19}$ cm^{-3} from the H_α red shift [6]. The discrepancy is caused by the significant temperature dependence of both widths and shifts, and the experimental uncertainty in temperature translates into the uncertainty in n_e. Most importantly, experimental widths of the H_α and H_β lines, and experimental shifts of the H_α line, show consistent results early in the plasma decay only when using the generalized theory to infer electron number density.

4.5 Diatomic molecular emission spectra

In the near-ultraviolet wavelength region of nominally 200–325 nm, the emission spectrum is composed of overlapped electronic transitions primarily of NO, N_2 and OH and, to a lesser degree, of O_2 and atomic species. The spectral analysis of superimposed spectra of such multiple species is most challenging, and the computer code for non-equilibrium air radiation (NEQAIR) is utilized in the spectral analysis of the transient micro-plasma. For

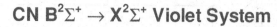

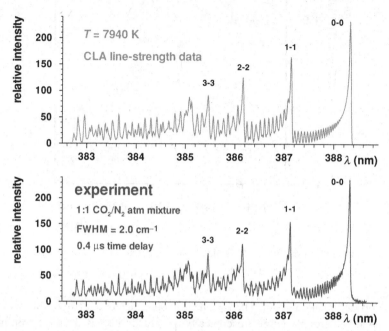

Figure 4.3. Measured and simulated emission spectrum. The measured set shows an average of 40 individual single-shot spectra; the simulated temperature obtained was $T = 7940 \pm 210$ K.

isolated molecular spectra, however, the use of so-called line-strength files for a particular molecular transition is recommended. The theoretical model and simulation used are based on computing the polarization- and angle-averaged, spontaneous-emission intensity of a specific, isolated molecular transition. The model has been tested, for example, for the CN violet system [7], the C_2 Swan system [8], and the first negative N_2^+ system [9]. It has also been applied recently in the analysis of OH emission spectra following laser-induced optical breakdown in air [10, 11], and in the analysis of Swan spectra following laser-induced breakdown at and near the surface of graphite [12, 13]. The spectroscopic temperature is obtained by fitting the entire measured emission spectrum.

Figure 4.3 shows a comparison of measured and computed spectra for the $\Delta v = 0$ sequence of the CN violet system for 2.0 cm^{-1} spectral resolution data. This spectrum was measured subsequent to nominal 1 ns pulse-width, 308 nm excimer laser-induced optical breakdown.

The emission intensity $\bar{I}$ produced by laser-induced breakdown is proportional to the transition strength $S(n'v'J', n''\,v''J'')$ and the number $N(n'v'J')$ of excited molecules by

$$\bar{I} = \frac{64\pi^4 c \tilde{v}^4}{3(4\pi\varepsilon_0)} S(n'v'J', n''v''J'')N(n'v'J'). \tag{4.7}$$

The set of single-primed quantum numbers n', v', and J' is associated with the electronic, vibrational, and rotational levels, respectively; the double-primed quantum numbers describe the lower level. The transition wavenumber $\tilde{v}$ is given by the term-value difference $F(n'v'J') - F(n''v''J'')$. The transition strength is factorized and written as the product of the electronic transition strength $S_{\text{electronic}}(n'v', n''v'')$, the Franck–Condon factor $q(v', n'')$, and the Hönl–London factor $S(J', J'')$:

$$S(n'v'J', n''v''J'') = S_{\text{electronic}}(n'v', n''v'')q(v', v'')S(J', J''). \tag{4.8}$$

For isolated molecular transitions, the Franck–Condon factors are found by numerically solving the radial Schrödinger equation, and the Hönl–London factors are obtained by numerical diagonalization of the rotational and fine-structure Hamiltonian. Simulation of a measured spectrum makes use of these results, and the results are applied to determine temperature.

For measurement of spectra, typically an array detector is used. The intensity that falls within one pixel is modeled by

$$I_{\text{pixel}} = C \sum_{v'J''v''J''} \tilde{v}^4 q(v', v'')S(J', J'') \exp\{-hcF(n'v'J')/(k_B T)\}, \tag{4.9}$$

where all the factors that do not vary for the observed spectrum are collected in the constant C; in addition I_{pixel} is multiplied (not explicitly indicated in equation (4.8)) by the slit-function $SF(\tilde{v}, \tilde{v}_{\text{pixel}})$. In the analysis, the equilibrium temperature T is obtained by use of nonlinear curve-fitting methods such as downhill simplex. For high-temperature and high-density breakdown spectra, and for typical gate-width times and gate-delays used in time-resolved measurements of the recombination spectra, thermal equilibrium of vibrational and rotational modes is assumed.

Figure 4.4 shows the simulated and measured spectra subsequent to laser-induced breakdown using Nd:YAG 1064 nm radiation of 7.5 ns pulse-width focused into a cell containing CO at 3.25 psi (168 Torr) above atmospheric pressure. These spectra are recognizable within the first microsecond after optical breakdown. Typically, for delays of 30 μs the Swan spectra become prominent, without presence of atomic carbon lines. A spectral resolution of 11 cm^{-1} was used, and a temperature in excess of 6000 K was fitted.

4.6 Simulation by use of the program NEQAIR

The program NEQAIR is widely used to predict and interpret optical spectra from air plasmas [34–41]. The majority of the code's applications concern atmospheric radiation, but it is also applied in simulation facilities and has other applications such as for the prediction of laser sustained plasmas or arcjet flows. The recent non-equilibrium air radiation code NEQAIR96 [42] includes significant improvements over the original version [34] and the updated versions [35, 36]. The program is a line-by-line and a line-of-sight code to compute the emission and absorption spectra for atomic and diatomic molecules and the transport of

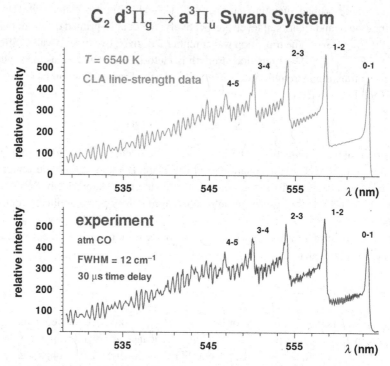

Figure 4.4. Measured and simulated $\Delta v = -1$ sequence of the C_2 Swan system.

radiation through a non-uniform gas mixture to a surface. The NEQAIR96 computer code shows improvements in accuracy and capability by including the rotational Hamiltonian matrix formalism for calculating rotational energy levels and Hönl–London factors.

Simulation of experimental spectra from air breakdown is, first, based on the use of calculated number densities of a dry air plasma at a pressure of 100 kPa [43] and, second, by computing the plasma composition by use of the NASA chemical equilibrium code [44, 45] (see Figure 4.6). Figure 4.5 exhibits results obtained 50 μs after optical breakdown in air.

For the simulation results shown in Figure 4.5, tabulated number densities are used. These number densities are calculated [43] by first characterizing, from a thermodynamic point of view, the chemical system by its temperature and pressure, and consequently, by minimizing the Gibbs free energy together with conservation of the chemical elements and the principle of electrical neutrality. Air is considered to be a mixture of nitrogen, oxygen and argon at room temperature, with all other species neglected. Starting from one mole of air at room temperature, the number densities of the species N, O, Ar, N^+, O^+, Ar^+, N_2, N_2^+, O_2, O_2^+, NO, NO^+, and the free electrons e, are given as functions of temperature. These data are used for computation of synthetic spectra; i.e. a Boltzmann distribution with

Table 4.1. *Neutral species number densities* (cm^{-3}) *for* $T = 6000$ K *and* $T = 6700$ K

T (K)	N	O	Ar	N_2	O_2	NO
6000	2.3×10^{17}	3.9×10^{17}	8.5×10^{15}	6.6×10^{17}	3.3×10^{14}	1.0×10^{16}
6700	4.6×10^{17}	3.1×10^{17}	6.8×10^{15}	3.6×10^{17}	7.8×10^{13}	4.5×10^{15}

Table 4.2. *Ionized species and free electron number densities* (cm^{-3}) *for* $T = 6000$ K *and* $T = 6700$ K

T (K)	N^+	O^+	Ar^+	N_2^+	O_2^+	NO^+	e
6000	0	0	0	0	0	2.7×10^{14}	2.7×10^{14}
6700	4.7×10^{13}	4.8×10^{17}	0	0	0	4.4×10^{14}	5.4×10^{14}

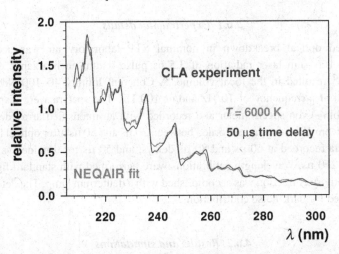

Figure 4.5. Measured and NEQAIR-simulated NO spectrum.

equal translational, rotational, vibrational and electron temperatures was assumed in our analysis.

Tables 4.1 and 4.2 summarize the temperature and number densities that were utilized as input for the computer code NEQAIR8; temperatures of $T = 6000$ K and $T = 6700$ K were found for delays of 50 µs and 25 µs, respectively.

For applications of the NEQAIR96 program in detailed analyses of laser-induced breakdown spectra, the computed files for the rotational energy levels are replaced with line-strength files that are generated for selected transitions of specific diatomic molecules. These modifications are required for an accurate prediction of wavelength positions of measured, highly excited molecular spectra. Alternatively, if a particular species dominates the emission spectrum, the code's background predictions can be combined in the data reduction with an individual, diatomic molecule for which an accurate line-strength file is available.

In the following sections, we will discuss the applicability and extension of the NEQAIR96 program that has been modified for computations on machines other than a CRAY C-90 computer, for example, for the Silicon Graphics Power Challenge L machine or a nominal IBM Personal Computer with a Microsoft or Linux Operating System. For the adaptation of the code, changes were made owing to different FORTRAN compiler implementations on these platforms [36]. In addition, Lambda doubling has been included for selected (externally provided) line-strength files [46, 47]. Results from NEQAIR96 computations are presented below to predict and simulate time-resolved measurements of recombination emission spectra in the near uv wavelength region, in particular between 305 nm and 322 nm [10, 11, 48–51].

4.6.1 Experimental details

Laser-induced optical breakdown in nominal STP laboratory air was accomplished by focusing 1064 nm laser radiation of 3.5 ns pulse width; an irradiance of typically 10 TW cm^{-2} resulted in the focal volume. A Coherent Infinity 40–100 Nd:YAG laser was operated at a frequency of 10 Hz and/or 100 Hz. The spectra were dispersed with a 0.64 m Jobin–Yvon spectrometer and recorded with an intensified array detector. The gate-widths were 10 μs for data recorded between 20 μs and 50 μs after optical breakdown, 20 μs for data recorded at 60 μs and 80 μs delays, and 50 μs for time delays of 100 μs, 200 μs, and 250 μs. Wavelength calibrations were performed with standard light sources and the sensitivity correction was accomplished with a deuterium lamp. The detector's data were corrected for dark-noise contribution.

4.6.2 Results and simulations

In the experimental investigation, superposition spectra of OH with significant contributions of the N$_2$ Second Positive system were recorded at delay times of typically 20–40 μs. Such combined spectra were analyzed by the use of the NEQAIR96 program to infer temperature and number density. The species number densities (input to NEQAIR96) were calculated by the use of the NASA chemical equilibrium code [44, 45].

Figure 4.6 shows the computed concentration of the species of interest in this work. Figure 4.7 shows the experimental and fitted spectrum at a delay of 30 μs from the optical breakdown event. The spectral resolution was 32 cm^{-1}. The major OH contributions are

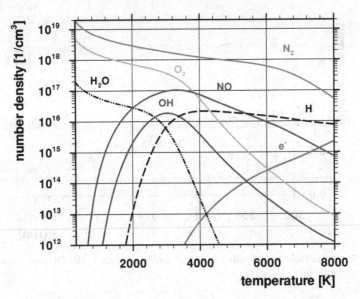

Figure 4.6. Selected air species concentrations versus temperature for atmospheric pressure.

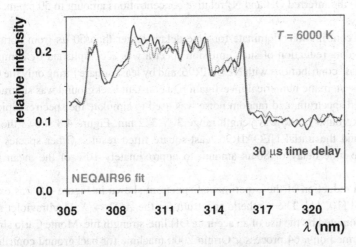

Figure 4.7. Comparison of measured (solid) and synthetic (dotted) emission spectra. The C_2 Second Positive and OH emissions dominate in the 305 nm to 322 nm wavelength range.

illustrated near 306 nm and 309 nm, and the major N_2 Second Positive contributions are illustrated near 314 nm and 316 nm in the figure. The obvious differences of the spectroscopic fit and of the recorded spectrum are the result of inaccuracies in computed wavelength positions and inaccuracies in species number densities. For the laboratory air, mole fractions of 0.78, 0.20, 0.004, and 0.016 for N_2, O_2, CO_2, and H_2O, respectively, were

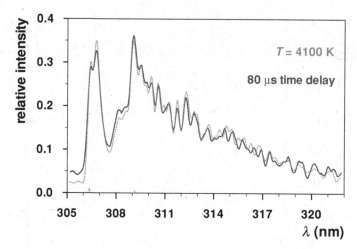

Figure 4.8. Measured (solid) and synthetic (dotted) emission spectra. The OH emissions dominate for time delays larger than 50 μs.

assumed at room temperature ($T = 298$ K) and atmospheric pressure ($p = 1$ atm). At 6000 K (Figure 4.6) the inferred OH and N_2 relative concentrations amount to 50 p.p.m. and 50%, respectively.

The OH contributions dominate for time delays larger than 50 μs from optical break-down. The data reduction of such emission spectra was accomplished by computing the "background" contributions with NEQAIR96, and by least-square fitting only the OH emissions. In a systematic numerical investigation, a constant background was subtracted from the recorded spectrum, and random noise was used to simulate the spectroscopic features of the background in the wavelength range 305–322 nm. Figure 4.8 shows the recorded spectrum and the initial NEQAIR96 least-square fitted results. Other species contributions at the time delay of 80 μs amount to approximately 10% of the mean measured intensity.

The data reduction of the experimental spectrum shown in Figure 4.8 was extensively investigated [10, 11]. The synthetic spectrum for the $A^2\Sigma \leftrightarrow X^2\Pi$ ultraviolet system of OH was computed by the use of an accurate OH line-strength file. Monte Carlo simulations were performed on a 64 processor Origin 2000 machine for background contributions of 10% ($x = 0.1$), 20% ($x = 0.2$), and 30% ($x = 0.3$) of the mean intensity modulated by a normally distributed random noise. Figure 4.9 shows the simulation results. As can be seen in the figure, the average temperatures are almost identical while the variances scale with the magnitude of the background contributions.

Comparison with the initial NEQAIR96 least-square fitting (see Figure 4.8) shows that the inferred spectroscopic temperatures differ by approximately 10%, or 400 K. Accordingly, the OH concentration differs by a factor of 3 at this time delay. The use of an accurate OH line-strength file yields a lower temperature for the data that were measured

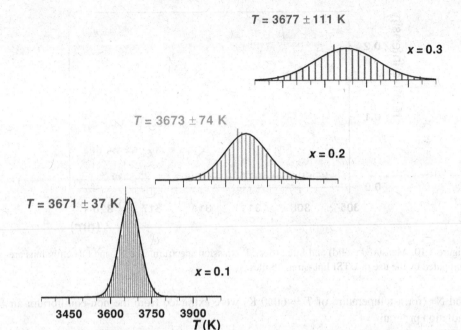

Figure 4.9. Temperature distributions for different error magnitudes obtained from Monte Carlo simulations in the wavelength range 305–322 nm, 80 μs after laser-induced optical breakdown of air.

80 μs after optical breakdown, and the number density for OH amounts to approximately 1.6×10^{16} cm^{-3}.

The application of the NEQAIR96 code to extract only the background contribution is preferred, owing to the uncertainties in the theoretical wavelength positions. Accurate line-strength files for the N_2 Second Positive system and for OH were used to further investigate the spectrum that was recorded at a delay of 30 μs. An improved spectral fit would result if the synthetic spectra were computed from accurate line-strength files for the OH, N_2 Second Positive system, and other species contributions. The spectral variations from other species cause wavelength inaccuracies in the fitting of the measured spectra.

The program NEQAIR96 uses a matrix representation of the diatomic Hamiltonian that is expressed in terms of the molecular parameters and then numerically diagonalized to give the term values. For the purpose of halving the dimension of each Hamiltonian matrix, Λ-doubling [52–56] terms are not included in the NEQAIR96 calculation of the matrix elements. Neglect of Λ-doubling does indeed halve the dimensions of the diatomic Hamiltonian matrix. However, in the establishment of the line-strength files for selected diatomic molecules, Λ-doubling is included, with subsequent adaptation and inclusion of the line-strengths in the NEQAIR program.

Figure 4.10 shows the results of fitting the measured spectrum with UTSI's line-strength files that were used in the NEQAIR96 code. The values for the partition functions of OH

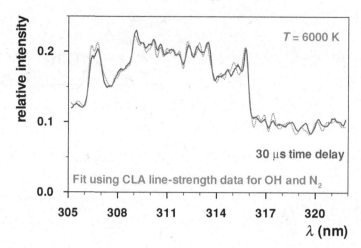

Figure 4.10. Measured (solid) and fitted (dotted) emission spectrum. The N_2 and OH emissions are computed by the use of UTSI line-strength files.

and N_2 (for a temperature of $T = 6000$ K) were extracted from the non-equilibrium air radiation program.

The results indicate significant potential of the non-equilibrium air radiation program for applications in optical breakdown plasma diagnostics and in combustion studies [57–59]. High-temperature molecular emissions from optical breakdown plasma appear to be predicted reasonably well, and the synthetic NEQAIR96 spectra are particularly useful for initial investigations. The analysis of optical breakdown spectra that show contributions from atomic emissions would require extensions to the program, for example, the inclusion of hydrogen Stark line-broadening. The indicated replacement of the line-strength files allows us accurately to predict selected molecular emissions.

4.7 Computational fluid dynamic simulations

Laser-induced optical breakdown generates fluid physics phenomena. The simulation of these phenomena following optical breakdown is challenging considering that the time scale to be covered spans 6 orders of magnitude, from nanosecond to millisecond, and the spatial scale spans 10 orders of magnitude, from 10^{-11} m^3 to 10^{-1} m^3. The developed computational model includes a kinetic mechanism that implements plasma equilibrium kinetics in ionized regions, and non-equilibrium, multi-step, finite-rate reactions in non-ionized regions. The computational fluid dynamics model time-accurately predicts species concentrations, free electron number density decay, blast wave formation and dynamics, and flow field interactions of the laser spark decay [15, 16].

Owing to the generation of a nominally dense plasma and the resulting high collision frequency, local thermal equilibrium is assumed to be satisfied subsequent to the termination

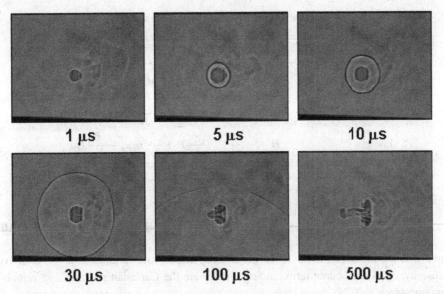

1 µs   5 µs   10 µs

30 µs   100 µs   500 µs

Figure 4.11. Shadow graph records for room-air optical breakdown induced by use of 50-mJ break-down pulses of Nd:YAG 1064 nm laser radiation. The laser beam propagation is from left to right. The shock wave in the picture labeled 100 µs delay is caused by reflection from the surface at the bottom of the image, indicated by the dark area.

of laser pulse. An asymmetric laser energy deposition profile is used to generate the initial condition for the numerical studies. Included are characteristics of the laser-induced breakdown process and results from focal volume investigations [21, 27, 62–68]. Effects of ionization and dissociation are included in the simulation [69] that is used to predict temperature, pressure and velocity profiles at selected times. The model involves solution of the two-dimensional axial-symmetric transport equations of mass, momentum, and energy. These equations are reproduced below.

Shadow graph measurement techniques [60, 61] have been applied to investigate post-breakdown effects. A short pulse light source is used to investigate the temporal development of the shock wave and the onset of typical fluid dynamics phenomena. Figure 4.11 shows shadow graphs recorded subsequent to laser-induced optical breakdown of air. These shadow graphs are generated using a 308 nm back-light source (XeCl excimer laser, 10 ns pulse-width) and by video-recording the images that are projected on a screen placed in the beam path of the back-light source. The pulsed laser is focused from the left for each of these images (36 mm × 48 mm).

In the modeling of the laser-spark decay, the CFD-ACE software package (CFD Research Corporation, Huntsville, AL) is used to solve the fluid equations. The CFD software package utilizes finite element methods to solve the mass, momentum, and energy transport equations [70]. The transport equations inherently satisfy their respective conservation laws, and have

Table 4.3. *Transport equations*

Transport equation	ϕ	Γ	S_ϕ
Mass	Y_α	0	$\dfrac{\partial}{\partial x_j} J_{\alpha j} + M_\alpha \omega_\alpha$
Momentum	u_i	0	$-\dfrac{\partial}{\partial x_i} P + \dfrac{\partial}{\partial x_j} \tau_{ij}$
Energy	H	$\dfrac{\kappa}{c_p}$	$\dfrac{\partial}{\partial T} P + \dfrac{\partial}{\partial x_i} \tau_{ij} u_j + \dfrac{\partial}{\partial x_j} J_{\alpha j} h_\alpha$

the general form

$$\frac{\partial}{\partial t}(\rho \phi) + \frac{\partial}{\partial x_j}(\rho u_j \phi) = \frac{\partial}{\partial x_j} \Gamma \frac{\partial}{\partial x_j} \phi + S_\phi, \tag{4.10}$$

where ρ is the mass density, ϕ is the generalized scalar flow variable, Γ is the effective diffusivity, S_ϕ is the source term, and x_j and u_j are the Cartesian position and velocity components.

Table 4.3 displays the evaluation of the generalized parameters for each of the transport equations.

The terms M_α, Y_α, and ω_α are respectively the mass, mass fraction, and molar production rate of species α, P is the gas pressure, T is the gas temperature, τ_{ij} is the viscous stress tensor, H is the total enthalpy, κ is the thermal conductivity, c_p is the specific heat, $J_{\alpha j}$ is the total diffusive mass flux for species α in the j-direction, and h_α is the specific enthalpy of species α. Assuming Newtonian fluids, the viscous stress tensor is reduced to a function of velocity gradients and the scalar dynamic viscosity (μ). The discretization of the partial differential equations (along with initial and boundary conditions) onto a computational grid leads to the governing finite difference equations (FDE). The FDE are solved for the flow variables by a variation of the Semi-Implicit Method of Pressure-Linked Equations Consistent (SIMPLEC) algorithm [70, 71].

The laser-spark decay model predicts the quasi-adiabatic expansion of the high-pressure plasma. This expansion results in the formation of an asymmetric shock front that becomes increasingly spherical over time. Shadow graphs are a common photographic technique used to record fluid flow patterns. Figure 4.12 shows the comparison of recorded and computed shadow graphs. The computed image (left) is generated by use of the relative luminosity, L, for a shadow graph [72]:

$$L = S \int_{z_0}^{z_1} \left(\frac{\partial^2}{\partial x^2} n + \frac{\partial^2}{\partial y^2} n \right) dz, \tag{4.11}$$

where n is the index of refraction and S is a scaling factor.

As the blast wave moves outward, the temperature and pressure of the inner gas region decrease. The pressure of the high-temperature inner gas decreases until it is low enough

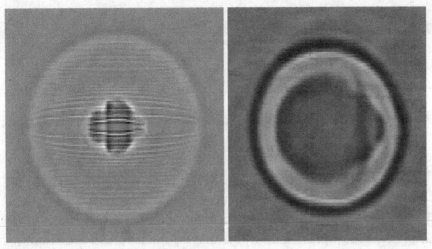

Figure 4.12. Simulated (left) and measured (right) shadow graphs 5 μs after optical breakdown of air.

to overcome the outward momentum of its periphery. The subsequent collapse of the two opposed fronts generates vortices that form a toroidal pattern about the optic axis. Because of the asymmetric initial conditions and subsequent asymmetric expansion, the low-pressure region also causes an asymmetric collapse along the optic axis. The spherical asymmetry in the laser-spark profiles leads to the collapsing front moving towards the laser side (see the left-hand part of Figure 4.12) being much stronger. Eventually the collapsing front moving toward the laser pushes through its counterpart. This flow pattern is characteristic of laser-spark decays. Computational and experimental shadow graphs of these flow fields are compared in Figure 4.13, showing a record of 4 mm × 4 mm at a delay of 100 μs after optical breakdown in air.

Extension to other mixtures of gases, for example mixtures of ammonia and oxygen, adds additional levels of complexity by the increase in the number of finite-rate reactions. The temperature-triggered finite-rate energy release is included in the modeling of the reaction processes. Noteworthy is that the computational fluid-dynamics model includes simultaneously the equilibrium plasma in the ionized regions of the kernel and the combustion reactions in the neutral regions. For example, time-dependent species concentrations are predicted by the computational model and, in turn, planar laser-induced fluorescence (PLIF) measurements show agreement with predictions of molecular species concentrations [21–26].

4.8 Summary

The experiments and simulations discussed in this chapter show the importance of computational physics in modeling post-breakdown phenomena. The computational fluid dynamic

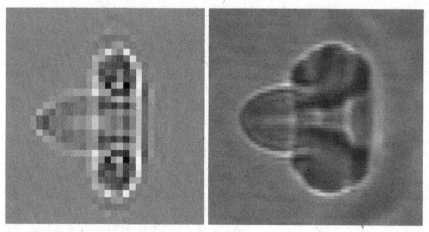

Figure 4.13. Simulated (left) and measured (right) shadow graphs 100 μs after optical breakdown of air.

model clearly indicates that the asymmetry of the laser energy deposition as well as ionization affects the fluid properties. A detailed characterization of the laser focusing is important for obtaining the initial conditions used in the computations.

Atomic and molecular spectroscopy allow one to characterize the plasma decay phenomena. The application of the NEQAIR program is shown to be useful in the study of laser-induced post-breakdown phenomena. Spectroscopic temperature measurements frequently comprise the analysis of overlapped spectra. Extensions are indicated for the NEQAIR program to include line-strength files.

The computationally predicted fluid phenomena agree with various flow patterns characteristic of laser spark decay by direct comparison with experimental shadowgraph records. These flow patterns develop from the asymmetric laser energy deposition caused by laser-induced optical breakdown processes, and are typically observed in investigations of laser-induced optical breakdown with LIBS.

The author thanks Dr. Y.-L. Chen, Dr. I. G. Dors, Dr. G. Guan, Dr. D. R. Keefer, Dr. J. W. L. Lewis, Dr. E. Oks, Dr. D. H. Plemmons, Dr. W. Qin, Dr. Y. Tang, Mr. J. O. Hornkohl, Mr. F. A. Schwartz, and Mr. N. W. Wright, for their interest and contributions to this work. Also, thanks to NASA Ames Research Center for making available the NEQAIR96 program, along with detailed comments by Dr. E. E. Whiting regarding the applications and extensions of the code.

This material is based upon work that is supported in part by the National Science Foundation under Grant No. CTS-9512489, in part by a grant of High Performance Computer (HPC) time from the Department of Defense HPC Center, AEDC at Arnold AFB, on the Origin 2000 computer, in part by UTSI and UTSI's Center for Laser Applications, and in part by UTSI's Space Grant.

4.9 References

[1] K. Grant and G. L. Paul, *Appl. Spectrosc.*, **8** (1990), 1439.

[2] P. D. Maker, R. W. Terhune and C. M. Savage, in *III International Conference on Quantum Electronics* (Paris, 1963).

[3] C. G. Parigger, J. W. L. Lewis and D. H. Plemmons, *J. Quant. Spectrosc. Radiat. Transfer*, **53** (1995), 249 (and references therein).

[4] C. G. Parigger, D. H. Plemmons and J. W. L. Lewis, *Appl. Opt.*, **34** (1995), 3325.

[5] C. G. Parigger and D. H. Plemmons, in *OSA Trends in Optics and Photonics (TOPS), Laser Induced Plasma Spectroscopy and Applications, OSA Technical Digest, Postconference Edition*, vol. 81 (Optical Society of America, Washington, DC, 2002), pp. 112–113.

[6] C. G. Parigger, D. H. Plemmons and E. Oks, *Appl. Opt.*, **42** (2003), 5992.

[7] J. O. Hornkohl, C. G. Parigger and J. W. L. Lewis, *J. Quant. Spectrosc. Radiat. Transfer*, **46** (1991), 405.

[8] C. G. Parigger, D. H. Plemmons, J. O. Hornkohl and J. W. L. Lewis, *J. Quant. Spectrosc. Radiat. Transfer*, **52** (1994), 707.

[9] C. G. Parigger, D. H. Plemmons, J. O. Hornkohl and J. W. L. Lewis, *Appl. Opt.*, **34** (1995), 3331.

[10] C. G. Parigger, G. Guan and J. O. Hornkohl, in *OSA Trends in Optics and Photonics (TOPS), Laser Induced Plasma Spectroscopy and Applications, OSA Technical Digest, Postconference Edition*, vol. 81 (Optical Society of America, Washington, DC, 2002), pp. 102–103.

[11] C. G. Parigger, G. Guan and J. O. Hornkohl, *Appl. Opt.*, **42** (2003), 5986.

[12] C. Parigger, J. Hornkohl, A. M. Keszler and L. Nemes, in *OSA Trends in Optics and Photonics (TOPS), Laser Induced Plasma Spectroscopy and Applications, OSA Technical Digest, Postconference Edition*, vol. 81 (Optical Society of America, Washington, DC, 2002), pp. 104–105.

[13] C. G. Parigger, J. O. Hornkohl, A. M. Keszler and L. Nemes, *Appl. Opt.*, **42** (2003), 6192.

[14] C. G. Parigger, D. H. Plemmons, J. W. L. Lewis, G. Guan and Y. L. Chen, private communication (1996), the University of Tennessee Space Institute, Tullahoma, TN.

[15] I. G. Dors and C. G. Parigger, in *OSA Trends in Optics and Photonics (TOPS), Laser Induced Plasma Spectroscopy and Applications, OSA Technical Digest, Postconference Edition*, vol. 81 (Optical Society of America, Washington, DC, 2002), pp. 78–79.

[16] I. G. Dors and C. G. Parigger, *Appl. Opt.*, **42** (2003), 5978.

[17] J. H. Lee and R. Knystautas, *AIAA Journal*, **7** (1969), 312.

[18] F. J. Weinberg and J. R. Wilson, *Proc. R. Soc. Lond.* A**321** (1971), 41.

[19] N. M. Witriol, B. E. Forch and A. W. Miziolek, *CPIA Publication 557*, Volume III (1990), pp. 213–217.

[20] P. D. Ronney, *Opt. Eng.*, **33** (1994), 510.

[21] D. H. Plemmons, Ph.D. dissertation, The University of Tennessee, Knoxville, TN (1996).

[22] Y.-L. Chen, Ph.D. dissertation, The University of Tennessee, Knoxville, TN (1998).

[23] G. Guan, Ph.D. dissertation, The University of Tennessee, Knoxville, TN (1999).

[24] I. G. Dors, Ph.D. dissertation, The University of Tennessee, Knoxville, TN (2000).

[25] W. Qin, private communication (2000), The University of Tennessee Space Institute, Tullahoma, TN.

[26] Y.-L. Chen and J. W. L. Lewis, *Opt. Express*, **9** (2001), 360.

[27] C. G. Parigger, Y. Tang, D. H. Plemmons and J. W. L. Lewis, *Appl. Opt.*, **36** (1997), 8214.

[28] A. E. Siegman, *Lasers* (Mill Valley, CA: University Science Books, 1986).

[29] M. Born and E. Wolf, *Principles of Optics* (Oxford: Pergamon Press, 1980).

[30] A. Nussbaum and R. A. Phillips, *Contemporary Optics for Scientists and Engineers* (Englewood Cliffs, NJ: Prentice-Hall, 1976).

[31] L. M. Smith, Ph.D. thesis, The University of Tennessee, Knoxville, TN (1988).

[32] H. R. Griem, *Spectral Line Broadening by Plasmas* (New York: Academic Press, 1974).

[33] Y. Ispolatov and E. Oks, *J. Quant. Spectrosc. Radiat. Transfer*, **51** (1994), 129.

[34] C. Park, *Technical Report NASA TM 86707*, NASA Ames Research Center, Moffet Field, CA (1985).

[35] C. O. Laux, Ph.D. thesis, Department of Mechanical Engineering, Stanford University (1993).

[36] E. E. Whiting, private communication (1995), Space Administration, Reacting Flow Environments Branch, Ames Research Center.

[37] C. Laux, R. Gessman and C. H. Kruger, in *Proceedings of 28th AIAA Thermophysics Conference*, AIAA paper 93–2802 (Orlando, FL, 1993).

[38] C. Laux, R. Gessman, B. Hilbert, and C. H. Kruger, in *Proceedings of 30th AIAA Thermophysics Conference*, AIAA paper 95–2124 (San Diego, CA, 1995).

[39] C. Laux, R. Gessman and C. H. Kruger, in *Proceedings of 26th AIAA Plasmadynamics and Lasers Conference*, AIAA paper 95–1989 (San Diego, CA, 1995).

[40] D. A. Levin, C. Laux and C. H. Kruger, in *Proceedings of 26th AIAA Plasmadynamics and Lasers Conference*, AIAA paper 95–1990 (San Diego, CA, 1995).

[41] G. Palumbo, E. E. W. R. A. Craig and C. Park, *J. Quant. Spectrosc. Radiat. Transfer*, **57** (1997), 207.

[42] E. Whiting, C. Park, Y. Liu, J. Arnold and J. Paterson, *Technical Report NASA RP-1389*, NASA Ames Research Center, Moffet Field, CA (1996).

[43] P. M. I. Boulous and E. Pfender, *Thermal Plasmas – Fundamentals and Applications* (New York: Plenum Press, 1994).

[44] S. Gordan and B. McBride, *Interim Revision NASA Report SP-273*, NASA Lewis Research Center (1976).

[45] B. McBride and S. Gordan, *Technical Report NASA Report RP-1311, Part I* (1994), and *NASA RP-1311, Part II*, NASA Lewis Research Center (1996).

[46] J. O. Hornkohl, G. Guan and I. Dors, private communication (1996), The University of Tennessee Space Institute, Tullahoma, TN.

[47] J. O. Hornkohl and C. Parigger, private communication (2002), The University of Tennessee Space Institute, Tullahoma, TN; J. O. Hornkohl and C. Parigger, Boltzman equilibrium spectrum program (BESP), http://view.utsi.edu/besp; J. O. Hornkohl, C. G. Parigger and L. Nenes, *Appl. Opt.* **44** (2005), 3686.

[48] C. G. Parigger, J. W. L. Lewis, D. H. Plemmons, G. Guan and J. O. Hornkohl, in *1996 OSA Technical Digest Series: Laser Applications to Chemical and Environmental Analysis*, vol. 3 (Washington, DC: Optical Society of America, 1996), pp. 82–84.

[49] C. G. Parigger, J. W. L. Lewis, D. H. Plemmons, G. Guan and J. O. Hornkohl, in *1996 OSA Technical Digest Series: Laser Applications to Chemical and Environmental Analysis*, vol. 3 (Washington, DC: Optical Society of America, 1996), pp. 85–87.

[50] G. Guan, C. G. Parigger, J. O. Hornkohl and J. W. L. Lewis, in *Technical Digest of the APS-SES meeting*, APS-SES paper BC.04 (Nashville, TN, 1997).

[51] J. Lewis, C. Parigger, J. Hornkohl and G. Guan, in *Proceedings of 37th AIAA Aerospace Sciences Meeting & Exhibit*, AIAA paper 99–723 (Reno, NV, 1999).

[52] G. Herzberg, *Spectra of Diatomic Molecules*, 2nd edition (New York: Van Nostrand, 1950).

[53] I. Kovacs, *Rotational Structure in the Spectra of Diatomic Molecules* (New York: American Elsevier, 1969).

[54] R. N. Zare, A. L. Schmeltekopf, W. J. Harrop and D. L. Albritton, *J. Mol. Spectrosc.*, **46** (1973), 37.

[55] J. H. V. Vleck, *Rev. Mod. Phys.*, **23** (1951), 213.

[56] R. N. Zare, *Angular Momentum* (New York: Wiley, 1988).

[57] J. E. Harrington, A. R. Noble, G. P. Smith, J. B. Jeffries and D. R. Crosley, in *Proceedings of Western States Section of the Combustion Institute*, WSS/CI paper 95F-193 (Stanford, CA, 1995).

[58] M. Rumminger, N. H. Heberle, R. W. Dibble, *et al.*, in *Proceedings of Western States Section of the Combustion Institute*, WSS/CI paper 95F-212 (Stanford, CA, 1995).

[59] J. A. Drakes, D. W. Pruitt, R. P. Howard and J. O. Hornkohl, *J. Quant. Spectrosc. Radiat. Transfer*, **57** (1997), 23.

[60] G. V. Sklizkov, in *Laser Handbook, Vol. 2*, part F3, edited by F. T. Arrechi and E. O. Schulz-DuBois (New York: North Holland Publishing Company, 1972), pp. 1545–1576.

[61] G. S. Settles, *Schlieren and Shadowgraph Techniques, Visualizing Phenomena in Transparent Media* (New York: Springer Verlag, 2001).

[62] D. C. Smith, in *Proceedings of 38th AIAA Aerospace Sciences Meeting & Exhibit*, AIAA paper 2000–716 (Reno, NV, 2000).

[63] G. M. Weyl, in *Laser-Induced Plasmas and Applications*, part 1, 1st edition, edited by L. J. Radziemski and D. A. Cremers (New York: Marcel Dekker, Inc., 1989), pp. 1–67.

[64] R. G. Root, in *Laser-Induced Plasmas and Applications*, part 2, 1st edition, edited by L. J. Radziemski and D. A. Cremers (New York: Marcel Dekker, Inc., 1989), pp. 69–103.

[65] C. G. Morgan, *Rep. Prog. Phys.*, **38** (1975), 621.

[66] C. G. Morgan, in *Radiative Processes in Discharge Plasmas*, edited by J. M. Proud and L. H. Luessen, NATO Advanced Study Institute on Radiative Processes in Discharge Plasmas (New York: Plenum Press, 1985), pp. 457–507.

[67] Y.-L. Chen, J. Lewis, and C. G. Parigger, *J. Quant. Spectrosc. Radiat. Transfer*, **67** (2000), 91.

[68] J. Stricker and J. G. Parker, *J. Appl. Phys.* (1982), pp. 851–855.

[69] I. G. Dors, C. G. Parigger and J. W. L. Lewis, in *Proceedings of 38th AIAA Aerospace Sciences Meeting & Exhibit*, AIAA paper 2000–717 (Reno, NV, 2000).

[70] *CFD-ACE Theory Manual*, 5th edition, Computational Fluid Dynamics Research Corporation, 215 Wynn Drive, Huntsville, AL 35805 (1998).

[71] P. J. VanDoormal and G. D. Raithby, *Numerical Heat Transfer*, **7** (1984), 147.

[72] A. S. Dubovik, *Photographic Recording of High-Speed Processes* (New York: Pergamon Press, 1968).

5

Analysis of aerosols by LIBS

Ulrich Panne

*Laboratory for Applied Laser Spectroscopy,
Institute of Hydrochemistry, Technical University Munich*

David Hahn

*Department of Mechanical and Aerospace Engineering,
University of Florida*

5.1 Introduction to aerosol science

5.1.1 Fundamentals of aerosols

Laser-induced breakdown spectroscopy (LIBS) is well suited for the analysis of aerosol particles because of the unique point sampling nature of the laser-induced plasma. The discrete plasma volume uniquely couples with the discrete nature of aerosol particles to enable a wide range of data analysis options, including spectral averaging, conditional spectral processing, and single-shot analysis. In this chapter, a detailed introduction to aerosol science and aerosol analysis is presented to frame the overall problem of LIBS-based aerosol analysis. A detailed analysis of the laser-induced breakdown process is focused on the gas-phase processes associated with plasma initiation and propagation. Quantitative aerosol analysis is presented in terms of the aerosol-sampling problem, followed by direct and indirect quantitative aerosol measurements. We conclude with a detailed discussion of LIBS applications to aerosol analysis and future directions in this challenging and important area.

Aerosols (Latin, Aer (air) and sole (solutions)) are particle ensembles of solid and/or liquid matter with characteristic dimensions in the nanometer to micrometer range suspended in a gaseous carrier gas. Common usage, however, refers to aerosols often only as the particulate component. For many processes involving semivolatile components, the gas phase is, however, inextricably linked to the particle composition. In relation to their number density, aerosols are ubiquitous; however, the mass of a single aerosol particle is often negligibly small. For example, a solid spherical particle of 100 nm diameter with a density of 2 g cm^{-3} has a total mass of 1 fg. Aerosols can be extremely heterogeneous in size and chemical composition owing to their external and internal mixing, so that under natural conditions it is rare for two identical particles to be found. Besides the average bulk composition, the surface composition is an important characteristic since it affects interfacial mass transfer and surface reactions. On a global average, the total concentration of the atmospheric

Laser-Induced Breakdown Spectroscopy: Fundamentals and Applications, ed. Andrzej W. Miziolek, Vincenzo Palleschi and Israel Schechter. Published by Cambridge University Press. © A. W. Miziolek, V. Palleschi and I. Schechter 2006. A. W. Miziolek's contributions are a work of the United States Government and are not protected by copyright in the United States.

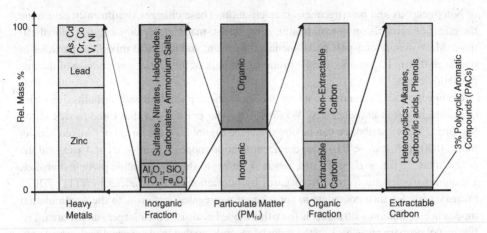

Figure 5.1. Typical composition of an ambient aerosol.

(i.e. troposphere) aerosols is in the order of 1 μg m^{-3}, while in urban areas concentrations on the order of 100 μg m^{-3} can be observed. For more detailed information the reader is referred to the following text books and review articles [1]–[7].

Real aerosols are a complex mixture from natural and anthropogenic sources, and are generated either via direct emission (primary particles) or via gas-to-particle conversion (secondary particles). Aerosol particles in the ambient atmosphere originate from several sources: wind-raised dust, construction sites and open fields, sea spray, industrial activity, emissions from animals and plants (microbial and fungal spores), traffic, volcanoes, forest fires and combustion processes, residues from the evaporation of sprays, mists and fogs, and photochemical conversion of gas to particles. Figure 5.1 depicts the typical composition of the tropospheric aerosol: a large fraction of this aerosol is of anthropogenic origin, i.e. from fossil-fuel emissions and other industrial activities, biomass burning, and agricultural emissions, and has increased dramatically over the past century. Fossil-fuel emissions are mainly emitted into the Northern Hemisphere, and emissions associated with biomass burning are mainly emitted into the tropics, with a seasonal variation that depends on the timing of the dry season and agricultural activities [8]. Chemical components include sulfates, ammonium, nitrates, chlorides, trace metals, carbonaceous materials, crustal elements, and water. The carbonaceous fraction consists of elemental carbon (black carbon, soot) predominantly from combustion sources, and organic carbon emitted directly or generated by condensation of compounds of low volatility. While the main inorganic components are known fairly well, a large percentage of the organic fraction is still not identified (see [9] and references therein).

In contrast to gases, which have mean atmospheric lifetimes between some seconds and several hundred years, the typical atmospheric residence time of aerosols is limited to some days owing to their inherently dynamic character. Aerosols not only undergo wet and dry deposition, but also change in particle size, number, and composition as a result of homogeneous and heterogeneous (photo)chemical and/or gas-particle reactions, agglomeration, or homogeneous and heterogeneous condensation. These changes significantly determine the effect of aerosols on human health, ecosystems, material degradation, and global climate. Macroscopic transport phenomena, convection, and turbulent mixing of particles in the range 0.001–10 μm are overall similar to the bulk of the carrier gas because of the fast relaxation times.

Despite the countless aerosol sources, the atmospheric particle mass distribution is usually discussed in terms introduced by Whitby [10], who pointed out that three modes in the lognormal mass distribution can be observed (see Figure 5.2): the nuclei or "Aitken" nuclei mode, $0.005 \, \mu m < d_p < 0.01 \, \mu m$, the accumulation mode, $0.1 \, \mu m < d_p < 1 \, \mu m$, and the coarse mode, $1 \, \mu m < d_p < 50 \, \mu m$, where d_p refers to the aerodynamic particle diameter. Ultrafine particles are formed from gases by nucleation or by condensation [11]. These primary particles then coagulate to form larger aggregates, but grow to the accumulation mode via coagulation with aerosols from this mode because of their larger surface area. The lifetime of primary particles is in the order of seconds, so that they are usually observed only in the vicinity of high-temperature combustion sources. The accumulation mode is populated via gas-to-particle conversion, condensation, chemical reactions, and coagulation. Particle growth is limited to a few micrometers because of deposition, so that the atmospheric lifetimes range in the order of days. The coarse mode therefore remains distinct from the "fine particle" modes, as particles are mainly generated by mechanical processes. Further, the fine and ultrafine aerosols differ in many aspects other than aerodynamic diameter, such as formation mechanisms, chemical composition, sources, physical behavior, and human exposure relationships, from the third mode, so that these differences alone are often used to justify their consideration as separate pollutants.

The nuclei mode is clearly observable only near sources of condensable gases, for example biogenic emissions in remote areas or homogeneous condensation of sulfuric acid. However, recent studies point to controlled and uncontrolled high-temperature sources of mainly anthropogenic origin, which contribute significantly to the nuclei mode and the accumulation mode. These sources are characterized by particle size distributions with small geometric standard deviations σ_g of 1.2–1.3. The origin of the major particle fraction in the accumulation mode is still unclear despite numerous field studies and extensive modeling [12–14]. The coagulation of particles from the nuclei mode cannot account for the bulk of the accumulation modes. Hence, direct anthropogenic input, aging and hygroscopic growth as well as cloud processing and heterogeneous conversion are a possible source or mechanism [15–17].

Aerosols can influence the radiation balance of the Earth through absorption or scattering of sunlight (direct effect), or by acting as cloud condensation nuclei (CCNs), which influence the formation, lifetime, and radiative properties of clouds (indirect effect). The role of this negative radiative forcing of aerosols in combination with the positive forcing of the greenhouse gases is generally recognized as the most uncertain effect of anthropogenic emission on climate [18].

Trace metals, the major target analytes for LIBS and other elemental analysis techniques, are ubiquitous in various raw materials, such as fossil fuels and metal ores, as well as in

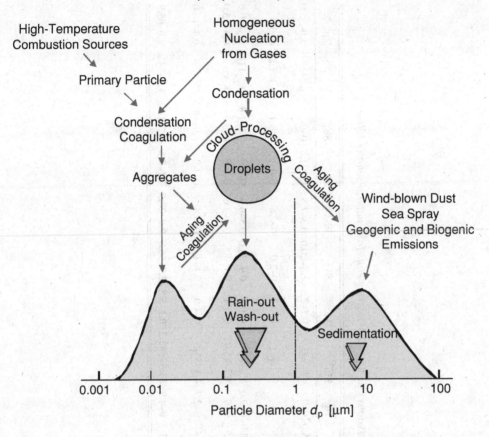

High-Temperature
Combustion Sources

Homogeneous
Nucleation
from Gases

Primary Particle

Condensation

Condensation
Coagulation

Cloud-Processing

Droplets

Aging
Coagulation

Wind-blown Dust
Sea Spray
Geogenic and Biogenic
Emissions

Aggregates

Aging
Coagulation

Rain-out
Wash-out

Sedimentation

0.001　0.01　0.1　1　10　100

Particle Diameter d_p [μm]

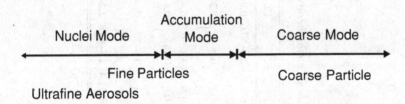

Nuclei Mode

Accumulation
Mode

Coarse Mode

Fine Particles

Coarse Particle

Ultrafine Aerosols

Figure 5.2. Sources and sinks of atmospheric particles according to the generalization by Whitby [10].

industrial products. It is not surprising that some trace metals evaporate entirely or partially from raw materials during the high-temperature production of industrial goods, combustion of fuels, and incineration of municipal and industrial wastes, thereby entering the ambient air with exhaust gases. Most of the time, aerosols from such sources have very distinct particle distributions with a low geometric standard deviation and a mean particle size in the order of 0.1 μm. When emitted to the atmosphere, trace metals are subject to transport within air

Table 5.1. *Worldwide atmospheric emissions (10^3 t y^{-1}) of heavy metals from natural sources [20–22]*

Source	As	Cd	Co	Cr	Cu	Hg	Mn	Mo	Ni	Pb	Sb	Se	V	Zn
Wind-borne soil aerosols	0.3–5.0	0–0.4	0.6–7.5	3.6–50.0	0.9–15.0	0–0.1	42–400.0	0.1–2.5	1.8–20.0	0.3–7.5	0.1–1.5	0–0.4	1.2–30.0	3.0–35.0
Sea-salt aerosol	0.2–3.1	0–0.1	0–0.1	0–1.4	0.2–6.9	—	0–1.7	0–0.4	0–2.6	0–2.8	0–1.1	0–1.1	0.1–7.2	0–0.9
Volcanoes	0.2–7.5	0.1–1.5	0–1.9	0.8–29.0	0.9–18.0	0–2.0	4.2–80.0	0–0.8	0.9–28.0	0.5–6.0	0–1.4	0.1–1.8	0.2–11.0	0.3–19.0
Wild forest fires	0–0.4	0–0.2	0–0.6	0–0.2	0.1–7.5	0–0.1	1.2–45.0	0–1.1	0.1–4.5	0.1–3.8	0–0.5	0–0.5	0–3.6	0.3–15.0
Biogenic processes	0.4–7.5	0–1.7	0–1.3	0.1–2.2	0.1–6.4	0–2.7	4.1–55.5	0–1.0	0.1–1.7	0–3.4	0–1.3	0.6–14.3	0.1–2.4	0.4–16.0
Total	**1.1–23.5**	**0.1–3.9**	**0.6–11.4**	**4.5–82.8**	**2.2–53**	**0–4.9**	**51.5–582.2**	**0.1–5.8**	**2.9–56.8**	**0.9–23.5**	**0.1–5.8**	**0.7–18.1**	**1.6–54.2**	**4.0–85.9**

masses and migration through the ecosystem, causing perturbations of their geochemical cycles not only on a local scale but also on regional and even global scales. Deposition of trace metals in areas surrounding the emission sources, as well as deposition en route during their long-range transport, can exceed the maximum permissible values. Further problems arise from enrichment within the biosphere, leading to significant biomagnification.

Natural sources are related primarily to the geological presence of trace metals in the crustal material and are transformed into aerosols during various natural physical, chemical, biological and meteorological processes. Release of trace metals during industrial production processes and disposal of wastes is regarded as anthropogenic emissions. Obviously, it is often difficult to differentiate between the natural and anthropogenic origin of trace metals, particularly when elements such as mercury have the potential to re-evaporate after being deposited into the aquatic or terrestrial surfaces. Measurements of sources and source strengths depend on the type of source and the elements in question. Note that, for many elements, natural emissions exceed those from anthropogenic sources. To determine the source of a given element in aerosols, three main approaches exist, namely direct measurements, calculation of enrichment factors, and source receptor modeling. Those techniques are well described by Fergusson [19].

It is generally assumed that the principal natural sources of trace metals include wind-borne soil particles, volcanoes, sea-salt spray, and wild forest fires. Table 5.1 provides a short summary of natural sources [20–22]. The emission factors for wind-blown dust are calculated from the concentration of each metal in different soils and the amount of continental dust annually brought into the atmosphere. Soil-derived dust accounts for over 50% of the total Cr, Mn, and V emissions, as well as for 20%–30% of the Cu, Mo, Ni, Pb, Sb, and Zn released annually to the atmosphere. The most extensively studied source appears to be volcanic emanations, which account for 40%–50% of the total natural Cd and Hg and 20%–40% of the total natural As, Cr, Cu, Ni, Pb, and Sb emitted every year. Organic aerosols are dominant in non-urban areas and 60% of the airborne trace metals in forested regions can be attributed to biogenic origin. Indeed, Nriagu [20] estimated that biogenic sources contribute, on average, over 50% of Se, Hg and Mo, and from 30% to 50% of the As, Cd, Cu, Mn, Pb, and Zn, to the total atmospheric emissions from all natural sources. Selenium in the marine aerosol can originate from its gaseous precursors as a result of gas-to-particle conversions, as can sulfur. Methylation processes in the aquatic and terrestrial environments result in the re-emission of Hg on fine particles. Sea-salt aerosols seem to account for <10% of atmospheric trace metals from natural sources. Sea-spray aerosols release primarily Cd, Cu, Ni, Pb and Zn through "bubble bursting" and As through gas exchange. In certain parts of the world, forest fires are the major emission sources and more than 10% of atmospheric Cu, Pb and Zn from natural sources can originate from this source. For forest wildfires, the emission factors are estimated from the average acreage that is burned and the concentrations of trace elements in forest stock.

Global, regional and local anthropogenic source processes are mainly related to the volatility of elements at the high temperatures of fossil-fuel combustion, and many other high-temperature industrial processes, particularly the extraction of non-ferrous metals from

Table 5.2. *Worldwide anthropogenic emissions (10^9 g y^{-1}) of heavy metal aerosols adapted from [20–22]*

Source	As	Cd	Cu	Ni	Pb	Se	Zn
Mining, non-ferrous metals	0.013	0.002	0.8	—	8.2	0.005	1.6
Primary non-ferrous metal production	15.2	4.71	20.8	9.4	76.5	0.28	106.7
Secondary non-ferrous metal production	—	0.60	0.33	0.2	0.8	—	9.5
Iron and steel production	4.2	0.07	5.9	1.2	50	0.01	35
Industrial applications	0.02	0.05	4.9	1.9	7.4	0.06	26
Coal combustion	0.55	0.06	4.7	0.7	14	0.68	15
Oil combustion (including gasoline)	0.004	0.003	0.74	27	273	0.06	0.1
Wood combustion	0.60	0.20	12	3.0	4.5	—	75
Waste incineration	0.43	1.40	5.3	3.4	8.9	—	37
Manufacture, phosphate fertilizer	2.66	0.21	0.6	0.6	0.05	—	1.8
Miscellaneous	—	—	—	—	5.9	—	6.7
Total	23.6	7.3	56	47	449	1.1	314

sulfides. To compare these emissions (see Table 5.2 for an overview) with trace element releases from natural sources, one must take into account the scale of pollutant perturbations. For Se, Hg, and Mn, global natural emissions exceed (for example) the total releases from anthropogenic sources. Combustion of fossil fuels to produce electricity and heat is the main source of anthropogenic emissions of atmospheric Be, Co, Hg, Mo, Ni, Sb, Se, Sn, and V [22] and an important source of As, Cr, Cu, Mn, and Zn. In general, the amount of emissions from a conventional thermal power plant depends on the content of trace metals in the fuels, the physical and chemical properties of trace metals during combustion, technological conditions of a burner, and the type and efficiency of emission control equipment. The last, in the form of electrostatic precipitators, wet scrubbers, and flue gas de-sulfurization, have led to a considerable decrease in emissions in Japan, Europe, and the United States in the past decade. The largest emissions of atmospheric As, Cd, Cu, In, and Zn arrive from the pyrometallurgical processes employed in the production of non-ferrous metals, such as Pb, Cu, and Zn. The type of technology employed in smelters, refineries and other operations, such as roasting, the content of trace metals in ores and scrap material, and the type and efficiency of emission control equipment are the most relevant parameters for the actual emissions. Other major anthropogenic sources of atmospheric trace metals include high-temperature processes in steel and iron manufacturing such as coke production, pig iron manufacturing, and steel production. Major aerosol emissions from cement plants result from high-temperature operations in kilns and driers. Trace metals, including Cd, Pb, As, Zn, Sb, Bi, Te, V, and Hg, are used as additives to produce or improve various products, particularly within the manufacturing, construction, and chemical industries. Incineration of municipal and industrial hazardous wastes, the next most common method of waste disposal

Table 5.3. *Source category list of the European Monitoring and Evaluation Program (EMEP)*

Source	Cd	Hg	As	Cr	Cu	Ni	Pb	Zn	Se
Public power, electricity/heat generation	●	③	③	●	③	③	●	●	③
Commercial, institutional, and residential combustion	●	③	●	●	●	●	●	●	●
Industrial combustion	●	③	●	●	●	●	●	●	●
Production processes	③	③	③	③	③	●	③	③	●
Extraction of fossil fuels/solvent use					●	●	●	●	
Distribution of fossil fuels/road transport	●				●	●	③	●	
Other mobile sources and machinery	●				●	●	③	●	
Waste treatment and disposal	③	③	●	●	●	●	●	③	●
Agriculture			●					●	
Nature		●	●		●		●	●	●

Minor source, ●; major source, ③

after land filling, is an important source of trace metal emissions to the atmosphere. In many countries, for example Sweden, Japan and Switzerland, incineration has even overtaken landfill as the predominant disposal option for municipal wastes, accounting for more than 85% of this waste stream. The largest sources of trace metals entering the environment from burning petroleum products are tetraethyl lead and other gasoline additives, diesel fuel combustion, metal compound additives for lubricants, worn metals that accumulate in spent lubricants, and automobile tires. The abandonment of leaded gasoline resulted in a recent significant reduction of global Pb emissions.

The most commonly inventoried heavy metals, i.e. priority metals for emission reductions, are As, Cd, Hg, Pb, and Zn. This is because of their effects on environmental and human health, as well as their ubiquitous appearance in the environment. As an example, Table 5.3 displays source categories corresponding to the European Monitoring and Evaluation Program (EMEP). In this context sources can be treated individually, i.e. such as power plants, refineries, waste incinerators and airports. Sources comprising large numbers of small emitters, for which the emission conditions are relatively similar, are treated collectively, for example vehicle emissions from road transport, railways, inland navigation, shipping, or aviation.

Aerosols can also have adverse health effects on humans, a fact mentioned by Pliny the Elder (AD 23–79), who referred to inhalation of "fatal dust." While modern technologies allowed a serious reduction in the emission in coarse aerosols and gases such as SO_2, many studies have by now demonstrated the relevance of fine and ultrafine aerosols to human health [23, 24]. This thoracic particle fraction is of special concern as these particles, presumably surface-enriched with contaminants, can penetrate deep into the lungs. Controversy remains at present regarding only the question if particles with $d_p < 100$ nm have an impact on human health, i.e. irritation of the pulmonary alveoli, lung inflammation, and hyperactivity, by their physical presence and/or their chemical composition [25]. Compelling evidence suggests that – among other components – water-soluble metal components of urban aerosols are important determinants of the respiratory effects observed during air pollution episodes. Metals, particularly reactive transition metals such as V, Zn, Fe, Cu, and possibly Hg, associated with submicrometer aerosol particles may directly initiate or exacerbate irritation of respiratory tissues by stimulating local cells to release reactive oxygen species (e.g. hydrogen peroxide and superoxide free radicals) and inflammatory mediators, such as cytokines [26]. Cytokines are protein molecules involved in the lung's immune response to bacteria and, apparently, to aerosol particles. They act either by attracting macrophages, which subsequently release reactive oxygen species, or by directly manufacturing such species.

The mass of various inorganic constituents, including first-series transition metals capable of producing reactive oxygen species, are predominately associated with primary aerosol emissions from high-temperature combustion sources (HTCS) [13]. Hence, future legislative thresholds for aerosol emission will have to consider not only the particle mass but also the particle concentration weighted by the chemical composition and size distribution. This is confirmed by the observation that the incidence and preponderance of respiratory diseases have increased in the past decade while (in contrast to the total particle concentration) the total emitted particle mass decreased. The knowledge of aerosol size distributions is essential because the particle size significantly affects ambient transport and deposition processes as well as the described uptake in the respiratory system. Moreover, elemental size distributions can give an indication of the source of the element [27, 28]. The typical characteristics of size distributions vary from element to element and from sampling location to sampling location, but generally they display a trimodal or bimodal distribution [14]. The calculation of enrichment factors points to additional sources in comparison with soil erosion, for example. Lannefors *et al.* [29] calculated enrichment factors for several elements relative to Ti, which can be considered to be mainly soil derived.

Not surprisingly, many tasks in aerosol analysis stem ultimately from the potential health effects of aerosols. The applications range from aerosol analysis in combustion to process analysis and control in various industrial production processes. Further needs for aerosol analysis result from the fact that aerosol processes themselves have the potential to create complex chemical materials including insulators, semiconductors, superconductors, metals and alloys, which are useful in producing multicomponent and artificially structured nanoparticles for electronic, optical and magnetic applications (see [30] and references therein). The principal ease of process control for material characteristics such as particle size, crystallinity, degree of agglomeration, porosity, chemical homogeneity, and stoichiometry makes the aerosol synthesis pathway very attractive. Most synthesis methods of nanoparticles in the gas phase are based on homogeneous nucleation in the gas phase and subsequent condensation and coagulation. Interestingly, laser ablation can also yield nanoparticles, but the formation mechanism does not involve a homogeneous nucleation step [31].

The characterization of aerosols is very relevant for industrial hygiene, i.e. aerosols at workplaces [32–34]. Although the aerosols can vary substantially in chemical composition and size distribution with the work process involved, many process are similar to the processes described for the free atmosphere. The major issue in occupational hygiene is the question of exposure, that is the time-averaged concentration of the agent under study at the relevant interface between the environment and the biological system (the worker). It is not only the concentration, but also other characteristics such as morphology and particle-size distribution, that are coming into focus in exposure threshold definitions.

Another application for particle analysis is clean room technology, which is becoming more and more important not only in the microelectronic industry but also in the pharmaceutical industry and related areas. Contamination control is a serious analytical problem in clean room technology and demands the detailed chemical analysis of particles for identifying sources. This is often a critical issue when establishing new production processes as the generation of aerosols from different production steps and materials can not be deduced a priori.

The assessment of indoor aerosols is often connected to phenomena such as "sick building syndrome," where heating, ventilation, and air conditioning can result in an increased exposure to potentially health relevant aerosols. The concentrations often easily exceed the outdoor concentrations.

5.1.2 Chemical analysis of aerosols

In the past decade, numerous studies have focused on the observation and prediction of the physical characteristics of atmospheric aerosols (scatter, absorption, agglomeration), while the size-resolved chemical analysis of aerosols, especially fine aerosols, remained unsatisfactory (see [4], [35]–[38] for an overview on instrumentation and techniques for atmospheric and occupational aerosol analysis). Ultrafine aerosols originating from nucleation processes are generated in a time scale on the order of 0.1–1 s, which determines the ultimate time resolution needed for an analysis. In general, the necessary temporal resolution increases with decreasing particle diameter. Not surprisingly, the temporal and size resolution of present analytical methods for the chemical composition is woefully inadequate compared with the available methods to probe number concentration, charge, or particle diameter. The latter have sufficient temporal resolution and sensitivity to cover particle diameters between 0.003 μm and 1 μm. In contrast, the chemical analysis is often based on conventional filter collection and subsequent laboratory based analysis because of the minute amounts of aerosol mass available. Although the chemical composition can be related in some cases to the observed physical parameters by a priori information on the aerosol sources, the integral nature of a filter-based measurement can often be no more than a crude approximation.

The conditions for sampling and analysis of atmospheric aerosols are less variable than the conditions encountered in performing on-line measurements in process analysis and industrial hygiene. The obvious implications for the interface concern construction and materials and whether monitoring is done by sampling by an *in situ* approach. Furthermore,

Table 5.4. *Toxic metal limits for industrial plants*

EU		USA (hazardous waste combustor rule)	
	Threshold (mg m^{-3})		Threshold (mg m^{-3})
Dust	10–50 [a]	dust	34
Cd + Tl	0.05	semi-volatiles (Cd, Pb)	0.240[b] / 0.120[c]
Hg	0.05	Hg	0.130[b] / 0.045[c]
Sb + As + Pb + Cr + Co + Cu + Mn + Ni + V	0.5	low volatiles (As, Be, Cr)	0.097

Notes: [a]depending on the type of plant; [b]existing sources; [c]new sources; [c]directive 2000/76/EC; [d]EPA, Method 29/2001.

for industrial applications, robustness and compactness of measuring equipment are another often decisive key factor. Often the processes demand working at high pressures and temperatures where the effects of strong background signals or absorption of exciting light beams by process gas have to be considered. Power plants may operate at atmospheric pressure, but many systems are pressurized up to 15–20 atm with temperatures at the measurement point in the range 600–1000 °C. In combustion systems, the main components of the flue or stack gas will be N_2, CO_2, O_2, H_2O, and, depending on the fuel, HCl with minor amounts of CO, NO_X, and SO_2. In addition, the particle load can vary enormously, again depending on the conditions at the chosen position of monitoring, for example directly after the combustor, after a cyclone, hot gas or particle filter or other cleaning or cooling sections of the plant. Sample monitoring methods have to consider care of sampling lines, which should be constructed of inert material and kept at process conditions, i.e. at temperature above the dew point of water and other species of interest. If an optical access can be installed at the measurement point, measurements can be made that are both on-line and *in situ*. However, in industrial-scale systems, the large dimensions and the often high pressures and temperatures involved put practical constraints on the design of such optical access. Specific demands are made on materials for flue gas piping, flanges, and windows. The large dimensions can result in considerable beam absorption, or the laser and the emission, respectively, break down outside the observed interaction volume. Correct and statistically significant sampling of large dimensions is often an issue when working with laser plasmas; appropriate scanning procedures and multiple plasma generation along the characteristic dimension of the measurement point are possible solutions. Experience has shown that *in situ* calibration, while the plant is actually in operation, is seldom feasible, so the instrument has to be calibrated externally and a validated calibration transfer, possibly supported by some reference analysis, has to be done. However, species concentration, e.g. O_2, can change in flue gas stack gases rapidly, which can result in variable interferences such as collisional quenching. Modern process control requires, in many cases, detection limits at the microgram per cubic meter level (see Table 5.4) with a high time resolution (1 min or less) and a sufficient dynamic range. For some applications, only the total species concentration present in the gas phase is needed, while discrimination of chemical species and/or particle vs. gas phase can be also of interest. Standard analysis methods usually operate in batch mode, which involves sampling over many hours, and then sending the samples for analysis. Such long sampling times mean that large, short-term changes in metal concentration are averaged out.

The "ideal" aerosol analyzer was described by Friedlander who, in his seminal 1970/1971 work, classified different aerosol measurement techniques on the type of information given by the instrument [39, 40]. Although today's aerosol analysis by mass spectrometry approaches the ideal of a perfect single particle counter, which should provide the complete size-resolved chemical composition of an aerosol in real time, most current methods are far from this ideal. Other analytical concepts depicted in Figure 5.3 have been realized by now, compromising by integration or utilizing discrete channel for time, size and/or composition.

Sampling of aerosols usually comprises at least a fractionation into coarse and fine particles, which can be achieved with a variety of dichotomous filter samplers. The temporal resolution is determined by the flow rate and the detection limits of the subsequent employed analytical technique. Typical sampling times range from 1 h to several days. Sampling always includes a number of sub-components that contribute to the overall accuracy with which a sample is taken. These components are: the sampling head, the transmission section, the particle size selector (not always present), the collecting or sensing region, calibrated flow monitoring and control, and the pump. An inlet should not be biased to a particle size and must be omni-directional and independent of wind speed for atmospheric aerosols. For proper design of the transfer section particle losses to the internal walls must be avoided; losses can occur over the whole size spectrum of particles, ranging from inertial and sedimentation losses for the large particles to diffusional losses for ultrafine particles and additional losses through deposition from electrostatic forces. For health-related sampling certain aerosol fractions, i.e. $PM_{2.5}/PM_{10}$, are required, so that some form of particle size selector must be employed to isolate the relevant fraction. In the 1980s, the PM_{10} control statutes established a cut-off criterion for sampling of hazardous aerosol. The PM_{10} threshold enforced a reduction of anthropogenic particles with a mass median of aerodynamic diameters less than 10 μm. Based on improved knowledge on the toxicological importance of fine particles the PM_{10} threshold was lowered in 1997 to a $PM_{2.5}$ cut-off, i.e. the sampler discriminates particle diameters above and below 2.5 μm. Cascade impactors (e.g. Berner impactor with 11 stages) allow an improvement insofar as they fractionate aerosols according to their aerodynamic size in several classes. However, sampling time relates to the number of stages, so that under atmospheric conditions sampling times in the order of days often have to be accepted. Particles are generally selected by aerodynamic means by using physical processes similar to those involved in the deposition of particles in the respiratory system. Gravitational sedimentation processes are used to select particles in horizontal and vertical elutriators, centrifugal sedimentation is used in cyclones, inertial forces are used in impactors, and porous foams employ a combination of both sedimentation and inertia forces.

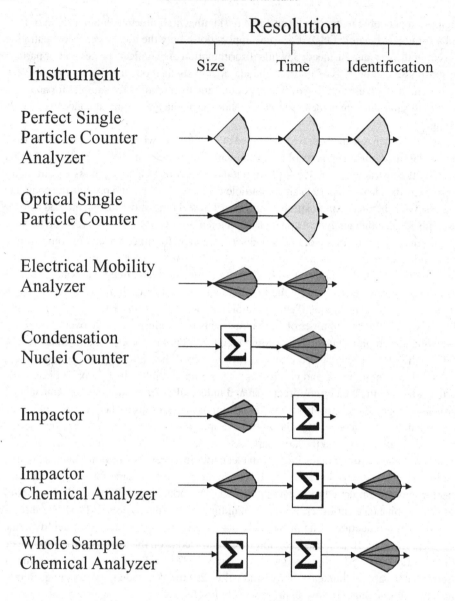

Figure 5.3. Classification of aerosol analyzers according to Friedlander [39, 40].

A filter is the most common means of collecting the aerosol sample in a form suitable for assessment. That assessment might include gravimetric weighing on an analytical balance before and after sampling to obtain the sampled mass. It might also include visual assessment using an optical or electron microscope and/or a whole range of analytical and chemical techniques. The choice of filter type for a given application depends greatly on how it is proposed to analyze the collected sample. Many different filter materials, with markedly different physical and chemical properties, are now available. These include fibrous (e.g. glass, quartz), membrane (e.g. cellulose nitrate, polycarbonate, Teflon), and sintered (e.g. silver) filters. Membrane filters have the advantage that they can retain particles effectively on their surface (good for microscopy), whereas fibrous filters have the advantage of providing in-depth particle collection and hence a high load-carrying capacity (good for gravimetric assessment). Such filters are available in a range of dimensions (e.g. 25–100 mm diameter) and pore sizes (e.g. 0.1–10 µm). Collection efficiency is usually close to 100% for particles in most size ranges of interest, although sometimes some reduction in efficiency might be traded against the lower pressure drop requirements of a filter with greater pore size. For some types of filter, electrostatic charge can present aerosol collection and handling problems – in which case, the use of a static eliminator may sometimes (but not always) provide a solution. For other types, weight variations due to moisture absorption can cause difficulty, especially when being used for the gravimetric assessment of low masses. The chemical requirements of filters depend on the nature of the proposed analysis. As already mentioned, weight stability is important for gravimetric assessment. If particle counting by optical microscopy is required, then the filters used must be capable of being rendered transparent (i.e. cleared). Direct on-filter determination of elemental composition can lead to several requirements from low elemental blanks to good transmission of the analyzing radiation. The common range of wet chemical methods includes ultrasonication, ashing, etc., each of which imposes a range of specific filter requirements.

A laboratory-based analysis of aerosols collected on filters starts typically in the case of elemental analysis with an appropriate digestion method, which is eventually microwave supported or, in the case of organic analytes, by extraction with suitable organic solvents. After a further clean-up the elemental composition can be determined via conventional techniques such as ion chromatography (IC), atomic absorption spectroscopy (AAS), inductively coupled plasma optical emission spectroscopy, (ICP-OES), or inductively coupled plasma mass spectrometry (ICP-MS). A speciation of heavy metals in aerosols can be performed in a simplified way through sequential leaching or with further chromatographic techniques such as liquid chromatography (LC) or capillary electrophoresis (CE). These techniques are mostly hampered by the time, cost, and labor-intensive procedures, which do not allow on-line and *in situ* measurements, and increase the turn-around time for appropriate measures. Furthermore, sampling can change the thermodynamic environment or gas-phase composition and thereby cause variations in the particle composition.

This process can be shortened if the direct analysis of filters is possible. X-ray fluorescence spectroscopy permits the detection of the most relevant heavy metals in the order of 10–50 ng cm^{-2}. With a suitable blank free quartz filter band [41] a quasi-continuous automatic sampling was also reported by Dannecker's group [42, 43]. The system can be employed for stack sampling at 150 °C with a flow of 5 m^3 h^{-1} and a sampling time of 0.5 h, with which the authors realized volume detection limits in the order 20–80 ng m^{-3} [44]. Portable systems with a radioisotope source excitation [43] revealed only detection limits in the order of 10–30 μg cm^{-2}. Laser-ablation ICP-MS is probably the most sensitive detection method for aerosols on filters and allows a fast and comprehensive analysis of filters with today's commercially available ablation systems [45–47]. Aerosols can be sampled not only on filters but also on other suitable materials. The most elegant way for subsequent AAS, ICP-MS, or TXRF analysis is the direct aerodynamic impaction or electrostatic precipitation into an electrothermal evaporation system or a suitable sample carrier [48–55]. While the spatial resolution of laser ablation of aerosols from filters permits only analysis of particles with micrometer-sized diameters, several other beam techniques for elemental analysis, such as EPMA (electron probe microanalysis), PIXE (proton induced X-ray emission), LAMMA (laser microprobe mass analyzer), or SIMS (secondary ion mass spectrometry), are frequently employed for microanalysis of aerosol particles with diameters below 1 μm ([3], [56] and references therein).

Several authors tested continuous atomization sources such as ICPs [57–65], microwave plasmas (MIP), glow discharges (GD) [66, 67], electrical sparks [68], or direct current (DC) plasmas [69, 70] for development of continuous emission monitors. All sources can, in principle, be used in combination with atomic emission spectroscopy (AES) or mass spectrometry. Often, problems arise from the change in the plasma parameters owing to incomplete vaporization or atomization and introduction of larger air volumes, especially with high water loadings, as is common in process analysis. The use of an ICP plasma, when compared with other plasmas, is not very cost effective for long-time monitoring owing to the necessary high consumption of Ar. In combination with a suitable aerosol sizer such as a differential mobility analyzer (DMA), size-resolved elemental analysis is possible [65, 71]. Microwave plasmas in combination with AES are probably the most promising candidate for a CEM, as they can be miniaturized and have low requirements concerning the infrastructure, i.e. power and gas consumption. The reported detection limits are in the order of 1 μg m^{-3} and, under suitable plasma conditions, metalloids can be also detected in the NIR [72–76]. The utilization of laser desorption and ionization for on-line analysis of atmospheric aerosols has certainly generated a recent impact on chemical analysis of aerosols due to the possibility of analysis of inorganic and organic aerosols [77–80]. However, the current generation is still hampered by several shortcomings, i.e. particle cut-off below 0.2 μm [81], only semi-quantitative analysis possible because of matrix effects, and considerable investment costs. Progress is expected by using different wavelengths for the desorption and ionization step [78, 82], or the use of an absorbing MALDI (matrix assisted laser desorption and ionization) matrix, which is condensed on the aerosols in an on-line and *in situ* fashion [83, 84]. The utilization of an aerodynamic lens permits a size analysis of particles entering the system [85] of diameters between 40 and 800 nm.

The application of LIBS for aerosol analysis offers inherent advantages that overcome many of the limitations described above. The LIBS technique has been developed in recent years as a novel means for the quantitative, direct measurement of particle size and composition of individual particulates, although many fundamental issues remain regarding the interactions of the laser-induced plasma with aerosol particles. As a starting point for LIBS-based aerosol analysis, the following sections provides a detailed analysis of the laser-induced breakdown process, as focused on the gas-phase processes associated with plasma initiation and propagation.

5.2 Laser-induced breakdown of gases

Along with the advent of the ruby laser more than 30 years ago, the first observation of a laser plasma in gases was reported by Maker *et al.* [86] and Meyerand and Haught [87]. Laser-induced breakdown transforms transparent gases into opaque, highly conducting plasmas. Since then, a considerable body of literature aimed at understanding the various aspects of gas and aerosol-induced breakdown has been written. For detailed information see the reviews [88]–[90] and corresponding chapters in the monographs [91]–[94] and references therein. In contrast to a breakdown on solids or liquids, the breakdown threshold in gases is in excess of 10^9 W cm^{-2} owing to the involved mechanism of multiphoton ionization (MPI) and/or cascade ionization [95].

While high-energy photons can ionize gases in single-photon interactions, and microwave radiation can initiate breakdown in relatively large volumes of gas, it is not obvious in what way a laser photon of low energy (1–2 eV), compared with the ionization potential of common gases, can generate a breakdown. The formation of a laser plasma in a neutral gas follows three distinct but overlapping stages: plasma ignition, plasma growth and interaction with the laser pulse (in case of laser nanosecond pulse width), and plasma development accompanied by shock wave generation and propagation in the surrounding gas. For a pulse width $> 10^{-6}$ s a laser-supported combustion wave (LSC) model is usually assumed as the absorbed energy results in a slow, subsonic expanding plasma, whereas for shorter pulses a laser-supported detonation wave (LSD) with a supersonic, rapidly expanding shock wave front is generated. For further details of the laser–plasma interaction and later plasma and shock wave development the reader is referred to the corresponding chapters on the basics of LIBS in this book. Plasma ignition comprises the growth in the free electron and ion concentration with arrival of the first laser photons. The growth stage is characterized by a fast amplification of free electrons and ions. The term breakdown is rather arbitrarily and loosely defined in the literature; often the luminous plasma emission or the acoustical detection of the shock wave is taken as the only criterion. In the following discussion this will imply an electron density $> 10^{13}$ cm^{-3} or a degree of ionization of about 10^{-3} which permits significant absorption and scattering of the incident laser beam and finally leads very fast to a fully developed plasma (n_e typically 10^{17}–10^{19} cm^{-3}). Note that, in the early stage when the electron density is below 10^{13} cm^{-3}, electron–electron collisions are not frequent enough to make the electron distribution function Maxwellian. With the onset of breakdown a rapid plasma development stage results in the formation of a highly ionized plasma in which – for nanosecond pulse widths – further absorption and photoionization occur. With the end of the laser pulse, the plasma gradually dies away as a result of radiation and conduction of thermal energy, diffusion, attachment, and recombination of ions and electrons, until local thermodynamic equilibrium with the surrounding gas is restored. Although precise measurements of the involved spatial and temporal beam parameter along the focal volumes remain difficult, the main processes for gas breakdown have now emerged from a large body of observations.

For nanosecond and microsecond pulse widths at pressures above $100\,\text{mbar}$ ($10^4\,\text{Pa}$) the breakdown proceeds at long wavelengths ($>1\,\mu\text{m}$) as a result of an electron avalanche or cascade by inverse bremsstrahlung, i.e. free–free absorption. In this way, electrons absorb photons in the presence of a neutral gas atom for momentum transfer and ultimately acquire sufficient energy for collisional ionization of gas atoms and generation of further electrons. At visible or ultraviolet wavelengths, highly excited states of the gas molecules or atoms can readily be photoionized over times much shorter than the typical nanosecond pulse widths. This reduces the observed breakdown thresholds considerably, and reduces the losses in electron density due to diffusion.

Under low-pressure conditions and sufficient irradiance, the breakdown is initialized via electrons from multiphoton absorption and ionization (MPI), while at later times the cascade process often overtakes the electron generation. Beam trapping and self-focusing in gases under special experimental conditions can also lead to enhanced ionization. In addition, MPI comes back into play when considering the source of the "first" electron for the cascade process. In contrast to MPI the cascade ionization is not self-sufficient, but requires at least one electron in the focal volume. Owing to natural local radioactivity, cosmic rays or ultraviolet radiation, ions occur naturally in the atmosphere at a rate concentration of 10^2–$10^3\,\text{cm}^{-3}$; free electrons are, however, immediately attached to O_2, yielding O_2^-. The mean lifetime of O_2^- (which can be treated as a free electron because of electron tunneling) is in the order of $10^{-7}\,\text{s}$, so that the probability of encountering an electron in an interaction volume of about $10^{-6}\,\text{m}^3$ during a laser pulse with nanosecond pulse width is rather negligible. With the assumption that all absorption cross sections σ_{abs} for an atom excited from m to the $(m+1)$th state are similar [89], the ionization rate per atom, W_m, can be simply estimated via

$$W_m = \frac{\sigma_{\text{abs}}^m I^m}{\nu^{m-1}(m-1)!} = AI^m, \tag{5.1}$$

where ν is the laser frequency, I is the irradiance, and A is a constant. If the onset of breakdown is coupled to a critical electron density $n_{\text{e,c}}$ in a defined volume V and pressure p, then the breakdown threshold F_{th} is given by

$$F_{\text{th}} = \frac{\nu}{\sigma_{\text{abs}}}\left(\frac{n_{\text{e,c}}(k-1)!}{N_0 p V \tau \nu}\right)^{1/m}, \tag{5.2}$$

where N_0 is the number of gas atoms/molecules in V, and τ is the laser pulse width. The breakdown threshold depends weakly on the gas pressure through $p^{-1/m}$. If the breakdown starts when a fraction δ, typically taken as 10^{-3}, of the $N_0 p V$ atoms present in the volume V is ionized, equation (5.2) can be simplified to

$$F_{th} \approx \frac{\nu}{\sigma_{abs}} \left(\frac{\delta m!}{\tau \nu} \right)^{1/m},$$

(5.3)

Equation (5.3) holds under the assumption that the lifetime of each virtual state is $1/\nu$ and the photon arrival time in the interaction volume follows a Poisson distribution. In air, an MPI-initiated breakdown at the fundamental wavelength of a Nd:YAG (1064 nm) is an 8- or 10-photon process for O_2 and N_2. With $m = 10$, $\tau = 10^{-8}$ s, $\delta = 10^{-3}$, $\sigma = 10^{-16}$ cm^2, $\nu = 2.8 \times 10^{14}$ s^{-1}, F_{th} yields 1.44×10^{30} photons s^{-1} cm^{-2}, corresponding to 267 GW cm^{-2}, which is in reasonable agreement with the reported experimental observations (see [96]–[101] for experimental studies of MPI). The simple expressions above do not account for the fact that at visible and ultraviolet wavelengths highly excited states can be readily photoionized over times much shorter than the laser pulse width. At the fourth harmonic of a Nd:YAG laser (266 nm), 1- and 2-photon ionization of excited states lowers the threshold significantly (typically in the order of $5-10$ GW cm^{-2}). Stark shift and broadening of intermediate levels can additionally bring the levels into resonance with the laser wavelength. Also, if a sufficient n_e is generated by MPI early enough in the pulse so as to affect the diffusion of electrons out of the focal volume, the diffusion becomes affected by the space charge of the ions remaining in the focal region. This ambipolar nature reduces the overall diffusion and the breakdown threshold will be lowered especially in experiments with small focal spots where diffusion losses can be important [102, 103]. Finally, it must be remembered that laser modes can lead to local fields which significantly exceed the average value over the focal volume. Hence the effective irradiance of reported experimental values is probably somewhat larger than the required irradiance.

At longer wavelengths the problem of generating the first electron in the absence of gaseous or aerosol impurities becomes more serious. Breakdown at 10.6 μm [104–106] can then increase to an irradiance level of 10^{12} W cm^{-2}, where the electric field induces tunneling of an electron through its potential barrier. However, the effect has been shown to be equivalent to multiphoton absorption [107].

While production of the first electron is crucial for the start of the breakdown, the irradiance threshold is governed by the cascade or avalanche growth of ionization, which is fed by absorption of the laser light. Electrons in a laser field will gain energy to ionize and increase in number through electron-neutral inverse bremsstrahlung (IB). Although microwave breakdown theory, developed in the 1950s [108], has been successfully employed to evaluate breakdown thresholds at wavelengths beyond 1 μm (for examples of the early work, see [109]–[112]), where IB is the dominant mechanism, data at visible and near-ultraviolet wavelengths are not so well understood [103, 113–115]. If electron-impact ionization were the dominant mechanism leading to gas breakdown, the breakdown threshold for a given gas at a given pressure would scale inversely with the IB absorption coefficient, that is as $\{\lambda^3 [1 - \exp(-hc/\lambda kT)]\}^{-1}$ or approximately λ^{-2}, where T_e is the electron temperature during the cascade. Experimental observations [116, 117] revealed that the breakdown

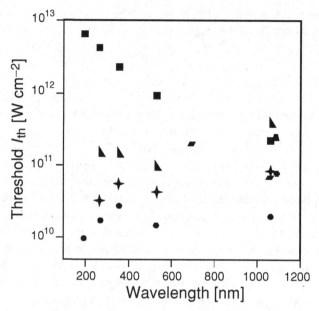

Figure 5.4. Breakdown threshold in air at different wavelengths (adapted from [118]; ▲ data from [119]; ■ modeled data from [93]; + data for nitrogen from [120]; ◢ data from [93]; ● data from [118]; ◆ data from [121]; ▲ data from [122]).

thresholds peaked in the middle of the visible spectrum owing to competition between IB and MPI of ground states and excited states. Figure 5.4 gives an overview of the observed breakdown threshold in air at different wavelengths [93, 118–122].

For electromagnetic frequencies where the photon energy $h\nu$ is sufficiently small (e.g. CO_2 laser) compared with the total free electron energy ε that quantum effects can be ignored, the absorption of energy by free electrons can be treated classically in a continuous way. Zel'dovich and Raizer showed in 1965 [123] that the quantum-mechanical and classical treatments lead to the same conclusions in this case. Because the total free electron energy (not the energy of the oscillations in the field, see below) is approximately equal to the ionization potential ϕ of the gas, this results in $h\nu/\phi \approx 1$. Otherwise the energy absorption is no longer continuous and has to be treated via quantum kinetic version of the Boltzmann equation.

The rate of energy increase for a free electron is classically given by

$$\frac{d\varepsilon}{dt} = \frac{e^2 E_0^2 \nu_c}{2m \left(\nu_c^2 + \omega^2\right)}, \tag{5.4}$$

where ε, e, and m are the electron energy, charge, and mass, respectively, ν_c is the momentum-transfer collision frequency of electrons with gas molecules, ω is the angular frequency of the laser light, and E_0 is the amplitude of the electric field. In this way, free electrons can have a net increase in energy through oscillation with the field. If the loss mechanisms are lower than the net energy increase an electron will eventually have enough energy for ionizing collisions. Note that ν_c is usually small compared with ω except at high pressures. For many gases ν_c is proportional to p and is largely independent of the electron energy, i.e. ν_c/p is in the order of 10^8 s^{-1} Pa^{-1} and is in the order of 10^{13} s^{-1} at 1 atm. Note that in the case of wavelengths in the visible, the photon energy is in the order of 1–2 eV, whereas the oscillation energy is of the order of 10^{-2} eV. Hence, in most of the collisions the electrons do not acquire energy from the field, and rarely instantaneously the energy $h\nu$.

The cascade process will lead to an exponential growth of electron density. Assuming a fully developed plasma at $n_e = 10^{17}$ cm^{-3}, about 40 generations are required to grow from an assumed initial value of $n_{e,0} = 1-10$ cm^{-3} in the focal volume. This number is not strongly dependent on the assumed value of n_0 as the electron concentration becomes large only near the end of the process; 99% of the ionization is produced in the last 7 generations. When the electron concentration exceeds 10^{13} cm^{-3}, i.e. the onset of the breakdown, electron–electron collisions will tend to populate the tail of the electron distribution function and this has a dramatic effect on the cascade rate.

Quantities such as the growth and losses from the cascade and the time to breakdown are determined by conditions at times when the electron concentration is small. Losses can be through several inelastic processes such as vibration and rotational excitation (for polyatomic gases), excitation of electronic levels of atoms and molecules, elastic collisions, attachment, recombination, and diffusion. The diffusion of electrons out of the focal volume is the slower the higher the gas density and the larger the dimensions of the focal region. Note that at low pressures the electrons diffuse rapidly from the very small regions in which large local fields exist because of the interaction between the laser modes. Electrons then become distributed over the entire focal volume and are, in general, subject to the action of the fields averaged over the volume. On the other hand, at high pressures diffusion is very slow and the cascade develops predominantly in places where the local fields exceed the average field.

In the case of large losses, it seems that $d\varepsilon/dt < 0$ becomes possible and that the electron energy can never reach the ionization potential, which is equivalent to a breakdown of the classical cascade model. However, owing to the inherent quantum character, there is always a finite probability for ionization. The reason is that the electron motion along the energy axis is made up of random jumps of finite magnitude and has a stochastic character, which is described by the quantum-mechanical model.

The master for the electron growth is given by

$$\frac{dn_e}{dt} = \nu_i n_e + W_m I^m N - \nu_a n_e - \nu_R n_e - \nu_D n_e, \tag{5.5}$$

where ν_i, ν_a, ν_R, and ν_D are the impact ionization, attachment, recombination, and diffusion rate coefficients, and N is the gas number density.

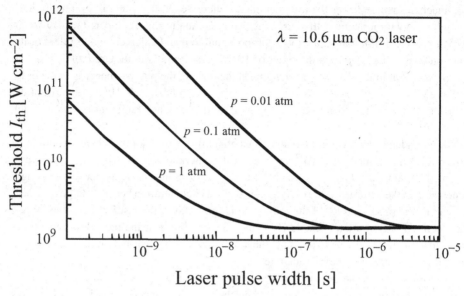

Figure 5.5. Dependence of the breakdown threshold in air upon the laser pulse width (adapted from [93]).

If we neglect all loss processes from equation (5.5), the threshold condition is reached if, during a squared laser pulse of duration τ, the energy gained by the electrons is equal to the ionization potential ϕ times the critical number of generations, i.e. about 40,

$$40\phi = \int_0^\tau \frac{d\varepsilon}{dt} dt. \tag{5.6}$$

Using equation (5.4) and $E_0^2/2 = I/(4\pi/c)$, the threshold F_{th} is given by

$$F_{th} = \frac{10\phi m\omega^2 c^2}{\pi e^2 \nu_c \tau}. \tag{5.7}$$

Obviously it has to be remembered that all energy-loss processes have been ignored and $\omega \ll \nu_c$ does not hold at very high pressures. Figure 5.5 displays the dependence of the breakdown threshold in air upon the laser pulse width according to equation (5.7).

If the cascade ionization rate ν_i cannot be derived through equation (5.4) as in gas breakdown below 1 µm, it has to be obtained via a Boltzmann code that calculates the quasi-steady-state distribution function in energy space of electrons that are heated through the laser pulse [113, 123–125]. An approximation which relates the IB absorption coefficient K_{IB} [126] to the momentum-transfer cross section σ_M [127, 128] is defined by

$$K_{IB} = \frac{4\pi e^2}{3mc\omega} \left[\frac{2(\varepsilon + h\nu)}{m} \right]^{1/2} \left[\frac{2\varepsilon + h\nu}{h\nu} \right] \sigma_M(\varepsilon + h\nu). \tag{5.8}$$

Note that ε is the electron energy before the absorption of a photon. For irradiances $I > 10^{10}$ W cm^{-2} the stimulated emission given by K_e,

$$K_e(\varepsilon) = K_{IB}(\varepsilon - h\nu)\left(\frac{\varepsilon - h\nu}{\varepsilon}\right)^{1/2}, \tag{5.9}$$

can be neglected and ν_i approximated through

$$\nu_i \simeq \frac{K_{IB}(\bar{\varepsilon})N}{\hat{\varepsilon}}I, \tag{5.10}$$

where $\bar{\varepsilon}$ is a fraction (0.6–0.75) of the ionization energy and $\hat{\varepsilon}$ is slightly higher than the ionization energy. Equation (5.10) indicates that ν_i scales with I and the gas density N. Diffusion effects are not important in MPI-dominated breakdown because electrons are generated from neutral particles remaining in the focal volume. However, in cascade ionization diffusion out of the focal volume can be quite important with focal diameters in the order of 10–50 μm [129]. Diffusion and cascade ionization are treated together via

$$\nu_{AV} = \nu_i - \nu_D \tag{5.11}$$

and assuming a top radial intensity profile for the laser and imposing a sink of electrons at the edge. In equation (5.5) ν_i grows in a reduced fashion through

$$\nu_{AV} = \nu_i - \frac{2.408D}{a^2}, \tag{5.12}$$

where D is the electron diffusion coefficient and a is the beam radius. At low electron densities, D can be derived from the kinetic gas theory via

$$D = \frac{l\nu_{gas}}{3N\sigma_{gas}}. \tag{5.13}$$

Here, l denotes the mean free path in the gas, ν_{gas} is the velocity of the gas molecules, and σ_{gas} is the scattering cross section. If $\nu_i \gg \nu_{AV}$, i.e. in the case of large diffusion losses, the breakdown threshold becomes $F_{th} \propto p^{-2}$ as D scales with p^{-1} and ν_i scales with p. For a beam diameter in the order of 10 μm at 1 atm, Rosen and Wegl estimated the change in breakdown threshold through diffusion to be between a factor of 2 and 32 for different gases [115]. The onset of the ambipolar diffusion (see above) can be estimated by

$$n_e \geq \frac{kT_e}{4\pi e^2 a^2}, \tag{5.14}$$

which means that the Debye length becomes smaller than the beam parameter a. The fact that, for many experimental situations, i.e. lower pressures, picosecond pulse widths, wavelengths <1 μm, and small focal spot diameters, a slight pressure dependence is still found, indicates a predominant cascade mechanism and points to the importance of the photoionization of excited states [130–135]. Although the theoretical modeling of the breakdown in gases has developed to quite a sophisticated level [103, 113], to elucidate the exact mechanism of the plasma formation in gases, a strict control of the experimental parameters such as focusing, laser modes, contamination of the gas and the cell is still needed, accompanied by a suitable plasma diagnostic [136, 137].

The study of interactions between laser light and aerosols began in the 1960s with the availability of the first lasers. Haught and Polk reported already in 1966 the first observations of particle-induced breakdown in gases; that is, a breakdown on an electrostatically levitated single 20 μm particle of LiH [138]. It became immediately clear that the propagation of lasers through the atmosphere and hence the operative efficiency of lasers for applications from nuclear fusion to military weapons research were intimately connected to aerosols (see [139]–[141] and references therein).

Once irradiated with the laser beam, an aerosol particle starts to absorb, which leads to heating and further melting, boiling, and gradual evaporation (sublimation) of particle material (see [142] for an elegant visualization through a molecular dynamics simulation). The process may not be uniform because the distribution of the electromagnetic field inside the particle is not necessarily uniform, especially with internally mixed particles. Hot spots in the nodes of the field can virtually explode the particle before the thermal conductivity smooths out the temperature distribution. For particles small in comparison with the laser wavelength, the absorptivity decreases with d_p. The particle heating depends strongly on the heat losses caused by contact with the carrier gas and evaporation. The breakdown is promoted through aerosols by heating the surrounding gas and providing the breakdown zone with additional electrons caused by a number of mechanisms: heat explosion, shock waves in the surrounding carrier gas, breakdown in evaporated matter, electron emission (thermo-, photo- and triboelectrons). The increase in the initial electron density can be up to 10^7 cm^{-3}. Naturally, this decreases the threshold and reduces the time for a full plasma development.

The presence of aerosol particles in the focal volume will lower the threshold for gas breakdown by several orders of magnitude (typically around 10^7 W cm^{-2} compared with breakdown in air at 10^{11} W cm^{-2}). Aerosol breakdown depends upon the pulse width, wavelength, focusing of the laser and particle size; Figure 5.6 shows a well known summary of early results from several authors for breakdown at 10.6 μm (CO$_2$ laser). The threshold usually scales with the particle diameter as d_p^{-1} to d_p^{-2}. For a small focal volume with a sufficient irradiance $V(I_{th})$ for breakdown of a certain particle with diameter d_p, the threshold will depend upon the particle size distribution $f(d_p)$ and the total particle number density n_p. The threshold will then follow the inequality

$$\int_{r_{th}}^{\infty} n_p V(I_{th}) f(d_p) \mathrm{d}d_p \geq 1. \tag{5.15}$$

Aerosol-induced breakdown has been described by several authors with different wavelengths between 10.6 μm and 0.248 μm [143–154]. For pulse widths in the order of some microseconds, a wavelength dependence of λ^{-1} was observed, while with nanosecond pulse widths a λ^{-2} scaling was found [151]. The observed increase of the threshold with a decrease of the focal beam diameter is explained through inclusion of larger particles, which produce a higher initial increase in electron density, in the interaction volume [155].

The breakdown of droplets is modified through the curved liquid–gas interface [141, 156–160]. The droplet can be envisioned as a lens that concentrates the incident light

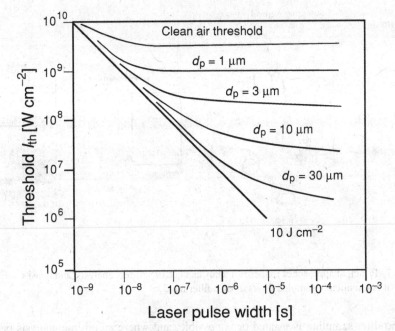

Figure 5.6. Dependence of the threshold intensity upon the laser pulse width for aerosol-induced breakdown (CO_2 laser, $\lambda = 10.6\ \mu m$, adapted from [150]).

wave to a localized region just within the droplet shadow face and focuses it to a localized region just outside the droplet shadow face. For large transparent droplets and moderate irradiance, the breakdown occurs just outside the shadow face at which the irradiance is highest. Whether the breakdown is initiated in the gas outside the droplet or within the droplet depends on parameters such as the breakdown threshold of the gas and liquid as well as the droplet morphology [161, 162]. An exact Lorenz–Mie formalism allows one to substantiate the simple conclusions from geometric optics, and has also quantified the amount of enhancement at all locations within the droplet, i.e. an internal field distribution and the corresponding enhancement factors of 100–1000 [163]. For the scaling of the threshold with the wavelength, Pinnick *et al.* found (for the harmonics of a Nd:YAG) thresholds between 4×10^7 and 3×10^9 W cm^{-2} for breakdown of 50 μm droplets [160]. The threshold scaled with decreasing wavelength as $\sqrt{\lambda}$. With shorter wavelengths an increased contribution of MPI and enhancement of internal fields was observed.

5.3 Analysis of aerosols by LIBS

In principle two different approaches for aerosol analysis via LIBS can be envisioned: (i) direct analysis in an on-line and *in situ* fashion, and (ii) quasi-continuous analysis of auto-matically sampled filters or on a filter band. Direct analysis is often used in process control

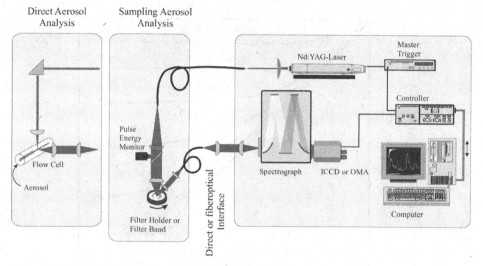

Figure 5.7. Principal approaches for aerosol analysis via LIBS: direct analysis, and quasi-continuous analysis of automatically sampled filters or on a filter band.

where no filter sampling is wanted or is possible, and where an on-line analysis permits fast turnaround times for process control. Filter sampling is more appropriate for long-term monitoring with extended sampling times usually in the order of 0.5–24 h; possible applications besides quasi-on-line monitoring of industrial emissions are workplace monitoring or characterization of environmental relevant aerosols. Figure 5.7 shows the experimental set-up for both types of aerosol analysis. In general the instrumentation is similar to other LIBS experiments (a Nd:YAG or excimer laser for plasma ignition), a high-resolution spectrograph with an intensified detector (CCD camera or OMA), and some kind of flow cell. For both filter sampling and the flow cell, the conditions described for proper sampling such as isokinetic inlet, temperature control to avoid condensation of water, etc., should be considered.

5.3.1 Aerosol sampling with LIBS

The analysis of aerosols with laser-induced breakdown spectroscopy differs significantly from other common analytical techniques such as ICP-AES, ICP-MS or MIP-AES with regard to the actual sample volume analyzed. These methods make use of a continuous analyte-carrying gas stream and temporal signal integration to yield an analyte signal that represents the average analyte mass within the total volume of carrier gas analyzed. Aerosol particles are likewise introduced into the steady-state plasma, where they are vaporized and contribute to the overall analyte response. These methods are sensitive to both gaseous and particulate analyte fractions; however, the resulting signal response provides no information as to the presence or relative contribution from either gaseous or particulate species. In contrast, LIBS-based analysis makes use of a series of laser-induced plasmas to interrogate a similar continuous analyte-carrying gas stream. By their very nature, however, the individual laser-induced plasmas constitute discrete sample volumes collected from within the overall carrier gas stream. If the resulting spectra from many laser pulses are ensemble averaged, the resulting analyte signal is representative of the average analyte mass within the carrier gas stream in a manner analogous to the aforementioned methods. However, a unique aspect of LIBS is that the individual spectra contain additional information about the discrete analyte nature of the sample stream. For aerosol analysis, the LIBS technique is well suited to utilize the discrete sampling nature of laser-induced plasmas, thereby enabling optimal sampling strategies, single-particle analyses, and aerosol sample concentration schemes. A number of LIBS implementation strategies for aerosol analysis are available and will be discussed in this chapter. An appropriate starting point for further development of LIBS aerosol analysis is the consideration of aerosol sampling rates, which are presented here following the treatment reported by Hahn and co-workers [164, 165].

Perhaps the most fundamental parameter with regard to aerosol analysis with the LIBS technique is the actual volume of the laser-induced plasma. The plasma volume is a complex function of the laser-beam geometry, focusing optics, irradiance, and gas-stream conditions. Complete details regarding plasma initiation, plasma properties, and plasma morphology are presented elsewhere in this book. In general, the plasma is formed along the beam path just in front of the laser beam focal spot, resulting in a shape corresponding to a truncated cone, as reported in 1983 by Radziemski *et al.* [166]. Plasma data based on plasma imaging studies, transmission measurements, and sampling considerations have been reported in the literature (see [165]–[169] and references in Section 5.2), with plasma volumes typically on the order of 10^{-3} to 10^{-4} cm^3. The plasma sampling calculations presented in this chapter make use of an effective plasma volume (V_{plasma}) of 10^{-3} cm^3, which corresponds to an equivalent spherical diameter of about 1.2 mm. The plasma volume and aerosol size distribution parameters can then be used to calculate the average number of aerosol particles expected within a single plasma volume, as well as the overall aerosol particle sampling rates.

For a population of homogenous aerosol particles, the overall mass loading (i.e. aerosol mass per unit volume) will be referred to as the aerosol mass concentration C_A, which is given by the expression

$$C_A = \rho \frac{\pi}{6} \bar{r}_{\text{vmd}}^3 N, \tag{5.16}$$

where ρ is the bulk aerosol particle density (mass/volume), N is the aerosol number density (particles/volume of gas), and $\bar{r}_{\text{vmd}}$ is the volume mean diameter of the aerosol particle distribution, which is defined in terms of the normalized aerosol particle size distribution function $p(r)$ through the relation

$$\bar{r}_{\text{vmd}} = \left\{ \int_{r=0}^{\infty} r^3 p(r) \mathrm{d}r \right\}^{1/3}. \tag{5.17}$$

The product of the plasma volume and the aerosol number density yields the average number of aerosol particles, μ, within each laser-induced plasma, namely

$$\mu = N \cdot V_{\text{plasma}}. \tag{5.18}$$

Equations (5.16) and (5.18) may be combined to define the average number of particles per plasma volume in terms of the aerosol mass concentration and the volume mean diameter

$$\mu = \frac{6 C_A V_{\text{plasma}}}{\pi \rho \bar{r}_{\text{vmd}}^3}. \tag{5.19}$$

For discrete aerosol particles, the probability distribution of the expected number of aerosol particles per plasma volume may be expressed using the Poisson distribution

$$P_n = \frac{\mu^n}{n!} \exp(-\mu), \tag{5.20}$$

where P_n defines the probability of finding n discrete aerosol particles within a given plasma volume. The aerosol sampling rate (R_A) may be defined as the percentage of laser-induced plasmas (i.e. laser pulses) expected to sample at least one particle. This sampling rate is directly calculated as 1 minus the probability of sampling zero particles, namely

$$R_A = 100 \cdot (1 - P_0) = 100 \cdot (1 - e^{-\mu}). \tag{5.21}$$

The aerosol sampling rate enables an examination of the LIBS-based aerosol analysis problem to be made in the context of discrete aerosol particles and a finite number of discrete plasma sample volumes. Using this model, it is useful to explore the aerosol sampling rates for relevant LIBS parameters and realistic aerosol size and concentration data, thereby elucidating key regimes suited to ensemble averaging or perhaps better suited to more sophisticated data analysis approaches owing to aerosol sampling limitations and spectral data with no information.

Aerosol sampling rates are presented in Figure 5.8 corresponding to aerosol mass concentration values of 5 μg m^{-3} and 100 μg m^{-3}, or approximately 4 p.p.b. and 85 p.p.b. (parts per billion) by mass, respectively, based on the density of ambient air. These values are consistent with many speciated aerosol loadings in ambient air or particulate emissions from industrial processes or waste treatment facilities. The calculations are based on a plasma volume of 1×10^{-3} cm^3 as discussed above and an average particle density of 2.5 g cm^{-3}. By using the volume mean diameter of the aerosol size distribution (equation (5.17)) for these calculations, no explicit particle distribution function is specified. The key feature of Figure 5.8 is the marked decrease in aerosol sampling rate with increasing volume mean diameter of the aerosol population. For the 5 μg m^{-3} mass concentration, the aerosol sample rate is about 10% for a mean diameter of 0.33 μm and falls to 1% for a diameter of 0.7 μm. Similarly for the 100 μg m^{-3} calculations, the 10% and 1% sampling rates correspond to volume mean diameters of 0.9 μm and 2.0 μm, respectively. Using the PM$_{2.5}$ cut-off as an example, LIBS-based analysis would generally be limited by very low aerosol sampling rates for all but the very finest of particles sizes (<100 nm). One of

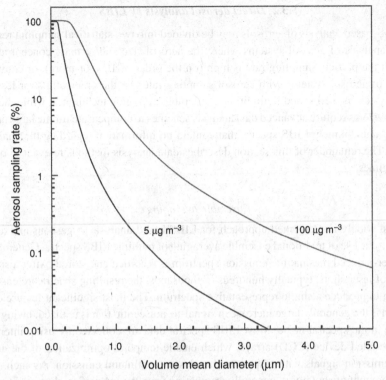

Figure 5.8. Aerosol sampling rates predicted as functions of aerosol size for two different aerosol mass loadings.

the most significant outcomes of low aerosol sampling rates with the LIBS technique is a potential decrease in method sensitivity (i.e. atomic emission intensity) for a given analyte species when ensemble averaging all collected spectra. For example, a 1% sampling rate corresponds to a single particle being sampled for each 100 laser pulses. If the analyte of interest is only contained within the aerosol particles, the analyte signal is diminished by a factor on the order of $1/10$ with ensemble averaging, assuming an increase in signal-to-noise of $\sqrt{100}$ is gained owing to a reduction in shot noise. For larger numbers of laser shots, the reduction in analyte signal scales directly as the sample rate, as ensemble averaging reaches an upper limit in the reduction of shot noise. Overall, the above calculations demonstrate that the LIBS-based sampling rates must be considered in the context of discrete aerosol particles and discrete plasma volumes. Accordingly, the aerosol mass concentrations and particle size distribution dictate the aerosol sampling rates, which can become exceedingly small for many realistic combinations such as ambient air measurements. In the following section, LIBS-based sampling strategies and data analysis schemes are formulated in the context of aerosol sampling rates.

5.3.2 *Direct aerosol analysis by LIBS*

The LIBS-based analysis of aerosols may be divided into two statistical sampling regimes, namely analysis of aerosol systems where the aerosol size and/or mass concentration is such that the particle sampling rate is high (on the order of 10% or more), or conversely analysis of aerosol systems with aerosol sampling rates on the order of 1% or less. The former system is well suited to traditional ensemble averaging techniques, while the latter aerosol systems require advanced data analysis schemes to compensate for the large amount of "null" data, namely LIBS spectra that contain no information regarding the analyte of interest. The remainder of this section describes data analysis methods relevant to each of these regimes.

Ensemble averaging

The most widely implemented approach for LIBS-based analysis of gaseous and aerosol systems is the use of traditional ensemble averaging of multiple LIBS spectra. Generally, for each laser-induced plasma, the emission spectrum is recorded and stored. After a suitable number of laser shots, typically hundreds or thousands, the resulting spectra are ensemble averaged to produce a single representative spectrum. The most significant feature of this approach is the generally large increase in signal-to-noise ratio that is realized, owing to the decrease in the spectral noise. Most LIBS spectral data are collected by using intensified charge-coupled device (CCD) arrays, which enable temporal optimization of the analyte atomic emission signals with respect to the plasma continuum emission. As such, a significant amount of shot noise, generally originating from the intensifier, characterizes each LIBS spectrum. An example of this behavior is illustrated in Figure 5.9, which contains a single-shot LIBS spectrum and a 100-shot ensemble-averaged LIBS spectrum recorded in ambient air. The prominent atomic emission line is the 247.86 nm C(I) emission peak due to the approximately 330 p.p.m. (part per million) level of carbon dioxide present in ambient air. The signal-to-noise ratio, as calculated based on the integrated carbon emission peak divided by the root-mean-square (r.m.s.) of the continuum emission adjacent to the carbon peak, is decreased by a factor of 6.2 when comparing the single-shot spectrum and the 100-shot average. Such data illustrate the use of spectral ensemble averaging with the LIBS technique.

The use of ensemble averaging for quantitative analysis of aerosol systems is limited to the evaluation of average analyte mass concentrations. Nonetheless, LIBS analysis offers the ability simultaneously to measure speciated mass concentrations for a number of analytes. Implementation for aerosol analysis is no different than that for analysis of gaseous species although several issues require additional attention. Specifically, (i) the total number of ensemble-averaged spectra must be large enough to ensure sufficient sampling of the aerosol population, (ii) the aerosol calibration source stream must be designed to ensure that the true LIBS analyte response is realized, and (iii) the targeted aerosol particle size distribution should lie within a size regime such that the largest particles can be totally vaporized within the laser-induced plasma. The exact number of required laser pulses to ensure adequate

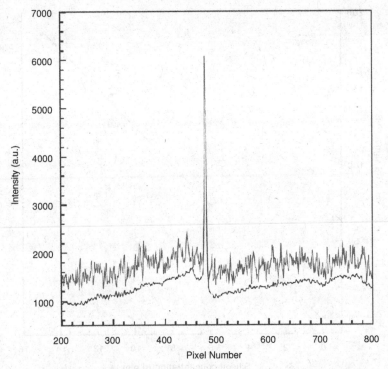

Figure 5.9. Representative single-shot (upper line) and 100-shot (lower line) ensemble-averaged LIBS spectra recorded in ambient air. The prominent line is the 247.86 nm C(I) emission line.

statistical sampling of an aerosol source is difficult to specify in general. Hahn and co-workers [165] performed Monte Carlo simulations in combination with experimental data to investigate the LIBS aerosol sampling problem in this context. It was concluded that approximately 20 aerosol particles were sufficient to characterize the average aerosol mass for rather broad distributions of submicrometer- to micrometer-sized aerosol distributions. Accordingly, a total of several hundred laser pulses is generally sufficient to determine aerosol mass concentrations using the guidelines discussed above, namely aerosol sampling rates greater than 10% for ensemble averaging.

Calibration schemes for LIBS-based aerosol analysis should be designed to produce an accurate analyte response for a known mass concentration. Although it is generally accepted that the resulting atomic emission from a laser-induced plasma is independent of the actual analyte source (i.e. atomic, molecular or aerosol), it is also expected that plasma non-homogeneity may affect the analyte response resulting from the spatial variation of discrete particles within the plasma. An ideal calibration approach would be the ensemble averaging of thousands of laser pulses in an aerosol stream composed of a high aerosol number density of submicrometer-sized particles. These conditions may be realized by nebulizing aqueous analyte solutions and subsequently drying the droplets in a gaseous

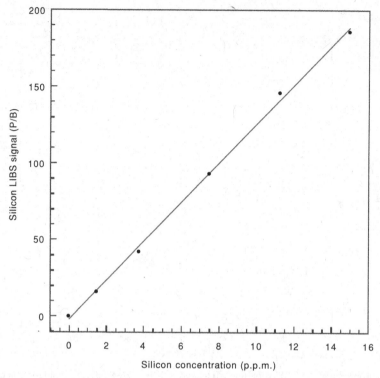

Figure 5.10. Typical LIBS calibration curve showing linear response of the 288.16 nm Si(I) emission line to changing concentration of submicrometer-sized silicon-rich aerosols.

flow stream [170]. This approach was used to produce the silicon (288.16 nm Si(I) atomic emission peak) calibration curve presented in Figure 5.10. This curve was produced with an aerosol stream of nominally 50 nm sized silicon-rich particles with a number density sufficiently high that each plasma volume contained on the order of 10^3 particles. The highly linear analyte response suggests complete vaporization of the silicon particles over this range of mass concentrations.

A final issue regarding quantitative analysis of aerosol systems concerns the upper particle size limit for complete particle dissociation and analyte signal size independence. Cremers and Radziemski [171] explored LIBS for the detection of beryllium particles deposited on filters corresponding to particle diameters of 50 nm, an ensemble collection of particles ranging from 0.5 μm to 5 μm, and nominally 15 μm sized particles. They used a cylindrical lens to focus a laser beam directly on the surface of the filters, thereby producing a plasma that engulfed the deposited beryllium particles, and subsequently collected the spectral emission. Cremers and Radziemski reported a different analyte response, manifest as different calibration curve slopes, for these three different particle-size classes, and concluded based on their experimental observations that incomplete particle vaporization

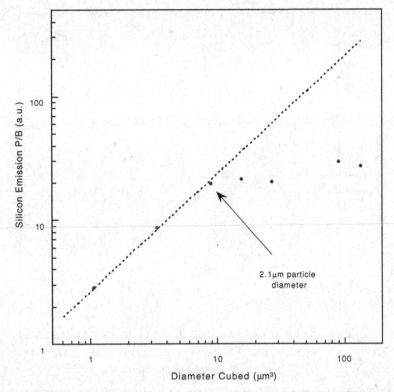

Figure 5.11. Linear and nonlinear response regions of monodisperse silica microspheres in a laser-induced plasma (after [173]).

occurred for particles with diameters greater than about 15 μm. In other studies, Radziemski and co-workers [166, 172] explored direct LIBS-based analysis of beryllium aerosols by using particles less than 10 μm in diameter, and noted in the latter study that such a particle size is consistent with complete particle vaporization. Several contemporary research papers have cited a 10 μm upper size limit for complete particle vaporization, most likely using the pioneering work of Radziemski *et al.* [166] as a useful guideline for quantitative LIBS analysis of aerosols. Given the recent interest in aerosol analysis with LIBS, the upper size limit was revisited in a study explicitly designed to quantify the aerosol response as a function of size.

Carranza and Hahn [173] investigated the laser-induced plasma vaporization of individual silica microspheres in an aerosolized air stream, which yielded an upper size limit for complete particle vaporization corresponding to a silica particle diameter of 2.1 μm for a laser pulse energy of 320 mJ, as determined by the deviation from a linear mass response of the silicon atomic emission signal. The results are summarized in Figure 5.11, which shows the silicon analyte response (P/B ratio) as a function of particle mass per plasma volume for single micrometer-sized microspheres. The most significant result is the clearly linear

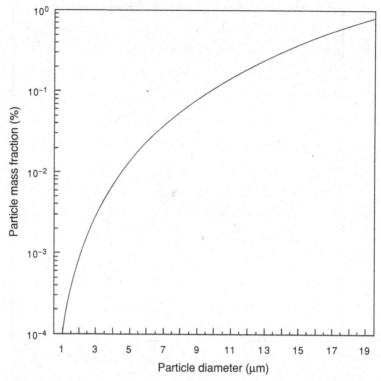

Figure 5.12. Estimated analyte mass fraction with respect to total plasma mass resulting from vaporization of a single particle of pure analyte.

relation between the silicon P/B ratio as a function of silicon particle mass (i.e. as diameter cubed) for the smallest three silica particle diameters (1.0 μm, 1.5 μm, and 2.1 μm), and the abrupt deviation from this linear trend for the particle diameters larger than 2.1 μm. For complete silica particle dissociation and vaporization, the resulting analyte signal (i.e. P/B) should scale as the analyte mass contained in the particle, hence as the diameter cubed. Accordingly, the break point in the data shown in Figure 5.11 represents the upper size limit in which an aerosolized silica particle is completely vaporized in the laser-induced plasma, for the experimental parameters reported in that study. It is noted that the limiting particle size of 2.1 μm is significantly below the frequently used 10 μm particle diameter limit, and should be taken into consideration for LIBS-based analysis of aerosol systems. Clearly additional experimental work and plasma modeling are needed to further determine the exact processes that govern particle vaporization, as well as to determine particle size limits for different laser pulse energies, wavelengths, focusing optics, and particle types.

To gain additional insight into the role of plasma–particle interactions, it is useful to consider the relative mass fraction due to an aerosol particle within a laser-induced plasma. Figure 5.12 presents the mass of a single aerosol particle (density of 2.5 g cm^{-3}) as a fraction of the mass of the gaseous plasma using the same plasma parameters as discussed above. The particle mass reaches 0.1% of the plasma mass at a diameter of about 10 μm, the often-cited upper limit for complete particle dissociation. Complete particle dissociation and subsequent heating of the particulate mass to the nominal plasma state require an amount of energy equal to the sum of the latent heat (i.e. heat of melting and heat of vaporization), the sensible heat (i.e. specific heat of gas-phase species), and the dissociation energies of the constituent species. Assuming comparable specific heats for various dissociated, gaseous species, the last fraction of energy will scale as the particle mass fraction with respect to total energy deposited into the plasma. Hence about 0.1% of the plasma energy will be consumed to heat the dissociated mass of a nominal 10 μm particle. However, heats of vaporization are typically two to three orders of magnitude greater than specific heat, hence the energy required to initially vaporize the particle is expected to be significant. Assuming a conservative value of the heat of vaporization a factor of 100 greater than the specific heat, the initial vaporization of the 10 μm diameter particle would consume a quantity of energy on the order of 10% of the energy required to heat the particle-free gaseous plasma. Clearly such an energy budget for particle vaporization is inconsistent with the primary tenet of LIBS-based aerosol analysis, namely that the analyte signal due to the presence of particulate-bound analyte fractions is simply additive to the overall plasma spectral emission.

More analyses and experimental data are required to understand the exact nature of plasma–particle interactions and the mechanisms responsible for such findings as the data presented in Figure 5.11. Additional work by Carranza and Hahn [174] demonstrated a linear analyte response for nanometer-sized silicon-based aerosols well beyond the equivalent mass contained in the 2.1 μm silica microsphere. This finding suggests that the plasma–particle vaporization process is controlled by parameters other than global energy conservation within the plasma, and that the consideration of process rates must be important. Therefore, plasma-induced processes such as thermophoretic forces and vapor expulsion may influence the vaporization dynamics in view of significant plasma temperature gradients. In view of the above data and discussions, it is reasonable to expect a linear mass response for LIBS-based aerosol analysis for submicrometer- to micrometer-sized aerosols, including a large fraction of the $PM_{2.5}$ class.

Conditional analysis

As discussed above, when the aerosol particle sampling rates become small, on the order of 1%, the inclusion of a large fraction of spectral data with no information with an ensemble-averaging approach can result in a greatly diminished analyte response. However, owing to the discrete nature of aerosols, the fraction of laser pulses that actually sample particles can contain significant analyte signal. Therefore, the use of a suitable conditional data analysis approach to identify the subset of analyte-rich spectral data corresponding to the presence of particles within given plasma volumes may greatly enhance the overall LIBS signal response. Such an approach was reported specifically for the analysis of aerosol particles based on the conditional analysis of plasma emission spectra on a pulse-to-pulse basis [165], while other researchers have used conditional analysis to increase the analyte response by rejecting both irregular spectra as well as spectra with poor analyte response [175]. In general, implementation of a conditional data analysis routine for LIBS-based aerosol analysis involves: (i) development of a threshold criterion for the detection of analyte-rich aerosol particles, i.e. "hits," within a given laser-induced plasma, and (ii) establishment of the total number of particle hits required to achieve a valid statistical sample of the aerosol population.

The total number of particle hits necessary adequately to assess the overall mass concentration of aerosols follows the same guidelines as discussed above regarding ensemble averaging, namely that on the order of 20 aerosol particles are sufficient to characterize the average aerosol mass for rather broad distributions of submicrometer-to micrometer-sized aerosol distributions. For a given average aerosol particle sampling rate, the total number of laser shots should then be selected to ensure at least 20–30 aerosol hits. Note, however, that to achieve an accurate measure of the aerosol sampling frequency, conditional data analysis routines should always be implemented with a fixed number of total laser pulses rather than acquiring pulses until a fixed number of hits is achieved [164]. Because the sampling of discrete aerosol particles is a random process, a rather large variance with respect to the number of total shots required to achieve a certain number of aerosol hits is expected. This will result in a large variation in the inferred aerosol sample rate (total hits/total shot), and therefore a large variation in the calculated aerosol mass concentration, as discussed in more detail below. This variance is avoided by terminating data collection after a fixed number of total shots. These two cases are analogous to determining the probability of rolling one number (i.e. one side) of a six-sided die by either rolling the die until the number is obtained, or rolling the die a fixed number of times and recording how many times the desired number is obtained. Clearly the latter approach is more accurate, in that the true probability of one-sixth is only obtained with the former approach if the desired number is rolled on the sixth try.

The remaining issue is determination of a suitable means to detect the presence of targeted analyte-rich aerosol particles, that is identification of particle hits for a given laser-induced plasma. A suitable threshold criterion should be selected taking into consideration the relatively large spectral noise associated with typical single-shot LIBS spectra, notably with intensified CCD detectors. Because the intensity of the laser-induced plasma emission is generally characterized by considerable variance on a shot-to-shot basis, it is desirable to normalize the analyte signal. A useful metric is the ratio of the analyte atomic emission peak (either integrated peak or peak intensity) to the intensity of the adjacent continuum emission. With this in mind, two approaches may be used for the identification of individual aerosol particles present within a given plasma [164], both of which make use of the analyte peak-to-base ratio. With the first approach, a peak-to-base threshold is selected such that the threshold value exceeds the maximum peak-to-base value obtained with the absence of any analyte species, in other words the threshold exceeds the normal spectral noise. This may be accomplished by increasing the threshold value, with no analyte present, until no particle "hits" are obtained for many thousands of laser pulses, hence the extreme fluctuations of the peak-to-base signal due to shot noise are insufficient to trigger an analyte hit. If the analyte cannot be readily removed from the source stream for this procedure, the threshold value may be obtained as based on an adjacent, surrogate spectral region of comparable intensity that is free from any analyte atomic emission lines. This latter approach is valid in that, over small spectral regions, the baseline continuum emission is relatively smooth and continuous, and the shot noise is random in nature. Regardless of how the threshold is determined, one drawback of this method is the potential loss of spectral data because of actual analyte hits that fall below the threshold criterion, which is necessarily strict owing to the large r.m.s. values of spectral noise on single-shot LIBS spectra. This limitation with respect to analyte discrimination may be reduced by using the second approach for threshold selection.

With the second approach, the threshold value as determined by using the above method is relaxed such that the analyte hit rate with no analyte present (i.e. false hit rate) is relatively low, on the order of 0.05% to 0.1%, but not exactly zero. This method enhances the sensitivity of the conditional analysis method and extends detection to smaller analyte-containing particles by not attempting to eliminate the extreme noise fluctuations. However, care must be exercised during data analysis to eliminate false analyte signals, in that the conditionally analyzed spectral data set contains spectra with no information (i.e. no actual analyte signal) that correspond only to the extreme shot-noise fluctuations. With true ensemble averaging of all spectra, the noise fluctuations about the analyte peak are averaged over all shots, thereby balancing all intensity fluctuations about the mean. This issue is readily avoided if additional analyte atomic emission peaks are available. If two or more emission lines are present on the recorded spectra, the additional atomic emission lines are used for quantitative analyte analysis, thereby eliminating any false hits (i.e. spectral noise) added to the analyte emission line used for triggering. This approach is valid because the shot noise is spectrally random, and it is statistically unlikely that the extreme noise values will occur simultaneously on multiple analyte emission lines. Note that the addition of a small number of spectra with no information to the ensemble average of spectra corresponding to particle hits does significantly impact the conditional analysis approach, in that the analyte signal may be proportionately diminished but the spectral r.m.s. noise is also reduced. More importantly, the inclusion of *random* spectra with no information, with respect to the alternative analyte emission lines, does not bias the analyte emission intensity owing to selective filtering of shot noise.

The methods discussed above enable the identification of LIBS spectra corresponding to discrete aerosol particles rich in the targeted analyte signal. This "preconcentration" of analyte-rich spectral data forms the basis of the conditional analysis approach to enhance the LIBS analyte response. The implementation of such a conditional data analysis scheme is relatively straightforward once a suitable method for the identification of analyte-peak-containing spectra, as described above, has been determined. The subset of spectra that are identified as containing analyte peaks that exceed the threshold criterion (i.e. the spectra corresponding to aerosol hits) are subsequently ensemble averaged to yield a single

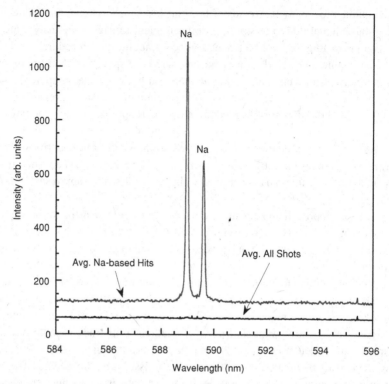

Figure 5.13. Ensemble-averaged spectrum of 13 200 laser shots (lower) and conditionally processed spectrum of a subset of 65 shots (upper) triggered on the presence of sodium. All spectra were recorded in ambient air in a 2 h period.

spectrum. For low aerosol sampling rates, the analyte signal is significantly enhanced in the conditionally analyzed spectrum owing to the rejection of spectral data without spectral analyte information. An example of this procedure is presented in Figure 5.13, which presents data corresponding to the detection of sodium-containing aerosol particles in ambient air. Two spectra are presented, namely the ensemble average of all 13 200 laser shots recorded during an approximately 2 h period, and the ensemble average of 65 spectra identified as containing sodium emission peaks using a conditional analysis routine triggered by the 589 nm sodium atomic emission line. The corresponding sodium-containing aerosol sampling rate was 0.49%, which resulted in a significant enhancement in the sodium peak signal-to-noise ratio when comparing the two spectra. The sampling rate of 0.49% results in an approximately 200-fold increase in the sodium peak-to-base ratio. The r.m.s. of the continuum emission intensity, a measure of signal noise, in the spectral regions adjacent to the sodium peaks increased by a factor of 2.5 when comparing the 65-shot ensemble average to the average of all 13 200 spectra. Taking all factors into account, the sodium emission signal, as measured by the signal-to-noise ratio, was enhanced by nearly two orders of magnitude with the conditional analysis routine, enabling excellent analyte sensitivity, when in contrast the traditional ensemble-averaged spectrum resulted in an essential non-detect.

With the above approach, the actual mass concentration of a given analyte species is determined by using a linear combination of the analyte signal as based on the subset of the conditionally analyzed spectra and the corresponding aerosol sampling (i.e. hit) rate. The spectrum based on the analyte hits is used in conjunction with a traditional LIBS calibration curve (e.g. analyte peak-to-base vs. known analyte concentration) to determine the corresponding equivalent concentration of the subset of analyte hits. Because this equivalent concentration is based on a subset of the total sampling data (i.e. all spectra), the actual or true analyte concentration is the product of this equivalent concentration and the aerosol sampling rate, namely

analyte concentration

$$= \text{(equivalent concentration of analyte hits)} \times \text{(total hits/total shots),} \quad (5.22)$$

where the latter term is the aerosol sampling rate. Note that as the sampling rate approaches 100%, the conditional analysis scheme reduces to a traditional ensemble-averaging scheme based on all laser shots. Furthermore, if two analyte peaks are used as discussed above (one for triggering and one for analysis), the effect of "false" analyte hits is minimal. A false hit reduces the intensity of the *analysis* emission peak but simultaneously increases the sampling rate, thereby producing the correct actual analyte concentration based on equation (5.22). Using this approach, the sodium concentration in the ambient air corresponding to the data in Figure 5.13 is 0.68 μg m^{-3}, or approximately 600 parts per trillion (p.p.t.) on a mass basis. As observed in Figure 5.13, such low sodium levels correspond to a non-detect with the simple ensemble-averaged spectrum.

Analysis of individual aerosol particles

Conditional data analysis was discussed above in the context of aerosol analysis as a means to enhance LIBS sensitivity for overall analyte mass concentration measurements by taking advantage of the discrete nature of aerosol particles. The adaptation of such conditional analysis schemes also makes feasible the detection and subsequent analysis of *single* aerosol particles corresponding to single laser pulses.

Quantitative analysis of the size and composition of single aerosol particles was reported in several recent papers [164, 165], and may be developed in terms of aerosol sampling considerations as discussed in equations (5.16)–(5.22). Consider the aerosol mass concentration for a monodisperse aerosol system which, in consideration of equation (5.16), is given by the expression

$$C_A = \rho \frac{\pi}{6} r^3 N, \quad (5.23)$$

where r is monodisperse aerosol diameter. Equation (5.22) may be recast as follows:

$$C_A = C_{\text{hit}} R_A, \quad (5.24)$$

where C_{hit} is the equivalent mass concentration of the subset of spectra corresponding to the particle hits and R_A is the aerosol particle sampling rate. The above two equations may be combined to yield an expression relating the equivalent mass concentration (i.e. LIBS response), the particle sampling rate, the particle mass, and the particle number density,

$$C_{hit} R_A = \rho \frac{\pi}{6} r^3 N. \tag{5.25}$$

For single aerosol analysis, the overall particle sampling rate must be sufficiently small to ensure that a single particle is within the plasma volume; hence the average number of particles per plasma volume μ (equation (5.18)) must be much less than unity. For small μ, the aerosol sampling rate given by equation (5.21) may be modified by using the series expansion $\exp(-\mu) = 1 - \mu$ to the order of μ^2, which yields $R_A = \mu$. Note that this expression uses the fractional sampling rate rather than a percentage; hence the factor of 100 is omitted. This relation may be combined with equation (5.18) to yield the aerosol sampling rate as

$$R_A = N \cdot V_{plasma}. \tag{5.26}$$

Equations (5.25) and (5.26) may be combined, and the number density N canceled, yielding the relation

$$C_{hit} V_{plasma} = \rho \frac{\pi}{6} r^3. \tag{5.27}$$

Noting that the right-hand side of this equation is the actual particle mass, this expression may be rearranged to yield the desired expression

$$C_{hit} = (\text{particle mass})/(\text{plasma volume}). \tag{5.28}$$

Equation (5.28) represents a direct relation between the equivalent mass concentration based on a calibrated LIBS analyte response, and the characteristic concentration represented by the actual particle mass in the discrete plasma volume. If the plasma volume is known, the above equation may be used to calculate the analyte mass in a single aerosol particle as the simple product of the plasma volume and the equivalent mass concentration, where the latter quantity is readily calculated from the corresponding single-shot spectrum and typical LIBS analyte calibration curve. The equivalent spherical diameter, r_{eq}, of the individual particle present in the plasma volume may be calculated from a modified form of equation (5.27), namely

$$r_{eq} = \left(\frac{6 C_{hit} V_{plasma}}{\pi \rho f} \right)^{1/3}, \tag{5.29}$$

where ρ is the particle bulk density (mass/volume), and f is the mass fraction of the analyte with respect to the overall bulk particle mass. For a pure, homogeneous particle f equals unity and the density is the actual density of the analyte.

The analysis of single-shot LIBS spectra for size, mass, and composition analysis corresponding to individual aerosol particles is straightforward using the above equations; however, the plasma volume must be known a priori. The plasma volume is readily measured

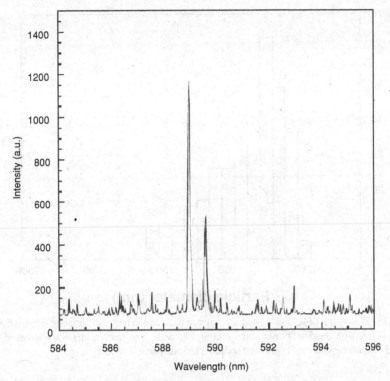

Figure 5.14. Representative single-shot LIBS spectrum recorded in ambient air. The spectrum was triggered using the 588.99 nm sodium emission peak.

by solving equation (5.29) with data for a characterized (i.e. known size and composition) monodisperse particle source. A suitable procedure is outlined following the approach reported by Carranza and Hahn [169]. Conditional data analysis may be used to collect spectra corresponding to single particle hits in a dilute, monodisperse aerosol stream. The equivalent mass concentration, C_A, based on the ensemble-averaged spectrum for all particle hits may be combined with the actual particle mass to determine the characteristic plasma volume. Once the plasma volume has been determined for a given set of LIBS parameters, this value may be used to calculate the analyte mass from any single LIBS spectrum using equation (5.29).

Using conditional-based analysis for detection of individual aerosols, a representative single-shot spectrum is presented in Figure 5.14. The spectrum corresponds to a single laser pulse recorded in an ambient air sampling stream, and represents one of the sodium particle hits corresponding to the conditionally analyzed data presented previously in Figure 5.13. The pronounced sodium doublet at 588.99 nm and 589.95 nm characterizes the spectrum. The submicrometer-sized sodium-containing particles were modeled as sodium chloride, consistent with sea salt derived atmospheric aerosols near coastal areas. The spectrum

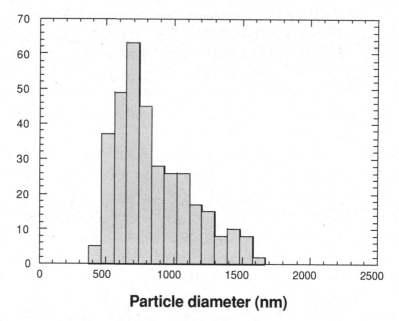

Particle diameter (nm)

Figure 5.15. Histogram of sodium-based aerosol size recorded in ambient air using single-shot LIBS spectra triggered on the 588.99 nm sodium emission peak. The particle diameter is based on the measured sodium mass using a spherical sodium chloride particle model.

corresponds to an absolute sodium mass of about 200 fg, with an equivalent spherical diameter of 750 nm as sodium chloride. A histogram of particle-size measurements based on single-shot analysis using the above algorithms is presented in Figure 5.15 corresponding to 399 sodium-based particle hits. The mean particle diameter is 840 nm with a modal diameter of 700 nm. The minimum recorded particle diameter of 450 nm represents the single-shot detection limit for sodium, which corresponds to a minimum detectable sodium mass of about 30 fg. These data demonstrate the overall sensitivity of the LIBS technique for single-particle analysis, as observed by the relative strength of the atomic emission lines on a shot-to-shot basis.

As discussed above, the ensemble-averaging approach (including data filtering algorithms) has been widely used as a tool to overcome the often extensive laser shot-to-shot spectral fluctuations, thereby improving the sensitivity of the LIBS technique. This approach, however, reaches a natural limit with the analysis of single-shot spectra. In contrast to ensemble averaging, single-shot LIBS analyses raise new questions regarding shot-to-shot fluctuations and precision. Although the issue of precision for LIBS-based analysis of aerosols remains a relatively undeveloped area of research, some preliminary data are presented in this chapter after the work of Carranza and Hahn [174].

The effect of laser pulse energy on the precision of the LIBS signal, as measured by the peak-to-base (P/B) and relative standard deviation (RSD) of the P/B of the 247.8 nm

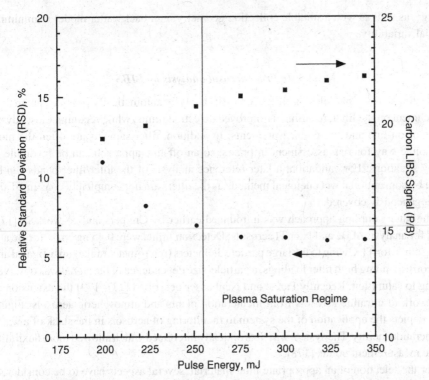

Figure 5.16. The effect of laser pulse energy on the 247.8 nm C(I) emission line recorded in purified air (after [174]).

C(I) atomic emission line, is illustrated in Figure 5.16. The source of carbon was carbon dioxide in a purified ambient air stream, which provided a homogeneous carbon source (~100 p.p.m. of atomic carbon) at the molecular level. To provide a valid comparison of carbon emission at different pulse energies, the carbon emission signal was optimized temporally for each different pulse energy, specifically using the ratio of the peak area of the carbon atomic line to the continuum emission intensity. The relative standard deviation of the carbon signal decreases with increasing laser pulse energy. Moreover, the rate of decrease from 200 mJ to about 255 mJ was significantly higher than for pulse energies in excess of 255 mJ. In fact, the RSD of the peak-to-base values is essentially constant above 250 mJ. For the LIBS parameters used (10 ns, 1064 nm laser pulse, and $f = 75$ mm focusing optic), the pulse energy of about 255 mJ corresponds to the saturation value for absorption of pulse energy by the plasma. The results suggest a natural breakpoint at the saturation energy, resulting in a higher degree of precision, for these LIBS parameters. A possible conclusion based on these preliminary experiments is that the enhanced precision obtained in the plasma saturation regime is the result of a decrease in shot-to-shot spatial variability of the plasma. Hence it is concluded that single-shot LIBS-based measurements should be made with sufficient laser pulse energy to achieve saturation with respect to absorbed pulse energy, as well as with a suitable collection geometry (e.g. backscatter mode) to minimize spatial variability.

5.3.3 Indirect aerosol analysis by LIBS

Collecting aerosol particles on filters for LIBS analysis automatically provides an analyte enrichment on the filter, resulting in improved detection limits, while retaining most advantages of on-line and *in situ* measurements. In addition, filter samples are often the most economic way for risk assessment in praxis, so an off-line approach can be favorable in many situations. For validation, a later reference analysis of the filter/filter bands can be done by conventional wet chemical methods as the filters are not completely consumed and homogeneously covered.

The filter sampling approach was introduced earlier by Cremers and co-workers [171, 176] for analysis of Be and later Tl aerosols. Detection limits were 0.45 ng cm^{-2} for Be and 40 ng cm^{-2} for Tl. Owing to the large particle diameters (micrometer range) investigated and the corresponding high filter loadings, a particle-size dependence of the signal was observed owing to saturation. Recently Panne and Neuhauser described [177–179] the extension of this work to ultrafine aerosols from incineration plants and atmospheric aerosols. Figure 5.17 depicts the application of the system to monitoring of aerosols in the stack of a waste incineration facility. The system was fiberoptically interfaced which allowed more flexibility at the measurement points [178].

For the selection of an appropriate filter material, several aspects have to be considered. Besides a sufficient temperature resistance (up to 200 °C) for utilization within an exhaust stream of incineration facilities, an adequate sampling efficiency of nearly 100% for ultrafine particles in combination with high flow rates is a prerequisite [180]. Furthermore, heavy metal blank values of the filter should be low [181] and for a large number of measurements the cost per single filter should be moderate. Figure 5.18a gives an impression of the overall ablation sampling on an aerosol filter, while Figure 5.18b and Figure 5.18c illustrate the differences in the ablation of a blank filter and a filter with a high mass load. Figure 5.18d displays the low mass loadings of typical filter sample taken at a waste incineration facility during routine sampling. Although the crater diameter is similar for both filters in Figures 5.18b,c, the depth of ablation is dependent on the mass loading. For low and moderate mass loadings, i.e. <10 μg cm^{-2}, nearly all aerosol mass was ablated with the first pulse, which is in good agreement with the observed penetration depth (below 100 μm) of aerosols on this type of filter [44]. For most modern industrial processes the observed aerosol mass is rather low because of the efficiency of the removal of particles in the micrometer range. In reality, for most end-of-pipe samples and atmospheric aerosol samples, the total mass loading for most real samples is below 1 μg cm^{-2}. Hence, the analysis can be restricted to a single pulse per filter spot. In this way linear calibrations were achieved for several heavy metal aerosols with typical detection limits in the order 10–400 ng cm^{-2}. In terms of concentration, this amounts for a sampled volume of 1 m^3 and a filter area of 30 mm^2 to detection limits between 0.1 μg m^{-3} and 1 μg m^{-3}. Longer sampling times at high flow as well as conditional analysis

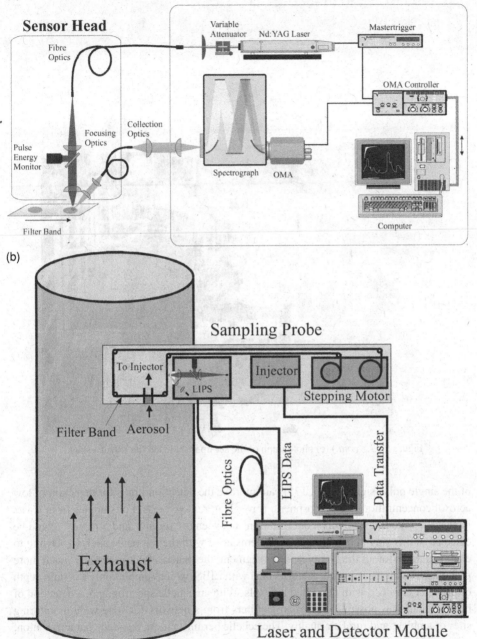

(a)

Sensor Head

Fibre Optics

Pulse Energy Monitor

Focusing Optics

Collection Optics

Filter Band

19"-Rack

Variable Attenuator

Nd:YAG Laser

Mastertrigger

OMA Controller

Spectrograph

OMA

Computer

(b)

Sampling Probe

To Injector

LIPS

Injector

Stepping Motor

Filter Band

Aerosol

Fibre Optics

LIPS Data

Data Transfer

Exhaust

Laser and Detector Module

Figure 5.17. Experimental set-up for LIPS analysis of heavy metal aerosols on filters. (a) Overview of the fiberoptically-coupled set-up; (b) installation with an automatic filter sampling device at an exhaust stack of a waste incineration (adapted from [177]); (*cont.*)

(c)

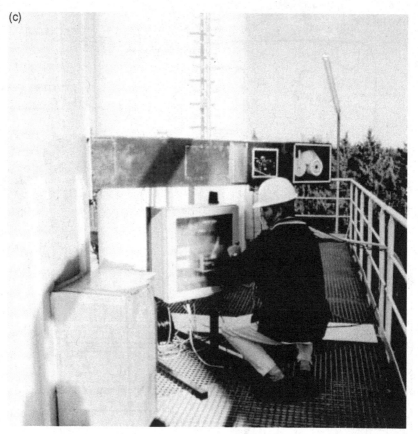

Figure 5.17. (*cont.*) (c) photograph of the set-up installed at the exhaust stack.

of the single pulse signals [165, 175] can improve the detection limits for detection of low aerosol concentration, e.g. atmospheric aerosols in remote areas. For samples from waste incineration facilities, volume concentrations between 0.1 μg m^{-3} and 5 μg m^{-3} could be detected in this way [179], although problems arise with the reference analysis. Owing to chemical digestion of the complete filter segment, the matrix effect is actually much more pronounced with the reference analysis than with LIBS, which samples only a certain depth of the filter enriched with deposited aerosols. With an echelle spectrograph, a detection of heavy metals was possible on numerous filters from a network for state-wide monitoring sites of ambient and rural air quality. The echelle permits a more detailed characterization, owing to the improved detection limits and the superior spectral resolution and spectral range. As an example, Figure 5.19 shows an irregular emission event which was detected on one of the filters from a town area. Besides cadmium (Figure 5.19a). (150 ng per filter, 20 ng cm^{-2}, 6 ng m^{-3}), lead was found in the same filter at elevated Pb (Figure 5.19b) concentrations (1.13 μg per filter, 150 ng cm^{-2}, 47 ng m^{-3}).

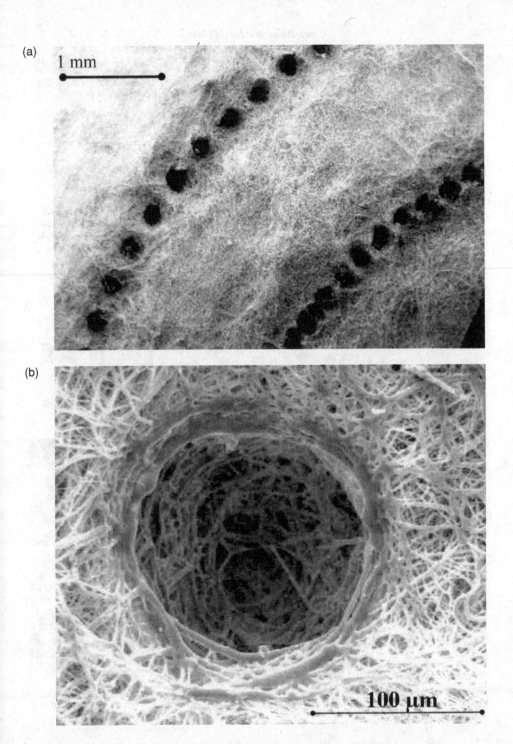

(a)

1 mm

(b)

Figure 5.18. Four SEM pictures of LIBS sampling on (a) a filter, (b) a crater on a blank filter (*cont.*)

(c)

(d)

Figure 5.18. (*cont.*) (c) a calibration filter with Cd(CH$_3$COO)$_2$ aerosols, and (d) a filter sampled at a waste incineration facility.

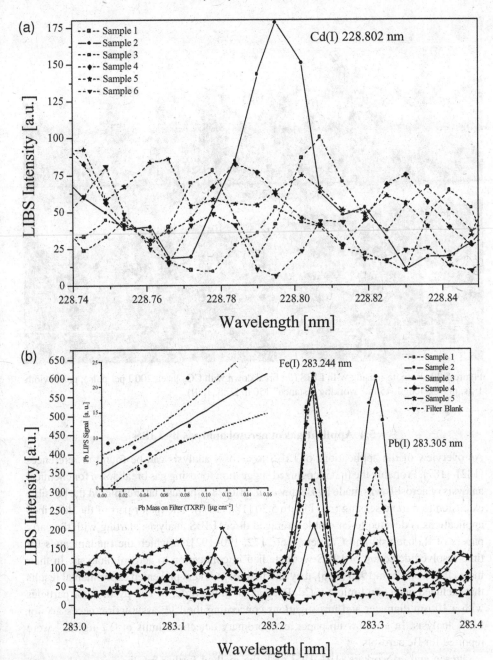

Figure 5.19. Cadmium (a) and lead (b) emission lines observed with the echelle system on filter samples from ambient environmental aerosol monitoring. The inset in (b) shows the LIPS signal vs. TXRF reference analysis (dashed lines: 95% confidence interval of the (solid) regression line, adapted from [179]).

Figure 5.20. Remote sensing with LIBS (air breakdown with CO_2 laser, 200 J per pulse, pulse width 1 μs, plasma length 15 m, "working distance" 150 m [188, 189]).

5.4 Applications of aerosol analysis by LIBS

An overview of the applications of LIBS to aerosol analysis can be found in references [182]–[187]. Probably the first recognized suggestion for using gas breakdown for chemical analysis of aerosols was made by Belaev *et al.* in 1978, who also later explored the possible extension to remote sensing (see Figure 5.20 [188, 189]). The major part of the described applications is devoted to ensemble-averaged direct LIBS analysis, starting with the early papers of Radziemski and Cremers [166, 172, 190–192], in which the fundamentals for time-resolved detection of LIBS were studied in connection to aerosol analysis. In their most cited work from 1983 [166], they and their co-workers presented experimental results that included detection limits for Cd, Pb, and Zn in the order of 200 μg m^{-3} for particles with a 10 μm diameter and investigations concerning the LTE assumption necessary for LIPS analysis. In a follow-up paper, extraordinary detection limits of 0.7 μg m^{-3} were reported for Be aerosols.

Ottesen and co-workers [193–196] took the method further into process analysis and demonstrated the use of LIBS for analysis of single coal particles in combustion systems. Because the particle size is in the order 50–100 μm, this approach featured a combination of particle sizing and LIBS to counter the observed incomplete evaporation. Further studies were focused on the quantitative analysis via inclusion of plasma diagnostics into the calibration, with the use of an internal standardization method (C(I) at 247.86 nm). The overall success was, however, subdued through severe matrix effects and the limitations of the available detector and spectrograph technology of that time which permitted only a sequential normalization through spectral scanning. Process analysis was also the objective of the work of Singh and co-workers [197–199]. They studied different optical arrangements at a coal-fired flow facility and achieved detection limits for different heavy metals in the range 1–600 $\mu g\ m^{-3}$. On-line detection of Cr aerosols for a fast monitoring of particulate emissions in an electroplating facility was reported by Neuhauser *et al.* [200]: the prototype was tested in cooperation with an independent laboratory, which determined the total chromium content. The system provided both the necessary time-resolution and detection limits (14 $\mu g\ m^{-3}$) for emission monitoring. A good correlation was found between the on-line LIBS measurement and the independent reference analysis. Similar results, however, by laboratory measurements have also been demonstrated [201, 202].

The results of the September 1997 DOE/EPA demonstration of continuous emission monitoring technologies provided a good indication of the performance of the available LIBS systems in the USA and the direct analysis approach [198, 203]. At a rotary kiln incinerator aerosols containing As, Be, Cd, Cr, Pb, Hg, and Sb were generated in concentrations is the range 15–100 $\mu g\ m^{-3}$ to test for the USA maximum achievable control technology (MACT) rules. Although the two LIBS systems demonstrated a fast response time, their overall precision and accuracy (between 50% and 150%) were rather low and could not meet the required 20% accuracy mandated by regulatory agencies. However, none of the other tested analyzers, i.e. on-line ICP-AES, MIP-AES, spark spectroscopy, and XRF on a filter band, was able to meet this required criterion. The spectral interferences observed for the two LIBS systems can be easily overcome with an echelle spectrograph, while the issues of precision and calibration remain more difficult. Batch-sampling on a filter/filter band was also recommended for LIBS to deal with the calibration and precision problems. Detection of mercury was a problem for all analyzers. Note, however, that concentrations in the range 15–100 $\mu g\ m^{-3}$ are rather high and can no longer be expected in the exhaust of systems with modern control technology (see Table 5.4). Later work by Hahn *et al.* and by Buckley *et al.* [204] revealed that LIBS with conditional analysis gave considerably better results in terms of detection limits and accuracy, including excellent agreement with independent extractive sampling, but also conventional direct analysis can be improved as demonstrated by Monts *et al.* [205].

In the past, several studies were devoted to analysis of liquid aerosols with particle diameters in the μm range. The obvious advantage is that the particles can be easily generated via nebulization, and particle-size distribution and chemical composition can be varied independently of each other [206, 207]. Not surprisingly, most applications deal with fundamental phenomena such as matrix effects due to droplet morphology and composition as well as determination of plasma parameters, i.e. n_e and T_e [158, 163, 208–211]. The detection limits were comparable to solid aerosols, i.e. in the order of 1–100 $\mu g\ m^{-3}$. Note that the accuracy can suffer in these experiments as the aerosol concentration is often directly inferred from the droplet generator or nebulizer without any reference analysis for validation.

Because of the detection limits in the order of micrograms per cubic meter, lower than conditional analysis, few studies were focused on atmospheric aerosols [212]. Carranza et al. measured ambient concentrations of Al, Ca, Mg, and Na as low as 5 ng m^{-3} using conditional signal processing, and where able to detect transient changes in magnesium- and aluminum-based aerosols due to the discharge of fireworks [213]. Nunez et al. [214] studied the detection of sulfuric acid, which is relevant for several atmospheric processes. A direct via the S(I) at 182.034 nm emission yielded a detection limit of 165 p.p.b. (by volume) after an integration time of 15 min. However, laser photofragmentation (LPF), after interaction of NaCl and H$_2$SO$_4$, gave improved detection limits of 46.5 p.p.b. (by volume) in only 10 s. Laser photofragmentation (LPF) and subsequent optical detection of atomic emission can be a very valuable alternative to LIBS in some cases. For alkali species, which are actively involved in the breakdown, corrosion, and erosion of power plant materials, LPF can be a sensitive method [215–217]. Similar observations were reported for the detection of Hg by Tong et al. [218, 219]. The metalloids and the halides are problematic elements for current LIBS systems owing to the low sensitivity of the detection systems and the low emission strength in the NIR. Consequently, a buffer gas (e.g. He) can be employed to increase the emissivity for the NIR lines. In this way, Tran et al. [220] demonstrated the detection of gaseous and particulate fluorides. For particulate fluorides in air, limits of detection were 9 mg m^{-3}, and for detection in He 0.5 mg m^{-3}, significant improvements were reported for filter-sampled aerosols, which allowed for a 5 μg m^{-3} detection limit in He after a 10 min sampling time at 10 l min^{-1}.

Another extension of the direct analysis approach is the use of the laser plasma only as an atomization reservoir, and subsequent single element analysis by excitation of atomic fluorescence with a second laser (LEAF, laser-enhanced atomic fluorescence). Neuhauser et al. demonstrated a detection limit in the ng m^{-3} range for ultrafine lead aerosols produced with a DMA [221]. However, the detection limit increased from 55 ng m^{-3} for a particle diameter of 48 nm to 130 ng m^{-3} for a particle diameter of 300 nm. The increasing detection limit with increasing particle diameter probably resulted from the incomplete atomization of larger particles in the colder periphery of the plasma. A suitable way for improving LIBS analysis of some of the relevant heavy metal species, i.e. Sn, Hg, and As, is the formation of the volatile hydrides [197, 222], and, in the case of Mg, the direct detection of the mainly gaseous species [223]. In all cases detection limits in the order of 50 μg m^{-3} were observed.

To determine the best conditions for the analytical performances of gas breakdown, several studies have examined the effects of ambient conditions on the laser plasma in different gases. Yalcin et al. [168, 224] investigated temporally and spatially resolved plasma temperature and electron density at atmospheric pressure in air. Good agreement was found between electron densities measured by using different atoms and emitted lines, between temperatures measured by using different elements, and between ionization and excitation temperatures for those elements. Changing the ambient gas, the laser energy, particulate levels, and humidity levels produced little variation in the temperature and electron density in the laser spark. Similar investigations concerning the spatial and temporal profiles of air breakdown were reported by Chen *et al.* [167], while other studies reported the wavelength dependence of gas breakdown and the plasma characteristics in different gases [118, 225–229].

5.5 Future directions

Possibilities for improvement of the direct analysis approach either are technology driven or are in the methodology. The advent of echelle spectrographs certainly extends the possibilities of coping with spectral interference and improving the approach to calibration. The combination of a kilohertz diode laser-pumped Nd:YAG and fast-readout ICCD cameras could improve the sampling efficiency and hence the overall analytical performance of the direct approach to LIBS aerosol analysis and conditional sampling by several orders of magnitude. The extension of echelle technology to the visible–ultraviolet range will permit the analysis of non-metals such as S, P, Cl, or Br, which are of considerable relevance for both process analysis and atmospheric aerosols. The direct approach can also benefit from aerodynamic lens systems, which not only allow the focusing of an aerosol stream for a short distance [85, 230–236], and hence a considerable enrichment, but also allow simultaneous sizing of ultrafine aerosols. Another interesting option to reduce possible matrix effects is on-line condensation of a matrix on the aerosols. This could reduce the variability of the plasma ignition and subsequent elemental emissions [83, 84]. Necessary detection limits on the order of $1 \, \mu g \, m^{-3}$ and lower make future applications to atmospheric sciences challenging. Process control is expected to be a significant LIBS application in the future. Owing to the improved emission control measures for all types of industrial emission, the requirements for the detection limits will be tightened in future and the emphasis will shift to ultrafine particles. In that case, conditional analysis will be a sine qua non condition to meet user requirements. The EPA/DOE test of 1997 showed that aerosol LIBS has similar problems to other LIBS applications in terms of accuracy and precision, so there is certainly room for considerable improvement in the general LIBS methodology. The filter-based approach offers some advantages for long-term monitoring of atmospheric aerosols, especially from remote or rural areas. Automatic filter band sampling could be easily combined with other aerosol characterization methods such as optical absorption for black carbon, Raman spectroscopy, and/or fluorescence spectroscopy. The latter could be also combined with the direct approach utilizing the fourth harmonic of a Nd:YAG for both plasma ignition and fluorescence spectroscopy (e.g. for bioaerosols). In this context, echelle spectrographs, which allow a combination of different techniques, i.e. Raman and LIBS, in a single instrument, could provide a decisive advantage.

5.6 References

[1] J. M. Prospero, R. J. Charlson, V. Mohnen *et al.*, *Rev. Geophys. Space Phys.*, **21** (1983), 1607–1630.
[2] J. H. Seinfeld, *Atmospheric Chemistry and Physics of Air Pollution* (New York: John Wiley & Sons, 1986).

[3] R. M. Harrison and R. E. Van Grieken, *Atmospheric Particles* (New York: John Wiley & Sons, 1998).

[4] K. R. Spurny, *Analytical Chemistry of Aerosols* (Boca Raton, FL: Lewis Publishers, 1999).

[5] P. C. Reist, *Aerosol Science and Technology* (New York: McGraw-Hill, 1993).

[6] W. C. Hinds, *Aerosol Technology* (New York: John Wiley & Sons, 1982).

[7] B. J. Finlayson-Pitts and J. N. Pitts, *Atmospheric Chemistry* (New York: John Wiley & Sons, 1986).

[8] J. E. Penner, D. Hegg and R. Leaitch, *Environ. Sci. Technol.*, **35** (2001), 332A–340A.

[9] M. C. Jacobson, H.-C. Hansson, K. J. Noone and R. J. Charlson, *Rev. Geophys.*, **38** (2000), 267–294.

[10] K. T. Whitby, *Atmos. Environ.*, **12** (1978), 135–159.

[11] O. Preining, *J. Aerosol Sci.*, **29** (1998), 481–495.

[12] P. F. Caffrey, J. M. Ondov, M. J. Zufall and C. I. Davidson, *Environ. Sci. Technol.*, **32** (1998), 1615–1622.

[13] J. M. Ondov and A. S. Wexler, *Environ. Sci. Technol.*, **32** (1998), 2547–2555.

[14] A. E. Suarez and J. M. Ondov, *Energy & Fuels*, **16** (2002), 562–568.

[15] L. S. Hughes, G. R. Cass, J. Gone, M. Ames and I. Olmez, *Environ. Sci. Technol.*, **32** (1998), 1153–1161.

[16] C. A. Pio, L. M. Castro, M. A. Cerqueira *et al.*, *Atmos. Environ.*, **30** (1996), 3309–3320.

[17] J. C. Chow, J. G. Watson, Z. Q. Lu *et al.*, *Atmos. Environ.*, **30** (1996), 2079–2112.

[18] S. N. Pandis, A. S. Wexler and J. H. Seinfeld, *J. Phys. Chem.*, **99** (1995), 9646–9659.

[19] J. E. Fergusson, *The Heavy Elements: Chemistry, Environmental Impact and Health Effects* (Oxford: Pergamon Press, 1991).

[20] J. O. Nriagu, *Nature*, **338** (1989), 47–49.

[21] J. M. Pacyna and E. G. Pacyna, *Environ. Rev.*, **9** (2001), 269–298.

[22] J. O. Nriagu and J. M. Pacyna, *Nature*, **333** (1988), 134–139.

[23] T. Reichhardt, *Environ. Sci. Technol.*, **29** (1995), A360–A364.

[24] C. A. Pope III, *Aerosol Sci. Technol.*, **32** (2000), 4–14.

[25] G. Oberdörster, R. M. Gelein, J. Ferrin and B. Weiss, *Inhalation Toxicol.*, **7** (1995), 111–124.

[26] B. Veronesi and M. Oortgiesen, *Neurotoxicol.*, **22** (2001), 795–810.

[27] H. Horvath, M. Kasahara and P. Pesava, *J. Aerosol Sci.*, **27** (1996), 417–435.

[28] A. G. Allen, E. Nemitz, J. P. Shi, R. M. Harrison and J. C. Greenwood, *Atmos. Environ.*, **35** (2001), 4581–4591.

[29] H. Lannefors, H. C. Hansson and L. Granat, *Atmos. Environ.*, **17** (1983), 87–101.

[30] F. E. Kruis, H. Fissan and A. Peled, *J. Aerosol Sci.*, **29** (1998), 511–535.

[31] S. E. Pratsinis, *Prog. Energ. Combust. Sci.*, **24** (1998), 197–219.

[32] W. H. Walton and J. H. Vincent, *Aerosol Sci. Technol.*, **28** (1998), 417–438.

[33] D. Y. H. Pui, *Analyst*, **121** (1996), 1215–1224.

[34] L. C. Kenny, *Analyst*, **121** (1996), 1233–1239.

[35] R. M. Harrison and R. Perry, *Handbook of Air Pollution Analysis* (London: Chapman and Hall, 1986).

[36] R. C. Flagan, in *Measurement Challenges in Atmospheric Chemistry*, L. Newman (Washington, DC: ACS, 1993), pp. 185–210.

[37] P. H. McMurry, *Atmos. Environ.*, **34** (2000), 1959–1999.

[38] P. Monkhouse, *Prog. Energ. Combust. Sci.*, **28** (2002), 331–381.

[39] J. Friedlander, *Aerosol Sci.*, **2** (1971), 331–340.

[40] J. Friedlander *Aerosol Sci.*, **1** (1970), 295–307.

[41] O. Haupt, B. Klaue, C. Schaefer and W. Dannecker, *X-Ray Spectrom.*, **24** (1995), 267–275.

[42] G. Steinhoff, O. Haupt and W. Dannecker, *Fresenius J. Anal. Chem.*, **366** (2000), 174–177.

[43] R. Harmel, O. Haupt and W. Dannecker, *Fresenius J. Anal. Chem.*, **366** (2000), 178–181.

[44] O. Haupt, K. Linnow, R. Harmel, C. Schaefer and W. Dannecker, *X-Ray Spectrom.*, **26** (1997), 79–84.

[45] C. F. Wang, C. J. Chin, S. K. Luo and L. C. Men, *Anal. Chim. Acta*, **389** (1999), 257–266.

[46] C. F. Wang, S. L. Jeng, C. C. Lin and P. C. Chiang, *Anal. Chim. Acta*, **368** (1998), 11–19.

[47] C. J. Chin, C. F. Wang and S. L. Jeng, *J. Anal. Atom. Spectrom.*, **14** (1999), 663–668.

[48] T. Buchkamp and G. Hermann, *Spectrochim. Acta*, **B 54** (1999), 657–668.

[49] J. Sneddon and Y.-I. Lee, *Microchem. J.*, **67** (2000), 201–205.

[50] C. Lüdke, E. Hoffmann, J. Skole and S. Artelt, *Fresenius J. Anal. Chem.*, **355** (1996), 261–263.

[51] C. Lüdke, E. Hoffmann and J. Skole, *Fresenius J. Anal. Chem.*, **359** (1997), 399–403.

[52] G. A. Petrucci, P. Cavalli and N. Omenetto, *Spectrochim. Acta*, **52** (1997), 1597–1615.

[53] M. Schmeling, R. Klockenkämper and D. Klockow, *Spectrochim. Acta*, **B 52** (1997), 985–994.

[54] M. Schmeling, *Spectrochim. Acta*, **B 56** (2001), 2127–2136.

[55] A. C. John, T. A. J. Kuhlbusch, H. Fissan and K. G. Schmidt, *Spectrochim. Acta*, **B 56** (2001), 2137–2146.

[56] R. Van Grieken and C. Xhoffer, *J. Anal. Atom. Spectrosc.*, **7** (1992), 81–88.

[57] H. Kawaguchi, N. Fukasawa and A. Mizuike, *Spectrochim. Acta*, **B41** (1986), 1277–1286.

[58] U. K. Bochert and W. Dannecker, *J. Aerosol Sci.*, **20** (1989), 1525–1528.

[59] D. P. Baldwin, D. S. Zamzow and A. P. D'Silva, *J. Air Waste Manage. Assoc.*, **45** (1995), 789–791.

[60] S. Kaneco, T. Nomizu, T. Tanaka, N. Mizutani and H. Kawaguchi, *Anal. Sci.*, **11** (1995), 835–840.

[61] C. C. Trassy and R. C. Diemiaszonek, *J. Anal. Atom. Spectrom.*, **10** (1995), 661–669.

[62] A. M. Gomes, J. P. Sarrette, L. Madon and A. Almi, *Spectrochim. Acta*, **B 51** (1996), 1695–1705.

[63] M. D. Seltzer and G. A. Meyer, *Environ. Sci. Technol.*, **31** (1997), 2665–2672.

[64] M. D. Seltzer, *Appl. Spectrosc.*, **52** (1998), 195–199.

[65] T. Myojo, M. Takaya and M. Ono-Ogasawara, *Aerosol Sci. Technol.*, **36** (2002), 76–83.

[66] W. Schelles, K. J. R. Maes, S. De Gendt and R. E., Van Grieken, *Anal. Chem.*, **68** (1996), 1136–42.

[67] R. K. Marcus, M. A. Dempster, T. E. Gibeau and E. M. Reynolds, *Anal. Chem.*, **71** (1999), 3061–3069.
[68] A. J. R. Hunter, S. J. Davis, L. G. Piper, K. W. Holtzclaw and M. E. Fraser, *Appl. Spectrosc.*, **54** (2000), 575–582.
[69] R. Oikari, M. Aho and R. Hernberg, *Energ. Fuel.*, **17** (2003), 87–94.
[70] R. Oikari, L. Botti and R. Hernberg, *Appl. Spectrosc.*, **56** (2002), 1453–1457.
[71] Y. Okada, J. Yabumoto and K. Takeuchi, *J. Aerosol Sci.*, **33** (2002), 961–965.
[72] P. P. Woskov, K. Hadidi, P. Thomas, K. Green and G. Flores, *Waste Manage.*, **20** (2000), 395–402.
[73] P. P. Woskov, D. Y. Rhee, P. Thomas, *et al.*, *Rev. Sci. Instrum.*, **67** (1996), 3700–3707.
[74] Y. Duan, Y. Su, Z. Jin and S. P. Abeln, *Anal. Chem.*, **72** (2000), 1672–1679.
[75] Q. Jin, C. Zhu, M. W. Borer and G. M. Hieftje, *Spectrochim. Acta*, **B 46** (1991), 417–430.
[76] H. Suto, M. Matsuura, K. Iinuma, S. Uchida and K. Takayama, *Rev. Sci. Instrum.*, **72** (2001), 4434–4441.
[77] C. A. Noble and K. A. Prather, *Mass Spectrom. Rev.*, **19** (2000), 248–274.
[78] A. L. Hunt and G. A. Petrucci, *Trends Anal. Chem.*, **21** (2002), 74–81.
[79] M. V. Johnston and A. S. Wexler, *Anal. Chem.*, **67** (1995), A721–A726.
[80] S. H. Wood and K. A. Prather, *Trends Anal. Chem.*, **17** (1998), 346–356.
[81] J. O. Allen, D. P. Fergenson, E. E. Gard *et al.*, *Environ. Sci. Technol.*, **34** (2000), 211–217.
[82] Z. Z. Ge, A. S. Wexler and M. V. Johnston, *Environ. Sci. Technol.*, **32** (1998), 3218–3223.
[83] S. N. Jackson and K. K. Murray, *Anal. Chem.*, **74** (2002), 4841–4844.
[84] D. B. Kane and M. V. Johnston, *Anal. Chem.*, **73** (2001), 5365–5369.
[85] G. A. Petrucci, P. B. Farnsworth, P. Cavalli and N. Omenetto, *Aerosol Sci. Technol.*, **33** (2000), 105–121.
[86] P. D. Maker, T. R. W. and S. C. M. (New York: Columbia University Press, 1963), pp. 1559–1565.
[87] R. G. Meyerand Jr. and A. F. Haught, *Phys. Rev. Lett.*, **11** (1963), 401–403.
[88] C. DeMichelis, *IEEE J. Quant. Electron.*, **5** (1969), 188–202.
[89] C. G. Grey Morgan, *Rep. Prog. Phys.*, **38** (1975), 621–665.
[90] N. Bloembergen, *J. Nonlin. Opt. Phys. Mat.*, **6** (1997), 377–385.
[91] T. P. Hughes, *Plasmas and Laser Light* (New York: Wiley, 1975).
[92] Y. B. Zel'dovich and Y. P. Raizer, *Physics of Shock Waves and High-Temperature Hydrodynamic Phenomena* (New York: Dover Publications, 2002).
[93] G. Bekefi, *Principles of Laser Plasmas* (New York: Wiley, 1976).
[94] J. F. Ready, *Effects of High-Power Laser Radiation* (New York: Academic Press, 1971).
[95] J. R. Bettis, *Appl. Opt.*, **31** (1992), 3448–3452.
[96] G. S. Voronov, *Zh. Eksperim. Teor. Fiz.*, **51** (1966), 1496–1498.
[97] G. S. Voronov, G. A. Delone and N. B. Delone, *Zh. Eksperim. Teor. Fiz., Pis'ma Redakts.*, **3** (1966), 480–483.
[98] P. Agostini, G. Barjot, G. Mainfray, C. Manus and J. Thebault, *IEEE J. Quant. Electron.*, **6** (1970), 782–788.
[99] P. Agostini, G. Barjot, J. F. Bonnal *et al.*, *IEEE J. Quant. Electron.*, **4** (1968), 667–669.

[100] G. Baravian and G. Sultan, *Physica, B+C*, **128** (1985), 343–352.

[101] A. L'Huillier *J. Phys.*, **C9** (1987), C9–415/C9–425.

[102] C. H. Chan, C. D. Moody and W. B. McKnight, *J. Appl. Phys.*, **44** (1973), 1179–1188.

[103] G. M. Weyl and D. Rosen, *Phys. Rev.*, **A 31** (1985), 2300–2313.

[104] E. Yablonovitch, *Appl. Phys. Lett.*, **23** (1973), 121–122.

[105] E. Yablonovitch, *Phys. Rev. Lett.*, **32** (1974), 1101–1104.

[106] E. Yablonovitch, *Phys. Rev.*, **A 10** (1974), 1888–1895.

[107] L. V. Keldysh, *Zh. Eksperim. Teor. Fiz.*, **47** (1964), 1945–1957.

[108] A. D. MacDonald, *Microwave Breakdown in Gases* (New York: Wiley, 1966).

[109] A. V. Phelps, in *Physics of Quantum Electronics*, edited by P. Kelley, B. Lax and P. Tannenwald (New York: McGraw-Hill, 1966), pp. 538–547.

[110] P. F. Browne, *Proc. Phys. Soc.*, **86** (1965), 1323–1332.

[111] J. K. Wright, *Proc. Phys. Soc.*, **84** (1964), 41–6.

[112] E. K. Damon and R. G. Tomlinson, *Appl. Opt.*, **2** (1963), 546–547.

[113] N. M. Kroll and K. M. Watson, *Phys. Rev.*, **A 5** (1972), 1883–905.

[114] G. M. Weyl, D. I. Rosen, J. Wilson and W. Seka, *Phys. Rev.*, **A 26** (1982), 1164–1167.

[115] D. I. Rosen and G. Weyl, *J. Phys.*, **D 20** (1987), 1264–1276.

[116] H. T. Buscher, R. Tomlinson and K. Damon, *Phys. Rev. Lett.*, **15** (1965), 847–849.

[117] K. C. Byron and G. J. Pert, *J. Phys.*, **D 12** (1979), 401–408.

[118] J. B. Simeonsson and A. W. Miziolek, *Appl. Phys.*, **B 59** (1994), 1–9.

[119] R. Tambay and R. K. Thareja, *J. Appl. Phys.*, **70** (1991), 2890–2892.

[120] J. P. Davis, A. L. Smith, C. Giranda and M. Squicciarini, *Appl. Opt.*, **30** (1991), 4358–4364.

[121] J. Stricker and J. G. Parker, *J. Appl. Phys.*, **53** (1982), 851–855.

[122] R. A. Armstrong, R. A. Lucht and W. T. Rawlins, *Appl. Opt.*, **22** (1983), 1573–1577.

[123] Y. B. Zel'dovich and Y. P. Raizer, *Zh. Eksperim. Teor. Fiz.*, **47** (1964), 1150–1161.

[124] V. Kas'yanov and A. Starostin, *Zh. Eksperim. Teor. Fiz.*, **48** (1965), 295–302.

[125] A. Dalgarno and N. F. Lane, *Astrophys. J.*, **145** (1966), 623–633.

[126] B. Fabre, *Phys. Lett.*, **A 30** (1969), 167–168.

[127] R. L. Taylor and G. Caledonia, *J. Quant. Spectrosc. Radiat.*, **9** (1969), 681–696.

[128] R. L. Taylor and G. Caledonia, *J. Quant. Spectrosc. Radiat.*, **9** (1969), 657–679.

[129] M. Young and M. Hercher, *J. Appl. Phys.*, **38** (1967), 4393–4400.

[130] R. J. Dewhurst, G. J. Pert and S. A. Ramsden, *J. Phys.*, **B 7** (1974), 2281–2290.

[131] R. J. Dewhurst, *J. Phys.*, **D 10** (1977), 283–289.

[132] C. L. M. Ireland and C. G. Morgan, *J. Phys.*, **D6** (1973), 720–729.

[133] Y. Gamal and M. Abdel Harith, *J. Phys.*, **D 14** (1981), 2209–2214.

[134] M. Hanafi, M. M. Omar and Y. E. E. D. Gamal, *Rad. Phys. Chem.*, **57** (2000), 11–20.

[135] L. M. Davis, L. Q. Li and D. R. Keefer, *J. Phys.*, **D 26** (1993), 222–230.

[136] C. L. M. Ireland, A. Yi, J. M. Aaron and C. G. Morgan, *Appl. Phys. Lett.*, **24** (1974), 175–177.

[137] H. Sobral, M. Villagran-Muniz, R. Navarro-Gonzalez and A. C. Raga, *Appl. Phys. Lett.*, **77** (2000), 3158–3160.

[138] A. F. Haught and D. H. Polk, *Phys. Fluids*, **9** (1966), 2047–2056.

[139] V. E. Zuev, A. A. Zemlyanov, Y. D. Kopytin and A. V. Kuzikovskii, *High-Power Laser Radiation in Atmospheric Aerosols* (Dordrecht: D. Reidel, 1984).
[140] A. A. Lushnikov and A. E. Negin, *J. Aerosol Sci.*, **24** (1993), 707–735.
[141] R. L. Armstrong, *Appl. Opt.*, **23** (1984), 148–155.
[142] T. A. Schoolcraft, G. S. Constable, L. V. Zhigilei and B. J. Garrison, *Anal. Chem.*, **72** (2000), 5143–5150.
[143] S. V. Zakharchenko, L. P. Semenov and A. M. Skripkin, in *Proceedings of the International Conference on Lasers* (1980), pp. 864–866.
[144] D. E. Lencioni and L. C. Pettingill, *J. Appl. Phys.*, **48** (1977), 1848–1851.
[145] D. E. Lencioni, *Appl. Phys. Lett.*, **25** (1974), 15–17.
[146] D. E. Poulain, D. R. Alexander, J. P. Barton, S. A. Schaub and J. Zhang, *J. Appl. Phys.*, **67** (1990), 2283–228.
[147] I. Y. Borets-Pervak and V. S. Vorob'ev *Quantum Electron.*, **23** (1993), 224–230.
[148] D. C. Smith and R. T. Brown, *J. Appl. Phys.*, **46** (1975), 1146–1154.
[149] D. C. Smith, *J. Appl. Phys.*, **48** (1977), 2217–2225.
[150] G. M. Weyl, *J. Phys.*, **D12** (1979), 33–49.
[151] D. C. Smith, *Opt. Eng.*, **20** (1981), 962–969.
[152] S. T. Amimoto, J. S. Whittier, F. G. Ronkowski *et al.*, *AIAA J.*, **22** (1984), 1108–1114.
[153] J. E. Lowder and H. Kleiman, *J. Appl. Phys.*, **44** (1973), 504–550.
[154] A. A. Boni and D. A. Meskan, *Opt. Comm.*, **14** (1975), 115–118.
[155] G. H. Canavan and P. E. Nielsen, *Appl. Phys. Lett.*, **22** (1973), 409–410.
[156] R. G. Pinnick, A. Biswas, J. D. Pendleton and R. L. Armstrong, *Appl. Opt.*, **31** (1992), 311–317.
[157] P. Chylek, M. A. Jarzembski, N. Y. Chou and R. G. Pinnick, *Appl. Phys. Lett.*, **49** (1986), 1475–1477.
[158] A. Biswas, H. Latifi, P. Shah, L. J. Radziemski and R. L. Armstrong, *Opt. Lett.*, **12** (1987), 313–315.
[159] A. Biswas, H. Latifi, L. J. Radziemski and R. L. Armstrong, *Appl. Opt.*, **27** (1988), 2386–2391.
[160] R. G. Pinnick, P. Chylek, M. Jarzembski *et al.*, *Appl. Opt.*, **27** (1988), 987–996.
[161] P. Chylek, M. A. Jarzembski, V. Srivastava *et al.*, *Appl. Opt.*, **26** (1987), 760–762.
[162] W. F. Hsieh, J. H. Eickmans and R. K. Chang, *J. Opt. Soc. Amer.*, **B4** (1987), 1816–1820.
[163] R. K. Chang, J. H. Eickmans, W. F. Hsieh *et al.*, *Appl. Opt.*, **27** (1988), 2377–2385.
[164] D. W. Hahn and M. M. Lunden, *Aerosol Sci. Technol.*, **33** (2000), 30–48.
[165] D. W. Hahn, W. L. Flower and K. R. Hencken, *Appl. Spectrosc.*, **51** (1997), 1836–1844.
[166] L. J. Radziemski, T. R. Loree, D. A. Cremers and N. M. Hoffman, *Anal. Chem.*, **55** (1983), 1246–1252.
[167] Y. L. Chen, J. W. L. Lewis and C. Parigger, *J. Quant. Spectrosc. Radiat.*, **67** (2000), 91–103.
[168] S. Yalcin, D. R. Crosley, G. P. Smith and G. W. Faris, *Appl. Phys.*, **B68** (1999), 121–130.
[169] J. E. Carranza and D. W. Hahn, *J. Anal. Atom. Spectrom.*, **17** (2002), 1534–1539.
[170] D. W. Hahn, J. E. Carranza, G. R. Arsenault, H. A. Johnsen and K. R. Hencken, *Rev. Sci. Instrum.*, **72** (2001), 3706–3713.
[171] D. A. Cremers and L. J. Radziemski, *Appl. Spectrosc.*, **39** (1985), 57–63.

[172] M. Essien, L. J. Radziemski and J. Sneddon, *J. Anal. Atom. Spectrom.*, **3** (1988), 985–988.

[173] J. E. Carranza and D. W. Hahn, *Anal. Chem.*, **74** (2002), 5450–5454.

[174] J. E. Carranza and D. W. Hahn, *Spectrochim. Acta*, **B57** (2002), 779–790.

[175] I. Schechter, *Anal. Sci. Technol.*, **8** (1995), 779–86.

[176] S. D. Arnold and D. A. Cremers, *Amer. Indust. Hyg. Assoc. J.*, **56** (1995), 1180–1186.

[177] R. E. Neuhauser, U. Panne and R. Niessner, *Anal. Chim. Acta*, **392** (1999), 47–54.

[178] R. E. Neuhauser, U. Panne and R. Niessner, *Appl. Spectrosc.*, **54** (2000), 923–927.

[179] U. Panne, R. E. Neuhauser, M. Theisen, H. Fink and R. Niessner, *Spectrochim. Acta*, **B56** (2001), 839–850.

[180] K. Willeke and P. A. Baron, *Aerosol Measurement* (New York: Van Nostrand Reinhold, 1993).

[181] T. Berg, O. Royset and E. Steinnes, *Atm. Environ.*, **27A** (1993), 2435–2439.

[182] M. Z. Martin, M. D. Cheng and R. C. Martin, *Aerosol Sci. Technol.*, **31** (1999), 409–421.

[183] L. J. Radziemski, *Microchem. J.*, **50** (1994), 218–234.

[184] L. J. Radziemski and D. A. Cremers (editors) *Laser-Induced Plasmas and Applications* (New York: Marcel Dekker, 1989).

[185] J. Sneddon, *Trends Anal. Chem.*, **7** (1988), 222–226.

[186] B. W. Smith, D. W. Hahn, E. Gibb, I. Gornushkin and J. D. Winefordner, *Kona*, **19** (2001), 25–33.

[187] D. A. Rusak, B. C. Castle, B. W. Smith and J. D. Winefordner, *CRC Crit. Rev. Anal. Chem.*, **27** (1997), 257–290.

[188] E. B. Belyaev, A. P. Godlevskii and Y. D. Kopytin, *Kvantov. Elektron.*, **5** (1978), 2594–2601.

[189] A. P. Godlevskii, Y. D. Kopytin, V. A. Korol'kov and Y. V. Ivanov, *Zh. Prikl. Spektrosk.*, **39** (1983), 734–740.

[190] L. J. Radziemski, D. A. Cremers and T. R. Loree, *Spectrochim. Acta*, **B38** (1983), 349–355.

[191] T. R. Loree and L. J. Radziemski, *Plasma Chem. Plasma Process.*, **1** (1981), 271–279.

[192] L. J. Radziemski and T. R. Loree, *Plasma Chem. Plasma Process.*, **1** (1981), 281–293.

[193] D. K. Ottesen, in *Advances in Coal Spectroscopy*, edited by H. L. C. Meuzelaar (New York: Plenum Press, 1992), pp. 91–118.

[194] D. K. Ottesen, J. C. F. Wang and L. J. Radziemski, *Appl. Spectrosc.*, **43** (1989), 967–976.

[195] D. K. Ottesen, L. L. Baxter, L. J. Radziemski and J. F. Burrows, *Energ. Fuel*, **5** (1991), 304.

[196] D. K. Ottesen, *Appl. Spectrosc.*, **46** (1991), 593–596.

[197] J. P. Singh, H. Zhang, F. Y. Yueh and K. P. Carney, *Appl. Spectrosc.*, **50** (1996), 764–773.

[198] H. Zhang, F. Y. Yueh and J. P. Singh, *Appl. Opt.*, **38** (1999), 1459–1466.

[199] J. P. Singh, F. Y. Yueh, H. S. Zhang and R. L. Cook, *Process Cont. Qual.*, **10** (1997), 247–258.

[200] R. E. Neuhauser, U. Panne, R. Niessner and P. Wilbring, *Fresenius J. Anal. Chem.*, **364** (1999), 720–726.

[201] L. W. Peng, W. L. Flower, K. R. Hencken *et al.*, *Process Cont. Qual.*, **7** (1995), 39–49.

[202] W. L. Flower, L. W. Peng, M. P. Bonin *et al.*, *Fuel Process. Technol.*, **39** (1994), 277–84.

[203] P. M. Lemieux, J. V. Ryan, N. B. French *et al.*, *Waste Manage.*, **18** (1998), 385–391.

[204] S. G. Buckley, H. A. Johnsen, K. R. Hencken and D. W. Hahn, *Waste Manage.*, **20** (2000), 455–462.

[205] D. L. Monts, J. P. Singh, Y. S. Abhilasha *et al.*, *Combust. Sci. Technol.*, **134** (1998), 103–126.

[206] H. A. Archontaki and S. R. Crouch, *Appl. Spectrosc.*, **42** (1988), 741–746.

[207] D. E. Poulain and D. R. Alexander, *Appl. Spectrosc.*, **49** (1995), 569–579.

[208] C. Parigger and J. W. L. Lewis, *Opt. Commun.*, **12** (1993), 163–173.

[209] K. C. Ng, N. L. Ayala, J. B. Simeonsson and J. D. Winefordner, *Anal. Chim. Acta*, **269** (1992), 123–128.

[210] J. H. Eickmans, W. F. Hsieh and R. K. Chang, *Appl. Opt.*, **26** (1987), 3721–3725.

[211] M. Martin and M. D. Cheng, *Appl. Spectrosc.*, **54** (2000), 1279–1285.

[212] M. Casini, M. A. Harith, V. Palleschi *et al.*, *Lasre Part. Beams*, **9** (1991), 633–639.

[213] J. E. Carranza, B. T. Fisher, G. D. Yoder and D. W. Hahn, *Spectrochim. Acta*, **B56** (2001), 851–864.

[214] M. H. Nunez, P. Cavalli, G. Petrucci and N. Omenetto, *Appl. Spectrosc.*, **54** (2000), 1805–1816.

[215] S. G. Buckley, C. S. McEnally, R. F. Sawyer, C. P. Koshland and D. Lucas, *Combust. Sci. Technol.*, **118** (1996), 169–188.

[216] S. G. Buckley, R. F. Sawyer, C. P. Koshland and D. Lucas, *Combust. Flame*, **128** (2002), 435–446.

[217] K. T. Hartinger, P. B. Monkhouse and J. Wolfrum, *Ber. Bunsenges. Phys. Chem.*, **97** (1993), 1731–1734.

[218] X. Tong, R. B. Barat and A. T. Poulos, *Rev. Sci. Instrum.*, **70** (1999), 4180–4184.

[219] X. Tong, R. B. Barat and A. T. Poulos, *Environ. Sci. Technol.*, **33** (1999), 3260–3263.

[220] M. Tran, Q. Sun, B. W. Smith and J. D. Winefordner, *Appl. Spectrosc.*, **55** (2001), 739–744.

[221] R. E. Neuhauser, U. Panne, R. Niessner *et al.*, *Anal. Chim. Acta*, **346** (1997), 37–48.

[222] E. A. P. Cheng, R. D. Fraser and J. G. Eden, *Appl. Spectrosc.*, **45** (1991), 949–952.

[223] C. Lazzari, M. De Rosa, S. Rastelli *et al.*, *Laser Part. Beam*, **12** (1994), 525–530.

[224] S. Yalcin, D. R. Crosley, G. P. Smith and G. W. Faris, *Hazard. Waste Hazard. Mater.*, **13** (1996), 51–61.

[225] Y. L. Chen, J. W. L. Lewis and C. Parigger, *J. Quant. Spectrosc. Radiat.*, **66** (2000), 41–53.

[226] T. X. Phuoc and F. P. White, *Combust. Flame*, **119** (1999), 203–216.

[227] T. X. Phuoc and C. M. White, *Opt. Commun.*, **181** (2000), 353–359.

[228] A. Sircar, R. K. Dwivedi and R. K. Thareja, *Appl. Phys. B*, **63** (1996), 623–627.

[229] R. J. Nordstrom, *Appl. Spectrosc.*, **49** (1995), 1490–1499.

[230] B. Dahneke, *Nature*, **244** (1973), 54–55.

[231] J. Schreiner, C. Voigt, K. Mauersberger, P. McMurry and P. Ziemann, *Aerosol Sci. Technol.*, **29** (1998), 50–56.

[232] P. Liu, P. J. Ziemann, D. B. Kittelson and P. H. McMurry, *Aerosol Sci. Technol*, **22** (1995), 293–313.

[233] P. Liu, P. J. Ziemann, D. B. Kittelson and P. H. McMurry, *Aerosol Sci. Technol.*, **22** (1995), 314–324.

[234] M.-D. Cheng, L. Karr, J. Kornuc *et al.*, *Microchem. J.*, **72** (2002), 209–219.

[235] M. D. Cheng, *Fuel Process. Technol.*, **65–66** (2000), 219–229.

[236] J.-W. Lee, M.-Y. Yi and S.-M. Lee, *J. Aerosol Sci.*, **34** (2002), 211–224.

6

Chemical imaging of surfaces using LIBS

J. M. Vadillo and J. J. Laserna

Department of Analytical Chemistry, University of Málaga

6.1 Introduction

The increasing demand of materials science for new and different kinds of information has resulted in radical changes in instrument capabilities during the past decade. Indeed, the central challenge to materials science of achieving an understanding of the continuum from the atomic level through nanomaterials to bulk properties of matter can be faced only with tools and methods of unprecedent analytical potential. Of particular relevance to this central problem has been the response given by chemical imaging methods. While a large number of imaging methods are available, the term chemical imaging refers to the information provided by some classes of imaging methods that can incorporate information on chemical identity to spatial dimensions. In this chapter we review the uses of LIBS as a chemical imaging tool. The several operating modes of LIBS imaging are presented and its properties of spatial resolution are discussed. The broad front of applications that has resulted from these properties is illustrated with examples taken from real world analytical situations.

It was not until relatively recent times that scientists came to understand the relationships between the structural elements of materials and their properties, empowering to a large degree the characteristics of materials, surfaces and interfaces [1]. To understand fully the surface of a solid material, analytical techniques are needed that not only distinguish the surface from the bulk of the solid, but also approaches are required that distinguish the properties of each [2]. Facing the investigation of surfaces is a multi-parameter problem such that a complete knowledge of the properties and reactivity of a surface often implies gaining information from multiple points of view including physical topography, chemical composition, chemical structure, atomic structure, electronic state, and a detailed description of bonding of molecules at the surface. Not surprisingly, no single technique can provide all these different pieces of information, and frequently more than one technique is required [3].

Laser ablation (LA) has experienced a growth in interest during the past few years as a direct analysis technique or as a solid sampling technique, where the particulated material is transported to a second excitation source where it is analyzed by different methods.

Laser-Induced Breakdown Spectroscopy: Fundamentals and Applications, ed. Andrzej W. Miziolek, Vincenzo Palleschi and Israel Schechter. Published by Cambridge University Press. © A. W. Miziolek, V. Palleschi and I. Schechter 2006. A. W. Miziolek's contributions are a work of the United States Government and are not protected by copyright in the United States.

Laser-induced breakdown spectrometry (LIBS) falls in the category of analytical methods based on atomic emission. Most applications of LIBS have so far concentrated on specimens where the concentration is uniform over the analyzed volume and have involved measurement of the average composition of a sample. In this type of measurement, an extended area of the sample is allowed to interact repetitively with the laser beam and the characteristics of the averaged emission spectrum are used to assign the composition of the sample. In this sense, LIBS does not differ significantly from classic arc/spark procedures, sharing important similarities in methodology and instrumentation. As recognized from the beginning [4], spatial localization is an inherent attribute of laser ablation and thus a potential exists for the use of this sample localization capability to provide chemical information confined to restricted areas of a sample. However, the capabilities of LIBS for surface and interface analysis have been under-estimated. In general, the fact that laser ablation can penetrate through a considerable amount of material gives LIBS a high degree of generality, but also seems to imply that it will be difficult to implement the characterization of surfaces and interfaces as the influence of the surface region of a sample on the light emitted by the plasma will be much weaker than that of the surrounding bulk [5]. More recently, refinement in laser ablation technology has demonstrated success in restricting the information generated to specific areas of the interior and the surface region of a sample. The knowledge gathered so far reveals that a number of analytical problems relevant to structured materials can find a practical solution in LIBS.

This chapter addresses some of the important considerations in designing and implementing LIBS for the description of the surface and subsurface composition of materials. Several possible operational modes of LIBS as a surface analytical tool, including those required for chemical image generation, are discussed. Since background topics such as the fundamentals of laser plasma formation, the interaction of lasers with solid surfaces, and the instrumental aspects of LIBS are presented in other chapters of this book, we do not cover these issues here. However, those aspects defining the performance characteristics of LIBS in terms of spatial resolution are discussed to some extent. The current availability of ultrashort laser pulses has reopened the interest in investigating the basic mechanisms of laser–surface interaction. However, this chapter deals only with nanosecond lasers, which are still the workhorse for laser ablation studies, and for which wide experience has been gained. Significant applications of LIBS for surface-oriented tasks are also discussed to illustrate the potential of the approach.

6.2 LIBS chemical imaging: operational modes

An impressive range of methods for imaging structures inside and at the surface of materials has been developed. The term chemical imaging has been coined to describe the information provided by some classes of these imaging methods. Many forms of imaging provide topographic details from the sample under study. However, production of chemical images is restricted to those methods that can incorporate information on chemical identity into spatial dimensions. The term chemical image thus refers to the spatial distribution of single chemical species within a multicomponent sample and generally involves the production of chemical information in two dimensions or in higher-dimensionality formats such as tomographic analysis. However, in a broad sense, monodimensional information such as depth profiles can be also considered as a form of image analysis. From this point of view different operational modes may be assigned in LIBS.

(a) Lateral distributions (scan analysis): this implies the acquisition of consecutive spectra while the sample is being ablated along a specific path. In this way, intensity–position plots are obtained. Extraction of data at specific wavelengths (detector pixels) allows monitoring different elements along the line scan.

(b) Depth-profiling: in this instance, the sample is exposed to consecutive laser shots in a static position. In this way, each acquired spectrum provides information on subsequent eroded layers of material. As in case (a), the choice of specific wavelengths permits the investigation of different elements as a function of depth.

(c) Mapping (area scan): this implies lateral scanning, row by row, over a selected surface. The output is a map or surface distribution of the species of interest. A combination of operational modes (b) and (c) is also known as tomographic analysis. Experimentally, this would consist of performing a complete depth-profiling analysis at the different positions that define a selected surface. In this instance, the use of multichannel detectors results in extremely large information sets, comparable to those obtained with other classical techniques such as SIMS.

The technique of compositional mapping involves performing the analysis at chosen coordinates in the sample and the subsequent representation of the resultant compositional maps in the form of images in which the gray or color scale is related directly to the intensity (or concentration) of the elements. The images directly convey the sense of the spatial distribution and may serve to establish compositional interrelationships of the elemental constituents in the sample. A sketch of how the process is performed in LIBS is presented in Figure 6.1. Basically, the complete sequence of events may be summarized in the following steps.

(a) Choice of the sample region to be analyzed (represented with a dashed rectangle in the upper portion of Figure 6.1). This process may be performed by previous inspection of the zone by a microscopy system.

(b) Selection of the spectral region. As laser ablation is a (micro)destructive technique, a previous knowledge of the elements of interest is important in order to choose an adequate spectral region because, in most cases, a combination of a grating spectrometer and a multichannel detector is used and the limited spectral coverage of the detector will determine the maximum number of elements to be analyzed. This limitation is completely overcome with the use of echelle spectrometers that cover the spectrum from the ultraviolet to the near infrared with constant spectral resolution at levels as high as $60\,000$ $(\lambda/\Delta\lambda)$. Despite this impressive resolution, the applicability of these systems for chemical imaging purposes is still restricted owing to the inherently slow data acquisition, which derives from the low readout rates of commercially available CCD chips. With present technology, acquisition rates are confined to a practical limit of about 0.25 Hz. However, recent developments [6] with echelle spectrometers of lower performance have been achieved in a system dispersing between 220 nm and 390 nm (a very suitable spectral region for analytical

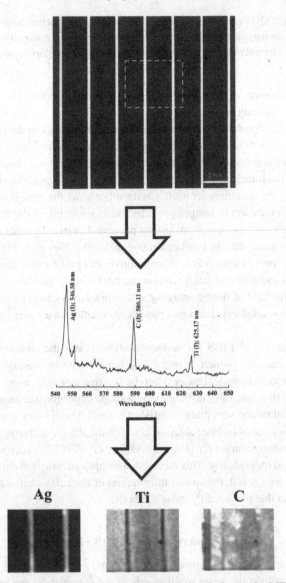

Figure 6.1. Sequence of events to generate chemical maps by LIBS.

purposes) with resolution of 16 000 ($\lambda/\Delta\lambda$) at 313 nm. The lack of moving parts in these echelle systems is an important factor to take into account as they are potentially free from repetitive wavelength calibration.

(c) Data acquisition. A different number of laser shots (depending on the lateral resolution required and the sampling area/volume) are acquired at a given sample position, the spectra are stored, and the sample is moved by means of a micrometric stage to another sampling point to start a new acquisition cycle.

(d) Data processing. This operation can build monochromatic images of the sample at any wavelength by retrieving from the computer memory the proper spectroscopic information for each location. The several operational modes described above are then available just by processing the data in the appropriate way.

Two important aspects of data processing in LIBS imaging should be mentioned. Multielemental chemical imaging usually results in large data sets. Hence, computerized data reduction becomes a significant step in the system performance. The data acquisition speed can be limited either by the sample positioning system, by the refreshing time of the detector, or by the laser repetition rate depending on the particular configuration being used. However, the throughput of the imaging system is often limited by the computational facilities instead of being limited by the spectrometer itself. On the other hand, the presentation of concentration information in the form of images provides the major utility of the imaging technique. It is often possible to recognize patterns in data presented as an image that would be difficult to derive from the same data in tabulated numerical form. However, the interpretation of information in the form of images is a major intuitive strength of the human mind. Although computer-assisted imaging and interpretation are beginning to provide automatic tools for image analysis, the field of digital imaging still depends very heavily on the observer as interpreter. Even for a dedicated computer the determination of a pattern automatically is a difficult task.

The main advantage of LIBS for surface analysis is not the quantitative accuracy of the chemical information (which is preferably done in the laboratory with other well-established chemical analysis techniques) but the fact that it enables users to survey a range of properties from the same sample, employing a single experimental set-up, while keeping the inherent flexibility and versatility of LIBS. Compared with other approaches used for surface investigation based on laser ablation in which the ablated material is transported to a second excitation/ionization source (LA-ICP-AES or LA-ICP-MS), each sampling position in LIBS is analyzed individually. This means that sample information derives from single-shot laser events. As a result, the spatial information of the LIBS chemical image is by far more accurate than that provided by those methods.

6.3 Spatial resolution in LIBS imaging

Spatial localization is an intrinsic property of LIBS as, usually, the laser is focused at the sample surface to obtain the beam irradiance required for ablation and plasma formation. The morphological parameters of the crater formed by a single laser shot (diameter, depth, and shape) define the three main properties associated to LIBS images, namely, lateral resolution, surface sensitivity, and depth resolution.

6.3.1 Lateral resolution

Lateral resolution refers to the capability of discriminating between atoms in adjacent positions on the surface. This property depends on the capability to focus the laser beam to a small spot. The minimum diameter to which a Gaussian beam can be focused with a single optical element is a function of the input beam parameters and is given by $d = k\lambda f D^{-1}$, where k is a constant, λ is the laser wavelength, f is the focal length of the focusing lens, and D is the diameter of the input beam. Thus the use of expanded, short-wavelength lasers, focused with short focal length lenses, produces smaller beam spots. Unfortunately, a number of factors contribute to degrading the lateral resolution in LIBS analysis, resulting in crater diameters larger than the beam waist. These parameters include imperfections in the focusing optics and the thermal diffusion length of the surface. Redeposition of ablated material in the crater rim also degrades lateral resolution.

In general an increase in laser fluence also tends to increase the crater size as more mass is ablated per laser pulse. Thus tightly focused, low-energy pulses produce the best lateral resolution. For spectral detection of the ablated material the minimum pulse energy to use is the plasma formation threshold or fluence needed for optical breakdown of the vapor. This parameter also depends on the surface characteristics and on several optical, thermophysical, and thermodynamical properties of the material. A survey of plasma threshold values for metals has been published [7]. Under optimal circumstances it is possible to obtain craters that are not much larger than the diameter of the beam spot. The practical limit to lateral resolution is found to be between 2 μm and 5 μm.

6.3.2 Surface sensitivity

This parameter describes the capability of a technique to provide data from the surface of the sample. It depends on the depth origin of the radiation detected and is mostly determined by the sampling depth, i.e. the average distance normal to the surface probed by the analysis technique. In LIBS the sampling depth is determined by the crater depth. The depth involved in single-shot laser ablation varies widely in solid materials, ranging from a few nanometers to several micrometers depending on the experimental conditions and material composition.

To ensure that LIBS is surface sensitive, conditions are chosen to ensure that the bulk signal is small compared with the surface signal, i.e. that the vast majority of the detected signal comes from the surface region of the sample. Three parameters can be considered when defining the surface sensitivity in LIBS, namely, laser pulse energy, angle of incidence of the laser beam on the surface, and beam focal conditions.

In general, an increase in pulse energy tends to decrease the surface sensitivity because a larger sample volume is ablated. The limiting sampling depth depends on the capability to detect an emission signal from the plasma on a single-laser-shot basis. For a spectral signal-to-noise ratio of 2, this limit approaches the ablated depth at the plasma formation threshold fluence. Under these circumstances the ablated depth per pulse has been calculated to be in the range ~1–8 nm [8, 9].

In LIBS the measured signal is strongest at normal incidence and decreases continuously as the angle of incidence moves toward grazing incidence. The reason for this effect is that the area of the beam at the surface increases continuously with increasing incidence,

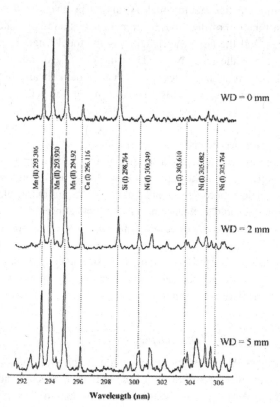

Figure 6.2. Comparison of single-shot LIP spectra of manganin on Si substrate at three different focal positions with ablation at 532 nm from a Nd:YAG laser. The thickness of the coating was *c.* 50 nm. WD is the distance from the sample to the lens focal position. Reprinted with permission from the American Chemical Society from reference [10].

resulting in an uniform decrease of beam irradiance at the surface. The surface sensitivity has been demonstrated to improve by a factor of 4 when changing the laser incidence angle from 0° to 60° [9]. Enough pulse energy has to be used for maintaining a sufficient signal level when working at large incidence angles.

Focal conditions of the laser beam affect the surface sensitivity of LIBS. For a given pulse energy, the crater depth increases as the laser is more tightly focused at the sample surface, with a subsequent degradation of the surface sensitivity. It is generally useful to locate the beam focal point at a distance above or below the sample surface in order to improve the surface sensitivity. The analysis of coatings is a good example of the importance of focal conditions. Figure 6.2 shows LIBS spectra derived from a manganin coating (an alloy of Cu 86%, Ni 2%, Mn 12%) deposited on a silicon wafer by pulsed laser deposition [10]. When the laser was focused at the sample surface and 2 mm above the surface, the corresponding spectra show lines from the manganin components and from silicon, indicating that both structural components are being ablated. Focusing the beam 5 mm above the surface results in a spectrum free from Si lines, i.e. the sampling depth is within the manganin coating thickness, resulting in good surface sensitivity, while still maintaining satisfactory spectral detection.

Good lateral resolution and simultaneous high surface sensitivity result in low signal level. Consequently there is a trade-off between spatial resolution and detection sensitivity. The degree of spatial resolution is determined by the required signal-to-noise ratio and the spatial distribution of the species of interest.

6.3.3 Depth resolution

This parameter measures the degree to which the experiment is able to define an abrupt interface and it is thus of interest in depth profiling of structured materials. In general, depth-profiling methods provide a profile that represents a distorted image of the true layer interface, i.e. the measured interface appears broader than the original in-depth distribution of composition. The broadening depends on a number of phenomena that are usually categorized into instrumental factors, beam–sample interactions, and sample characteristics. The broadening process is usually quantified by means of the depth resolution.

Depth resolution, as shown in Figure 6.3, is defined as the depth range over which the signal decreases (or increases) by a specified amount when profiling an ideally sharp interface between two media. By convention, the depth resolution corresponds to the depth range over which an 84% to 16% change in the full signal is measured. In LIBS, calculation of depth resolution requires the conversion of laser shots into depth by estimation of the ablation rate. A straightforward method for this task is to measure the average ablation rate, AAR (nm pulse^{-1}), over a profile, from the number of laser pulses required to reach the interface, $p_{0.5}$, and the thickness of the layer whose profile is being investigated, d,

$$\text{AAR} = d\,(p_{0.5})^{-1}. \tag{6.1}$$

The depth resolution, Δz (nm), is then calculated from the profile and AAR,

$$\Delta z = \Delta p\,\text{AAR} = \Delta p\,d\,(p_{0.5})^{-1}, \tag{6.2}$$

where Δp is the number of laser shots necessary to reach 84% and 16% of the signal. Equation (6.2) predicts that the best depth resolution (lowest Δz) is obtained from conditions in which the ablation rate is lowest. However, experiments with metal layers have demonstrated that the best values are obtained at moderate irradiance levels [11].

Depth resolution is also limited by the roughness of the original surface and, likely, by the formation of structures at the bottom of the crater. If the surface of the material to be analyzed is not flat to start with, it will not be possible to erode a well-defined crater and so a good depth resolution will not be possible. In addition, most materials develop rough structures (ripples) as laser ablation proceeds. The patterns are often unrelated to the beam profile and appear even if the incident beam is smooth and the surface is flat. These

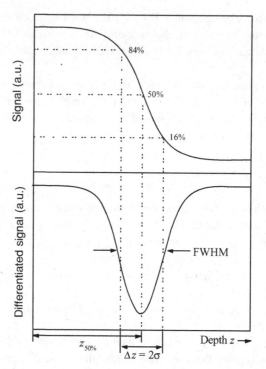

Figure 6.3. Definition of depth resolution for a Gaussian resolution function approximation [11]. Reproduced by permission of The Royal Society of Chemistry.

effects are usually caused by light scattering at material imperfections or dust particles. The mechanism involves material melting, deformation, and, after irradiation, resolidification, making the deformation permanent. The shape and extent of surface deformation depend on the absorbed fluence as well as on the material. Development of microroughness in the bottom of the crater can degrade the depth resolution of the profile but this effect has still to be demonstrated experimentally.

6.4 Applications of LIBS imaging

6.4.1 Scan analysis (lateral distributions)

Scan analysis is performed in LIBS by orienting the laser perpendicular to the direction of surface motion and then analyzing a series of spaced points, producing a composition profile. The number and spacing of the points are selected by considering the extent of the compositional variation and the spatial resolution of the analysis and the representativeness of the analysis. In general, a small crater is desirable and the experimental conditions (laser fluence, lens-to-sample distance, etc.) must be optimized. A useful "on-the-fly" determination of the spatial resolution without making use of visual examination of the crater may

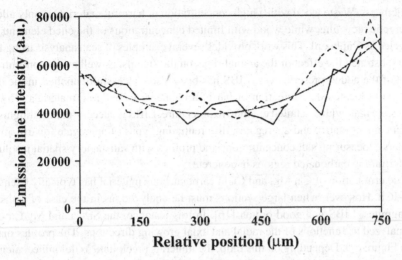

Figure 6.4. Variation in the Mg(I) emission line at 285.2 nm as a function of the spot position in a non-homogenized sample. The RSD on the net intensities was around 20%, but the evolution of the signal along the profile is not random. Distance between consecutive shots = 40 μm. Reprinted from reference [13] with permission from Elsevier Science.

be accurately determined by measuring the ratio of intensities of a given element measured at the same wavelength from adjacent peaks on the surface [12]. From a given distance between points this ratio is unity, indicating the complete absence of overlapping between craters.

Progress in focusing optics has resulted in successive improvements in lateral resolution. Mauchien and co-workers have been particularly active in the coupling of laser beams, especially the quadrupled Nd:YAG, into microscope objectives. Using reflecting objectives, lateral resolution in the range 6–8 μm has been routinely achieved [13]. This level of resolution, highly satisfactory from a microanalytical point of view, is larger by a factor of only 2 than the theoretical value expected considering the numerical aperture of the objective used and the laser wavelength. Among different experiments performed by the authors with a variety of samples, it is especially interesting to see one clear demonstration of the reproducibility of the technique at the microscopic level. For this purpose, two parallel line scans were performed along a distance of 750 μm with a spacing of 30 μm between shots. In every case, the intensity corresponded to a single-shot event at each position. As shown in Figure 6.4, the intensity RSD between points is large (about 20%), but the evolution of the signal along the profile was clearly non-random. It was concluded that the observed variations truly represent concentration changes at the microscopic level.

Apart from microanalysis of alloys or microelectronic materials, difficult applications to common surface analysis techniques have been demonstrated with LIBS. One example is the determination of wood preservatives, involving mainly Cu, Cr, B, Hg, Sn, and As [14]. In this application, the authors have built a transportable system to perform the analysis of these elements. Waste wood with a high concentration of harmful materials is only allowed for energetic recycling, while wood with limited concentrations of the cited elements may be recycled as chipboard. This work provides several examples of scan analysis using LIBS and demonstrated the effect of the annual rings on the signals, as well as the concentration profiles of the added preservatives. LIBS has been also successfully applied in the direct and spatially located determination of harmful elements in building materials [15]. The authors state that, with a detailed qualitative procedure, LIBS is successful in distinguishing between cement matrix and aggregates, differentiating types of aggregate (quartz, marble and others), measuring salt concentrations and profiles with satisfactory spatial resolution, and determining carbonated regions in concrete.

The determination of Ca, Mg, and Ca in carbonaceous material has typically been done by LA-ICP. However, when large sections must be analyzed, as in the case of stalactites or stalagmites, LIBS is a good option [16]. In this instance, the Sr/Ca and Mg/Ca ratios were analyzed as functions of the radial and axial growing directions. The profiles may be seen in Figure 6.5. Depending on this ratio, the minerals precipitate as dolomite, calcite, or aragonite, as determined by the solubility coefficient of the present elements at the formation temperature. In this sense, LIBS may be a complementary tool to perform paleoclimatic studies.

6.4.2 Depth profiling

One of the most difficult tasks in LIBS depth profiling involves converting the number of laser shots into depth [17]. Several approaches using stylus profilometers [18, 19], differential weight [20], or direct measurement by optical microscopy [17] have been tried. A new approach [13] involves the preparation of replicas of the craters by impregnating the sample with resins and performing posterior alkaline attack of the sample. Under these conditions, the resin may be prepared to be completely analyzed by SEM. A straightforward method consists of dividing the total crater depth by the number of laser shots [21, 22]. This calculation yields the ablation rate averaged over the measured profile. So far, there are no well-developed approaches for quantification in LIBS depth profiling, although several attempts are in progress [17]. Further effort is required to understand the erosion processes and develop a practical approach for this application.

In spite of these problems, several excellent works concerning depth-profile analysis may be found in the literature. In the first attempts, a simple comparison between spectra from different laser shots was presented [23–27]. However, new and more complex samples have been studied in order to demonstrate that LIBS is competitive enough with other established techniques. The resolution of multilayered samples is one application successfully resolved by LIBS [20]. In this application, a sample consisting of a commercial brass (Zn/Cu) that was deposited with Cu, Ni and Cr platings following an electrochemical process was analyzed. The LIBS analysis matched with the expected sample structure. One especially interesting application was the fast in-depth determination of phosphorus in photonic-grade silicon, an application typically restricted to SIMS [19]. LIBS is proposed as an alternative method,

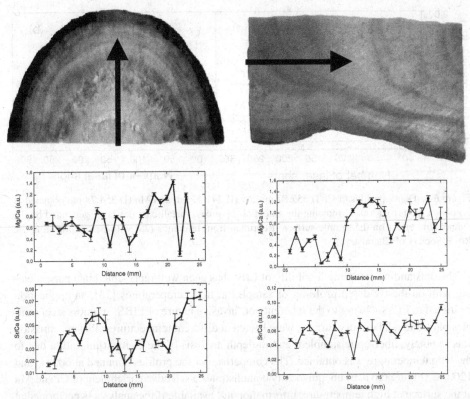

Figure 6.5. Spatially resolved variation of the Mg/Ca and Sr/Ca ratios as functions of the radial (left) and axial (right) growing directions in sections of stalactites. Each sample was rastered for 25 mm at 1 mm steps. The error bars correspond to the averaging of five scans in adjacent positions. Reprinted from reference [16] with permission from Springer-Verlag.

providing simplicity of operation and rapid data acquisition. For this material, the limiting sampling depth (non-optimized) was about 1.2 μm pulse^{-1}, allowing the determination of P depth as a function of the diffusion steps in the manufacturing process.

For applications requiring high surface sensitivity or the determination of submicrometric layers, different approaches must be found. The use of a collimated beam has probably been the most straightforward way to do it [8]. By exposing the sample to the collimated beam of an excimer laser, the irradiance on the sample was high enough to generate a plasma. In order to reach the plasma threshold, the energy per pulse was large, to compensate for the increment in spot area. The craters obtained present a regular shape with flat-bottom profiles, comparable to those obtained with a glow-discharge. An average sampling depth of 8 nm pulse^{-1} in the analysis of a 12 μm thick Zn layer of galvanized steel was reported. It is remarkable that the complete in-depth multi-elemental analysis of the sample was concluded in 3.3 min, a clear advantage when comparing the time required in other techniques with lower sputtering rates.

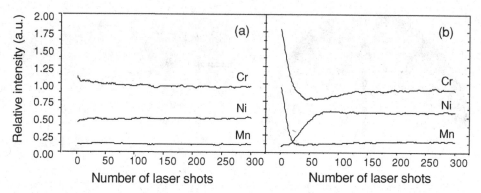

Figure 6.6. Depth profiles of Cr(I) 357.87 nm, Ni (I) 341.47 nm and Mn (I) 354.78 nm obtained at (a) 600 °C and (b) 1200 °C showing the extent of the mixed metallic oxides layer generated by the temperature effect on the sample surface. Reprinted from reference [28] with permission from the Royal Society of Chemistry.

The outstanding sampling flexibility of LIBS has been well represented in a paper dealing with on-the-fly depth profiling of samples at high temperatures [28], an application restricted to LIBS. Owing to the inherent non-invasive nature of LIBS, stainless steel samples placed in a laboratory furnace were heated at different temperatures. Plasma emission was remotely, fiber-optic sampled, and in-depth analysis of the element migration due to the high temperature was obtained. The comparison of the profiles obtained at 600 °C and 1200 °C (Figure 6.6 a, b) unequivocally demonstrates a selective enrichment of Cr and Mn at the surface at high temperature, information not available if the analysis is performed at room temperature.

Apart from improvements in the depth resolution, work has been directed to obtaining quantitative in-depth analysis. In a recent article [17], its authors have been able to translate the signal vs. laser shots axis into concentration vs. thickness for a galvanneal-coated steel. The thickness axis was calibrated by measuring the crater depth using optical microscopy, and certified samples were used to obtain the concentrations values.

6.4.3 Mapping

The first published compositional mapping article [29] applied the method in the analysis of coating coverage, coatweight distribution, and three-dimensional distribution of various pigments of paper coatings by their elemental composition (mainly Al, Ca, Mg and Si). Macroscopic areas of about 1 cm² were analyzed with a lateral resolution better than 250 μm. This work was performed by using two photomultiplier tubes and so only two elements were determined for each laser shot. The distribution of Si/Ca signal ratios of eight consecutive ablation layers of a multicoated paper provided evidence for a change in coating composition after the fourth pulse. An interesting contribution of this paper is the comparison between the obtained LIBS images and those obtained using laser-induced fluorescence, as shown

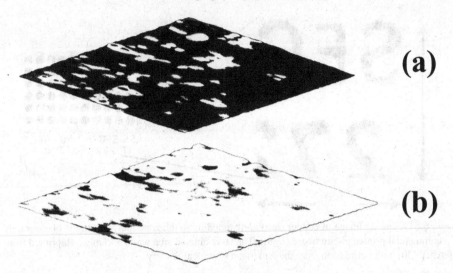

Figure 6.7. Demonstration of the mirror image correlation between a fluorescence image (a) and a LIBS image (b) of a paper coating. The field of view of the figure is 10×10 mm^2. Adapted from reference [29].

in Figure 6.7. The fluorescence emission at 422 nm was measured because areas with high content of organic constituents (that is, low pigment concentration) exhibited high fluorescence signals. As shown, there is an excellent correlation between the two images, showing the agreement between the two techniques.

An interesting application involving compositional mapping of copper conductor patterns from printed circuit board has been published [30]. Monitoring of the Cu emission at 510.6 nm was performed with a single photomultiplier tube. With this configuration, compositional LIBS in the sample showed unequivocally some Cu contamination around the conductor area. This effect may seen in Figure 6.8 as black dots between the pattern in the zoomed inset. The authors claimed that the contaminated salt probably resulted from an incomplete washing step during manufacturing, and could lead to short circuits in the electronic device. Interestingly, the authors found lower contamination in the upper zone of the analyzed printed board, indicating that the board was dried upright without a complete washing procedure.

The appealing alternatives offered by multi-elemental detectors for chemical imaging LIBS have been described [31]. The paper shows two-dimensional surfaces for the C distribution on solar cells derived from the first and fourth laser shots (Figure 6.9). The sampled area was a rectangle of 3.5×0.8 mm^2 with a distance between shots of 50 µm and 200 µm on the Y and X axes, respectively, owing to the oval shape of the nitrogen laser beam used. Five shots were individually recorded at each sample position, resulting in 1500 spectra acquired in the analysis area. The analysis was performed in a region that included two silver charge collectors of the solar cell (fingers) where the carbon content is higher because of the manufacturing process. As observed, during the first pulse, carbon contamination on

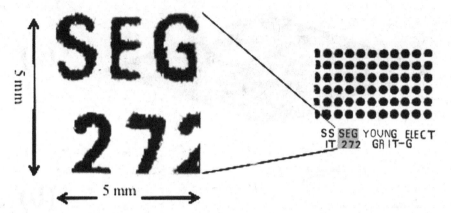

Figure 6.8. Mapping image of copper (left) corresponding to the zone indicated in the photography of a commercial printed circuit board (right). The laser-ablated area was 5×5 mm^2. Reprinted from reference [30] with permission from the American Chemical Society.

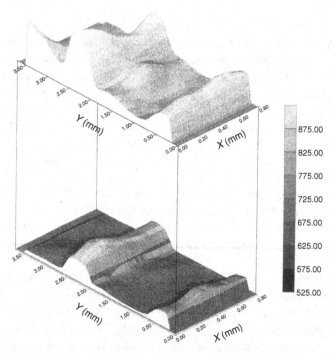

Figure 6.9. Spatial distribution of carbon on a silicon photovoltaic cell after one (top) and four (bottom) shots on a single surface position. The gray scale indicates the intensity level. Reprinted from reference [31] with permission from Springer-Verlag.

the surface is evident by comparison between the maps corresponding to the first (top) and fourth (bottom) pulses.

A fully automated system for performing depth profiling and compositional mapping of silicon solar cells has been described [12, 32]. The system consists of an N_2 laser fired by a pulse generator, which controls a pair of linear motorized stages as well as the opening of the shutter of the CCD detector. The specific home-written software allows control over the number of shots at each sample position, the step size between adjacent sample positions, and net distance traveled at each axis. Recall that Figure 6.1 shows compositional maps for Ag, Ti and C obtained from the first laser pulse on a 4×3 mm^2 surface. Additionally, tomographic maps corresponding to carbon at different depths are shown in the same article, demonstrating the analytical power of compositional LIBS. In this particular application (Figure 6.10), the images of carbon distribution were obtained from five consecutive laser shots. Carbon content between silver fingers decreases steadily with depth up to 39 nm, while it remained virtually constant in the silver fingers. However, at a depth of 52 nm, the carbon distribution between fingers changes suddenly (see the bottom left zone). This change results because the TiO_2 coating has been completely removed and the silicon substrate has been reached. Since the silver fingers penetrate into the silicon, the carbon content in the fingers remains unchanged at a depth of 60 nm.

A deeper study that focused on carbon distribution on the solar cell raw material (silicon wafers) has been performed [33]. The authors state that LIBS images revealed a higher carbon content, distributed more uniformly on the surface than deeper in the wafer. The application of imaging LIBS to determine the quality of raw material (silicon wafers) supplied by two independent manufacturers has been also demonstrated [34]. The study attempted to understand the relationship between the presence of these impurities with some faulty integrated circuits and the resulting low-efficiency solar cells. In this work the authors compared the distribution of Cu, Al, and Ca by compositional LIBS. A spectral window centered at 315 nm was selected as it allowed the simultaneous detection of the three elements. The samples were analyzed on a 20×20 mm^2 surface, the results obtained by imaging LIBS being more representative of the sample than those obtained by conventional scanning electron microscopy.

A detailed work that focused on the three-dimensional distribution of aluminum in solar cells has recently been published [35]. Aluminum is currently being used in the back surface of solar cells as a way to avoid photocurrent losses. Unfortunately, aluminum may diffuse into the silicon wafer, decreasing the energy conversion as it sinks the photovoltaic conversion. In this paper LIBS was used to provide silver and aluminum compositional maps as well as to determine the aluminum diffusion in solar cells. The analysis was performed on the surface and at depth in order to provide a complete three-dimensional visualization of the structure. A sequence of 51 locations in the axial direction with 20 shots in each location provided a total of 1020 spectra in a single data-acquisition sequence. This rastering was done over 28 parallel paths in the longitudinal direction, resulting in a total of 28 560 multi-elemental spectra for a total scanned area of 3×2.5 mm^2. An additional degree of information was provided by obtaining in-depth information corresponding to different

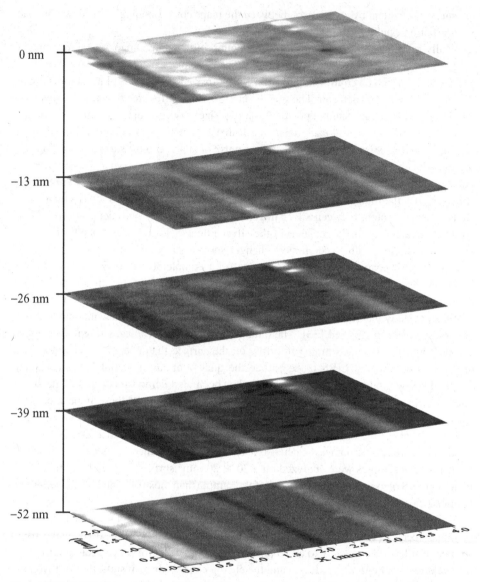

Figure 6.10. LIBS images of the spatial distribution of carbon (C(I) line at 598.106 nm) at several depths rendered from five consecutive laser shots. Reprinted from reference [12] with permission from the American Chemical Society.

pulses hitting the sample on the same location. In this sense, complete three-dimensional information is provided just by processing conveniently the results obtained. The power of this visual information is well exemplified in Figure 6.11, where the three-dimensional chemical maps corresponding to aluminum (top) and the aluminum diffusion zone (bottom) are shown. Much more specific information about a sample may be obtained by imaging

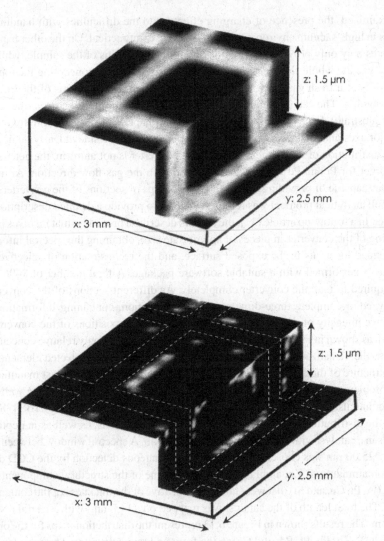

Figure 6.11. Three-dimensional representation of a LIBS elemental map of aluminum (top) and the aluminum diffusion into the silicon (bottom). The transverse section corresponds to depth maps generated from the spectra obtained after 20 laser shots in the same position. Reprinted from reference [35] with permission from Elsevier Science.

LIBS in a fraction of the time and experimental resources required when using conventional surface analytical techniques.

Compositional mapping of automobile three-way catalytic converters is another recent application where LIBS has found a very interesting niche. The measurement capabilities for the PGMs in converters are limited as a result of the complexity of the converter structure. Surface analysis techniques have been used for examining the catalyst surface, but several problems caused by the extensive sample preparation needed, the small sampling areas examined, the presence of charging effects, and the difficulties with handling large samples in high-vacuum environments make their use impractical. On the other hand, these techniques may only provide information about microdomains of the sample, while LIBS may analyze the whole sample. In the first of a series of papers concerning the topic [36], the analysis of a fresh gasoline catalytic converter is described in terms of the Pt, Pd, and Rh distribution. The catalyst analyzed consisted of two monolith substrates (the front and the rear substrate). Each substrate had an ellipsoidal shape with 80×170 mm for the minor and major axes and a length of 64 mm. The conclusions of the study clearly demonstrated that the distribution of the elements in the fresh converter is not uniform, the heterogeneity being larger for Pt and Pd than for Rh, and mainly in the gas-flow direction. At this step, LIBS was capable of providing two-dimensional maps of sections of the converters. However, a full analytical protocol was required in order to provide a detailed description of the converter. In a follow-up article [37], the authors describe a procedure that involves physical sectioning of the converter in pieces of appropriate size, obtaining the spectral information from defined locations in the exposed surface, and the reconstruction of selective chemical images performed with a suitable software package. A total number of 8496 images were required to map the converter completely. As different sections of the converter can be analyzed, a complete three-dimensional representation, containing information about the relative intensity of the elements of interest at specific positions of the converter, was obtained, as shown in Figure 6.12. To date, the images represent only relative concentration. However, the visualization in three dimensions of the location of selected elements in the whole structure of the converter is a most useful information to the catalyst manufacturers.

The problem of chemical deactivation by deposition of heavy metals in the walls of the converter has also been faced by the same authors using chemical imaging LIBS [38]. They studied the distribution of P, Zn, and Pb along the gas flow axis, as well as in depth in the channels in a catalyst with a lifetime of about 30 000 km. A spectral window between 209.00 nm and 225.00 nm was chosen at it provided simultaneous detection by the CCD detector of the contaminants (P, Zn, and Pb) as well as of some of the structural components of the catalyst (Pt, Pt, Ce, and Si) that were taken as indicative of the thickness of the contaminated surface. The total length of the catalyst was mapped out (143 mm) with a lateral resolution of 1.7 mm. The results shown in Figure 6.13 represent the distribution maps for the elements of interest (P, Zn, Pb, Pt, Pd, and Ce) in the front and rear substrates. The gray-level scale was established by assigning the color white to a value of 16% of the steady signal for each element, with different gray shades up to the maximum intensity. Given the criteria, the dark zones indicate the spatial location of a given element. As observed, P, Zn, and Pb are present throughout the converter, although they are preferentially located in the 10–20 mm closer to the engine. It is interesting as well to remark on the clear migration of these elements to deeper locations in the substrate in the same zone. However, Pt, Pd and Ce remain defined in a layer of constant thickness all the way along the converter.

New applications involving screen-printed electrodes [39] and biomaterials [40] have recently been reported. In the latter application, the authors describe low-spatial-resolution two-dimensional histograms of potentially toxic elements (Sr, Pb, and Al) within bones,

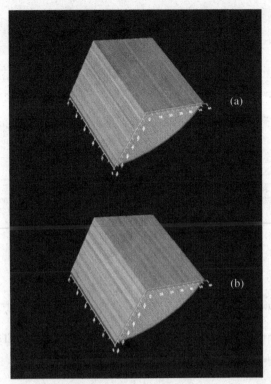

Figure 6.12. Three-dimensional LIBS image of the spatial distribution of palladium in the substrate of a fresh gasoline-engined automobile catalytic converter. Reprinted from reference [37] with permission from Elsevier Science.

teeth and dental materials. The choice of these particular elements is motivated by their importance related to medical, biological, and environmental exposure. In order to perform the quantitative determinations, $Ca\,CO_3$ pellets spiked conveniently with the studied elements were made. Different applications were performed, including single-point analysis, lateral distributions over selected paths (i.e. across caries-infected zones), and low-resolution (1 mm step) two-dimensional maps for Sr. Despite the low resolution, the authors demonstrated that the Sr concentration diminished from an average value of about 300 p.p.m. A similar experiment performed using a slice of a tibia bone demonstrated a higher overall Sr concentration (about 900 p.p.m.). However, the concentration decreased towards the bone marrow channel from where the bone receives its nutrients for build-up.

As mentioned above, compositional mapping LIBS is difficult, mainly because of the lack of suitable standards. In this sense, LIBS shares the same problems as many surface analysis techniques where relative sensitivity factor and matrix corrections must be introduced for each matrix–element pair to perform quantitative approaches. In compositional mapping LIBS, a quantitative approach was demonstrated for rock samples [41]. The authors used a frequency-doubled Nd:YAG to study element distribution mapping of polished rock

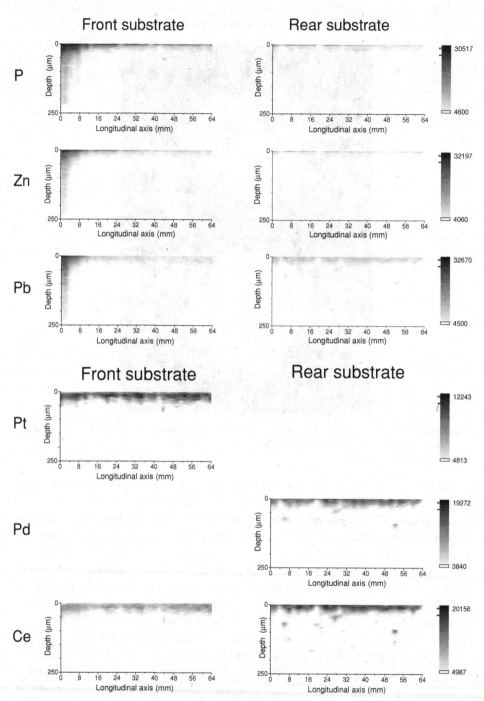

Figure 6.13. LIBS maps of P, Zn, Pb, Pt, Pd and Ce in the front and rear substrates of an aged three-way catalytic converter. The images shown represent depth profiles along the gas-flow axis in the converter. The gray scale indicated the relative abundance expressed as intensity levels. Adapted from reference [38]. (*cont.*)

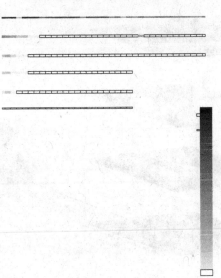

Figure 6.13. (*cont.*)

sections. An *X–Y* stage was used to move the sample, and an element image of 50 × 50 mm was made in 30 min. The element concentration distribution of Ba, Cu, Fe, Mn, Pb, Si, and Sr in polished sections was achieved by analyzing standard granite rock samples, pelletized at 1 MPa to overcome matrix matching properties. The limits of detection for Ba, Cu, Fe, Mn, Si, and Sr were 100 p.p.m., 5 p.p.m., 50 p.p.m., 10 p.p.m., 2000 p.p.m., and 50 p.p.m., respectively. The authors stated that LIBS may be thought of as an alternative to electron probe microanalysis (EPMA) in terms of its greater sensitivity, although the method still presents important calibration problems. A new quantitative attempt has been recently published [10]. In this work the authors present quantitative surface characterization of a 50 nm thickness manganin (an alloy consisting of Cu 86%, Ni 2%, and Mn 12%) layer applied on an Si substrate by pulsed laser deposition. In order to perform the quantitative approach, the authors assumed that the sum of Cu, Ni and Mn concentration is 100%, and that the emissitivity for each element (understood as the amount of light collected for the detector per amount of ablated material) is similar. Consequently, differences in atomic concentration across the coating could indicate the spatial stoichiometric change in the manganin elements deposited on the Si substrate. Selective compositional mapping of the simultanously collected signals revealed significant heterogeneity in the distribution of manganin elements on the Si substrate, as shown in Figure 6.14. One interesting aspect of this paper was checking the deviation from the theoretical stoichiometry of the deposited manganin layer. Homogeneity in the manganin layer was guessed for those points for which the composition of the three elements experienced simultaneously a deviation lower than 10%. Data processing demonstrated that only limited regions in the coating exhibited the stoichiometry of the sputtered manganin target. Excellent matching was observed between

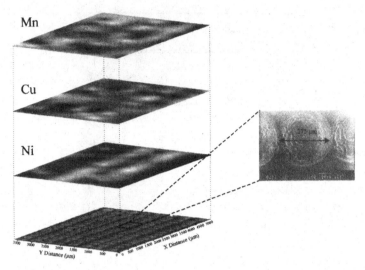

Figure 6.14. LIBS images of the spatial distribution of Mn, Cu, and Ni. A digitized image of the sample surface analyzed by LIBS is also shown. Data were acquired by accumulating four laser shots on each sampling position. The inset shows a scanning electron micrograph of the crater formed. Reprinted from reference [10] with permission from the American Chemical Society.

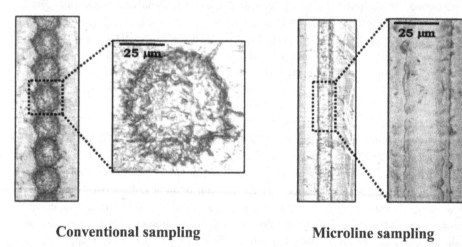

Conventional sampling   **Microline sampling**

Figure 6.15. Details of the discrete craters produced with a conventional LIBS set-up (left) or the continuous ribbon when using the microline configuration (right).

LIBS and SEM, indicating that, after careful choice of the operating conditions, LIBS offers at least the same structural information as SEM, with much lower experimental complexity.

All cited approaches to compositional mapping have in common the same experimental configuration, where the compositional maps have been achieved by point-to-point sampling using a spherical lens focusing system, at the cost of increased measurement time and data storage requirements. Under these conditions, an array of sampling spots is drawn in the sample as indicated in Figure 6.15 (left). The lateral resolution will be dictated by the crater diameter and the lateral displacement between adjacent pulses to avoid either overlapping or mixing. In the figure shown, as the practical lateral resolution was 50 μm, 7280 shots are required to scan a 3.2×5.6 mm^2 surface. At a laser repetition rate of 2 Hz, about 60 min of data collection are implied. A new alternative [42] involves exploding the two-dimensional capability of CCD detectors. In this approach, coined as microline-imaging LIBS (MILIBS), the laser is focused by using a cylindrical lens, and generating a microline in the sample. By collecting the plasma formed in a direction parallel to the spectrometer entrance slit, each ablated location in the sample will generate a signal at a defined height in the slit and, in this way, the lateral resolution will be defined by the pixel size. As shown in Figure 6.15 (right), the surface is continuously ablated, producing a ribbon in the surface with a thickness of about 20 μm. The only displacement required to scan a surface is between parallel ribbons, decreasing significantly the acquisition time. In comparison with conventional LIBS, the authors demonstrated a 100-fold decrease in the data-acquisition step with the microline approach. The fast scan speed has been fully exploited in the elemental mapping of microinclusions in stainless steel. An example is shown in Figure 6.16. A microinclusion with a diameter of about 900 μm was completely mapped out for six elements (Mg, Ca, Si, Al, Fe, and Ti) to only 64 laser pulses. The spacing between microlines was 25 μm, corresponding to a total analyzed area of 1625×3790 μm^2. The black areas indicate the zones with lower intensity (concentration) of the element of interest.

6.5 Concluding remarks and outlook

At the micrometer level of resolution, LIBS provides a powerful tool for imaging differences in atomic composition in a material. This level of detail is essential to defining processes and products in many areas of modern research and industry. When considering the sampling depth, the achieved levels of a few nanometers are outstanding for such simple instrumentation and operating conditions as those required in LIBS. Remarkable progress made during the past decade has resulted in a broad front of applications covering large added-value materials such as semiconductors, coatings, catalysts, and microelectronic devices. Also, direct analysis of non-conducting materials is very promising, with applications demonstrated for wood, rocks, biomaterials and ceramics.

In comparison with other methods, LIBS chemical imaging can still be considered as an emerging technique. Extensive development is needed in fields such as effective control of beam quality parameters, generation of reliable ultrashort laser pulses, and production of standards, not forgetting the development of commercial instrumentation. Quantitative mapping methods are required to extend its capabilities to other important applications. However, at the moment, the majority of LIBS investigations are qualitative with respect to the spatial distribution of species of a particular system. In the future, quantitative information needs to be obtained routinely by relative methods already implemented in surface-analysis techniques or by applying one of the several available LIBS algorithms that use no

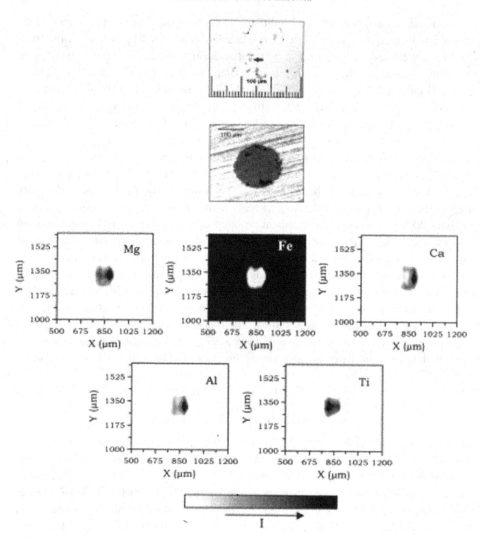

Figure 6.16. Ultrafast mapping of a 900 μm diameter inclusion in stainless steel using microline LIBS.

calibration. This is highly desirable, in view of the generalized lack of reference samples for use in materials analysis. Although still uncommon in the surface-analysis laboratory, LIBS may sometimes yield information that no other technique is able to provide.

The authors thank the Spanish National Research and Development Plan for granting successive projects that have made possible their work in this field (Projects PB94-1477, PB97-1107, AMB98-1361-CE). Also, the collaboration of all other authors from the group at the University of Málaga cited in the relevant references is gratefully acknowledged.

6.6 References

[1] W. D. Callister, Jr., *Materials Science and Engineering*: *An Introduction*, third edition (New York: Wiley, 1996).

[2] J. C. Vickerman (editor), *Surface Analysis. The Principal Techniques* (New York: Wiley, 1997).

[3] M. Grasserbauer, G. Friedbacher, H. Hutter and G. Stingeder, Trends in surface and interface analysis. *Fresenius J. Anal. Chem.*, **346** (1993), 594–603.

[4] D. Günther, S. E. Jackson and H. P. Longerich, Laser ablation and arc/spark solid sample introduction into inductively coupled plasma mass spectrometry. *Spectrochim. Acta*, **B 54** (1999), 381–409.

[5] T. F. Heinz and G. A. Reider, Surface studies with optical second-harmonic generation. *TrAC*, **8** (1989), 235–242.

[6] S. Florek, C. Haisch, M. Okruss and H. S. Becker-Ross, A new, versatile echelle spectrometer relevant to laser induced plasma applications. *Spectrochim. Acta*, **B 56** (2001), 1027–1034.

[7] L. M. Cabalín and J. J. Laserna, Experimental determination of laser induced breakdown thresholds of metals under nanosecond Q-switched laser operation. *Spectrochim. Acta*, **B 53** (1998), 723–730.

[8] J. M. Vadillo, C. C. García, S. Palanco and J. J. Laserna, Nanometric range depth-resolved analysis of coated-steels using laser-induced breakdown spectrometry with a 308 nm collimated beam. *J. Anal. Atom. Spectrom.*, **13** (1998), 793–797.

[9] C. C. García, M. Corral, J. M. Vadillo and J. J. Laserna, Angle-resolved laser-induced breakdown spectrometry for depth profiling of coated materials. *Appl. Spectrosc.*, **54** (2000), 1027–1031.

[10] L. M. Cabalín and J. J. Laserna, Surface stoichiometry of manganin coatings prepared by pulsed laser deposition as described by laser-induced breakdown spectrometry. *Anal. Chem.*, **73** (2001), 1120–1125.

[11] M. P. Mateo, J. M. Vadillo and J. J. Laserna, Irradiance dependent depth profiling of layered materials using laser induced plasma spectrometry. *J. Anal. Atom. Spectrom.*, **16** (2001), 1317–1321.

[12] D. Romero and J. J. Laserna, Multielemental chemical imaging using laser-induced breakdown spectrometry. *Anal. Chem.*, **69** (1997), 2871–2876.

[13] C. Geertsen, J. L. Lacour, P. Mauchien and L. Pierrard, Evaluation of laser ablation optical emission spectrometry for microanalysis in aluminum samples. *Spectrochim. Acta*, **B 51** (1996), 1403–1416.

[14] A. Morak, R. Unkroth, K. Sauerbrey and K. Schneider. Fast analysis of inorganic wood preservatives using laser-induced plasma emission spectrometry. *Field Anal. Chem. Technol.*, **3** (1999), 185–192.

[15] H. Wiggenhauser, D. Schaurich and G. Wilsch, LIBS for non-destructive testing of element distributions on surfaces. *NDT&E Internat.*, **31** (1998), 307–313.

[16] J. M. Vadillo, I. Vadillo, F. Carrasco and J. J. Laserna, Spatial distribution profiles of magnesium and strontium in speleothems using laser-induced breakdown spectrometry. *Fresenius J. Anal. Chem.*, **361** (1998), 119–123.

[17] L. St-Onge and M. Sabsabi, Towards quantitative depth-profile analysis using laser-induced plasma spectroscopy: investigation of galvannealed coatings on steel. *Spectrochim. Acta*, **B 55** (2000), 299–308.

[18] D. R. Anderson, C. W. McLeod, T. English and A. T. Smith, Depth profile studies using laser-induced plasma emission spectrometry. *Appl. Spectrosc.*, **49** (1995), 691–701.

[19] M. Milán, P. Lucena, L. M. Cabalín and J. J. Laserna, Depth profiling of phosphorus in photonic-grade silicon using laser-induced breakdown spectrometry. *Appl. Spectrosc.*, **52** (1998), 444–448.

[20] J. M. Vadillo and J. J. Laserna, Depth-resolved analysis of multilayered samples by laser-induced breakdown spectrometry. *J. Anal. Atom. Spectrom.*, **12** (1997), 859–862.

[21] J. M. Vadillo, J. M. Fernández, C. Rodríguez, and J. J. Laserna, Depth-resolved analysis by laser-induced breakdown spectrometry at reduced pressure. *Surf. Interface Anal.*, **26** (1998), 995–1000.

[22] L. M. Cabalín, D. Romero, J. M. Baena and J. J. Laserna, Saturation effects in the laser ablation of stainless steel in air at atmospheric pressure. *Fresenius J. Anal. Chem.*, **365** (1999), 404–408.

[23] C. J. Lorenzen, C. Carlhoff, U. Hahn and M. Jogwich, Applications of laser-induced emission spectral analysis for industrial process and quality control. *J. Anal. Atom. Spectrom.*, **7** (1992), 1029–1035.

[24] J. J. Laserna, N. Calvo and L. M. Cabalín, Imaging and space-resolved spectroscopy in the Xe–Cl-laser ablation of noble metals with charge-coupled device detection. *Anal. Chim. Acta*, **289** (1993), 113–117.

[25] M. Hidalgo, F. Martín and J. J. Laserna, Laser-induced breakdown spectrometry of titanium dioxide antireflection coatings in photovoltaic cells. *Anal. Chem.*, **68** (1996), 1095–1100.

[26] J. M. Vadillo and J. J. Laserna, Single-shot laser-induced breakdown spectrometry: lateral and depth studies of solid samples. *Recent Develop. Appl. Spectrosc.*, **1** (1996), 73–77.

[27] P. V. Maravelaji, V. Zafiropulos, V. Kilikoglou, M. Kalaitzaki and C. Fotakis, Laser-induced breakdown spectroscopy as a diagnostic technique for the laser cleaning of marble. *Appl. Spectrosc.*, **52** (1997), 41–53.

[28] S. Palanco, L. M. Cabalín, D. Romero and J. J. Laserna, Infrared laser ablation and atomic emission spectrometry of stainless steel at high-temperatures. *J. Anal. Atom. Spectrom.*, **12** (1999), 1883–1889.

[29] H. J. Häkkänen and J. E. I. Korppi-Tommola, UV-laser plasma study of elemental distributions of paper coatings. *Appl. Spectrosc.*, **49** (1995), 1721–1728.

[30] T. Kim, C. T. Lin and Y. Yoon, Compositional mapping by laser-induced breakdown spectroscopy. *J. Phys. Chem.*, **B 102** (1998), 4284–4287.

[31] J. M. Vadillo, S. Palanco, M. D. Romero and J. J. Laserna, Applications of laser-induced breakdown spectrometry (LIBS) in surface analysis. *Fresenius J. Anal. Chem.*, **355** (1996), 909–912.

[32] J. J. Laserna, F. Martín, S. Pots and M. Hidalgo, Método computerizado de análisis mediante espectroscopía de plasmas producidos por láser para el control de calidad de células solares. *Spanish patent no. P9702565* (1997).

[33] M. D. Romero and J. J. Laserna, Surface and tomographic distribution of carbon impurities in photonic-grade silicon using laser-induced breakdown spectrometry. *J. Anal. Atom. Spectrom.*, **13** (1998), 557–560.

[34] M. D. Romero, J. M. Fernández and J. J. Laserna, Distribution of metal impurities in silicon wafers using imaging-mode multi-elemental laser-induced breakdown spectrometry. *J. Anal. Atom. Spectrom.*, **14** (1999), 199–204.

[35] M. D. Romero and J. J. Laserna, A microanalytical study of aluminium diffusion in photovoltaic cells using imaging-mode laser induced breakdown spectrometry. *Spectrochim. Acta*, **B 55** (2000), 1241–1248.

[36] P. Lucena, J. M. Vadillo and J. J. Laserna, Mapping of platinum group metals in automotive exhaust three-way catalysts using laser-induced breakdown spectrometry. *Anal. Chem.*, **71** (1999), 4385–4391.

[37] P. Lucena and J. J. Laserna, Three-dimensional distribution analysis of platinum, palladium and rhodium in auto catalytic converters using imaging-mode laser-induced breakdown spectrometry. *Spectrochim. Acta*, **B 56** (2001), 177–185.

[38] P. Lucena, J. M. Vadillo and J. J. Laserna, Compositional mapping of poisoning elements in automobile three-way catalytic converters by using laser-induced breakdown spectrometry. *Appl. Spectrosc.*, **55** (2001), 267–272.

[39] J. Amador-Hernández, J. M. Fernández-Romero and M. D. Luque de Castro, Three-dimensional analysis of screen-printed electrodes by laser induced breakdown spectrometry and pattern recognition. *Anal. Chim. Acta*, **435** (2001), 227–238.

[40] O. Samek, D. C. S. Beddows, H. H. Telle *et al.*, Quantitative laser-induced breakdown spectroscopy of calcified tissue samples. *Spectrochim. Acta*, **B 56** (2001), 865–875.

[41] Y. Y. Yoon, T. S. Kim, K. S. Chung, K. Y. Lee and G. H. Lee, Applications of laser induced plasma spectroscopy to the analysis of rock samples. *Analyst*, **122** (1997), 1223–1227.

[42] M. P. Mateo, S. Palanco, J. M. Vadillo and J. J. Laserna, Fast atomic mapping of heterogeneous surfaces using microline-imaging laser-induced breakdown spectrometry. *Appl. Spectrosc.*, **54** (2000), 1429–1434.

7

Biomedical applications of LIBS

Helmut H. Telle

Department of Physics, University of Wales Swansea

Ota Samek

Department of Physical Engineering, Technical University of Brno

7.1 Introduction

In this chapter, applications of laser-induced breakdown spectroscopy (LIBS) to the analysis of biological and medical samples are outlined. The range of samples includes calcified tissue materials (e.g. teeth, bones, sea shells), soft tissue materials (e.g. human skin, plant parts like leaves and wood), and – to a limited extent – bio-fluids (e.g. blood). Specifically, applications providing spatial resolution information (lateral and/or depth) of elemental concentration distributions are discussed; no other technique provides such information with the ease of the method of LIBS. Strong efforts have been made to link this information with nutritional and environmental influences. Throughout the survey, the problem of reference standards, required for fully quantitative analysis, is addressed.

The technique of LIBS offers a simple and fast method of elemental analysis. Any material – solid, liquid or gaseous – can be analyzed, and no (or only very little) sample preparation is needed, as is being highlighted in other chapters of this book. Detection limits for solid samples are, in favorable cases, of the order of a few parts per million (p.p.m.). While this may be inferior to other methods of analysis one has to keep in mind that LIBS analysis can be undertaken directly from the solid sample, often *in situ* and at "remote" locations, and even *in vivo* analysis may be possible, when dealing with living organisms.

When considering the role of mineral nutrition and metabolism in the context of maintaining human health, the knowledge of the presence or absence of key elements is of vital importance. For example, the accumulation of various elements in excess concentration – and these may indeed only be extremely low – may have toxic effects; such elements include for example aluminum (Al), cadmium (Cd), mercury (Hg), and lead (Pb), to name but a few. On the other hand, the presence of other trace elements may be of vital importance to "health," including, for example, calcium (Ca), magnesium (Mg), and iron (Fe).

While the success of LIBS in the qualitative and quantitative analysis of solids is now unquestioned, a search of the published literature reveals that only very few investigations address non-ferrous materials. In particular, publications about using LIBS on bio-matrix materials are relatively few. Overall, this is not very surprising, for a range of reasons. First, biological tissue samples – "hard" calcified tissue and "soft" cell materials – are normally less "tough" in their texture than metals or minerals. Thus, the ablation process "destroys" the sample area much more rapidly, which results in poor statistics and reproducibility. Second, in many cases, biological samples are rather inhomogeneous. Again, this makes selectivity, statistics, and reproducibility of results difficult. Third, traces of specific elements in the bio-mass are often of most interest at concentration levels that are close to or below the detection limits of LIBS. And finally, more often than not, molecular species are of vital importance, and normally these are beyond the capabilities of LIBS.

Despite the shortcomings just summarized, we think that LIBS is starting to play a role of increasing importance in the analysis of biological samples. We will describe a range of examples, ranging from mineralized specimens through cell tissue to bio-fluids, which will clearly demonstrate the capabilities of LIBS for these types of material. The choice of examples mirrors their importance to biological, medical, and environmental problems. In general, clear links between the experimental findings for the concentration and distribution of particular elements, and certain external (environmental) conditions, can be identified.

7.2 Investigation of calcified tissue materials

Mineralized tissues, i.e. bones and teeth – and other bio-minerals – have been found to be excellent "archives" related to living habits, nutrition and exposure to changing environmental conditions [1]. Bones and teeth are found to maintain much of the "biological signature" from the living phase over a long time, revealing, for example, the uptake of contaminants from the surrounding environment during certain periods [2]. Traditionally, two major approaches have been followed for quantitative, compositional analysis of calcified tissue.

First, most commonly, a *dissolution approach* is used. The whole or part of a specimen is dissolved and analyzed by using methods based on atomic absorption spectroscopy (AAS) [3–5]. However, in this approach, information about the relation between concentration and spatial location is destroyed. This may not be an issue if only the total amount of an element is required; but, on the other hand, it does not allow us to follow elemental accumulation over time (related to the spatial position within the matrix), which is useful for example in nutritional studies.

Second, the elemental analysis is performed with *spatial distribution*; for this, selected parts of the specimen are raster scanned. These methods include inductively coupled plasma mass spectrometry (ICP-MS) [6, 7]; X-ray fluorescence microprobe [8]; and most recently laser-induced breakdown spectroscopy (LIBS) [9, 10].

In addition, in support of elemental analysis, *speciation* may be carried out to determine the nature of the "composite" material of bone and teeth that is dominated by extremely fine-sized crystalline (grain sizes on the order of tens of nanometers) carbonated hydroxyapatite. For this, Raman spectroscopic and laser scanning confocal microscopic analysis techniques have been successfully applied, also providing some degree of spatial distribution information [11].

LIBS compositional analysis has been used to determine the temporal/spatial evolution of trace element concentrations in bones and teeth, and whether tooth material is healthy or not [10, 12–15]. Continuously tracking the spectral analysis during the progress of ablation, information about the spatial (lateral and depth) distribution of elements in teeth and bones can be obtained *in vitro*, or even *in vivo*. It is possible to link the quantitative results from LIBS analysis to environmental influences (*in vitro* studies of tooth or bone sample cross sections) and dental disease states (*in vivo* monitoring).

Since it is likely that in the near future the clinical application of short-pulsed lasers may emerge as a real alternative to mechanical dental treatment, the provision of precise monitoring seems mandatory. The results from studies of the present authors and of other groups [16, 17] could provide the basis for such a monitoring procedure. An additional bonus of LIBS with spatial resolution capability is that it could provide an easy means for short- and long-term exposure characterization, related to environmental and nutritional factors [18].

Here we address a selection of different types of samples, including a range of teeth (first teeth of infants; second teeth of children; and teeth of adults), bones (primarily pieces of tibia and femur bones), and other calcified bio-samples (sea shells).

7.2.1 *Methodology*

Traditionally, qualitative and quantitative LIBS analysis has been applied to the analysis of solid samples, and hence the study of teeth should not be outside the general knowledge base of LIBS. The realization of LIBS for *in vitro* analysis of trace amounts in dental samples is straightforward: in principle any set-up used for LIBS analysis of solids would be suitable. We would like to note here that the ablation of dental samples using pulsed lasers (laser drilling) and simultaneously monitoring the plasma emission exactly mimics the principle behind the technique of LIBS. Exploitation of this for monitoring purposes was first proposed by Niemz and co-workers [19–21].

Here we will concentrate mostly on an implementation that is potentially applicable to *in vivo* analysis and monitoring of teeth during the process of drilling in teeth. Most of the measurements described here were conducted using a fiber-coupled assembly, which is capable both of delivering the laser radiation to the sample and of collecting the light emitted from the microplasma [22].

In teeth and bones the number of spectral lines being observed is greatly reduced, in contrast to metal matrices. In general only the strongest resonance lines of any trace element are observed besides the abundant Ca(I) and Ca(II) lines of the matrix. This may be perceived as beneficial for easy spectral identification. Also, in favorable cases multi-element analysis can easily be realized since wider ranges can be covered in individual spectral segments. In general, standard multiple grating spectrometers (150 600 and 2400 grooves mm^{-1} gratings in an instrument with $f = 500$ mm) with array detectors (typically 512 or 1024 pixels width) allow us to record spectral segments of about 160 nm, 40 nm and 10 nm, with resolution 2.0 nm, 0.5 nm and 0.125 nm, respectively. While the wider spectral ranges seem to favor the simultaneous observation of a multitude of elements, the spectral resolution may pose a problem for element detection and quantification. Specifically, this problem is encountered when the strongest (resonance) lines of a trace element overlap with those of calcium matrix lines. For example, under medium spectral resolution (i.e. $\Delta\lambda \cong 0.5$ nm), the Al lines at 394.4 nm and 396.1 nm, which are frequently used in aluminum trace analysis, cannot be fully resolved. In particular, at high laser pulse energies these Al lines are strongly masked by the Stark broadened Ca(II) lines at 393.3 nm and 396.8 nm (typically, the trace amount of Al in teeth is smaller than the amount of Ca in teeth by a factor of at least 100). Thus, alternative lines need to be utilized, which do not suffer from line overlap but which may be of lesser intensity in their respective spectral segment. Note, for example, that in the calibration measurements for Al content in teeth [23], the Al lines at 308.2 nm and 309.3 nm, and the Ca(I) line at 300.7 nm, were used, which are fully free from interference.

In most laser-based analytical methods, the reproducibility of quantitative measurements strongly depends on maintaining uniform, stable experimental conditions. For LIBS analysis this means that for accurate estimation of element concentrations, using the calibration curves obtained from reference samples (for the problem of reference samples see below), the plasma parameters have to be kept as constant as possible. For relatively "soft" matrices, like calcified tissue, this is even more important than for metallic samples. By measuring the electron density and plasma temperature on line, the plasma reproducibility can be assessed, in principle. However, this is not always straightforward since the rapid "drilling" into the sample for repetitive single-position exposure is accompanied by a gradual change in plasma conditions. In general, a compromise has to be found between keeping the plasma conditions reasonably constant on the one hand and sampling over a sufficient number of laser pulses for statistical averaging on the other hand, while at the same time maintaining spatial resolution if so required.

7.2.2 The analysis of teeth

The basic matrices of calcified tissue of animals and humans are similar, one of their major building blocks being hydroxyapatite, but they exhibit some distinct differences. For teeth the overall composition is as follows.

Enamel: this is the hardest substance of the body, about 95% hydroxyapatite, 4% water, and 1% organic matter [24]. Hydroxyapatite is a mineralized compound with the chemical formula $Ca_{10}(PO_4)_6(OH)_2$. Its sub-structure consists of crystallites that form enamel prisms (with diameter of about 5 μm). The crystal lattice is intruded by impurities of trace elements.

Dentine: this is not as hard as enamel and consists of about 70% hydroxyapatite, 20% organic matter (largely collagen fibres), and 10% water.

Because the major constituent of the tooth's crystalline enamel and dentine structure is hydroxyapatite, one should observe strong emission lines from the elements calcium, Ca, and phosphorus, P, these being indicative for the tooth matrix. Indeed, sequences of

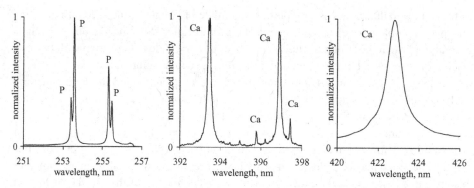

Figure 7.1. Segments of LIBS spectra from the enamel layer of a healthy adult tooth, exhibiting (resonance) lines from the matrix elements phosphorus and calcium.

characteristic lines are easily identified (see Figure 7.1). The other two elements in the crystal structure, oxygen, O, and hydrogen, H, are normally less conclusive if the samples are investigated in ambient air, which contains molecular oxygen and water vapor. Note the strong Stark broadening of the Ca(I) and Ca(II) lines, which may pose a problem if (weak) lines of trace elements sit on the shoulder of the calcium lines.

Spatial mapping of elemental content

Three examples are shown here; they are described in more detail in the work of Samek and co-workers [10, 23]. First, simple lateral (one-dimensional) scanning of samples was carried out on a fully virgin specimen. Then samples were cut into slices of about 1.5 mm thickness, and subsequently the cross sections of these samples were scanned to generate one-dimensional line scans and two-dimensional maps of elemental distributions.

As a first example of a one-dimensional measurement, a scan along the outside of a tooth, extracted during dental surgery, is shown in Figure 7.2a. The specific tooth sample has a bone fragment still attached to it, which makes its investigation an interesting exercise since both tooth and bone matrices are encountered in a single sample. In this specific example, we attempted to quantify the content of Sr in the sample. The ablation sites are separated by about 1 mm. Clearly, one can see that the Sr concentration in the bone fragment is higher than in the tooth, and that a large increase is observed at the cement–enamel junction. The absolute concentration values agree well with those reported in the literature [25].

A second example shows a scan across a slice of an adult tooth, cut horizontally, in steps of about 1 mm from the enamel to the pulp/root canal (Figure 7.2b) for an element/location and element/time profile for Al. The growth and calcification process of teeth evolves over time. Hence, measuring the concentrations of elements as functions of position in the tooth will provide a time evolution scale for a particular trace element [26]. In the figure the distribution of Al is not uniform across the profile. Because of the very slow accumulation/migration/depletion of elements, this feature could be exploited to generate element/time profiles. Thus one could determine the body burden of elements for people

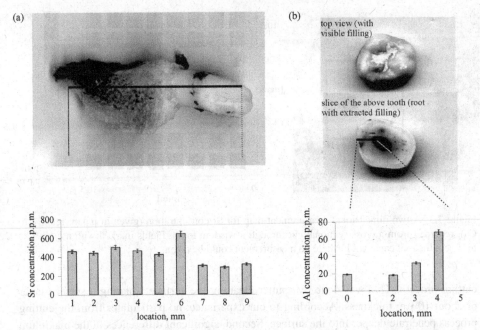

Figure 7.2. One-dimensional scans of teeth to determine trace element concentrations (measurement interval of 1mm). (a) Sr content in a tooth of an adult, measured along its length; (b) Al content in a tooth of an adult containing a filling, measured across a horizontal slice through the tooth; the concentration values are given in parts per million (p.p.m.), relative to Ca.

living in polluted areas, or – as in this case – that evidently Al migrated from the filling with high Al content outward into the dentine of the tooth. For this specific tooth sample the concentration of Al in the enamel layer was very high as well (measured from the side). This can be explained by the fact that the person from whom this particular sample was obtained had brushed his or her teeth with a "whitener" toothpaste, containing substantial amounts of Al_2O_3 [9]; the issue of toothpastes is addressed in more detail further below.

In Figure 7.3, a two-dimensional scan across a slice through a wisdom tooth is shown. The cross-sectional area of about 10×10 mm^2 is rastered in steps of 1.5 mm in each direction (the laser spot size on target was about 200 μm). In the photograph, the root canals of the tooth are quite clearly visible, and their presence is mimicked in the map by a hugely reduced to nearly absent Sr signal. Overall, the Sr content toward the edges of the tooth is elevated; on average about 250–350 p.p.m. of Sr are encountered. We stress that, in addition to the absolute overall value, these measurements clearly reveal changes in the spatial distribution of Sr. It should be noted that our values are in a good agreement with those obtained by Perez-Jordan *et al.* [25], and are consistent with the average values measured in a series of more than 100 tooth sample specimens.

Two things should be noted for the cross-section samples. First, care was taken to elimi-nate contributions from any possible contamination across the surface originating from the

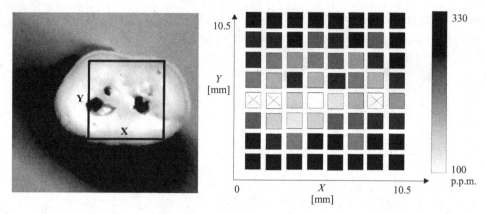

Figure 7.3. A two-dimensional measurement map for Sr concentration (given in p.p.m. relative to Ca), recorded from a cross-sectional cut through a wisdom tooth. Fields marked with a cross (×) represent the root canal void, where no measurement could be taken.

cutting process. This was done by sputter cleaning the surface, ablating a surface layer of about 10 µm thickness. According to our experience, no particulates from the cutting process penetrate deeper into the surface. Second, significant differences in the maximum spectral intensity on and off dental fillings were observed. This is caused by the different composition of the matrix materials. This information could potentially be exploited in monitoring the progress of the drilling process to remove any old filling material.

Although they are not shown here, we also generated two-dimensional maps for other trace elements, including Al, C, Mg, Na, and Zn. Unfortunately, for one other potentially toxic metal, Pb, no two-dimensional maps could be generated; the LIBS detection limit for Pb was not sufficient to detect its low average concentrations of about 6 p.p.m. in dentine [5]. However, we observed Pb lines when ablating the enamel layer of some of the teeth in our sample selection; the related concentration values were of the order 150 p.p.m. This is in broad agreement with Budd *et al.* [1]; they observed high Pb concentrations, similar to those measured by us, in the very thin layer of enamel (about 30–50 µm).

Determination of carious and heathy tissues

One major advantage LIBS analysis offers over methods traditionally applied to the analysis of calcified tissue is that the spatial information for the distribution of elements in teeth and bones is preserved. This was demonstrated in the previous section for samples for which elemental one- and two-dimensional maps of teeth were generated. Not only is this of interest for following the distribution of elements (for example to trace the temporal evolution of uptake of elements, from a polluted environment), but in future this may be of practical use for *in vivo* real-time applications.

As an example, we demonstrated that it is possible to distinguish unequivocally between healthy and caries-infected teeth. Hence, LIBS analysis could be implemented and used in

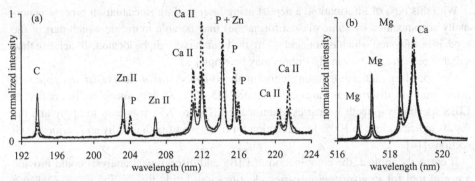

Figure 7.4. Selected LIBS spectra from the enamel part of a tooth, recorded at a location affected by caries (— full line trace) and at a sound, unaffected location (... dotted line trace). In the caries-affected section in (a) Zn and C increase against Ca and P, and in (b) Mg increases against Ca.

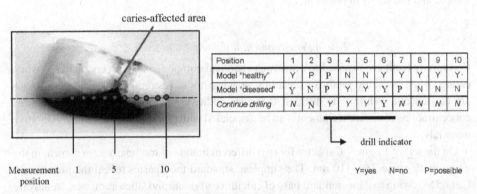

Position	1	2	3	4	5	6	7	8	9	10
Model "healthy"	Y	P	P	N	N	Y	Y	Y	Y	Y
Model "diseased"	Y	N	P	Y	Y	Y	P	N	N	N
Continue drilling	*N*	*N*	*Y*	*Y*	*Y*	*Y*	*N*	*N*	*N*	*N*

drill indicator

Measurement 1 5 10   Y=yes   N=no   P=possible
position

Figure 7.5. Relationship of pattern recognition analysis results across a caries-affected tooth (measurement locations spaced by about 1 mm). An "all-clear" indication is given if both models return the correct Y/N combination.

dental drilling using lasers, a technique that is increasingly being tested in dental laboratories and is reaching maturity.

The most frequent pathological condition of teeth is decay, or caries infection; the enamel becomes de-mineralized in a few days, and hard enamel is turned into "porous" material. Calcium bound to the hydroxyapatite is ionized and subsequently washed out from the tooth, being replaced by other elements such as magnesium or zinc. To demonstrate this point, we performed line scans across slices of a caries-infected tooth. In the examples shown in Figure 7.4, the differences in the spectra recorded for healthy and infected parts of the tooth are quite striking. Overall, dramatic changes in the Ba, C, K, Li, Mg, Na, P, Sr, and Zn peaks in comparison with Ca peak intensities were observed (spectral segments containing the alkali and most alkaline earth elements are not shown here).

With this type of information, a dentist using laser drilling (the ablation process is normally accompanied by some plasma formation) may be able to decide which part of the tooth is healthy and which is not, and where the boundary might be located. To achieve this, suitable pattern recognition algorithms may be employed.

One specific pattern recognition algorithm, the so-called *Mahalanobis Distance* approach often found in discriminant analysis methods [15, 27, 28], was successfully applied to LIBS spectra from teeth for caries identification (Figure 7.5). It is easy to apply and the results indicate close to 100% identification for materials of both healthy and carious tooth sections [14].

If correctly applied, the combination of LIBS and discriminant analysis could provide a useful tool for *in vivo/in vitro* caries identification during the drilling process when a luminous plasma is created by short laser pulses. Positive/negative identification could be signaled, for example, by a short sound when the transition from caries to healthy tooth is identified by the pattern recognition algorithm. This would allow the dentist to follow normal drilling routines while obtaining automatic, real-time information about the composition of sample and the status of ablation.

Tracing environmental influences on teeth

Commonly teeth are exposed to numerous chemical compounds, being encountered when eating, when brushing teeth for cleaning, or when tooth restorative/corrective processes are carried out. While it is difficult to assess additives in food, normally well-defined concentrations of different elements can be associated with toothpastes and dental restorative materials.

On the left of Figure 7.6 spectra for two different brands of toothpastes are shown, in the wavelength range 389–410 nm. The simplest, standard toothpastes reveal the presence of Ca and Si, owing to their standard base of calcium carbonate and silica abrasives. Relatively high contents of Al and Ti in addition to Ca were found in some toothpastes that contained "whitening" additives; alumina (Al_2O_3) and anatase (TiO_2) are commonly used. Thus one may suspect that, by using toothpastes that contain such whitening additives, migration of Al and Ti onto and into the tooth can be found. While the use of different toothpastes could clearly be distinguished when investigating relevant tooth samples, it should be noted that we found measurable concentrations of these metals only in the dentine layer of healthy teeth.

We would like to make one further point. Most toothpastes contain anti-caries agents, normally in the form of sodium fluoride, NaF. Hence one should be able, in principle, to follow the success of this anti-caries agent. Unfortunately, monitoring Na is not very satisfactory – although we observed significant Na signals – since its presence as a result of eating salted food (added or natural NaCl) may be rather large and mask the contribution due to NaF. Thus far we have failed to find suitable emission lines for F to address the problem of the protective influence of fluorine.

In dental treatment of tooth decay, fillings of amalgam were used routinely, which contain different mixtures of metals; alternatively, various other filling materials have become

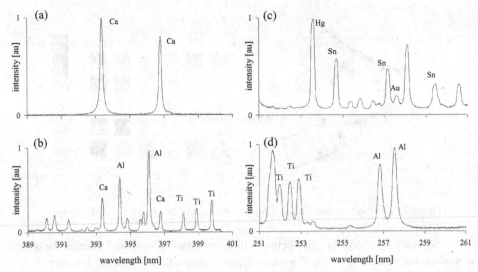

Figure 7.6. Segments of LIBS spectra of toothpastes (a) without and (b) with whitening additive; and filling materials (c) amalgam and (d) composite glass resin; selected elemental lines are indicated.

popular because of the concerns about amalgam and its high content of mercury. In a recent study by Samek *et al.* [9], four different types of teeth-filling material were investigated to identify the presence of potentially toxic elements; these may migrate into the tooth through long exposure by its presence in the filling. The spectral range 251–261 nm is displayed for two filling materials on the right of Figure 7.6.

The specific range was selected because it contains strong emission lines from potentially toxic elements, namely Al, Ti, Hg, and Sn. Differences in the filling material are quite evident, and may be used to identify specific types of fillings; no attempt at quantitative elemental analysis has been made at this stage of the investigation.

7.2.3 *The analysis of bones*

As with the markedly different composition of teeth's enamel and dentine, the detailed composition for various types of bone exhibits significant variations. The typical main chemical compounds are hydroxyapatite (50%–60%) – as for teeth; water (15%–20%); carbonates (~5%); phosphates (~1%); collagen (~20%); and proteins (~1%).

Numerous elements in the concentration range from below parts per billion up to the percentage region are encountered in bones. The actual concentrations often provide information on deficiency or disease states, or whether poisoning or contamination has occurred. The most widely encountered trace elements include Al, Ba, K, Li, Mg, Mn, Na, Pb, and Sr. Some of these, for example Al and Pb, are perceived as potentially toxic elements.

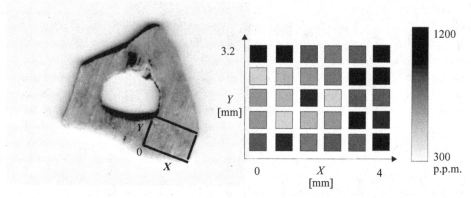

Figure 7.7. Two-dimensional scan of Sr concentration from a section of a tibia bone.

It is normally easy to distinguish between different types of bone, and whether the bone was of animal or human origin. For example, clear distinction in spectral intensity of Sr lines reflects different dietary habits – a meat-based diet that is often connected with Man contains less Sr than the vegetarian diet in some animals [29].

Here we address part of an investigation in which cross-section segments of bones were mapped for the determination of Sr content (Figure 7.7). In the particular case described here, a small section of about 3×4 mm^2 of a tibia bone is rastered, in steps of 0.8 mm in each direction; the laser spot size on target was about 200 μm. The concentration values for Sr in the particular tibia sample were near the upper limit of the normal overall range of concentration in human bones [2, 29]. On the other hand, we measured much lower values for some femur samples, using both the method of LIBS and, for reference, the method of RIMS [30]. Note that the concentration changes towards the bone marrow channel from where the bone receives its nutrients for build-up.

7.2.4 The analysis of other calcified bio-samples

Studies of marine life have been used in environmental monitoring [31]. Marine species accumulate elements first in their soft tissues, from where they precipitate into the calcified parts. Thus, elemental composition of calcified tissue will reflect the water composition, including pollutants. Specifically, the "hard" shells of mussels provide a time-base record of pollution. The shell samples discussed here were sampled from Swansea Bay, a shallow bay on the South Wales coast.

Without going into much detail (the study is still ongoing), we have analyzed a range of different shells, collected at specific locations in Swansea Bay, using standard LIBS on both natural shells (spatial information maintained) and pellets pressed from ground shells (spatial information lost). As expected, large variations in element concentration were encountered for different sample specimens. For example, it is well known that the

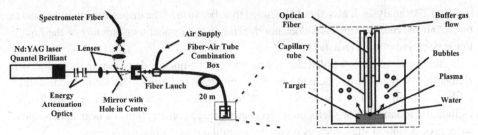

Figure 7.8. LIBS set-up for underwater experiments.

content of dissolved salt in sea water influences the Sr concentration, and we noted marked differences for shells collected from near the outlet of a fresh water river. In general, the Sr concentrations we determined were in the range 1500–2400 p.p.m., in the outer part of the shells. This is in broad agreement with findings of Raith *et al.* [31], who used the technique of LA-ICP-MS. We also followed the concentration variation of intake of Al, K, and Li (from effluents in a small river), and a few heavy metals whose presence in the water of Swansea Bay is notorious, as a consequence of extensive mining of these elements in the area until the mid 1950s. Concentration levels ranging from values close to the detection limits for LIBS (a few parts per million) through to values which have to be labeled as "heavily polluted."

The calcified parts of marine species can, in principle, even be analyzed directly under water, in their native environment. We proposed a novel approach to determine the composition of samples submerged in liquids [32]. The set-up used in those experiments is shown in Figure 7.8, and in its principle is very similar to the fiber-coupled LIBS system used in other parts of the studies of calcified tissue [33].

In the set-up, a single fiber (up to 20 m in length), with a core diameter $d_{core} = 550$ μm, was used. It both delivered the Nd:YAG laser radiation to the target material, and collected the plasma emission for subsequent analysis. Furthermore, no sophisticated fiber-coupling procedure or imaging techniques (only bare fiber-end near the target) were used. The fiber was fed through flexible plastic tubing. At the target end of the fiber, the probe consisted of a simple glass capillary tube of length 30 mm, fitted into the end of the plastic tubing; the fiber-end was held within the capillary tube at a suitable distance away from the sample surface (normally $D \cong 1.5$–2.0 mm). The inside diameter of the capillary ($d_{capillary} = 1$ mm) was slightly larger than the outer (cladding) diameter of the fiber ($d_{fiber} = 0.75$ mm). A buffer gas (ordinary air or dry nitrogen) was blown through the annular passage between the fiber cladding and the inside of the capillary tube, resulting in a bleed stream of gas displacing the water at the position of plasma generation.

Because of the close proximity of the fiber-end to the sample surface, sufficient light could be collected from the plasma by the optical fiber-end, without the need for a lens, to be suitable for quantitative spectral analysis. Note that for moderate buffer gas flow rates the temporal evolution of the plasma is similar to that found in ambient air, as used in routine on-line LIBS analysis. It also should be noted that, because of the displacement of the water by an "air bubble," the analysis does not depend on the optical transparency of the liquid to the laser and plasma radiation.

In summary, we demonstrated an easy-to-apply technique for determining (quantitative) elemental composition of calcified samples submerged in liquids. The technique is applicable to real-time and *in situ* monitoring of difficult-to-reach under-water species. These include the monitoring of shells on under-water bridge constructions or near under-water sources of pollution sites, such as urban waste outlets and oilrigs, for example.

7.2.5 On the problem of calibration standards

One major problem in the analysis of calcified tissue samples is that suitable reference standards, required for quantification of results, may not be available. It is next to impossible to prepare reference standards based on the hydroxyapatite majority compound encountered in calcified tissue. Hence, after numerous tests we opted to use $CaCO_3$ as the base matrix material. The overall physical properties of pellets pressed from $CaCO_3$ are roughly comparable to those of hydroxyapatite. However, it should be noted that the pellets were slightly more brittle than the biological specimen because of the absence of the biological growth mechanism. It would have been even more realistic if a phosphorus-carrying compound like $Ca_3(PO)_4$ had been added, but we encountered problems with homogenizing a mixture sufficiently to avoid substantial local variations in the Ca and P distributions.

In our tests, Al, Sr, and Pb were added simultaneously to the pellets in the form of $Al(NO)_3 \cdot 9H_2O$, $SrCO_3$, and $PbCO_3 \cdot Pb(OH)_2$, to reduce the number of individual samples and to allow for cross calibration. The relative element concentrations were adjusted in the range 100–10 000 p.p.m. relative to the Ca content of the matrix [23]. The relevant calibration curves for the three elements are shown in Figure 7.9.

The detection limits were estimated using the common 3σ rule, and we obtained 15 p.p.m., 95 p.p.m., and 30 p.p.m. for Al, Pb, and Sr, respectively. Also included in the figure are a number of data points for some of the tooth and bone samples discussed earlier in Sections 7.2.2 and 7.2.3. While at present we do not claim high precision of the calibration procedure (this is largely because of the incomplete characterization of the ingredients used in forming the reference pellets), the returned concentration values from real samples are well within expectation.

We would like to make one final remark here concerning reference samples based on a $CaCO_3$ matrix. Rusak *et al.* [34] found a strong dependence of LIBS spectra on the water content in pellets made from $CaCO_3$, and they showed that it is essential to account for the water content if accurate quantitative LIBS analysis is to be achieved. Thus, coupled with the fact that calcified tissue normally contains considerable amounts of water of crystallization, care has to be taken if precise, quantitative results are required. However, in a number of the potential applications outlined above, absolute measurements are not an issue but relative ratios are sufficient to guarantee success.

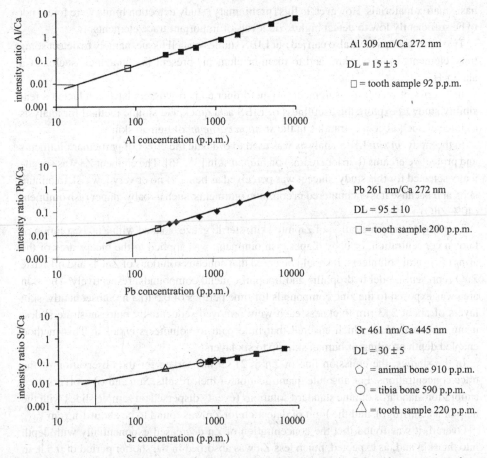

Figure 7.9. Calibration curves for Al, Pb and Sr in a $CaCO_3$ pellet matrix. Selected measurement data for some tooth and bone samples are included (open symbols).

7.3 Investigation of "soft" tissue materials with cell structure

7.3.1 Analysis of human and animal tissue

It should be noted from the outset that only very few LIBS studies of "soft" tissue have been carried out to date. This should not come as a surprise since by its matrix nature human and animal cell tissue is very rapidly destroyed by the laser ablation process and, because of the generally very high water content, spectral emission from elements in the tissue is severely affected. Recently, our research team has carried out a few measurements to analyze soft tissues, namely nails and skin [18]. Again, Ca was an abundant component in the spectra but there appears to be much less Ca in skin (soft cell tissue) and nail (corotid tissue) samples in comparison with teeth (calcified tissue). This is not unexpected, given the nature of the basic matrix materials. However, in this (preliminary) study detection limits were found not to be sufficiently low to detect high/toxic levels of important trace elements.

Two other groups have also carried out LIBS studies on cell tissue, namely to detect some trace elements in or on skin, and to monitor elements present in abundance, such as Ca and/or H.

In a series of recent measurements, Winefordner and co-workers have conducted a feasibility study to explore the usefulness of LIBS as a quick and simple method for analysis of trace elemental concentrations in the *stratum corneum* of human skin.

In the study, *in situ* LIBS analysis was used to evaluate the effect of particular ointments and protective creams (barrier creams) on human skin [35, 36]. The element Zn was specifically selected for this study, since it was perceived as being of no or very low risk to human skin, and because it is encountered in common ointments, such as baby diaper rash ointment (40% ZnO).

To evaluate the penetration of Zn into skin, sterile gauze, soaked with $ZnCl_2$ solution of known concentration, or baby diaper rash ointment, was applied to the biceps area of the arms of several volunteers. It should be noted that aqueous solutions of $ZnCl_2$ and oil-paste ZnO represent model hydrophilic and lipophilic metal compounds, respectively. The skin area was exposed to the zinc compounds for time periods of 0.5 to 3 h. Subsequently, skin layers of about 2–3 μm thickness each were removed sequentially onto substrate slides, using a special glue, which ensured that little pain to volunteers entailed. This method enabled depth profiling of human skin up to six layers.

In the studies, the emission line of Zn at 213.9 nm served for the observation of the trace concentrations. For absolute quantification of their results, Sun and co-workers used calibration standards of zinc standard solutions (evenly dispersed on sample slides with the help of a centrifuge), and the limit of detection for Zn was found to be about 1 $ng\,cm^{-2}$.

Overall, it was found that the concentration of Zn decreased exponentially with depth into the skin and, as expected, much less Zn was absorbed in the shorter period of 0.5 h, in comparison with the long exposure of 3 h.

It should be noted that normal, healthy skin exhibits natural Zn concentrations of about 30 $ng\,cm^{-2}$ while abnormal skin (e.g. atopic dry skin) may show concentration levels of >200 $ng\,cm^{-2}$. Thus, the results shown in Figure 7.10 are most likely statistically relevant only for the first three skin layers.

In addition to the actual penetration of Zn into the skin, the effect of protective barrier creams on the Zn exposure was studied [36]. The skin including the barrier cream was exposed once again to hydrophilic and lipophilic Zn compounds. Some of the results from this study are summarized in Figure 7.10. The results clearly revealed that the protective creams had a definite effect on the take-up of Zn by the skin, with the protection against the lipophilic ZnO being more significant. The authors also point out that their results were only deemed statistically relevant for the longer exposures of 3 h.

In principle, the study showed that LIBS may provide an *in situ* method in screening barrier cream candidates and in the formulation of barrier cream vehicles, and to explore the general assimilation behavior of skin to exposure of ointments.

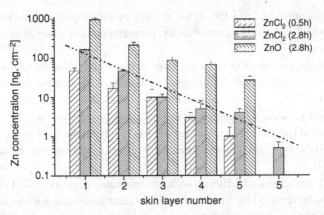

Figure 7.10. The Zn signal from LIBS depth profiling for successive human skin layers exposed to aqueous ZnCl$_2$ solution and ZnO oil-paste, for periods of 0.5 h and 2.75 h contact time. The dot–dash line indicates the fit to measurement of 2.75 h exposure. Data from Sun *et al.* [35].

Another interesting application of LIBS on soft tissues has been reported in which corneal hydration was monitored [37]; samples of human and rabbit corneas were used in the study. Measurements revealed that spectral intensities of Ca and H lines depended strongly on the water content in the cornea tissue. The intensity ratio parameter $I(H_\beta/Ca)$ was introduced for the construction of a calibration curve to quantitatively estimate the cornea's hydration. In the reported experiments use was made of a standard Nd:YAG laser. It should be noted that for actual clinical realization of this application a repeat study would be required, this time using an ArF excimer laser ($\lambda = 193$ nm), which is used in clinical cornea treatments.

Finally, in a preliminary study, the potential of LIBS in hair tissue mineral analysis (HTMA) was assessed [38]; HTMA is one of the most common analysis techniques in the assessment of nutritional deficiencies, in risk screening for metal poisoning, and in forensics. Only minute amounts of analyte were required (a single hair of a few millimeters in length) to identify hair from individual subjects. While at present the analysis of elements is only qualitative, efforts are under way to quantify the methodology.

7.3.2 Analysis of plants

Elemental analysis in food and agricultural products, and other botanical specimens, is necessary and important. Unfortunately, most analytical methods for determining the concentration of elements in plants require digestion of the sample in acid solutions before analysis. This is a time-consuming preparation method, and in addition any spatial distribution information is normally lost. While this may not necessarily be required for general analysis, it may be useful, or even paramount, to trace the time evolution of take-up and accumulation in specific areas of the sample. In this respect, the desire for spatial distribution information is not different from that for calcified tissue that was addressed above.

With the growing success of LIBS in the analysis of solid materials, and its known ease of use, speed, and direct sampling capabilities, the method is attracting increased attention for the analysis of plant samples.

One of the major problems in the analysis of plants – and, more generally, for all biological tissue – is the availability of reference standards. Although for calcified specimen reference standard matrices with quantifiable trace element concentrations can be simulated, as outlined in Section 7.2.5, this is normally not possible for soft tissue materials, including botanical specimens. Here one mostly has to rely on the availability of calibrated reference materials, for example those provided by the National Institute of Standards and Technology (NIST).

A successful application of LIBS to such pre-calibrated standards and unknown plant materials has been reported by Sun *et al.* [39]. In their study, the authors demonstrated that with very little sample preparation (e.g. grinding of leaves into a powder to apply to a sample slide) LIBS provided quantitative analytical results, which mostly were in excellent agreement with other methods, including inductively coupled plasma mass spectrometry (ICP-MS). The results were obtained for a variety of leaves (apple, peach, spinach, tomato), pine needles, and moss. Overall, from the measurement of the calibration standards, detection limits for a number of important elements (Al, Ca, Cu, Fe, Mg, Mn, P, and Zn) were found to be in the range 0.1–25 p.p.m. This is quite respectable, bearing in mind the minimal sample preparation required to obtain these results, using a rather unsophisticated LIBS system.

In this study, no efforts were made to measure the samples with spatial resolution. Investigations of leaves and tree samples with spatial information maintained have been carried out at the Institute for Spectrochemistry and Applied Spectroscopy (ISAS) in Berlin, Germany. The studies aimed at linking localized inhomogeneity of trace elements in the plant sample to differences in atmospheric pollutants, soil chemistry, or other factors stimulating or reducing growth. It should be noted that the majority of their studies were based on laser-ablation inductively coupled plasma mass spectrometry (LA-ICP-MS), rather than on simple LIBS.

Here we briefly note two studies addressing the spatial determination of elements in green leaves and tree rings [40, 41]. In both studies, spatial maps of a range of elements (Al, Ba, Ca, Cr, Cu, Fe, Mg, Mn, Ni, Pb and Sr) were recorded which revealed the response of the plant to changing environmental conditions. The reported detection limits for these elements were of the order 0.02–5 p.p.m.; as expected for ICP-MS measurements, this is about an order of magnitude better than for LIBS.

For the study of leaves [40], samples were collected during three periods of the growth cycle (May, June, and October). The results clearly revealed a link to the biological growth cycle. In addition, the spatial distributions of groups of element with similar migration behavior revealed differences between healthy and pollution-damaged trees, and thus provided information on pollution levels and indicated pollution-affected changes in nutrient flows. It should be noted that calibration samples were prepared using cotton cellulose doped with multi-element standard solutions, referenced against a peach leaf standard.

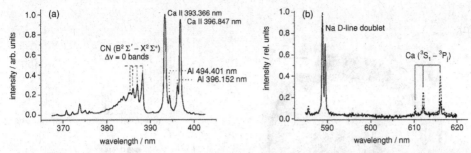

Figure 7.11. Typical LIBS spectra of wood samples; (a) pine tree sample from the Tawe valley; (b) laburnum tree sample from the Clyne valley. Prominent elemental and molecular features are indicated.

To reveal the link between plant tissue chemistry, soil chemistry, bedrock chemistry and environmental changes, cross sections through trees were studied, and the concentration of elements was measured as a (radial) function of tree rings, i.e. age. Furthermore, mapping the complete tree ring over 360° (circular scanning) highlighted considerable fluctuations in elemental concentrations in wood compression regions, which can be linked to predominant wind direction and frequency. Clearly, the samples under study confirmed links to pollution sources; the concentrations of fast-migrating pollution elements changed dramatically while those of general nutrients remained largely constant for the time period associated with the tree ring count. Once again, artificial calibration standards had to be made since wood standards were not available. These comprised pellets pressed from cellulose powder (cellulose is the major constituent of wood and has a composition similar to most types of woods); the pellets were doped with multi-element standard solutions.

As noted, the two studies of plant samples described here exploited LA-ICP-MS for analysis. To our knowledge, there is no published work on direct LIBS analysis with spatial resolution of such samples. In order to demonstrate the feasibility of quantitative LIBS analysis of trace element concentrations for tree rings, we commenced a study of trees in former pollution areas in and around Swansea [42]. The experiments were performed using a standard LIBS apparatus based on a Nd:YAG laser, operating at 1064 nm and providing pulse energies of about 15–20 mJ. Two typical LIBS spectra used in the analysis for trace elements are shown in Figure 7.11.

In the first example (Figure 7.11a), a sample of a pine tree from the Tawe valley (to the east of Swansea) is shown. This area was a former mining area with soil still loaded with heavy-metal ore remnants being washed out after prolonged rainfall periods. In this particular spectral segment the presence of toxic pollutant Al is seen beside the standard nutrient Ca. The molecular band of CN serves as an internal calibration standard; CN is being formed in a reaction between ablated carbon atoms and nitrogen in the ambient air. Its intensity did not change when the LIBS operating conditions were kept reasonably constant. Such molecular bands were found throughout the full spectral range used for analysis.

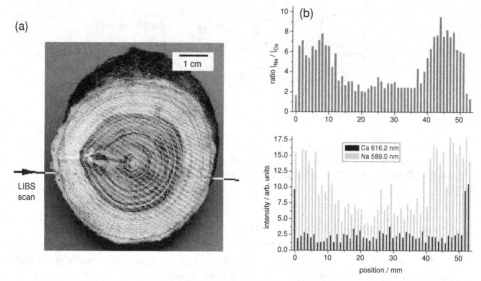

Figure 7.12. The LIBS analysis of a branch of a laburnum. (a) Cross section with LIBS crater trace; (b) relative line intensities, and their ratios, for Ca and Na, as functions of position across the tree branch.

The second example (Figure 7.11b) is a laburnum tree from the Clyne valley (to the west of Swansea), near the estuary of the Clyne river into Swansea Bay. This area also is a former mining area; the river still carries effluents from an industrial estate further upstream. Here the change in elemental composition in 2 different tree rings is evident. Note that the display is perceptive: the spectra from the 2 tree rings were normalized to the highest peak, i.e. the Na peaks. However, it is the Ca concentration that is relatively constant through the range of about 20 tree rings (see Figure 7.12). A clear link between the relative concentration levels of Na and the different coloring of tree stem cross section can be observed. The darkish brown (dark gray in the picture) region in the center exhibits low concentrations of Na, while in the light beige (light gray in the picture) section, typical for softer wood in the stem, a higher Na concentration is encountered. Evidently, some environmental change has occurred that influenced the nutrients and the health of growth (the specific tree was recently uprooted). The increase in Na content can possibly be attributed to a change in salinity in the estuary region due to dredging of the river bed, or to effluents in the river (we measured rather high concentrations of K and Na being carried by the river), or a combination of both.

We stress that thus far all measurements for the tree samples discussed here are relative measurements. Reference samples for absolute calibration are being prepared, following the same procedures outlined by Hoffmann *et al.* [40].

Finally, we would like to note one recent investigation into the use of a femtosecond (fs) laser source for LIBS of biological samples with high spatial resolution [43]. In this study the spatial distribution the nutrient element Ca in sunflower seedlings was investigated. The experimental system included a Ti:sapphire femtosecond laser, operating at a center wavelength of about 790 nm, and for comparative studies a nanosecond laser source (Nd:YAG, third harmonic at 355 nm). The laser radiation was sent through a dichroic beam splitter within a microscope to achieve collinear beam geometries. This property was important to "hit" specific, identified tissue features with very high spatial resolution. Detection of the plasma emission light was made in the usual manner through a spectrometer, coupled to the microscope, and a gated intensified CCD detector.

The analysis was performed in depth (a few laser pulse exposures penetrated through the peripheral cell wall of the sunflower stem) and analysis for the presence of Ca was performed inside the peripheral cell wall of 1.5–2.0 µm, with an axial resolution of around 140 nm, dictated by the NA of the microscope objective being used. It should be noted that the experiments are still in their infancy but have the prospect of three-dimensional element-specific imaging of biological specimen, with high spatial resolution.

7.3.3 Bio-monitoring in soft-tissue organisms

Broadly speaking, living organisms respond to surplus ("pollution") or deficiency of specific elements or molecular compounds in their environment, and these changes are reflected within their tissue. This was highlighted in Section 7.2 for bones, teeth, and shellfish, whose calcified tissue served as a long-term archive on nutrition and environmental changes. However, for short-term monitoring a much faster response to elemental/molecular concentration levels is required. In general, such a short-time-scale response is reflected in the accumulation of elements or morpho-structural modifications in organisms with a fast metabolism and a relatively short lifespan. Typical organisms that are used for this purpose are lichens [44]. Very recently, LIBS analysis was applied to lichens to trace the uptake of a range of heavy metals from the environment into their biomass [45]. Briefly, lichens were collected from environmental locations, in accordance with official guidelines for their sampling (in the reported study by the *Italian National Agency for Environmental Protection*), and sample pellets were prepared from homogenized lichen thallus. These pellets were then probed by LIBS for the presence of a range of important nutrient- and pollution-related elements. Clear links between industrial and human activities could be shown in the samples collected from a range of sites, providing geographical distribution maps.

7.4 Investigation of bio-fluids

The detection and quantification of many light elements and heavy metals within liquid samples is pertinent to industrial processing, waste management and environmental monitoring. A plethora of analytical techniques is used for this, including inductively coupled plasma atomic-emission spectroscopy (ICP-AES), this being closest to LIBS in its concept. For example, Simpkins *et al.* used ICP-AES and ICP-MS to analyze for a wide range of trace elements in orange juices and related products, reliably detecting sub-parts-per-million quantities [46].

To date, LIBS approaches have achieved only limited success when applied to sensitive, quantitative analysis of liquids; test arrangements have utilized static liquid pools, aerosols or droplets, or liquid streams. In general, for *in situ* measurements detection limits of 1–100 p.p.m. for light metal elements were realized, while heavy metals are rather more elusive, some of them having been found to be undetectable even at concentration levels as high as 1000 p.p.m. On the other hand, the simplicity of LIBS with its normally minimal sample preparation requirements offers great potential for rapid, real-time and *in situ* analysis.

A very promising approach for sensitive and reliable LIBS analysis of liquid samples has been made by Van der Wal *et al.* [47]. In their approach the analyte solution is evaporated upon an amorphous graphite substrate; this is followed by LIBS analysis of the substrate surface. In this manner, the trace-element identification within the liquid is transformed into a solid surface analysis measurement, with the advantage of the much greater flexibility and high sensitivity associated with LIBS solid surface analysis. In addition, for calibration purposes, use can be made of commercially available standard solutions, and the same sample substrates may be used in other analytical methods for cross-referencing. In comparison with other implementations of LIBS of liquid samples with limits of detection (LOD) normally larger than 1 p.p.m., the authors achieved LODs of 10–100 p.p.b. for many of the important elements in aqueous samples, being consistently superior by two to three orders of magnitude. Even for heavy elements, like Cd, Hg, and Pb, a LOD of (well) below 10 p.p.m. was realized. It should be noted that under the experimental conditions described by the authors a single LIBS analysis samples about 1–2 μl out of a 1 ml liquid volume initially deposited; this is competitive with, for example, graphite furnace atomic absorption spectroscopy used for trace-element analysis. The probed equivalent volumes of liquid and detection limits equate to the detection of sub-picogram quantities of specific elements.

In summary, LIBS analysis of liquid samples, dried on carbon substrates, offers greatly improved detection limits over LIBS applied to the *in situ* liquid itself. On the other hand, the method does require some sample preparation and thus may not be deemed to be an *in situ* measurement. However, the preparation time is minimal, and the volume of sample is minuscule. Consequently, the method may easily be applied to screening of liquid environmental and biological samples, including the analysis of urine or blood samples, but most importantly the analytical results are nearly "instantaneous."

In Swansea we have followed a similar approach to develop a liquid-to-surface transfer LIBS analysis procedure which lends itself to the screening of bio-fluid samples. In this implementation, the liquid (blood) sample is transferred to a standard filter paper (instead of the amorphous graphite substrate used by Van der Wal *et al.* [47]). The particular example highlighted here is that of the potential screening of blood for traces of Rb by LIBS, as a possible analysis technique to trace the effect of illegal doping drugs [48]. For example, certain performance-enhancing drugs lead to an increase in the count of red blood cells (erythrocytes). The method proposed here exploits the fact that at least 90% of Rb measured in whole blood is located in the erythrocytes [49]. When orally administering RbCl in aqueous solution, Rb will show up in the bloodstream after a relatively short time period

(~30 min). Because Rb is mainly bound to the erythrocytes, any Rb signal above the (normally low) background concentration can be related to the athlete's red blood cell count. A semi-quantitative result can be achieved, in principle, in less than a few minutes, by LIBS analysis of a minuscule quantity of blood deposited on filter paper, without time-consuming sample preparation. This would be in contrast to the tedious blood cell counting procedures currently used in drug abuse testing, which take a few hours to complete.

The liquid solution sample is deposited onto a piece of standard filter paper, over an area of about 5 cm^2, requiring normally much less than 1 cm^3 of liquid. The paper is dried and then attached to a rapidly rotating disk, which also can be moved laterally. The substrate is then exposed to the LIBS laser (a standard Nd:YAG laser operating at 1064 nm) and the plasma emission is analyzed in the standard way (the resonance lines of Rb at 780.03 nm and 794.76 nm are used for spectroscopic analysis). This procedure assures that the LIBS analysis statistically samples a large fraction of the applied liquid solution (note that the diameter of the focal spot is about 750 μm). Calibration measurements were made using aqueous solutions of RbCl or RbNO$_3$ (to simulate different pH values of the solution). Detection limits of the order 5–10 p.p.m. were determined; these were slightly disappointing when compared with measurements within liquid jets [33], and with the general results of liquid-to-surface transfer LIBS [47]. On the other hand, on closer inspection, these values do not represent too much of a surprise. Because of our simple "soak-and-dry" sample administering of solution, much less deposited liquid volume equivalent is probed. Calculating the loading in nanograms per square centimeter, the smallest observed signals were roughly in line with the values cited by Van der Wal *et al.* who found lower limits of detection of the order 100 p.p.b. [47].

In an attempt to improve on our LIBS detection limits, which might be insufficiently sensitive for the task, a secondary probe in the form of LIBS plus LIFS was introduced. This approach is known to improve on detection limits in the laser-generated plasma, at the expense of having to add a tunable laser to the measurement set-up. In our experiments, this secondary laser was a pulsed, tunable Ti:sapphire laser, synchronized with a delay of 20–40 μs to the plasma-generation laser. The plasma plume was exposed to photons at 780.03 nm (Rb *D2*-transition) and the Rb fluorescence was observed on the 794.76 nm *D1*-transition. By using the combination of LIBS plus LIFS, the detection limit for Rb could be lowered by about an order of magnitude; while not necessarily essential for the expected concentrations of Rb in blood, it could be crucial in the case that samples are extremely small. Further discussion can be found in the work of Al-Jeffery and Telle [48]. Some results from the LIBS and LIBS-combined-with-LIFS test measurements are summarized in Figure 7.13. Note that a few test measurements were also performed on "blank" and "spiked" blood serum (as expected, traces of Ca and K were observed, see also trace 3 in Figure 7.13).

No final assessment has yet been made as to whether the method will ultimately be suitable for the intended application of testing for taking illicit drugs. While the results from the preliminary test measurements look promising, procedural delays in dealing with whole blood samples and potential clinical trials have held up our studies. Whole blood

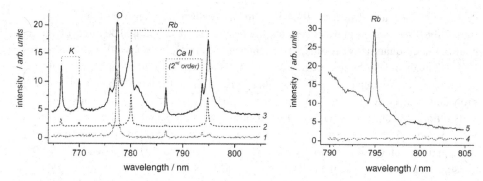

Figure 7.13. T LIBS and LIFS spectra of Rb traces on filter paper, loaded with aqueous solutions of RbCl/RbNO$_3$. Trace 1 – LIBS spectrum of virgin filter paper; trace 2 – LIBS spectrum of filter paper doped with 1000 p.p.m. Rb solution (ablation at $E_A = 12$ mJ); trace 3 – as trace 2 but for higher ablation energy, $E_A = 35$ mJ); trace 4 – LIBS signal after 40 µs delay; trace 5 – LIFS spectrum of filter paper doped with 1000 p.p.m. Rb solution (ablation at $E_A = 35$ mJ, probe at $E_P = 7.5$ mJ).

measurements are to commence shortly, first to determine the natural abundance of Rb, and then to conduct trials on whether short-term uptake of Rb into the bloodstream can be detected after orally administering RbCl solutions.

7.5 Investigation of microscopic bio-samples

As a final point we wish to address a topic that has received increased interest in recent years, namely the analysis of single microscopic particles, aerosols and cells. In particular, bio-aerosols (bacteria, fungi, viruses, pollen) have attracted wide attention, since they are found nearly everywhere, indoors and outdoors. While their concentrations are normally low, only minute amounts of (inhaled) bio-aerosols may cause disease or toxic/allergic reactions.

There have been a few campaigns over recent years to establish possible links between allergies and the presence of certain aerosol-type particles and nano-/micro-particulates carried in the atmosphere and in directed air streams (e.g. air conditioning ducts). A variety of techniques have been used for these studies, including characterization by fluorescence spectroscopy [50]; by Raman spectroscopy [51]; by ICP-MS [52]; by mass spectrometry methods [53]; and by on-line LIBS [54]. By and large, these methods are established techniques for measuring "bulk" materials; however, attention is turning to the on-line analysis of individual particles and the whole question regarding "bulk" vs. "single particle analysis" is being addressed.

Chemical characterization by Raman spectroscopy has been carried out both on-line (by trapping, for example, single pollen particles in an electro-dynamic balance, coupled to a Raman spectrometer) and off-line (for example pollen samples deposited on a substrate and analyzed using a Raman microprobe). Characteristic spectra were recorded for a range of common allergy-related pollen. While many of the spectral features are rather similar, which is not surprising given the common building blocks of pollen, there were nevertheless sufficiently large differences in the Raman spectra for unambiguous identification [51]. Such unique identification was much less straightforward in fluorescence measurements [50].

In a variety of epidemiological studies, attempts were made to establish geographic links between inhaled ambient aerosols and public health, using experimental studies based on physical particle counting and chemical bulk filter analyses using ICP-MS [52].

A much more versatile mass spectrometric analysis technique is based on a new breed of instruments, which allow simultaneous recording of positive and negative ions for the chemical identification of individual particles, together with measurement of their aerodynamic diameter. The method is termed aerosol time-of-flight mass spectrometry – ATOFMS – and is based on laser desorption ionization [55]. The method has been commercialized (e.g. TSI3800 ATOFMS), and the instrumentation easily lends itself to the study of ambient aerosols in real time and *in situ*. Higher time resolution and chemical detail can be achieved than with older bulk filter analysis methods.

Work in the field of aerosols using LIBS as an analysis method has been done by Carranza *et al.*, who studied ambient (inorganic) aerosols by "one-shot" analysis [54]. The LIBS atomic emission spectra can be used to infer the chemical and physical properties of individual particles. However, LIBS might not necessarily be the most appropriate method for detailed analysis, since very often the organic molecular components are of interest and not the presence of individual atomic species.

In addition to performing these experiments directly in an air stream (i.e. in real time), it is generally possible to apply standard sampling methods for such aerosol particulates, namely their accumulation on filters. This would permit us to carry out direct comparative cross-calibration measurements on the same sample by different analytical methods.

With respect to the topic of this chapter, here we wish to draw the reader's attention to some very preliminary experiments we carried out in our laboratory on pollen sampled on metal/glass substrates or standard filter paper [56]. To our knowledge, no LIBS work on biological micro-samples has yet been published, but some preliminary results are available only from our group and one other research group. Our measurements were performed on samples that were prepared in the form of "pastes," normally administered to glass substrates. A thin layer of the pollen paste was left to dry on the substrate, and was measured at a number of random locations, normally hitting on average 2–3 pollen particles (the sample was moved by a stepper motor in the two lateral dimensions). This approach is far from the desired final goal of single particle detection and identification, but the initial goal was to try and establish a "reference" base for such single-particle, single-laser-pulse analyses.

The analysis was performed using the same set-up as described in Section 7.2 for the two-dimensional raster analysis of calcified tissue materials; the laser pulse energy was in the range 20–30 mJ, focused to a spot size of approximately 0.2 mm^2. The gating of the spectrometer detector was set for 1 µs delay after the laser pulse, integrating the plasma intensity over a period of 4 µs. All spectra were recorded for a single-laser-pulse exposure, and hence absolute intensities varied substantially from laser pulse to laser pulse. However, the relative intensity pattern for spectral lines from individual sample types did not vary much (in general by only 10%). Typical example spectra for pollen of two different lilies

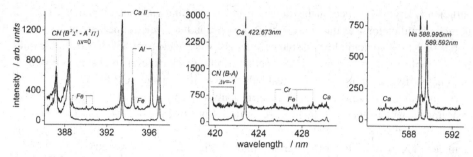

Figure 7.14. Typical LIBS spectra (single laser pulse exposure) of pollen samples. In each spectrum the upper trace is for the "yellow" tiger lily and the bottom trace is for the "white" smajeens lily. Prominent elemental and molecular features are indicated. Traces are normalized to the intensity of the CN(B–A,0–0) band and offset from each other for clarity; for further details see text.

are shown in Figure 7.14. Quite evident in all spectra was the presence of molecular bands, which in the range 300–500 nm could largely be attributed to emission from vibrational sequences of CN (B $^2\Sigma^+$ – X $^2\Sigma^+$). The CN radical is formed in a reaction of carbon liberated from the organic pollen sample with nitrogen in the LIBS plasma; we observed its formation for any organic cell material (see also Section 7.3.2 and Figure 7.11). While at times annoyingly masking atomic lines of interest, these emission bands can be used for internal standardization, since the overall matrix of hydrocarbons is reasonably constant for different types of cell material. Hence, in the spectra displayed in Figure 7.14 the relative spectral intensities of different pollen types were normalized to these molecular features.

Clearly, spectral differences can be recognized for the two samples belonging to the same family of flower pollen, namely lilies. While the presence of Ca and Na has to be attributed to standard "nutrient" uptake, nevertheless strong differences for the two lilies are observed. However, these are not yet sufficient for full identifications. Other elemental lines help in this respect. For example, the strong presence of Cr lines (see the middle spectrum in Figure 7.14) for the yellow tiger lily is associated with the color of the species – often the yellow color in plants is caused by chromium compounds. It should be stressed that it was rather easy to distinguish between different lily types, but that pollen from other plants often exhibited similar elemental patterns. Thus inter-species distinction was rather more difficult. In order to elucidate whether unique identification would be achievable by adding a complementary spectroscopic technique, we carried out Raman measurements in addition to this LIBS analysis, to attempt cross validation of differences between families of pollen.

These Raman measurements were performed on the same samples, using a Raman Scope 2000 (Renishaw). The differences in the spectra were substantial, and the observed spectra exhibited Raman band features rather similar to Raman spectra of different pollen families measured, for example features typical to carotenoids [57]. Examples of Raman spectra for a range of lily pollen and pollen from a different species (a marguerite) are shown in Figure 7.15. Two Raman band features are common to all species, and only subtle differences in

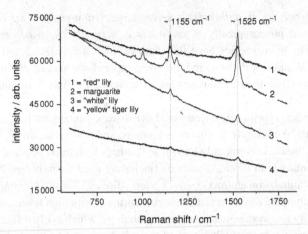

Figure 7.15. Partial Raman spectra of single pollen cells from a variety of lilies, and a marguerite. The Raman peaks sit on a background of strong fluorescence; prominent common features are marked (for further details see text).

the Raman features are observed. On the other hand, the fluorescence background varies significantly for the different species. However, as in the case of the LIBS spectra, unique association between species and spectral trace was problematic: while inter-species identification is now possible, distinguishing members within a species group remains very difficult. It is quite clear from these preliminary measurements that a single spectroscopic analytical method may prove to be insufficient for unequivocal identification, but that a combination of techniques is required.

As a follow-up to these preliminary studies on the feasibility of being able to distinguish between various pollens with sequential Raman and LIBS analyses, a more systematic study is under way to achieve analysis of single pollen particles [58]; this work is being done in collaboration with a group in our Medical School to assess allergic reactions to specific pollen as well. It should be noted that our initial studies were carried out at a period of the year when allergenic pollen activity is relatively low; future work will be carried out to specifically address these pollens, including grass pollen and the pollen of certain trees. We also envisage linking these measurements with on-going studies of bio-aerosols in other groups, using different analytical techniques.

A study very similar in nature to the above analysis, i.e. to determine chemical differences in single bio-particles, is being pursued by researchers in the USA, in conjunction with the anthrax attacks in the aftermath of September 11, 2001. It became clear during those cases of exposure to biological warfare agents that a fast and *in situ* analytical method was urgently needed. The method of LIBS was contemplated as one of the possible solutions to identify anthrax spores. Again, the measurement of atomic components might not be seen as the most logical approach – other specimens might exhibit very similar stochiometric composition, and their molecular and steric signatures might be more indicative for conclusive identification and the detection of minute amounts.

Bacillus anthracis (or *B. anthracis*) is a very large, Gram-positive, spore-forming rod. Genotypically and phenotypically *B. anthracis* is very similar to *B. cereus* and to *B. thuringiensis*. The three species have the same cellular size (1.0–1.2 µm in width, and 3–5 µm in length) and morphology, and form oval spores located centrally in a non-swollen sporangium. Therefore, their dimensions are such that in principle LIBS analysis for individual bacilli is feasible.

Bacillus anthracis is primarily encountered as a disease-causing bacillus in domesticated and wild animals, particularly herbivorous animals, such as cattle, sheep, horses, mules, and goats. The disease is commonly known as "anthrax." Humans become infected when brought into contact with diseased animals (including their flesh, bones, hides, hair and excrement). As "inhalation anthrax" (wool-sorters' disease), most commonly from inhalation of spore-containing dust, the disease begins abruptly with high fever and chest pain. It progresses rapidly to a systemic hemorrhagic pathology, which is often fatal.

Bacillus cereus is a normal inhabitant of the soil, but it can be regularly isolated from foods such as grains and spices. In humans, *B. cereus* may cause food-borne intoxications (as opposed to infections). It resembles either *Staphylococcus aureus* food poisoning in its symptoms and incubation period, or that caused by *Clostridium perfringens*. In either type, the illness usually lasts less than 24 h after onset.

Bacillus thuringiensis is distinguished from *B. cereus* by its pathogenicity to, for example, Lepidopteran insects (moths and caterpillars) and by production of an intracellular parasporal crystal in association with spore formation. The bacteria and protein crystals are sold as "Bt" insecticide, which is used for the biological control of certain garden and crop pests. It is also used in non-pathogenic and non-toxic animal species tests, and for sterilization procedure testing.

It is clear from this brief summary of these bacilli that it is vital in an analysis to be able to distinguish anthrax from its less pathogenic, or harmless, brethren, which might be encountered on a much more frequent basis.

The LIBS experiments conducted recently were aimed at establishing exactly this analysis capability. It should be noted that the preliminary measurements have not yet been carried out for anthrax itself but have for three "close cousins" *Bacillus cereus*, *Bacillus globigii*, and *Bacillus thuringiensis* [59]. The three types of spore were prepared on a porous silver substrate. For each of the spore-loaded disks, five single laser pulses were administered, at random locations, and the five LIBS spectra (associated with the individual bacilli types) were subsequently averaged. When comparing these averaged spectra, numerous spectral regions could be identified that exhibited significant differences between the three types of bacilli; furthermore, principal-component analysis of the spectra clearly allowed full distinction against other biological micro-particles. In a similar study on bacteria, Morel *et al.* investigated the influence of the culture medium on the LIBS signatures of a variety of biological agents, including *Bacillus globigii*, *Escherichia coli* and *Staphylococcus aureus*, all being related to pathogenic agents [60]. While these results certainly have to be seen as preliminary – specifically, no single-particle analysis has been attempted yet – they are nevertheless encouraging, suggesting that LIBS might be developed into an analytical tool for very rapid identification of certain bio-hazards.

7.6 Concluding remarks

Although the number of LIBS studies of biological and medical samples is on the increase, it has to be said that this particular application is still very much in its infancy. This is not unexpected since the acceptance of any novel use of an analytical technique is normally slow unless the benefits are undoubted and it is perceived as being superior to other techniques. In the analysis of biomedical samples, for a large fraction of cases, the identification and quantification of molecular species are required; LIBS normally lends itself only to the analysis of atomic components. But even if element analysis is required, some other techniques may provide detection limits superior to that of LIBS: in the best-case scenario LIBS is able to reach limits of detection of the order 0.1 p.p.m., which may be insufficient for some applications. However, LIBS comes into its own if spatial information on the analyte is required, or if *in situ* and/or real-time measurements are of benefit. Then LIBS may be the only analytical technique available.

It is clear from the examples outlined in this chapter that LIBS definitely has a role to play in the analysis of biomedical samples, and that the results justify the efforts. Specifically this seems to be true for the examples in which precise spatial concentration distribution information (lateral and/or depth) could be gathered.

And, last but not least, the very much reduced sample preparation efforts may give LIBS the edge over other techniques, where speed of analysis is of the essence, or where preparation procedures affect the results, provided its limitations in sensitivity can be tolerated.

7.7 References

[1] P. Budd, J. Montgomery, A. Cox, P. Krause, A. Barreiro and R. G. Thomas, The distribution of lead within ancient and modern human teeth: implications for long-term and historical exposure monitoring. *Sci. Total Environ.*, **220** (1998), 21–36.

[2] K. C. Stamoulis, P. A. Assimakopoulos, K. G. Ioannides, E. Johnson and P. N. Soucacos, Strontium-90 concentration measurements in human bones and teeth in Greece. *Sci. Total Environ.*, **229** (1999), 165–182.

[3] H. M. Tvinnereim, R. Eide, T. Riise *et al.*, Trace elements in primary teeth from six areas in Hungary. *Internat. J. Environ. Studies*, **50** (1996), 267–275.

[4] B. Nowak, Accumulation of metals in the teeth of inhabitants of two towns in the south of Poland. *J. Trace Elem. Med. Biol.*, **12** (1999), 211–216.

[5] V. Spevackova and J. Smid, Determination of lead in teeth of children for monitoring purposes by electro-thermal atomic absorption spectrometry. *Spectrochim. Acta B*, **54** (1999), 865–871.

[6] K. Grunke, H. J. Stark, R. Wennrich and U. Franck, Determination of traces of heavy metals (Mn, Cu, Zn, Cd and Pb) in micro samples of teeth material by ETV-ICP-MS. *Fres. J. Anal. Chem.*, **354** (1996), 633–635.

[7] A. Cox, F. Keenan, M. Cooke and J. Appleton, Trace element profiling of dental tissues using laser ablation – inductively coupled plasma – mass spectrometry. *Fresenius J. Anal. Chem.*, **354** (1996), 254–258.

[8] T. Pinheiro, M. L. Carvalho, C. Casaca *et al.*, Microprobe analysis of teeth by synchrotron radiation: environmental contamination. *Nucl. Instrum. Meth. B*, **158** (1999), 393–398.

[9] O. Samek, D. C. S. Beddows, H. H. Telle *et al.*, Quantitative analysis of trace metal accumulation in teeth using laser-induced breakdown spectroscopy. *Appl. Phys. A*, **69** (Suppl.) (1999), S179–S182.

[10] O. Samek, M. Liska, J. Kaiser *et al.*, Clinical application of laser-induced breakdown spectroscopy to the analysis of teeth and dental materials. *J. Clin. Las. Med. Surg.*, **18** (2000), 281–289.

[11] J. J. Freeman, B. Wopenka, M. J. Silva and J. D. Pasteris, Raman spectroscopic detection of changes in bioapatite in mouse femora as a function of age and in-vitro fluoride treatment. *Calcified Tissue Internat.*, **68** (2001), 156–162.

[12] E. Hoffmann, H. Stephanowitz, E. Ullrich *et al.*, Investigation of mercury migration in human teeth using spatially resolved analysis by laser ablation-ICP-MS. *J. Anal. Atom. Spectrom.*, **15** (2000), 663–667.

[13] T. Jeffries, Quintupled YAG probes 200-year-old teeth to uncover ancient diet details. *Opt. Las. Eur.*, **84** (2001), 15.

[14] O. Samek, H. H. Telle and D. C. S. Beddows, Laser spectroscopy: a tool for real-time, *in vitro* and *in vivo* identification of carious teeth. *Oral Health*, **1** (1) (2001), 1–9. http://www.biomedcentral.com/1472-6831/1/1.

[15] O. Samek, V. Krzyzanek, D. C. S. Beddows *et al.*, Material identification using laser spectroscopy and pattern recognition algorithms. *Lecture Notes on Computer Science*, **2124** (2001), 443–450.

[16] T. Pioch and J. Matthias, Mercury vapour release from dental amalgam after laser treatment. *Eur. J. Oral Sci.*, **106** (1998), 600–602.

[17] J. Kruger, W. Kautek and H. Newesely, Femtosecond-pulse laser ablation of dental hydroxyapatite and single-crystalline fluoroapatite. *Appl. Phys. A*, **69** (Suppl.) (1999), S403-S407.

[18] O. Samek, M. Liska, J. Kaiser *et al.*, Laser ablation for mineral analysis in the human body: integration of LIFS with LIBS. In *Biomedical Sensors, Fibers and Optical Delivery Systems*, ed. F. Baldini, N. I. Croitoru, M. Frenz *et al.*, *Proc. SPIE*, **3570** (1998), 263–271.

[19] M. H. Niemz, Investigation and spectral analysis of the plasma-induced ablation mechanism of dental hydroxyapatite. *Appl. Phys. B*, **58** (1994), 273–281

[20] M. H. Niemz, Cavity preparation with the Nd:YLF picosecond laser. *J. Dent. Res.*, **74** (1995), 1194–1199.

[21] L. Willms, A. Herschel, M. H. Niemz and T. Pioch, Preparation of dental hard tissue with picosecond laser pulses. *Las. Med. Sci.*, **11** (1996), 45–51.

[22] D. C. S. Beddows, H. Kondo, G. G. Morris and H. H. Telle, Remote laser-induced breakdown spectroscopy using a novel single-fiber arrangement. *CLEO/Europe – EQEC '98*, (Glasgow: Technical Digest. 1998), p. 237.

[23] O. Samek, D. C. S. Beddows, H. H. Telle *et al.*, Quantitative laser-induced breakdown spectroscopy analysis of calcified tissue samples. *Spectrochim. Acta B*, **56** (2001), 865–875.

[24] M. H. Niemz, *Laser–Tissue Interactions* (Berlin: Springer Verlag, 1996).

[25] M. Y. Perez-Jordan, A. Salvador and M. Guardia, Determination of Sr, K, Mg and Na in human teeth by atomic spectrometry using a microwave-assisted digestion in a flow system. *Anal. Lett.*, **31** (1998), 867–877.

[26] K. M. Lee, J. Appleton, M. Cooke, F. Keenan and K. Sawicka-Kapusta, Use of laser ablation inductively coupled plasma mass spectrometry to provide element versus time profiles in teeth. *Anal. Chim. Acta*, **395** (1999), 179–185.

[27] J. Amador-Hernandez, J. M. Fernandez-Romero and M. D. Luque de Castro, In-depth characterization of screen-printed electrodes by laser-induced breakdown spectroscopy and pattern recognition. *Surf. Interface Anal.*, **31** (2001), 313–320.

[28] D. C. S. Beddows, Industrial application of remote and *in situ* laser induced breakdown spectroscopy. PhD thesis: University of Wales Swansea, United Kingdom (2000).

[29] G. P. Sighinolfi, S. Sartono and Y. G. Artoli, Chemical and mineralogical studies on hominid remains from Sangiran, Central Java (Indonesia). *J. Hum. Evol.*, **24** (1993), 57–68.

[30] E. Vandeweert, J. Bastiaansen, V. Philipsen *et al.*, The detection of Sr sputtered from metallic and biological matrices by double-resonant photoionization mass spectrometry. In *RIS 2000: Laser Ionization and Applications Incorporating RIS* (ed. J. E. Parks and J. P. Young), *AIP Conf. Proc.* **584** (2001), 301–304.

[31] A. Raith, W. T. Perkins, N. J. G. Pearce and T. E. Jeffries, Environmental monitoring on shellfish using UV laser ablation ICP-MS. *Fresenius J. Anal. Chem.*, **355** (1996), 789–792.

[32] D. C. S. Beddows, O. Samek, M. Liška, and H. H. Telle, Single-pulse laser-induced breakdown spectroscopy of samples submerged in water using a single-fibre light delivery system. *Spectrochim. Acta B*, **57** (2002), 1461–1471.

[33] O. Samek, D. C. S. Beddows, J. Kaiser *et al.*, The application of laser induced breakdown spectroscopy to *in situ* analysis of liquid samples. *Opt. Eng.*, **39** (2000), 2248–2262.

[34] D. A. Rusak, M. Clara, E. E. Austin *et al.*, Investigation of the effect of target water content on a laser-induced plasma. *Appl. Spectrosc.*, **51** (1997), 1628–1631.

[35] Q. Sun, M. Tran, B. W. Smith and J. D. Winefordner, Zinc analysis in human skin by laser induced-breakdown spectroscopy. *Talanta*, **52** (2000), 293–300.

[36] Q. Sun, M. Tran, B. Smith and J. D. Winefordner, In-situ evaluation of barrier-cream performance on human skin using laser-induced breakdown spectroscopy. *Contact Dermatitis*, **43** (2000), 259–263.

[37] I. G. Pallikaris, H. S. Ginis, G. A. Kounis *et al.*, Corneal hydration monitored by laser-induced breakdown spectroscopy. *J. Refractive Surg.*, **14** (1998), 655–660.

[38] M. Corsi, G. Cristoforetti, M. Hidalgo *et al.*, Application of laser-induced breakdown spectroscopy technique to hair tissue mineral analysis. *Appl. Opt.*, **42** (2003), 6133–6137.

[39] Q. Sun, M. Tran, B. W. Smith and J. D. Winefordner, Direct determination of P, Al, Ca, Cu, Mn, Zn, Mg and Fe in plant materials by laser-induced plasma spectroscopy. *Can. J. Anal. Sci. Spectrosc.*, **44** (1999), 164–170.

[40] E. Hoffmann, C. Lüdke, J. Skole *et al.*, Spatial determination of elements in green leaves of oak trees (*Quercus robur*) by laser ablation-ICP-MS. *Fresenius J. Anal. Chem.*, **367** (2000), 579–585.

[41] T. Prohaska, C. Stadlbauer, R. Wimmer *et al.*, Investigation of element variability in tree rings of young Norway spruce by laser-ablation-ICPMS. *Sci. Total Environ.*, **219** (1998), 29–39.

[42] Y. Civell, LIBS analysis of biological samples to trace environmental influences. B.Sc. Project Report: University of Wales Swansea, United Kingdom (2001).

[43] A. Assion, M. Wollehaupt, L. Haag *et al.*, Femto-second laser-induced breakdown spectrometry for Ca^{2+} analysis of biological samples with high spatial resolution. *Appl. Phys. B*, **77** (2003), 391–397.

[44] P. L. Nimis, S. Andreussi and E. Pittao, The performance of two lichen species as bio-accumulators of trace metals. *Sci Total Environ.*, **275**, (2001) 43–51.

[45] A. Tozzi, R. Barate, G. Cristoforetti *et al.*, LIBS analysis of lichens as bio-indicators of environmental pollution. In *Laser-Induced Plasma Spectroscopy and Applications – LIBS2002* (25–28 November 2002, Orlando, FL, USA), Technical Digest (2002), pp. 119–121.

[46] W. A. Simpkins, H. Louie, M. Wu, M. Harrison and D. Goldberg, Trace elements in Australian orange juice and other products. *Food Chem.*, **71** (2000), 423–433.

[47] R. L. Van der Wal, T. M. Ticich, J. R. West and P. A. Householder, Trace metal detection by laser-induced breakdown spectroscopy. *Appl. Spectrosc.*, **53** (1999), 1226–1236.

[48] M. O. Al-Jeffery and H. H. Telle, LIBS and LIFS for rapid detection of Rb traces in blood. In *Optical Biopsy IV* (ed. R. R. Alfano), *Proc. SPIE*, **4613** (2002), 152–161.

[49] R. J. Davie, Rubidium. In *Handbook on Metals in Clinical and Analytical Chemistry*, ed. H. G. Seiler, A. Sigel and H. Sigel (New York: Marcel Dekker, 1994), pp. 543–547.

[50] R. G. Pinnick, S. C. Hill, P. Nachman *et al.*, Aerosol fluorescence spectrum analyser for rapid measurement of single micrometer-sized airborne biological particles. *Aerosol. Sci. Technol.*, **28** (1998), 95–104.

[51] M. L. Laucks, G. Roll, G. Schweiger and E. J. Davis, Physical and chemical (Raman) characterisation of bio-aerosols – pollen. *J. Aerosol Sci.*, **31** (2000), 307–319.

[52] W. C. Hinds, *Aerosol Technology: Properties, Behaviour, and Measurement of Airborne Particles*, second edition (New York: John Wiley and Sons, 1999).

[53] D. T. Suess and K. A. Prather, Mass spectrometry of aerosols. *Chem. Rev.*, **99** (1999), 3007–3035.

[54] J. E. Carranza, B. T. Fisher, G. D. Yoder and D. W. Hahn, On-line analysis of ambient air aerosols using laser-induced breakdown spectroscopy. *Spectrochim. Acta B*, **56** (2001), 851–864.

[55] E. Gard, J. E. Mayor, B. D. Morrical *et al.*, Real-time analysis of individual atmospheric aerosol particles: design and performance of a portable ATOFMS. *Anal. Chem.* **69** (1997), 4083–4091.

[56] W. Murrey, Micro-LIBS analysis of plants: leaves and pollen. B.Sc. Project Report: Department of Physics, University of Wales Swansea, United Kingdom (2002).

[57] B. Schrader, H. H. Klump, K. Schenzel and H. Schulz, Non-destructive NIR FT Raman analysis of plants. *J. Molec. Struct.*, **509** (1999), 201–212.

[58] A. Boyain, D. C. S. Beddows, B. Griffiths and H. H. Telle, Single-pollen analysis by laser-induced breakdown spectroscopy and Raman microscopy. *Appl. Opt.*, **42** (2003), 6119–6132.

[59] A. C. Samuels, F. C. DeLucia, K. L. McNesby and A. W. Miziolek, Laser-induced breakdown spectroscopy of bacterial spores, molds, pollens, and protein: initial studies of discrimination potential. *Appl. Opt.*, **42** (2003), 6205–6209.

[60] S. Morel, N. Leone, P. Adam and J. Amouroux, Detection of bacteria by time-resolved laser-induced breakdown spectroscopy. *Appl. Opt.*, **42** (2003), 6184–6191.

8

LIBS for the analysis of pharmaceutical materials

Simon Béchard and Yves Mouget

Pharma Laser Inc., Boucherville, Canada

8.1 Introduction

The uniform distribution of pharmaceutical materials such as drugs, lubricants, and other components used in the formulation of pharmaceutical powder blends and tablet dosage forms is critical in order to achieve optimal product performance. Ideally, each of the several million tablets produced during the manufacture of a batch should contain the same amount of each component. In practice, however, this goal is difficult to achieve because the pharmaceutical solids that are blended together generally have wide particle-size distributions as well as dissimilar bulk properties, such as density, shape, specific surface area, and energy. The analytical technologies that are currently available for the determination of the components in these tablet dosage forms are time consuming, i.e. 10–30 min per sample, can be laborious, and are not readily amenable to at-line or on-line process monitoring. About 70% of the marketed drug products are formulated as tablet or capsule dosage forms where the drug molecules contain one or more chlorine, sulfur, fluorine, sodium, potassium, magnesium, phosphorus, or calcium elements as integral parts of their structure. Laser-induced breakdown spectroscopy (LIBS) was found to be a good analytical technique to measure quickly the surface and internal distribution of many pharmaceutical materials, i.e. drugs and excipients, used in the formulation of solid dosage forms. The determination of this distribution is valuable to pharmaceutical research and development since it leads to a better understanding of formulations and manufacturing processes as well as improvements in at-line process monitoring. As well, the analysis time, which is generally in the order of 30 s or less per sample, facilitates the evaluation of a statistically significant number of samples.

The objective of this chapter is to provide the reader with an overview of the use of LIBS for the analysis of pharmaceutical samples. It is not the intention to cover fundamental aspects of LIBS, and the interested reader is referred to other chapters. A description of the needs of the pharmaceutical industry as well as the nature of pharmaceutical matrices will first be addressed. Then, a description of the instrumentation developed specifically for the

Laser-Induced Breakdown Spectroscopy: Fundamentals and Applications, ed. Andrzej W. Miziolek, Vincenzo Palleschi and Israel Schechter. Published by Cambridge University Press. © A. W. Miziolek, V. Palleschi and I. Schechter 2006. A. W. Miziolek's contributions are a work of the United States Government and are not protected by copyright in the United States.

analysis of pharmaceutical materials will be presented together with typical applications data.

Pharmaceutical tablets and capsules represent approximately 70% of the drug products marketed worldwide. In addition to the drug molecule, other substances that do not possess any pharmacological activity enter into the composition of pharmaceutical formulations. These excipients must be added to impart specific functionalities to drug powders before their compaction into millions of tablets, each typically weighing 50–1000 mg. Ideally, each dosage form should be of exactly the same weight and composition, which is a need for optimal drug product safety and efficacy. In practice, this goal is difficult to accomplish because the pharmaceutical powders that are mixed together to achieve this uniform composition generally have wide particle-size distributions, i.e. 10–200 µm, as well as dissimilar bulk properties such as density, shape, specific surface area, and energy. In addition, these differences make them subject to segregation or de-mixing after the blending process.

Pharmaceutical companies perform various physical and chemical tests at different stages of manufacture as well as on finished dosage forms in support of developmental and commercial batches. These tests must demonstrate that a given manufacturing process delivers a drug product that meets consistent pre-established attributes. High-performance liquid chromatography or HPLC [1] is by far the primary methodology used worldwide to provide information on the uniformity of drugs in solid dosage forms. The technique involves extracting the drug from its matrix by dissolving or disintegrating it in an appropriate solvent, removing the insoluble materials, and introducing the solutes into the mobile phase of the chromatographic system. During the chromatography, the compound of interest is separated from other soluble components before its detection and quantitation. Despite major technological advances made to the technology over the past two decades, i.e. improved automation through computer controlled instrumentation, higher performance columns and detectors, improved data acquisition, treatment and management, little progress has been made to improve sample throughput, which is still in the order of 10–30 min per sample when the sample preparation steps are included. As a result, the technology remains a laboratory tool that does not suit the needs of formulation, process development or manufacturing groups very well. Indeed, it can take anywhere from a day to a week to obtain the results of an HPLC analysis after a sample is collected. Therefore, any analytical technology that can perform a rapid determination of the uniformity of drugs as well as other components in solid dosage forms is of significant importance to the pharmaceutical industry.

It is estimated that 65% of the marketed drug molecules contain one or more chlorine, sulfur, fluorine, sodium, potassium, magnesium, phosphorus, or calcium elements specific to the drug moiety. These elements are present either as an integral part of the molecular structure or as a salt and are often absent from the excipients or matrices, which generally contain only carbon, hydrogen and oxygen. Nonetheless, certain excipients that have a critical impact on the product performance, such as magnesium stearate (which is used as a lubricant), cross-linked polymers containing sodium (which are used as disintegrants), and titanium dioxide (which is used for the production of opaque film coatings), are of significant interest. However, because of the lack of reliable and rapid analytical technologies, which often require extensive sample preparation steps, these particular excipients are often not examined.

Laser-induced breakdown spectroscopy (LIBS) is an elemental analytical technique that can detect the presence and the amount of the elements present in various types of sample and is therefore ideally suited for their quick analysis to provide qualitative and quantitative information. The use of the technology applied to the analysis of pharmaceutical materials and other solid organic compounds has recently received growing attention [2–13].

The laser ablation technique samples only a very small portion of a sample. This feature can provide a wealth of information, but will also determine the approach to be used for sampling, acquiring, and analyzing the data. Significant data require several laser pulses on the same sample. By calculating the integrated signal resulting from each laser pulse, an overall average value for a particular element in the sample can be determined. If the element is unique to the compound of interest, then the concentration of this compound can be established through the use of appropriate standards. Surface and internal distribution information can also be obtained either by selectively averaging peak intensity data collected across the surface, or by averaging the data for one particular target site on the sample. The absence of any sample preparation, the short analysis time, and the fully automated operation of the technology provide the capability for at-line or on-line measurement of drug content in dosage forms.

The development of the technology applied to the analysis of pharmaceutical materials began in 1996 in conjunction with the National Research Council Canada. After demonstration of the technology in the laboratory, prototype units were designed, fabricated, and delivered to pharmaceutical companies for advanced testing. As a result of these beta tests, software and hardware improvements were implemented leading to the introduction of the first commercial instrument, the PharmaLIBS^{TM} 200.

8.2 Needs of the pharmaceutical industry

A typical capsule or tablet formulation contains at least four or five powder ingredients such as the active drug, one or two compaction agents such as cellulose and lactose, a lubricant such as magnesium stearate that prevents the adhesion of the blend to the processing equipment, and a disintegrant that generally consists of a sodium salt of cross-linked polymers, which accelerates the disintegration of the dosage forms in the stomach. Additional ingredients such as buffers, coloring agents, and antioxidants may be required, depending on the properties of the active drug. Film coatings also are frequently applied on tablets in order to mask the taste, to shield the active compound from light, to protect a drug from degradation when in the presence of stomach fluids, or to modulate the release of the active drug from the dosage form. These coatings generally contain titanium dioxide as an opacifier.

The typical manufacturing process involves loading a vessel with the powders and blending them for a predetermined period of time. In order to assess the uniformity of the active drug in the blend, multiple powder samples are removed at different locations from the vessel and compacted into wafers. The samples then are sent to the laboratory for analysis, this being most frequently performed by high-performance (-pressure) liquid chromatography (HPLC). Providing the HPLC results are acceptable, i.e. the process is complete and the resulting blend is uniform, the powder blend is encapsulated in hard gelatin capsules or compacted on a tablet press at a rate of several thousand units per minute. During the encapsulation or tableting process, samples are collected at predetermined time intervals and sent to the analytical laboratory for various qualitative as well as quantitative analyses. The sample load submitted to the quality control laboratory can therefore be significant for each batch of tablets, especially during the validation stage of the manufacturing process. Therefore, any analytical technology that can reduce the load of samples to be analyzed by HPLC is of interest for the pharmaceutical industry.

It is well known among formulation and process development groups that the amount of time required to develop and scale up a formulation and manufacturing process is directly related to the speed with which the analytical data become available. A fast turn-over rate of the samples in the laboratory can help to reduce the formulation and development time, and thus the time required to put a drug on the market. It also provides the opportunity to acquire greater knowledge on the formulation and development processes, which can be limited by the inability of the current technologies to analyze a statistically significant number of samples in a reasonable timeframe. Indeed, with the exception of a few companies that have developed or adapted near-infrared (NIR) spectroscopy in order to obtain critical process information, the vast majority of companies still rely on physical measurements such as pressure, tablet weight, thickness, and breaking strength to monitor their processes. Therefore, there is a strong economic justification to develop analytical tools that can provide close to real-time chemical information during the manufacture of drug products, whether during their development stage or after market entry.

LIBS offers unique solutions to the problems encountered during formulation and process development, as well as the production of pharmaceuticals. The main advantage of the technology is that a greater amount of data will be available faster without the need for any sample preparation. In addition, surface and internal distribution of components can be determined owing to its good spatial resolution.

8.3 Comparison of LIBS with the current technologies

As mentioned above, pharmaceutical companies perform various chemical tests at different stages of manufacture as well as on finished products. The most prevalent analytical tool for these tests is HPLC [1]. The introduction of HPLC as a new analytical technology in the 1970s revolutionized the drug development arena. Since the technique deals with samples that are solutions, which are by definition more homogeneous than solid mixes, it provides excellent accuracy, precision, and reproducibility. It also offers a level of selectivity and sensitivity that could not previously be achieved with thin-layer chromatography and ultraviolet spectroscopy, with detection devices that can provide limits of quantitations of micrograms per milliliter to picograms per milliliter. In addition, the chromatographic step allows the separation of synthesis by-products as well as degradation products from

Table 8.1. *Advantages and disadvantages of HPLC technology*

Advantages	Disadvantages
Accuracy	Extraction required
Precision	Analysis time of 10–30 min per sample
Specificity	Use organic solvents
Selectivity	Method development required
Sensitivity	At-line and on-line testing require higher throughput
Reproducibility	Destructive
Robustness	Labor intensive
Availability	Many consumables

Table 8.2. *Advantages and disadvantages of the PharmaLIBS technology*

Advantages	Disadvantages
Fast	Partially destructive
No sample preparation	Matrix effects
Specific to the target element	No selectivity
Sensitivity down to 10 p.p.m. depending on the target element	
Spatial resolution of *c*. 0.5 mm	
Analysis of metals	
Multi-elemental analysis	
Surface and three-dimensional mapping	

the parent compound and cannot be achieved by LIBS or NIR. Its robustness has made it easily amenable to computer control. Today, HPLC is used to provide information on the chemical stability as well as concentration of drugs in samples collected at various stages of processing and in finished dosage forms. The technique, like any other technique, possesses disadvantages. The most significant of these is the analysis time, which can range from 10 to 30 min per sample. The sample preparation and method development is inherently labor intensive since it involves the dissolution of the dosage form in a solvent and the optimization of chromatographic conditions. An analysis by HPLC is therefore expensive and difficult to adapt to at-line, and to an even greater degree on-line, for process monitoring where several thousands tablets are produced every minute. Consequently, chemical information on a statistically significant number of samples cannot easily be provided (see Tables 8.1 and 8.2).

Another technique introduced in the 1980s is near-infrared spectroscopy (NIR), which is now widely accepted and used to provide qualitative information on pharmaceutical

Table 8.3. *Advantages and disadvantages of NIR technology*

Advantages	Disadvantages
Fast	Sensitivity down to *c.* 2%–5% by weight of drug
No sample preparation	Not specific
Non-destructive	Chemometrics required
Surface mapping	Matrix effects
	Transferability

materials [14]. The technology is based on the fact that organic compounds contain structural units that absorb infrared radiation at various intensities for given frequencies. The technique can be used in a reflectance mode, where the compounds at the surface of a tablet are examined, or in a transmittance mode, where a complete picture of all the compounds in a tablet can be obtained. Because NIR is a spectroscopic technique, it is fast, requires no sample preparation, and is non-destructive. However, for a multi-component sample such as a pharmaceutical tablet, NIR generates a composite spectrum requiring chemometric analysis before performing quantitation. This implies the maintenance of a spectral library of every compound in its pure form that could be present in the sample. A composite spectrum also greatly reduces the sensitivity of the technique, and as a result the technique is not widely accepted as a quantitative tool (see Table 8.3).

Alternative methods such as thermal effusivity [15] and frequency-domain photon migration measurements [16] have been proposed in order to resolve particular problems during the production of pharmaceuticals. The spatial discrimination of X-ray diffraction and scanning electron microscopy have also been proposed for various applications but, currently, none of these techniques is suitable for everyday use. LIBS is a technique that can greatly assist the aforementioned methods by providing a fast elemental analysis with spatial discrimination. Tables 8.1–8.3 provide lists of the advantages and disadvantages of HPLC, LIBS and NIR technologies, respectively.

8.4 Components of a LIBS instrument for applications in the pharmaceutical industry

Several issues must be addressed when designing and manufacturing an instrument for the pharmaceutical industry. These issues concern the choice of the components best suited for the analysis of the particular chemicals, the need for automation and high sample throughput, as well as other features that are frequently summed up by the words robustness, self-contained, minimal maintenance, and most importantly ease of use.

As discussed in the other chapters of this book, LIBS has been used in a large number of applications, each with its own particular range of compounds or elements and matrices of interest. As a result, significant differences in the type of laser, spectrograph and detector can exist between the various LIBS instruments used for specific applications. At the outset,

Table 8.4. *Excipients most commonly used for the manufacture of pharmaceutical solid dosage forms*

Compound	Empirical formula	Carbon (%)
Microcrystalline cellulose	$(C_6H_{10}O_5)_n$ $n \approx 220$	44.4
Pregelatinized starch	$(C_6H_{10}O_5)_n$ $n \approx 300–1000$	44.4
Lactose	$C_{12}H_{22}O_{11}$	42.1
Magnesium stearate	$C_{36}H_{70}MgO_4$	73.1

it was important to identify the type of compounds and matrices that are most frequently used during the manufacture of pharmaceutical solid dosage forms. When examining the composition and corresponding LIBS spectra of the most common excipients or inactive ingredients used in pharmaceutical formulations, it turns out that they are mainly composed of carbon, hydrogen, and oxygen, as shown in Table 8.4. Their emission spectra from 190 nm to 850 nm are therefore very similar, with slight variations in the emission lines and background intensities (Figure 8.1). This facilitates the distinction of the emission lines from other elements unique to the compound of interest. The spectra shown in Figure 8.1 contain emission lines due to the elements present in the sample as well as the ambient atmosphere. Although the signal-to-background ratio could be improved by using helium rather than air as an ambient gas, or by evacuating the chamber and operating in a vacuum environment, sufficient sensitivity was observed in air for most elements in the concentration range of interest, which is typically 0.1%–3% by weight. The ablation chamber was therefore designed so that the process is conducted under ambient atmosphere where the dust produced by the laser ablation is removed from the chamber by an HEPA filtered exhaust system.

The method used to mount the samples and their location with respect to the exhaust system were found to contribute significantly to the shot-to-shot, and thus site-to-site and tablet-to-tablet, signal variability. In order to provide the same environment for each sample, an XY-rotating sample stage was developed. Three stepper motors accurately position each sample according to the target coordinates specified in the software. The sample tray can hold up to 26 samples in individual sample holders. These sample holders are machined so as to ensure reproducible positioning of the samples.

The detection system was selected for specific elements frequently encountered in the chemical composition of pharmaceutical drugs, i.e. S and Cl. These two elements have emission lines below 200 nm or above 800 nm. Because the instrument operates under ambient atmosphere, the region below 200 nm is not accessible and therefore requires the use of a detection system that can operate up to 1000 nm since S and Cl emission lines are detected at 921 and 837 nm, respectively. Indeed, most analytes of interest were found to have emission lines in the 500–1000 nm spectral domain. In order to obtain acceptable sensitivity

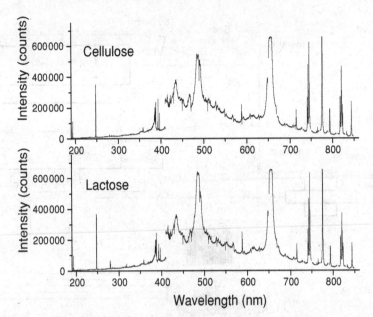

Figure 8.1. Cellulose (upper) and lactose (lower) LIBS emission spectra from 190 nm to 850 nm. The spectra were obtained using a Nd:YAG laser operating at 1064 nm, 150 mJ per pulse, and 2 Hz.

in that spectral range, an interline readout charge-coupled device (CCD) detector was used. This device, although non-intensified, showed considerable improvements in sensitivity over intensified detectors in the near-infrared region.

The current design of the PharmaLIBS™ 200 instrument (Figures 8.2 and 8.3) consists of a Q-switched Nd:YAG pulsed laser operating at the fundamental wavelength of 1064 nm. A beam attenuation system consisting of a half-wave plate and a polarizing beam splitter is placed immediately after the laser to permit the adjustment of the incident laser energy from 0 to 200 mJ. The laser can operate at a repetition rate of 10 Hz and delivers pulses with a width of approximately 5 ns. A 30 cm focal length plano-convex lens focuses the laser beam approximately 7 mm beyond the target surface in order to minimize the possibility of initiating air breakdown before the sample. The resulting spot size on the tablet surface is approximately 200 µm in diameter. A sampling window is located before the focusing optics. This window reflects 4% of the laser beam energy onto a joulemeter.

The light emitted by the plasma is focused by the ellipsoid mirror onto a fiber optic bundle, which in turn transfers the light to a 300 mm focal length spectrometer of Czerny–Turner configuration. The ellipsoid mirror possesses a hole through its center, collinear with one of the focal points. The laser is steered through this hole onto the sample surface. This arrangement greatly reduces variations in signal due to roughness, imperfections or markings on the sample surface. The fiber optic bundle has a 1 mm circular aperture at the input end and a linear aperture at the spectrometer end. The linear aperture output end of the fiber optic bundle is lined up with the input slit of the spectrometer. The spectrometer gratings

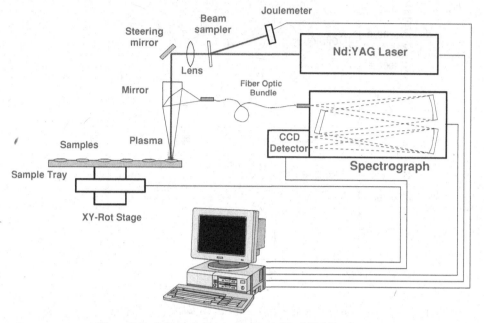

Figure 8.2. Schematic representation of the LIBS instrument.

Figure 8.3. Photograph of the PharmaLIBS™ 200 (left); rotating tray with 26 sample holders (right).

consist of a 1200 g mm^{-1} grating (blaze 750 nm), a 1200 g mm^{-1} grating (holographic visible), and a 600 g mm^{-1} grating (blaze 1 μm).

The result is a mobile, self-contained, and sealed LIBS instrument that does not require special equipment to operate. The software controls the functionality of the instrument allowing for automatic operation, site-definition and site-targeting on the sample of interest as well as data processing.

Figure 8.4. Photograph showing a pharmaceutical tablet after ablation with 1 (center) to 13 number of laser pulses per site (Nd:YAG laser operating at 1064 nm, 150 mJ per pulse).

8.5 Applications of LIBS to the analysis of pharmaceutical materials

8.5.1 Signal characteristics

The laser is focused on a spot of approximately 200 μm in diameter on the sample surface and each pulse (100–175 mJ per pulse) typically removes only a few micrograms from the sample. When several laser pulses are targeted at the same location (Figure 8.4) the laser is literally drilling through the sample, thus producing craters. Because the particle-size distribution of pharmaceutical materials is typically in the range 10–200 μm, considerable signal intensity variation is observed when looking at shot-to shot values, as shown in Figure 8.5. However, when averaging signal intensity values over several sites and/or several tablets, the precision increases significantly. Figure 8.6 illustrates the effect of varying the number of shots per site and/or the number of sites per tablet on the tablet-to-tablet %RSD values for tablet samples containing 10% by weight of chlorpheniramine maleate, a chlorine-containing compound, in a cellulose matrix. The figure shows that, to achieve better precision, it is more important to sample a larger number of sites than to perform more shots per site.

8.5.2 Blend and tablet uniformity

The problems associated with the blending of powders have always been significant in the manufacture of pharmaceutical tablets. The current methodology to assess blend uniformity involves sampling the powder bed followed by compaction of the sample into a wafer that is then sent to the analytical laboratory for analysis by HPLC.

In order to assess the ability of LIBS to monitor a blending process and determine blend uniformity, a mixture of 50:50 lactose and microcrystalline cellulose was blended for 10 min in a low shear mixer rotating at 35 r.p.m. Subsequently, chlorpheniramine maleate (a chlorine-containing drug), croscarmellose sodium (a sodium-containing disintegrant), and magnesium stearate (lubricant) were added at 10%, 3%, and 0.5% by weight, respectively. The disintegrant and lubricant are of the type typically found in a large number of formulations and they were used here at representative concentrations in the powder blend. The blender was rotated and powder samples were collected after a pre-determined number of rotations from five locations in the blender. The powder samples, *c.* 300 mg in weight, were compacted using a hydraulic press into round flat tablets 12.5 mm in diameter, which were subsequently analyzed by LIBS. The wavelengths chosen were 517 nm for Mg, 589 nm for

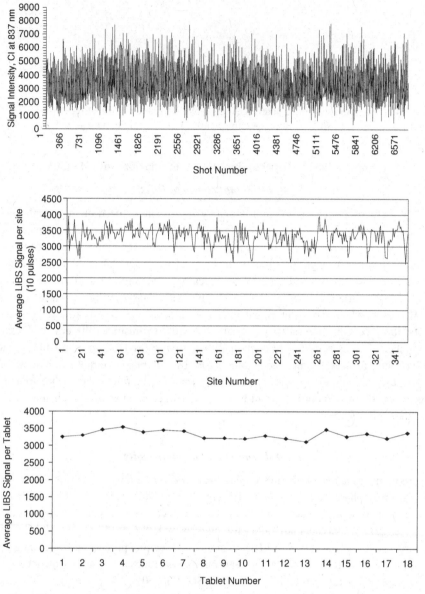

Figure 8.5. Evolution of the LIBS signal intensity with respect to shot number, %RSD 10%–20% (upper); site to site, %RSD 5%–10% (middle); and tablet to tablet, %RSD 1%–4% (bottom).

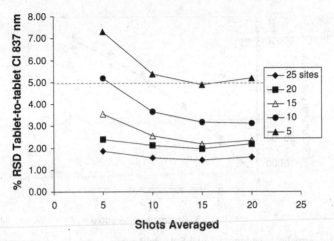

Figure 8.6. Evolution of the tablet-to-tablet %RSD values with respect to number of averaged shots per site and the number of averaged sites.

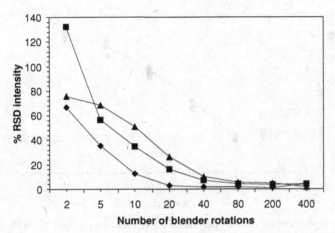

Figure 8.7. Plot of the net peak area %RSD for (■) magnesium stearate, (♦) W croscarmellose sodium, and (▲) chlorpheniramine maleate as functions of the number of blender rotations.

Na, and 837 nm for Cl. Figure 8.7 shows the tablet %RSD for each analyte with respect to blending time expressed as number of rotations. Each datapoint represent the mean LIBS signal from five tablets. These results clearly show that LIBS technology can provide rapid answers with respect to blender selection, blending endpoint, and blend uniformity. These types of data cannot be obtained easily with other conventional technologies.

An examination of the relation of emission intensity to concentration was performed by preparing three series of calibration standards for the three analytes. The plots in Figures 8.8–8.10 show the calibration curves for concentrations of 1%–5% croscarmellose

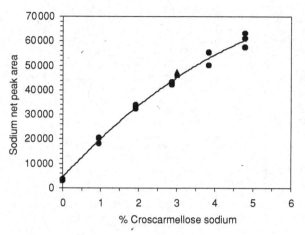

Figure 8.8. Croscarmellose sodium (Ac-Di-Sol) calibration curve for concentrations 1%–5%. Symbols: (•) croscarmellose sodium standards; (▲) QC samples.

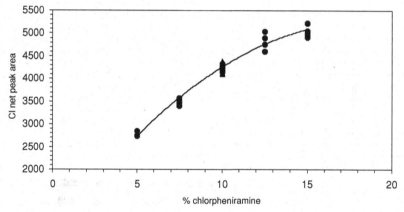

Figure 8.9. Chlorpheniramine maleate calibration curve for concentrations 5%–15%. Symbols: (•) chlorpheniramine maleate standards; (▲) QC samples.

sodium, 5%–15% chlorpheniramine maleate, and 0.1%–1% magnesium stearate per tablet, respectively. Calibration standards were prepared by adding each component to 50:50 mixtures of lactose and microcrystalline cellulose. The powders were mixed for 5 min with a mortar and pestle, and then 300 mg aliquots were sampled and compressed into 12.5 mm round flat tablets. Quality control samples were prepared in a blender according to the procedure described for the blend uniformity study. The nominal concentrations were 3%, 10%, and 0.5% by weight for the disintegrant, active drug, and lubricant, respectively. In this case, however, the blender underwent 400 rotations before samples were collected and

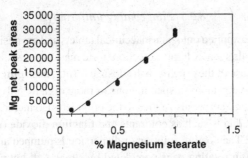

Figure 8.10. Magnesium stearate calibration curve for concentrations 0.1%–1%. Symbols: (•) magnesium stearate standards; (▲) QC samples.

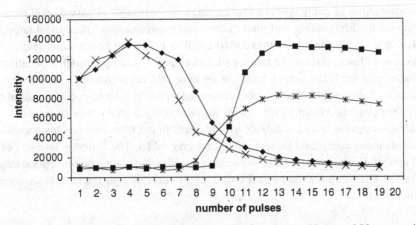

Figure 8.11. Titanium and sulfur signals as functions of laser pulses on 10 mg and 20 mg enterically coated omeprazole tablets. Symbols: (♦) Ti (20 mg); (■) S (20 mg); (×) Ti (10 mg); (✳) S (10 mg). Ten tablets of each of the 10 mg and 20 mg dose strengths were analyzed.

compacted into tablets. The laser was set at 125 mJ and 2 Hz with a delay and exposure of 1 μs and 3 μs, respectively. The mean net peak areas for each tablet (3 to 5 per concentration) are reported for each concentration. Results showed the QC samples to be within 2% of the nominal values. Similar results have been observed for other compounds; however, the response factor and curve fit vary with the matrix composition. In practice, calibration standards are generally prepared at 75%–125% of the nominal concentration or label claim using powder blends sampled from production and diluted with placebo material to achieve the appropriate concentrations. Another method is to blend the formulation ingredients using a mortar and pestle and compact wafers out of the mixtures. Linear regressions performed over narrow concentration ranges such as 75%–125% of the nominal value were generally found significant.

8.5.3 Film coating uniformity

Film coatings are often applied onto pharmaceutical tablets and are designed and formulated to achieve specific goals such as light protection, taste masking, resistance to gastric fluids, or modulating the release of the drug from the dosage form. These film coating formulations are prepared by dissolving and/or dispersing one or more organic polymer(s) in aqueous or organic solvents. Plasticizers are also added to the polymer(s) to increase their elasticity and flexibility. Other ingredients such as colorants and titanium dioxide (used as an opacifier) are also often present. The coating solution or dispersion is pumped and atomized onto the bed of tablets located in rotating pans percolated by streams of hot air to dry the coating as it is being deposited. The film coating is therefore applied by successive deposition of plasticized polymer droplets onto core tablets until a desired endpoint is reached. Depending on the application, film coating thickness can vary between 30 μm to 200 μm. Because of the functionalities of many types of film coatings, it is desirable to have a rapid means of determining the film coating uniformity across and between tablets. Analytical targets for LIBS are often present in the coatings and core tablets and can be monitored simultaneously as functions of laser pulses in the same spatial location, i.e. drilling through the coating and into the core tablet. When several locations are analyzed, the uniformity can be assessed.

Figure 8.11 shows mean titanium and sulfur signal profiles with respect to the number of laser pulses generated from 10 mg and 20 mg enterically coated omeprazole tablets. The formulation contains titanium dioxide as a pigment in the film coating. Omeprazole is a sulfur-containing compound present only in the core tablet. The emission intensity of the sulfur atom (921 nm) increases after the tenth laser pulse, indicating the penetration into the core tablet. The mean titanium signal (566 nm) also decreases as the laser beam samples the core tablet.

8.5.4 Dose strength identification

LIBS can also be used to examine and identify a tablet dose strength during a production run. The ability to monitor different dose strengts was evaluated using commercially available bromazepam tablets. Dose strengths of 1.5 mg, 3 mg, and 6 mg of bromazepam per tablet were selected, which correspond to 0.75%, 1.5%, and 3% of bromazepam by weight, respectively. The bromine emission was monitored at 827 nm because the molecular structure of bromazepam contains a bromine atom. The average signal of five tablets of each dose strength is plotted in Figure 8.12. A good correlation coefficient for the linear regression suggests that a single point calibration could be used to monitor various doses of bromazepam in this concentration range.

8.5.5 Drug and component mapping

A unique feature of the LIBS technology is the very small target size being sampled. The spot size ablated by the laser is approximately 200 μm in diameter. If a 500 μm distance

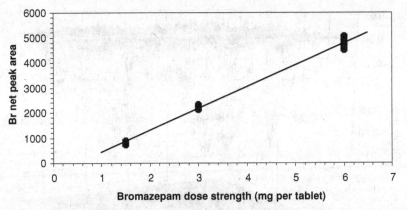

Figure 8.12. Bromine mean signal with respect to bromazepam dose strengths of 1.5 mg, 3 mg, and 6 mg per tablet.

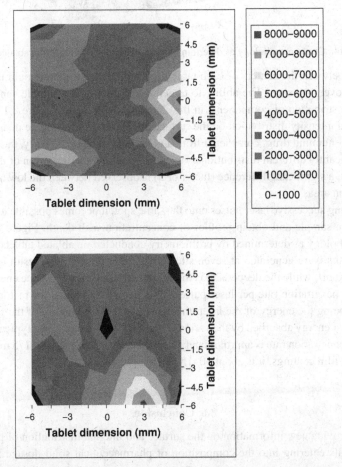

Figure 8.13. Plots showing the chlorpheniramine distribution on the surface of two flat tablets 12.5 mm in diameter compacted from a non-uniform blend.

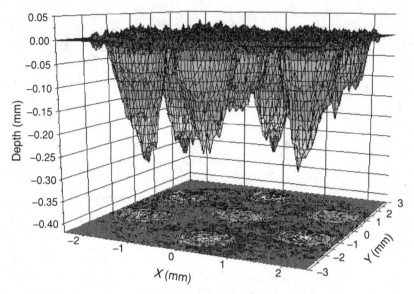

Figure 8.14. Crater morphology as determined by profilometry conducted on ablated tablets.

is selected between the centers of each target, a 6 mm diameter tablet could therefore be sampled at over 140 sites. The ability to target specific areas of a sample can be used to assess the distribution of components in the sample. For example, Figure 8.13 illustrates the Cl signal intensity distribution on the surface of two chlorpheniramine maleate tablets, a chlorine-containing drug, compacted from a non-uniform blend. Clearly, the two tablets show significantly different distributions. When examining the distribution of the drug in a single tablet, a significant difference (factor of 3) is observed between the low and the high concentration areas.

By focusing successive laser pulses onto the same spot, it becomes possible to probe into the core of a sample and obtain profiles of concentration with depth. Figure 8.14 shows crater morphology as determined by profilometry conducted on ablated tablets. Craters of various depths were generated at seven sites. The center crater is the result of two laser pulses at 100 mJ, while the deeper site is the result of 10 pulses at the same energy setting. The sample penetration rate per laser pulse will depend on a number of factors, the most significant being the energy of the laser, the density and granulometry of the sample, and the amount of energy absorbed by the various components in the sample. Typically, on core tablets, the penetration rate is approximately 50–100 μm per pulse at 100–175 mJ per pulse, whereas, on film coatings, it is about 10–15 μm per pulse.

8.6 Conclusions

LIBS can provide new information on the surface and internal distribution of pharmaceutical materials entering into the composition of pharmaceutical solid dosage forms. The technology does not require any sample preparation, samples can be analyzed quickly in less than one minute, and it is specific to the element(s) contained in the chemical structure of drugs and/or other functional materials. Qualitative and quantitative determinations can then be obtained rapidly on drug formulations during the developmental stage or commercial manufacturing.

8.7 References

[1] L. R. Snyder, J. J. Kirkland and J. L. Glajch, *Practical HPLC Method Development* (New York: Wiley, 1997).

[2] M. Sabsabi and J. Bussière, United States Patent 5,781,289 (1998).

[3] J. Lademann, H. J. Weigmann, H. Schaefer, G. Mueller and W. Sterry, *Skin Pharmacol. Appl. Skin Physiol.*, **13** (5) (2000), 258–264.

[4] S. Béchard and R. Sing, United States Patent Pending 09/607,093 (2000).

[5] S. Béchard, *Pharmaceut. Formulation Qual.*, **3** (2), (2001), 37–40.

[6] L. St-Onge, R. Sing, S. Béchard and M. Sabsabi, *Appl. Phys. A*, **69** (Suppl.) (1999), S913–S916.

[7] M. Tran, Q. Sun, B. W. Smith and J. D. Winefordner, *Appl. Spectrosc.*, **55** (6) (2001), 739–744.

[8] Y. Mouget, R. Sing and S. Béchard, *PITTCON 2001 Proc.* (2001), 2066P.

[9] Y. Mouget, R. Sing and S. Béchard, *PITTCON 2001 Proc.* (2001), 2067P.

[10] Y. Mouget, R. Sing and S. Béchard, *PITTCON 2001 Proc.* (2001), 297.

[11] R. Sing, E. Kwong and S. Béchard, *PharmSci*, **2** (4) (2000), 3403.

[12] R. Sing and S. Béchard, *1st International Conference on Laser Induced Plasma Spectroscopy and Applications Proceedings* (2000), p. 15.

[13] R. Sing and R. Béchard, *1st International Conference on Laser Induced Plasma Spectroscopy and Applications Proceedings* (2000), p. 58.

[14] M. Mckelvy, T. R. Britt, B. L. Davis *et al.*, *Anal. Chem.*, **70** (1998), 119R–177R.

[15] S. R. Dipali, C. Chandler, P. Adusumilli *et al.*, *PharmSci*, **3** (3) (2001), abstract.

[16] R. R. Shinde, G. V. Balgi, S. L. Nail and E. M. Sevick-Muraca, *J. Pharmac. Sci.*, **88** (10) (1999), 959–966.

9

Cultural heritage applications of LIBS

Demetrios Anglos

Institute of Electronic Structure and Laser, Foundation for Research and Technology–Hellas, Heraklion

John C. Miller[*]

Life Sciences Division, Oak Ridge National Laboratory, USA

9.1 Introduction

Laser-induced breakdown spectroscopy (LIBS) has emerged in the past ten years as a very promising technique for the analysis and characterization of a broad variety of objects of cultural heritage including painted artworks, icons, polychromes, pottery, sculpture, metal, glass, and stone objects. The analytical capabilities of LIBS with respect to its applications in the fields of archaeology, art history, and art conservation are discussed. Basic instrumentation requirements are briefly described followed by an analysis of critical parameters and methodological approaches. Illuminating examples from the application of LIBS in specific problems are presented demonstrating the potential of the technique in becoming a useful analytical tool for the characterization of works of art and archaeological objects.

The aim of this chapter is to provide an overview of the application of LIBS to the characterization of objects of cultural heritage. The basic principles and analytical features of the technique will be presented in relation to the specific fields of application and several illuminating examples will be given.

Following a brief introduction to the current use of physical and chemical techniques for addressing analytical questions and problems in the fields of art history, conservation, and archaeology, the basic principles of LIBS and its principal analytical advantages, within the context of the analysis of objects of art, are presented. A basic presentation of instrumentation issues is given, followed by a discussion of critical experimental parameters and methodological approaches that differ according to the analytical problem at hand. Specific examples demonstrating the use of LIBS in addressing real problems of the analysis of objects or materials of cultural heritage are presented that indicate the broad range of analytical questions addressed within the fields of archaeology, art history, and art conservation. The use of LIBS in conjunction with other laser analytical techniques such as laser-induced fluorescence (LIF) spectroscopy, Raman microscopy, and laser ablation-mass

[*] Present address: Chemical Sciences, Geosciences and Biosciences Division, Basic Energy Sciences, Office of Science SC-14 Germantown Building, US Department of Energy, 1000 Independence Avenue, SW Washington, DC 20585-1290, USA.

Laser-Induced Breakdown Spectroscopy: Fundamentals and Applications, ed. Andrzej W. Miziolek, Vincenzo Palleschi and Israel Schechter. Published by Cambridge University Press. © A. W. Miziolek, V. Palleschi and I. Schechter 2006. A. W. Miziolek's contributions are a work of the United States Government and are not protected by copyright in the United States.

spectrometry, is also described, showing the analytical advantages of combining complementary techniques. Final comments based on the authors' experience and a brief discussion on the future prospects of LIBS in cultural heritage are given at the end of the chapter.

9.2 Art and analytical chemistry

Archaeologists, art historians, and art conservators have long realized the need to know the chemical composition of the materials used, as perhaps the most important information for characterizing archaeological objects or works of art. This allows an understanding of what they were made of, how they were made, and when and possibly where they were made. Thus the use of proper analytical methods has become a necessary component of the systematic analysis of objects of cultural heritage, leading to important information on the chemical constituents of materials examined [1–5]. In fact, analytical applications in archeological scientific research have led to the creation of a distinct branch of analytical chemistry and archaeological science, called archaeometry.

The development of powerful analytical instrumentation has undoubtedly led to important advances in the field of cultural heritage analysis as shown not only by the increasing number of research papers and books on the subject, but also by the investments made by major museums and institutions in new analytical instrumentation. Examples and indicative literature references [6–48] are given in Table 9.1, illustrating the use of a broad spectrum of analytical techniques in addressing problems in archaeometry and artwork analysis.

9.3 Why LIBS in cultural heritage?

Laser-induced breakdown spectroscopy is an elemental analysis technique based on the characteristic atomic emission recorded from a laser-produced plasma [49, 50]. However, quite a few elemental analysis techniques, which have long been established in materials science and analytical chemistry, enjoy recognition in the fields of archaeometry and artwork analysis providing valuable qualitative and/or quantitative information often with detection limits at the level of a few parts per billion (p.p.b.) (Table 9.1).

A question put forward in this respect is "why LIBS in cultural heritage?" given the presence of a rather wide range of competitors in the field, including X-ray based, optical spectroscopic and mass spectrometric techniques, which feature excellent analytical capabilities. The answer to this question will hopefully become obvious by the end of this chapter, especially through the test cases that are presented. However, some key points, indicating the potential of LIBS in applications related to cultural heritage, will be outlined here.

LIBS, also known as laser-induced plasma spectroscopy (LIPS) is a straightforward and simple analytical technique, which can be employed even by non-specialized users. It is a rapid analysis technique providing results in very short time, practically instantaneously, after the analysis. Also, it is applicable *in situ* – that is, on the object itself – and under

Table 9.1. *Analytical techniques in cultural heritage*

Analytical method	Applications	References
UV–visible absorbance/reflectance spectroscopy	Analysis of inorganic pigments	[6–9]
Fluorescence emission spectroscopy	Pigment analysis	[10–13]
Infra-red (FTIR) spectroscopy	Paint analysis (pigments, binders)	[14–16]
Raman spectroscopy/microscopy	Inorganic and organic pigments, binder, and varnish analysis	[7, 17–23]
Inductively Coupled Plasma – Optical Emission Spectrometry (ICP-OES)	Major and trace element analysis of metals and minerals	[24]
Inductively Coupled Plasma – Mass Spectrometry (ICP-MS)	Trace element and isotope analysis of metals and minerals	[25, 26]
Scanning Electron Microscopy (SEM, ED X-ray microanalysis)	Mapping and elemental analysis of pigments, pottery, metals, and minerals	[27,28]
X-ray Fluorescence spectrometry (XRF)	Elemental analysis of pigments, metals, and minerals	[29–32]
X-ray Diffraction (XRD)	Pigment analysis	[27, 33, 34]
Particle-induced X-ray emission (PIXE)	Major and trace element analysis of pigments, pottery metals, and minerals	[35–37]
Neutron Activation Analysis (NAA)	Analysis of major and trace elements in pigments, pottery and minerals. Provenance.	[38, 39]
Secondary Ion Mass Spectrometry (SIMS)	Elemental analysis of pigments, pottery, metals, alloys, and minerals	[40]
Gas chromatography (GC), Gas chromatography – Mass spectrometry (GC-MS)	Analysis of organic components such as binders, varnishes, etc.	[41–43]
Thermoluminescence spectroscopy	Dating and provenance	[44]
Electron spin resonance (ESR) spectroscopy	Dating and provenance	[45]
Radiocarbon dating	Dating and provenance	[46]
Isotope analysis	Dating and provenance	[47, 48]

certain conditions is nearly non-destructive. In the context of analysis related to objects of cultural heritage the above features are considered very important. The simplicity of the technique and its speed permit the analysis of a relatively large number of objects in a short time. For example, archaeological excavations often produce a large number of artifacts and their timely characterization or simply screening is required. The *in situ* use of the technique eliminates the need for sampling, a time consuming and sometimes damaging process. This is important because the sensitivity and value of most works of art and archaeological objects often preclude sampling, thus excluding the use of several laboratory analytical techniques.

For the same obvious reason, non-destructive techniques are preferred over destructive ones and even though LIBS is not a strictly non-destructive technique it is considered minimally invasive given the tight area of interaction of the laser pulse with the sample surface. On the basis of these features and research done to date, LIBS appears as a useful alternative to other sophisticated techniques for obtaining information on the elemental composition of materials in cultural heritage objects. The potential of LIBS in this field is shown by several research papers that have appeared in the last few years, describing its use for the analysis of works of art and objects of archaeological importance [51–66]. Earlier reports can also be found on the use of laser microspectral analysis in the determination of the elemental content of metal, pottery, and paint samples from different objects [67, 68]. The analytical capabilities of LIBS are dealt with in more detail in the following sections and in the examples presented.

9.4 Physical principles

Laser-induced breakdown spectroscopy is a well-known atomic emission spectroscopic technique, developed shortly after the discovery of the laser in 1960, that provides compositional information on the elemental content of materials [69]. It has been used in a wide variety of analytical applications for the qualitative, semi-quantitative and quantitative analysis of materials [70–73]. LIBS relies on the generation of a micro-plasma plume by focusing a laser pulse with a few nanoseconds duration on the surface to be analyzed. Upon cooling, the plasma emits fluorescence characteristic of the neutral or ionized atoms in the plume, which are in principle representative of the elemental content of the solid surface probed. The resulting spectrally and temporally resolved recording of the emitted fluorescence comprises the LIBS spectrum whose proper analysis/interpretation yields the compositional information regarding the material examined. More specifically, the characteristic, sharp atomic emission peaks in the spectrum lead to the identification of the elements contained in the minute amount of material ablated, reflecting the local elemental composition of the sample. The peak intensity or the integrated emission can, in principle, be associated with the number density of each emitting species in the plume and this, in turn, with the concentration of specific elements in the ablated material. The integrated intensity of an emission spectral line from a single species present in the plasma, in a state of local thermodynamic equilibrium, is given in equation (9.1), as a function that relates the emission line intensity to the number density of the species, relevant spectroscopic parameters and the plasma electron temperature:

$$I_\omega^{ik} = N \frac{h\omega}{8\pi^2} \frac{g_i A_{ik} e^{-(E_i/k_B T_e)}}{Z(T_e)}, \tag{9.1}$$

where I_ω^{ik} is the intensity of emission line at frequency ω corresponding to a transition from the upper quantum state i to the lower k; N is the species number density in the point of observation within the plasma; h is Plank's constant; g_i is the statistical weight of state i; A_{ik} is the transition probability for spontaneous emission from energy level i to k; E_i is

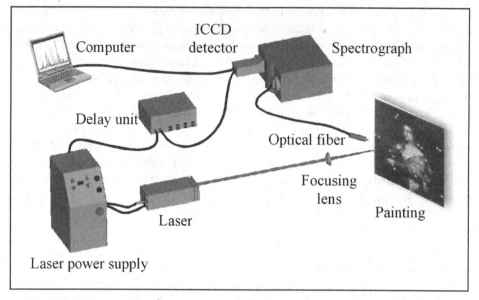

Figure 9.1. Diagrammatic representation of typical instrumentation used in a LIBS experiment.

the energy of quantum state i with respect to the ground state of the emitting species; k_B is Boltzmann's constant; T_e is the plasma electron temperature; and $Z(T_e)$ is the partition function for the quantum states of the emitting species.

On the basis of equation (9.1), the number density for each emitting species can be associated with the intensity of its spectral lines, thus providing a link between signal intensity and elemental concentration. A discussion on quantitative analysis by LIBS follows in Section 9.6.4.

9.5 Instrumentation

As already mentioned, the main ingredients of LIBS analysis include laser-induced plasma formation and detection and recording of the plasma emission. The instrumentation required for LIBS is straightforward and the basic components are shown in a diagrammatic representation of a typical experimental setup (Figure 9.1).

The laser most commonly used is the nanosecond, Q-switched Nd:YAG operating at its fundamental frequency corresponding to 1064 nm, or at its harmonics at 532 nm, 355 nm, and 266 nm. Excimer lasers emitting at 193 nm, 248 nm, or 308 nm have also been used. Laser wavelength is critical as it is one of the major parameters that determine the coupling of the irradiation to the material surface analyzed. Focusing of the laser beam is done through a convergent lens of appropriate focal length, typically between 50 and 500 mm. It can be much longer in cases of remote analysis [74] or very short in cases of analysis done through a microscope objective [75]. Typical values of the laser pulse energy lie in the range of 1–30 mJ, which translate to energy density values on the sample surface in the range of 1–50 J cm^{-2}. These values are varied by adjusting the incident energy or the working distance (the latter affects the laser spot size on the sample). Beam quality is a critical parameter in achieving tight focusing, which is important both for minimizing the affected surface area and for obtaining good spatial resolution.

The collection of the emitted light is done by using lenses (either through the focusing lens or through a separate lens located at an angle with respect to the irradiating beam) or directly through an optical fiber placed near the plume. Lens systems lead to improved collection efficiency but optical fibers offer simplicity in signal collection.

The spectral analysis is done in a grating spectrograph, which, coupled to a one-dimensional array or two-dimensional charge coupled device (CCD)-type detector, permits the simultaneous recording of a wide spectral range. Monochromators coupled to photo-multiplier tube detectors are not practical because of the single wavelength detection mode of operation, which requires scanning in order to record a proper spectrum. This implies exposure of the object to a large number of pulses, which in many cases is simply not wise as it can lead to extensive damage. Employing small- or medium-size standard spectrographs gives one the option of recording a broad range of the spectrum (typically 100–400 nm) with medium- to low-resolution gratings or a narrow part of the emission spectrum with the use of high-resolution gratings. Obviously the gain in the former case is the ability to maximize spectral coverage and detect more elements or more lines for each element; however, even in relatively simple spectra, lines can be overlapping and spectral information rather confusing. In the latter case, the high-resolution spectrum provides a clean record-ing of even very closely lying emission lines, permitting the unambiguous discrimination between different elements. However, at the same time the limited spectral coverage can result in the "loss" of the emission of other elements. To obtain both high resolution and wide spectral coverage, multiple spectrographs can be used, which obviously affects the cost of the instrumentation [76, 77]. Recently, significant technological progress has been achieved by employing spectrographs based on echelle gratings, which, when coupled to two-dimensional CCD-type detectors, provide wide spectral coverage simultaneously with excellent spectral resolution [78–81]. Figure 9.2 shows spectra obtained on regular spectro-graphs at low and high resolution and on an echelle spectrograph illustrating these effects.

Detection nowadays is based solely on intensified diode array or intensified CCD detec-tors. These detection devices offer high sensitivity with variable gain. They also permit adjustable gating in order to achieve discrimination of the atomic emission signal from the continuum background that is present at early times after the laser pulse.

Algorithms for rapid spectral analysis and peak recognition are straightforward and have been developed, while simple databases covering certain types of materials (e.g. pigments) and the elements used to identify their presence have been proposed. An important issue in this respect is the development of a user-friendly interface and software, particularly when the instrumentation is to be used by non-specialized personnel.

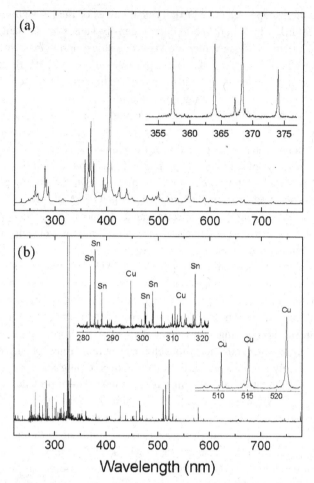

Figure 9.2. (a) LIBS spectra of lead white pigment recorded at low spectral resolution (spectrograph: 0.19 m; grating: 300 grooves mm^{-1}) and high spectral resolution (inset; spectrograph: 0.32 m; grating: 1800 grooves mm^{-1}). (b) LIBS spectrum of bronze recorded with an echelle spectrograph (reproduced from reference [63]).

Compact or medium-sized LIBS systems, which lend themselves to portability, have appeared in the market or as prototypes [82–84]. With this development more tests will be possible by archaeologists, historians, and conservators interested in employing LIBS analysis. Technological advances are expected to lead to lowering of the cost of such units, which may still be prohibitive for small- and medium-scale laboratories.

9.6 Analytical parameters and methodology

Several important factors governing LIBS analysis relate to both material and instrument parameters and depend on the specific analytical questions asked. To ensure success of the analysis, an appropriate methodological approach must be followed and optimum working parameters must be determined. This is even more important in the field of cultural heritage as the analysis often deals with objects of great historical and artistic value and any modification of the surface analyzed has to be limited to the minimum possible level.

9.6.1 Material parameters

From the historical and artistic point of view the value of the object is a critical factor, and this often imposes limitations in the analysis. The approach for analyzing a valuable painting can be very different from the one used for examining a pottery sherd or metal scrap from an archaeological excavation. For example, in the case of a sensitive miniature painting the analysis of pigments has to be performed so that selected points, representative of the pigments used, are examined under working conditions, which ensure that any adverse effects experienced by the painted surface are minimal.

Apart from the obvious dependence on the value of the object, certain physical and chemical material properties are critical for the analysis. First of all, the optical properties of the surface probed (absorptivity) determine the material–irradiation interaction, namely the ablation process [85–87]. Obviously different materials have different ablation and plasma formation threshold values for a given irradiation wavelength. For instance, while irradiation parameters, at the fundamental frequency of Nd:YAG laser (1064 nm), may be appropriate for a metal object, they can be inadequate for another type of material, such as a sample of pottery or glass, which is much more weakly absorbing. Differences in optical properties even across the surface of the same object can lead to significant signal variations.

The macroscopic nature of the object's surface is also important to the extent that it can affect the result of the qualitative or quantitative analysis. So in the analysis of ancient metal objects care has to be taken to avoid analyzing environmental deposits on the surface or corrosion products instead of the metal itself. In that case, the result can be significantly different and not representative of the composition of the bulk metal. Furthermore, surface fragility may pose the danger of causing damage to a more extended surface area than that irradiated. This is because the pressure shock-wave that accompanies ablation can lead to the detachment of poorly adhering parts of the surface at or around the spot analyzed. Therefore, a thorough examination of the areas to be analyzed is highly recommended in order to avoid unwanted side effects of the laser pulse on the surface of the object.

9.6.2 Irradiation-detection parameters

Important parameters associated with the irradiation include laser wavelength, pulse energy, and duration. As already discussed, the laser wavelength is a critical factor in the ablation process because, along with the material optical properties, it determines the ablation threshold. The energy per laser pulse, in turn, determines the intensity of the emitted signal.

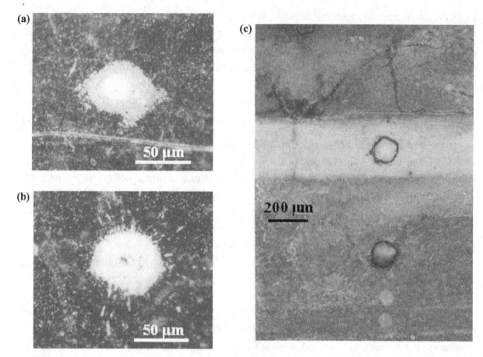

Figure 9.3. Craters formed in measurements carried out on a nineteenth-century daguerreotype upon irradiation with (a) one laser pulse and (b) ten laser pulses at 1064 nm; (c) craters formed on Byzantine icon upon irradiation with 50 (top) and 40 (bottom) pulses at 355 nm.

Therefore, given the wide variety of objects and the complexity of surfaces, it is essential to know the energy density threshold values for ablation of different materials and irradiation wavelengths in order to establish optimal parameters for each case. In general, because of the sensitivity and value of the objects examined, work is carried out at the lowest possible power density regimes (0.2–2 GW cm^{-2}). However, when working at energy density values close to the ablation threshold, care should be taken to ensure that the emission is truly representative of the sample's composition and one must avoid the regime where selective ablation could be occurring.

Proper focusing contributes to a greater signal intensity as it optimizes the laser energy density for a given pulse energy while at the same time it limits the effect of the laser pulse on the surface of the object. The minimization of the beam waist depends largely on the laser beam quality, the proper alignment and the quality of the optical components. For a laser beam having a Gaussian profile, a beam waist as small as a few micrometers is anticipated by theory. In practice, the beam diameter achieved at the focal plane of the lens is usually an order of magnitude higher. Figure 9.3a,b shows examples of craters formed with one and ten 1064 nm laser pulses, during the LIBS analysis of a daguerreotype placed at the focal plane of the focusing lens. Crater diameters were measured to be in the range of

70–80 µm. The craters' surface profiles show a deeper central part reflecting the intensity distribution of the laser beam (for the particular laser used) and a thermally affected zone in the periphery. Around the area irradiated with ten pulses, metal droplets are observed while the crater profile (measured with a profilemeter) indicates the presence of a rim formed by material melting and redeposition. For the case of a painting, craters produced in a depth-profiling study (see Section 9.6.3) employing multiple pulse irradiation (355 nm) are shown (Figure 9.3c) for two different points on the painted surface. As a result of the large number of pulses per spot the craters are wider than those observed with a single pulse. In both model and realistic samples, irradiation with one or two pulses results in average crater diameters less than 100 µm, the exact diameter depending upon laser beam profile and focusing, the specific ablation characteristics of the sample (pigment and medium) and the local inhomogeneity of the surface.

Optimization of detection parameters such as signal collection, spectral resolution, delay-time, and gate pulse duration is very important. This is a consequence of the low pulse energy values, employed particularly in valuable objects, yielding relatively weak plasma emission. For this reason, relatively short delay times, even as short as 100 ns, are employed in certain cases of weak emission in order to keep the signal intensity at sufficient levels to obtain satisfactory spectra. Typically the time delay between the laser pulse and the gating pulse ranges from 500 to 1000 ns. This range has been shown to be adequate to suppress the continuum background emission and minimize Stark broadening of spectral lines. The time window (gate) for detection is usually in the range of 500–1000 ns.

Also, the selection of the proper spectral range and resolution so that detection of certain elements is optimum is essential for recording clean spectra, thus optimizing the analytical information obtained from each spectrum. The advantage of the echelle spectrographs, which provide wide spectral coverage without sacrificing resolution, is obviously very important in this respect.

9.6.3 *LIBS analysis of objects of cultural heritage (qualitative analysis)*

As briefly described in Section 9.5, the LIBS spectrum provides immediate information on the identity of the elements present in the spot examined. The analysis is carried out on the object itself (*in situ*) and requires no sample removal or sample preparation. The qualitative and semi-quantitative information provided in such an analysis can lead to valuable results, for example the identification of pigments used on a painting or the determination of the components in a metal alloy. In view of the value and sensitivity of art objects, the analysis should be carried out under such conditions that a single laser pulse measurement is adequate for providing the information needed. The single-pulse experiment is important for minimizing material removal from the object's surface and also avoiding possible side effects of the laser radiation such as light- or heat-induced pigment discoloration [88, 89]. While averaging (employing multiple pulses) in general improves signal quality, it should be avoided because it can result in excessive material removal. Multiple pulses can even result in the sampling of successive material layers of varying composition, thus producing

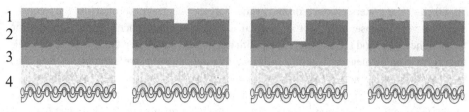

Figure 9.4. Schematic representation of a multi-layer paint, indicating the basic principle of the LIBS depth profile analysis. Progressive material removal as a result of laser ablation results in elemental mapping of the successive layers of varnish (1), paint layers (2, 3), and primer (4).

a more complex superposition of spectral information. Despite the fact that LIBS is not strictly a non-destructive technique, it can be characterized as micro-destructive or nearly non-destructive since tight focusing of the laser beam leads to very small affected areas (formation of microscopic craters, see Section 9.6.2), which in most cases are invisible to the naked eye.

The tight focusing of the laser beam is a very important feature of LIBS not only because it results in minimization of surface damage but also because it gives rise to superb spatial resolution. Thus it is possible, by using a single focusing lens, to discriminate between neighboring features and analyze very fine paint lines thinner than 100 μm [90, 91]. In fact, recent developments in micro-LIBS analysis show that much smaller dimensions can be probed, extending the range of spatial resolution to a few micrometers [75].

In the same context, LIBS has the ability to perform depth profile analysis. This can be very important if layered materials are probed [92, 93]. Such cases are common in painted works of art, which are always composed of multiple paint layers. Also, probing of environmental deposits on stone or marble or corrosion products on metal surfaces is another situation in which depth profiling can provide important details of the cross-sectional distribution of different species. Conventional stratigraphic analysis is done by examining, under an optical or electron microscope, the cross-section of a properly prepared sample, removed from the object. By using LIBS, an alternative approach to depth profile analysis is achieved. It essentially avoids removal of sample from the object while performing, in effect, minute sampling *in situ*. Each laser pulse in the process of analysis ablates a thin layer of material and, as a result, the next pulse probes a fresh surface. Therefore, after delivering a certain number of laser pulses on the same spot and recording the spectra produced by each pulse, the analyst has the ability to monitor the elemental content of successive layers (Figure 9.4). In the case of a painting, typical thickness values for paint layers range from 5 μm to 50 μm. The depth resolution depends on the characteristics of the ablation process as already discussed. Measurements in model samples and real paintings have shown that on average the ablation depth per pulse ranges from 0.5 μm to 2 μm. However, as mentioned earlier, local imperfections within the paint layers often cause the removal of larger amounts of material as a result of the shock wave induced by the ablation process. In such cases craters as deep as 10 μm have been observed.

Typically, depth resolution studies are qualitative in nature owing to the difficulty of accurate calibration of the procedure. The ideal laser beam would have a "top-hat" profile and all of the intensity of each succeeding laser shot would impinge on fresh surface at a new depth. However, the profile of a typical laser is, at best, Gaussian and often may have an irregular profile or even "hot spots." Under these circumstances, each successive laser pulse will sample the sides of the evolving crater in addition to the bottom of the hole. Some fresh surface will also be sampled as the hole diameter typically increases with the number of pulses. So the observed intensities of the components will be a complex mixture of signals from different depths, which is hard to calibrate. At higher laser power values, some partial melting may also complicate the picture. In spite of these problems with the analysis of elemental composition in the interior of samples, if the thickness of a layer is determined by noting the "breakthrough point," where a new element appears, for instance, the results can be quite accurate and reproducible. Several examples of depth profile analysis are shown in the next section.

9.6.4 Quantitative analysis

In many cases quantitative information about the composition of the material analyzed is important as, for example, in the determination of minor and trace elements in glass objects, coins, metallic objects or minerals. Such data, apart from material characterization and classification, can provide provenance information.

An important issue with LIBS is the capability of the technique to produce reliable, quantitative information. In this respect, a lot of work has been published where methodologies have been proposed for getting quantitative results, which rely mainly on calibration curves for selected analytes produced with samples of known composition [49, 50, 70–73, 94, 95]. Such an approach is, in principle, transferable to analysis of art objects; however, the following considerations have to be seriously taken into account. The vastly different types of material encountered and their highly varying composition would require a large number of calibration samples to cover all cases, which would obviously complicate the analysis. Furthermore, in order to minimize the effects of material heterogeneity and other matrix effects [95] on the analytical result, averaging of several pulses is required, which may not be possible in all cases, especially for valuable and/or fragile objects.

A promising approach to the problem appears to be the use of the calibration free LIBS (CF-LIBS) method, which can, in principle, provide quantitative elemental analysis with no need of calibration curves or internal standards [96, 58–62]. It relies on the assumption that stoichiometric ablation takes place and that an optically thin plasma is produced, which is at local thermodynamic equilibrium (LTE approximation). From the practical point of view, it requires recording of the emission intensity at high resolution and across a wide spectral range for identifying all or most of the elements present in the sample and for improving the accuracy in the calculations. According to the CF-LIBS method, the intensity of each emission line from a LIBS spectrum is represented as a Boltzmann plot, producing a set of parallel straight lines (one for each species; different states of ionization for each element are treated as separate species) whose slope yields the plasma electron temperature and the intercept relates to the logarithm of species concentration within an instrumental factor (equation (9.1)). The Saha–Boltzmann equation is used, if needed, to calculate species concentrations for a given element in adjacent states of ionization. Elimination of the unknown instrumental factor in a final normalization step, based on the requirement that the sum over all species concentrations equals unity, leads to the calculation of the concentration of each species within the sample. It has to be noted, however, that, despite the advantages of the CF-LIBS approach, data interpretation must still be done with caution. This is especially true in cases of highly heterogeneous substrates (e.g. pigment mixtures), where analytical results are representative only of the local microscopic composition of the sample and not of the average composition, which is of most interest.

9.7 Examples of LIBS analysis in art and archaeology

Specific examples involving analysis of real objects are presented to illustrate the application of LIBS to various types of analytical problems encountered in art history, conservation, and archaeological analysis.

9.7.1 Analysis of pigments

Painting has been used in all types of art forms from antiquity to modern times. Easel and wall paintings, wood and metal polychromes, illuminated manuscripts, and pottery represent art or craft forms where pigments have been used [97–99]. Determining the identity of pigments is of importance for several reasons. For example, in the case of painted works of art, it can help art historians to characterize the available materials and to understand techniques and effects used by the artist in achieving the result in the final work. Similarly, in the case of ancient painted pottery or frescoes, pigment characterization may lead to increased understanding of the materials and technology available to the craftsmen. This knowledge may even relate to the origin of the materials, suggesting local or remote sources that might indicate communication and trade between sites. In addition, pigment identification in painted works of art can be often significant in providing dating information on the basis of the known history of the pigment manufacturing or in assessing the state of preservation of a painted work of art in preparation for proper restoration.

Physically the pigments are insoluble matter in the form of small particles/grains (with diameters in the range of a few micrometers), which are dispersed in a matrix normally called the binding medium or the carrier, forming a relatively viscous paste, the paint [97]. The matrix can be a drying oil (oil paint), an acrylic polymer (acrylic paint), or lime (fresco paint). After the paint is applied on the appropriate substrate, the medium gradually solidifies because of chemical reactions (e.g. cross-polymerization of the fatty acid chains of linseed oil in the case of oil painting, or transformation of calcium hydroxide to calcium carbonate in the case of wall paintings) producing solid matrices that trap and stabilize the pigment grains. A schematic illustration of a painted structure composed of two paint layers on top of a ground layer on canvas is shown in Figure 9.4. The ground layer (often called primer or preparation layer) serves as a background on which the drawing is made while it also acts as a barrier between the paint and the support (e.g. canvas, wood, wall). In easel painting, for example, it is normally a thick (*c.* 0.5–2 mm) layer of a white paint, gypsum or chalk, in animal glue. The varnish, applied finally on top of the paint, functions as a transparent, protective layer, which also improves the color and gloss of paintings owing to proper refractive index matching. Varnishes are natural or synthetic organic resins that are applied onto the paint in the form of solution in a volatile organic solvent, yielding a hard, glassy film on the surface of the painting after evaporation of the solvent.

With the exception of a limited number of organic color substances, most pigments used in paintings from antiquity to modern times are inorganic compounds. The main reasons for this have been the higher chemical and photochemical stability of inorganic pigments compared with organic ones, and the easy availability of most inorganic pigments in the form of natural minerals or as products of relatively simple chemistry. In contrast, lengthy and delicate procedures were often involved in extracting organic pigments from insects or plants. Table 9.2 shows, arranged by color, several pigments used in painting. Their chemical structure and common names are given along with chronological information on their use.

The different elemental composition of inorganic pigments is reflected in the corresponding LIBS spectra and this enables their discrimination based on the characteristic atomic emission peaks recorded. Major analytical emission lines, used to identify the presence of specific elements in pigments and other materials through their corresponding LIBS spectrum, are shown in Table 9.3. The positive identification of a certain pigment results from correlating the spectral data with the color of the paint analyzed. In this respect, information from an art historian or conservator about the possible pigments anticipated is essential in the analysis of the work. In many cases, the characteristic elements can easily lead to determination of the pigment used. For example, the presence of mercury undoubtedly suggests the use of the red pigment vermilion (HgS, also known as cinnabar in its natural mineral form). The presence of lead in a white paint is definitive proof of the use of lead white ($Pb(OH)_2 \cdot 2PbCO_3$). However, in reality, mixtures of pigments are often used by the artist to achieve the desired result of color and shade. For example, it is not uncommon for a green paint to be a result of admixing a green pigment with a white one or even a yellow pigment with a blue one. The situation of pigment mixtures gives rise to more complex LIBS spectra but the presence of individual pigments can usually be determined based on the elements found. In such a case, microscopic examination, when possible, is of great help in guiding the analysis of the LIBS spectral data, if pigment grains can be optically discriminated.

Typical spectra from model samples of two common pigments, cadmium red and lithopone, are shown in Figure 9.5a and b indicating how the different spectral emission lines can lead to identification of each pigment on the basis of the specific elements detected (Cd and Ba, Zn respectively). The use of single- or multi-component pigment model samples is essential in establishing proper measurement conditions before carrying out an analysis of a real object.

Table 9.2. *List of selected inorganic pigments analyzed by LIBS and elements identified*

Pigment name	Chemical composition / formula	Identified elements	Origin and chronological information[a]
Lead White	$Pb(OH)_2 \cdot 2PbCO_3$	Pb	Synthetic, pre-500 BC
Titanium White	TiO_2	Ti	Synthetic, 1920
Zinc White	ZnO	Zn	Synthetic, 1834
Lithopone	$ZnS \cdot BaSO_4$	Ba, Zn, (Ca)	Synthetic, 1874
Chalk	$CaCO_3$	Ca	Mineral (Calcite)
Barytes, Barium sulfate	$BaSO_4$	Ba	Synthetic, early nineteenth century
Gypsum	$CaSO_4 \cdot 2H_2O$	Ca	Mineral
Cadmium Yellow	CdS or CdS $(ZnS \cdot BaSO_4)$	Cd, Zn, (Ba)	Synthetic, 1829
Chrome Yellow	$PbCrO_4$	Cr, Pb (Ca)	Synthetic, 1818
Cobalt Yellow	$2K_3(Co(NO_2)_6) \cdot 3H_2O$	Co	Synthetic, 1861
Orpiment	As_2S_3	As	Mineral
Naples Yellow	$Pb_2Sb_2O_7$	Pb, Sb	Synthetic, Egypt *c.* 1500 BC
Lead Tin Yellow	Pb_2SnO_4 (Type I)	Pb, Sn	Synthetic, *c.* 1300
Strontium Yellow	$SrCrO_4$	Sr, Cr	Synthetic, early nineteenth century
Barium Yellow	$BaCrO_4$	Ba, Cr	Synthetic, early nineteenth century
Yellow Ochre	$Fe_2O_3 \cdot nH_2O$, SiO_2, Al_2O_3	Fe, Si, Al	Mineral
Cadmium Red	$CdS_xSe_{(1-x)}$	Cd	Synthetic, *c.* 1910
Cinnabar/ Vermilion	HgS	Hg	Mineral/synthetic, eighth century
Red Ochre	Fe_2O_3 (Al_2O_3)	Fe, (Al)	Mineral
Realgar	As_2S_2	As	Mineral
Mars Red (Hematite)	Fe_2O_3	Fe	Synthetic, middle nineteenth century
Red Lead (Minium)	Pb_3O_4 ($2PbO \cdot PbO_4$)	Pb	Synthetic
Lapis Lazuli/Ultramarine	$Na_7Al_6Si_6O_{24}S_3$	Al, Si, Na	Mineral/synthetic, 1828
Egyptian Blue	$CaCuSi_4O_{10}$	Cu, Si, Ca	Synthetic, Egypt *c.* 3100 BC
Cobalt Blue	$CoO \cdot Al_2O_3$	Co, Al, Na	Synthetic, 1802
Cerulean Blue	$CoO \cdot nSnO_2$	Co, Sn	Synthetic, 1860
Prussian Blue	$Fe_4[Fe(CN)_6]_3 \cdot nH_2O$	Fe, (Ca)	Synthetic, 1704
Azurite	$2CuCO_3 \cdot Cu(OH)_2$	Cu, (Si)	Mineral
Malachite	$CuCO_3 \cdot Cu(OH)_2$	Cu, (Si)	Mineral
Viridian Green	$Cr_2O_3 \cdot 2H_2O$	Cr	Synthetic, 1838
Emerald Green	$Cu(CH_3COO)_2 \cdot 3Cu(AsO_2)_2$	Cu, As	Synthetic, 1814
Verdigris	$Cu(CH_3COO)_2$	Cu	Synthetic, antiquity
Ivory Black (Bone Black)	$C + Ca_3(PO_4)_2$	Ca, P	Charring of ivory, antiquity
Manganese Black	MnO	Mn	Mineral
Magnetite/Mars Black	Fe_3O_4	Fe	Mineral/synthetic, middle nineteenth century

[a]Origin information distinguishes between natural (minerals, in use since antiquity) and synthetic pigments. Chronological information indicates when pigments were first used (historical sources) or became commercially available (modern pigments).

Table 9.3. *Major analytical emission lines of elements*

Element	Wavelength (nm)[a]
Ag	272.18 (I), 282.44 (I), 328.07 (I), 338.29 (I), 487.41 (I), 520.91 (I), 546.55 (I)
Al	308.21 (I), 309.27 (I), 394.40 (I), 396.15 (I)
As	274.50 (I), 278.02 (I), 286.04 (I), 289.87 (I)
Au	267.60 (I), 274.82 (I), 312.28 (I)
Ba	225.47 (II), 230.42 (II), 233.53(II), 389.18 (II), 413.07 (II), 455.40 (II), 493.41 (II), 553.55 (I), 614.17 (II), 649.69 (II)
C	247.85 (I)
Ca	315.89 (II), (317.93, 318.13(II)), 393.37 (II), 396.85 (II), 422.67 (I), (526.17–527.03(I))
Cd	228.80 (I), 288.08(I), 298.06 (I), 340.36 (I), 346.62 (I), 346.77 (I), 361.05 (I), 467.81 (I), 479.99 (I), 508.58 (I), 643.85 (I)
Co	(240.72–241.53 (I)), (242.49–243.90 (I)), numerous lines in the range 340–360, 389.41 (I), 399.53 (I), 412.13 (I)
Cr	357.87 (I), 359.35 (I), 360.53 (I), 391.92 (I), 396.37 (I), 396.97 (I), 397.67 (I), 425.43 (I), 427.48 (I), 428.97 (I), (520.45, 520.60, 520.84 (I))
Cu	261.84 (I), 276.64 (I), 324.75 (I), 327.40 (I), 510.55 (I), 515.32 (I), 521.82 (I)
Fe	Numerous emission lines (from neutral and ionized atoms) appear throughout the spectrum. Features at 273–275 nm, 373–376 nm, 381–384 nm and 404–407 nm are characteristic of iron.
Hg	253.65 (I), 296.73 (I), (312.57–313.18 (I)), 365.02 (I), 404.66 (I), 435.83 (I), 546.07 (I)
K	766.49 (I), 769.90 (I)
Mg	279.55 (II), 280.27 (II), 285.21 (I), (382.93–383.83 (I)), 517.27 (I), 518.36 (I)
Mn	257.61 (II), 259.37 (II), (279.48–280.10 (I)), 293.31 (II), 293.93 (II), 294.92 (II), 380.67 (I), 382.35 (I), (403.08–403.57 (I)), 404.14 (I)
Na	330.13 (II), 589.00 (I), 589.59 (I)
Ni	239.45 (II), 241.61 (II), 243.77 (II), 251.09 (II), (300.25, 300.36 (I)), 301.20 (I), (310.16, 310.19 (I)), 313.41 (I), 338.06 (I), 341.48 (I), 344.63 (I), (345.85, 346.16 (I)), 349.30 (I), 351.50 (I), 352.45 (I), 356.64 (I), 359.77 (I), 361.05 (I), 361.94 (I)
P	(253.40, 253.56 (I)), (255.32, 255.49 (I))
Pb	280.20 (I), 282.32 (I), 283.30 (I), 287.33 (I), 357.27 (I), 363.96 (I), 367.15 (I), 368.35 (I), 373.99 (I), (405.78, 406.21 (I)), 416.80 (I), 500.54 (I), 722.90 (I)
Sb	206.83 (I), 217.58 (I), 231.15 (I), 252.85 (I), 259.80 (I), 265.26 (I), 277.00 (I), 287.79 (I)
Si	263.13 (I), 288.16 (I), 390.55 (I)
Sn	235.48 (I), (242.17, 242.95 (I)), 284.00 (I), 286.33 (I), 300.91 (I), 303.41 (I), 317.50 (I), 326.23 (I)
Sr	407.77 (II), 421.55 (II), 460.73 (I)
Ti	Numerous emission lines (from neutral and ionized atoms) appear throughout the spectrum. Features at 323–325 nm, 332–339 nm, 372–376 nm, 390–400 nm, 428–432 nm, 444–446 nm, 451–454 nm and 498–502 nm are characteristic of titanium.
Zn	213.86 (I), 328.23 (I), (330.26, 330.29 (I)), (334.50–334.59 (I)), 468.01 (I), 472.21 (I), 481.05 (I), 636.23 (I)

[a] Wavelengths in parentheses indicate two or more lines not resolved in low-resolution spectra but still characteristic of the elements. The wavelengths refer to emission from neutral atoms when followed by (I) and to emission from singly charged ions when followed by (II).

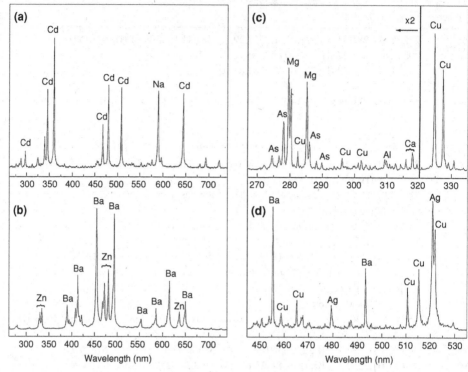

Figure 9.5. LIBS spectra of (a) cadmium red model paint sample, (b) lithopone model paint sample, (c) green paint on miniature (see Figure 9.6a) and (d) white paint on a daguerreotype (see Figure 9.6b).

A real pigment analysis case is represented by the study of the miniature (nineteenth century, France), shown in Figure 9.6a [55]. This miniature is an example in which the size of the object and sensitivity of the painted surface (thin and weakly adhering paint) dictated the examination of a minimal number of spots and the use of no more than three laser pulses per spot to minimize any damage to the paint. Examination of the green paint used extensively on the miniature shows intense emission due to copper and weak emission due to arsenic (Figure 9.5c). This suggests the use of a green pigment based on a copper–arsenic compound, such as either emerald green ($Cu(CH_3COO)_2 \cdot 3Cu(AsO_2)_2$) or Scheele's green ($Cu(AsO_2)_2$). An alternative possibility could be the use of a mixture composed of the blue pigment azurite (copper compound) and the yellow pigment orpiment (arsenic compound). Additional examination of the paint under an optical microscope verified the identification as emerald green on the basis of its characteristic appearance (small spherulites).

In a similar case several painted daguerreotypes were examined in order to identify the pigments used. Daguerreotypes were the first forms of photographs and were prevalent between 1839 and 1860, before photography based on paper became practical [100]. The

Figure 9.6. Paintings analyzed by LIBS. (a) Nineteenth-century miniature on ivory; (b) nineteenth-century colored daguerreotype; (c) eighteenth-century oil painting (copy of Palma Vecchio's "La Bella"); (d) Byzantine icon (St. Nicholas, nineteenth century, Russia).

images consist of silver particles photochemically produced on silver-coated copper plates. The plates were sometimes colored afterwards to highlight certain features by applying a very thin paint layer on the surface. As an example, the white paint used on the daguerreotype (nineteenth century, USA) shown in Figure 9.6b, was analyzed and the LIBS spectrum showed the presence of emission lines due to barium, suggesting the use of barium white (Figure 9.5d). Given the delicate nature and thickness of the paint layer ($c.1$ μm), a single laser pulse was employed for the analysis, showing that reliable results can be obtained with careful selection of working parameters [101].

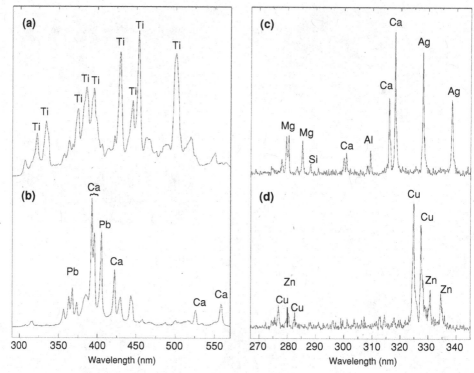

Figure 9.7. LIBS spectra of (a) titanium white paint on restoration and (b) original lead white paint on an eighteenth-century oil painting (Figure 9.6c). LIBS spectra of (c) silver foil and (d) restoration on silver foil used in Byzantine icon of St. Catherine (seventeenth century).

In the context of pigment analysis, LIBS can be also employed to identify the type of prior restoration performed on paintings by discriminating between the original paint and that used in the restoration. In certain cases, the time a painting was made or an intervention took place can be indirectly estimated on the basis of known dates when synthetic pigments became available. In one case the restoration carried out on several parts of an oil painting (Figure 9.6c) was examined and found to contain mainly titanium white, a modern pigment, in contrast to the original paint, which was composed mainly of lead white (Figure 9.7a,b) [51, 53]. This result suggests a retouching, done on the original painting after 1920, as titanium white became commercially available after that time. In a similar case the "gold" background used on a Byzantine icon (St. Catherine, seventeenth century) was examined by carrying out several spot analyses, which revealed dramatic local differences apparently resulting from a past restoration. The original "gold" foil was actually shown to be a silver foil on the basis of characteristic emission peaks due to Ag in the LIBS spectrum (Figure 9.7c). The golden appearance was achieved by coating the silver foil with a yellow varnish. Probing a restored area by LIBS revealed the use of a copper foil in place of the damaged silver foil (Figure 9.7d).

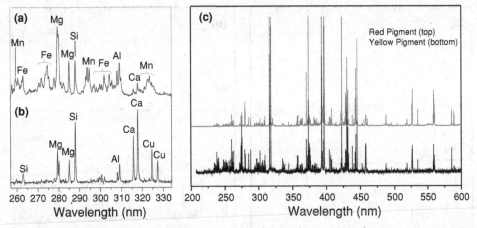

Figure 9.8. LIBS spectra from (a) black paint on pottery sherd, (b) blue fresco paint sample, and (c) red and yellow fresco paint samples. Part (c) is reproduced from reference [59].

In the context of archaeological objects, pigment analysis is important as well. For example, the study of pigments and painting techniques reveals information on technological progress in ancient cities. Such data can even be used to discriminate objects on the basis of their production site and reveal possible links between such sites. Examples of the application of LIBS in ancient paint analysis are shown in Figure 9.8. The black paint on a fragment of a Minoan ceramic sherd (twelfth century BC, Eastern Crete) was analyzed and found to contain manganese in addition to the usual clay components (Al, Si, and Mg). Detection of Mn suggests the use of MnO_2, a black pigment used commonly in antiquity. In a different case, the blue paint on a fresco fragment (Egypt) was examined and the LIBS spectrum showed the presence of Mg, Si, Ca, Cu, and Al. The spectral data suggest the use of Egyptian blue ($CuCaSi_4O_{10}$), which is further supported by comparison with a known sample of the pigment. Having the ability to quantify the spectral information is obviously important, especially in the case of compositionally similar paints. An example is shown in Figure 9.8c from the LIBS analysis of two different Roman fresco samples (second century, St. Albans, UK) performed on an echelle spectrograph system [58]. The wide spectral range covered, and the high resolution achieved, are obvious in the spectra. By applying the CF-LIBS analysis it became possible to determine the elemental content of a red and a yellow paint, detecting elements with concentrations as low as 0.1%.

In several cases LIBS has been successful in revealing stratigraphic information on paint layers by means of depth profile analysis [51, 52, 54, 56, 58, 59, 61]. In the case of a Byzantine icon examined (nineteenth century, St. Nicholas, egg tempera on silver foil; Figure 9.6d) depth profile analysis mapped the different paint layers and the silver foil in between the brown paint (iron-based pigment, possibly red ochre) and the ground layer (gypsum) (Figure 9.9a). Similar analysis on a different spot on the same icon revealed a thin zinc white overpaint layer on the original lead white paint [54]. Examination of the red

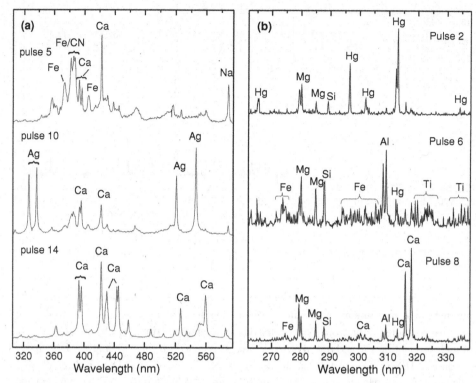

Figure 9.9. Depth profile studies on Byzantine icons: (a) St. Nicholas (Figure 9.6d); (b) Three Saints.

paint on another icon (eithteenth century, Three Saints, egg tempera) showed the use of the red pigment vermilion (HgS) on a dark background containing iron (possibly a red ochre or a sienna) (Figure 9.9b). The ground layer is rich in calcium, suggesting the likely use of gypsum or chalk. These results demonstrate the ability of LIBS to extract information on successive paint layers by means of depth profile analysis. This approach is not limited to painted structures, but is general for surfaces that have a multi-layer structure or elemental concentration gradients.

9.7.2 Analysis of pottery

Ceramic objects are the most common remnants of ancient life, uncovered in numerous excavations, having been used as storage containers, serving dishes, and votive figurines, among other uses. They are made of clay, which is often decorated by paint, depending on the use and quality of the object. In the analysis of ceramic sherds, questions are related to the characterization of pigments, the determination of the elemental composition of the clay, and the characterization of surface encrustation. Analysis of clay inclusions can be

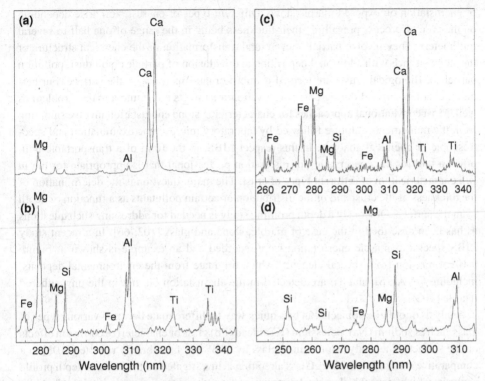

Figure 9.10. LIBS spectra from (a) white and (b) dark pottery inclusions, (c) marble encrustation (dendritic crust), and (d) ancient glass.

of importance if these inclusions are characteristic of the clay source or the technique employed. An indicative example is shown in Figure 9.10a,b, where LIBS spectra from two different types of microscopic inclusion are shown. The dark inclusion appears to be based on clay with a high content of iron (possibly from magnetite, Fe_3O_4, a black mineral), while the white inclusion shows a high content of calcium, most likely in the form of calcite. Quantitative elemental analysis of pottery sherds is important in differentiating between various types of clay from the same or different excavation sites, aiding archaeologists in classifying objects and drawing conclusions about materials and techniques that relate to the socio-economic status of populations.

9.7.3 Analysis of marble, stone, glass, and geological samples

Marble and stone materials have been used extensively in monuments, sculpture, and tools. Their elemental analysis related to geological data can be helpful in determining the source of the material. An important application of LIBS to the analysis of marble or stone monuments has to do with the investigation of environmental pollution effects. It is known that atmospheric pollution through physical and chemical interactions leads to the formation of encrustation on exposed marble/stone. Different types of encrustation arise depending on the type of process prevailing, their thickness being in the range of one half to several millimeters. They involve partial or even total transformation of the chemical structure of the stone surface with additional deposition and inclusion of particles (soil dust, pollution particles). Biological crusts are formed if micro-organisms grow on the surface. Furthermore, coatings applied during past conservation treatments constitute a major problem as well. Most conventional approaches for characterizing stone encrustation involve sampling from the monument/sculpture followed by laboratory microscopic examination and spectroscopic or chemical analysis. In this respect LIBS, in the form of a transportable unit, offers a tool for the conservation or restoration professional that is appropriate for the *in situ* characterization of different types of crust. The major question is the determination of the thickness of the crust and of the distribution of certain pollutants as a function of depth from the surface. Obviously a depth profiling study is needed for addressing such questions as has been described in the case of marble, stone and glass [102, 66]. In a recent study LIBS spectra of marble encrustation were recorded and an example is shown in Figure 9.10c. Emission from several elements which originate from the environmental deposits, including Fe, Al, Si, and Ti, are detected in the encrustation but not in the unweathered marble [102].

Analysis of old glass objects can be a quick way to differentiate between various types of glass such as sodium (Na), potassium (K) or lead (Pb) glass or characterize corrosion effects on the surface. The analysis of stained glass by LIBS has been briefly reported [103] in a comparative study against XRF. Glass alteration is investigated on the basis of depth profile analysis, which maps the elemental distribution from the outer surface layers into the bulk. Related work on LIBS analysis of glass has been performed in the context of investigating the process of laser cleaning of stained glass (Figure 9.10d) [66].

Geological materials were used in early times to manufacture simple tools and to create sculpture. Their elemental composition, particularly that of minor and trace elements, has been related to provenance of the object. In this respect, such analysis can be of importance in uncovering, for example, the sources of raw materials used to make certain objects by comparison of compositional patterns between the object in question and the original raw materials. Quick screening can be done on the basis of qualitative analysis, if differences in elemental content are significant [104]. For more detailed studies a systematic quantitative analysis of major and minor components is obviously needed.

9.7.4 Analysis of metals

Objects made of metal or metal alloys, including sculpture, tools, weapons, home utensils, and jewelry, have been widely used for different purposes since metallurgy was invented. The main materials used include copper and bronze – that is copper–tin alloys (in the Bronze Age) – and later iron. Other metals used include lead and tin, while precious metals such as silver or gold alloys have been used in jewelry and for decorating different objects. Also, metal alloys have been extensively used in coinage. The first analytical question concerning

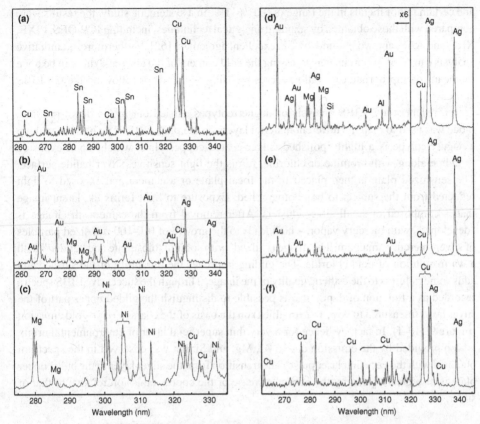

Figure 9.11. LIBS spectra from (a) a Minoan bronze object, (b) a Minoan golden jewel, and (c) a twentieth-century coin. Depth profile analysis of a nineteenth-century daguerreotype after irradiation with (d) one, (e) three and (f) ten laser pulses at 1064 nm.

a metal object is to identify the type of metal or metal alloy. In addition, determination of the quantitative content of the various metals and of any trace elements can lead to a more complete characterization of the objects, yielding information on technological know-how and possibly provenance.

Examples of spectra collected from various metal objects including a Minoan bronze tool and a part from a piece of Minoan golden jewelry (from Eastern Crete, Greece) and a twentieth-century coin are shown (Figure 9.11a–c). The quick screening of ancient metal objects on the basis of their different elemental contents can be achieved by LIBS in an efficient way. For example, discriminating between copper and bronze (copper–tin alloy) objects is of importance for the classification of archaeological findings. On the other hand, quantitative LIBS analysis based on the CF-LIBS method was used to analyze samples from two Renaissance bronze sculptures in Florence, Italy, leading to the determination of the percentage content of the major alloy components (Cu–Sn–Pb and Cu–Sn–Pb–Zn) and certain minor metals in the range of 0.01%–1%. In a systematic study, the results were compared with those obtained by standard analytical techniques, including ICP-OES, PIXE, XRF, and AAS, and were found to be in excellent agreement [62]. Furthermore, quantitative analysis can be of importance in assessing the gold content of the alloy and this can be done by means of appropriate calibration curves by using standards or following the CF-LIBS approach [80].

In a recent study, LIBS analysis of daguerreotypes carried out in the depth-profiling mode was used to map the different material layers on the plate [101]. The basis of making a daguerreotype is a highly polished, silver-coated copper plate, which, upon exposure to iodine along with bromine or chlorine, forms the light-sensitive, silver halide surface. The sensitized plate is then placed in the focal plane of a camera and exposed to light reflected from the subject to be photographed. Exposure to light forms the latent image that is composed of small silver clusters. After removal from the camera, the image is "developed" with mercury vapor, which leads to the growth of 0.1–100 μm sized particles of silver–mercury amalgam. The image is "fixed' with sodium thiosulfate and "gilded" with a warm solution of gold chloride. The gilding improves the durability of the surface and adds some richness to the esthetic quality of the image. Through the successive LIBS spectra recorded as a function of depth, it was possible to distinguish the gilded upper part of the silver layer (measured to be *c*. 1–2 μm thick) on the basis of the characteristic gold emission (Figure 9.11d–f). In fact, evidence for a very thin superficial layer of environmental origin is also provided by the emission due to Al, Mg, and Si that was observed in the spectrum obtained with the very first laser pulse. The transition from the silver layer to the bulk copper plate is also clearly seen by the abrupt increase of the copper emission relative to that of silver.

9.7.5 Biological samples

This is an area of potential interest to archaeological research focused on human remains aiming to identify certain diseases, nutritional habits or deficiencies or the presence of potentially toxic elements [105–107]. Recent work [108] on the analysis of calcified tissues has demonstrated the capability of LIBS to quantitatively detect traces of aluminum (Al), strontium (Sr), and lead (Pb) in human bones or teeth, and furthermore to map the distribution of these elements across the surface of the calcified tissue samples. These results indicate strongly the potential advantages of LIBS in the analysis of bioarchaeological samples.

9.7.6 Control of laser cleaning

A major breakthrough in art conservation has been brought about in the past two decades by the introduction of laser-based techniques for cleaning art objects ranging from marble, stone, metal sculpture, and stained glass to paintings, icons, and paper [109]. The cleaning process relies on the controlled removal of contaminant or other unwanted layers from the surface of the object treated. This is effected by means of laser ablation, which results from the interaction of focused nanosecond laser pulses with the substrate. The process depends strongly on the material properties (absorptivity, surface roughness, mechanical stability) and irradiation parameters (wavelength, energy density, pulse duration). There are extensive studies on the use of laser ablation cleaning, which are beyond the scope of this chapter; however, one critical question regarding the success of the cleaning process relates to LIBS. It is highly important, when carrying out any type of cleaning methodology, to know where to stop the process. That is, to be able to assess reliably to what extent the contamination layer has been removed. Such control of laser cleaning can, in certain cases, be achieved by monitoring the optical emission resulting from material ablation. In essence, the LIBS measurement is carried out simultaneously with the laser cleaning. As discussed in Section 9.6.3, successive laser pulses on the same area can provide compositional information about the material layers. If distinct differences between the LIBS spectra of the contamination layer to be removed and the cleaned surface exist, then it is, in principle, possible to control the process of cleaning by a simple algorithm implemented through a computer. This is a delicate approach and thus careful preliminary tests of working parameters with respect to the specific case in hand have to be done beforehand in order to define properly the end point of the cleaning. Such control of laser cleaning has been demonstrated in several cases including the removal of overpaint from frescos [64] or encrustation from marble [65, 110] or glass [66]. An example is shown in Figure 9.12 indicating clear spectral changes as an overpaint layer is removed, which allow effective, on-line control of the process. Cleaning is then stopped before exposing the original paint layer to the laser irradiation in order to avoid any pigment modification.

9.8 LIBS in combinations with other techniques

The combination of more than one technique in addressing analytical questions can be advantageous, as use of complementary approaches could yield more complete information on materials found on an object. Strengths of one technique could complement weaknesses of another. The use of LIBS in various combinations with other laser analytical techniques such as Raman microscopy, laser-induced fluorescence (LIF) spectroscopy, and laser time-of-flight mass spectrometry is described below. Examples of combining results from LIBS analysis with other analytical techniques such as diffuse reflectance, hyper-spectral imaging [57] and chromatography have also been reported [111] but are not detailed herein.

9.8.1 LIBS and Raman microscopy

Raman spectroscopy provides molecular species information complementary to the elemental analysis data obtained in a LIBS analysis. The Raman effect relies on the inelastic scattering of light from molecules or materials and is sensitive to vibrational transitions, which in many cases are characteristic of the substrates analyzed. Particularly in the case of pigments, studies have shown that Raman spectra constitute a unique spectral fingerprint for each pigment [17–21, 23, 112–114]. Raman microscopy is an excellent technique for the

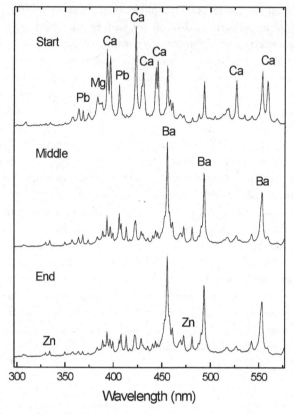

Figure 9.12. LIBS spectra revealing the cross-sectional elemental composition of overpainted plaster used as a criterion for controlling the laser removal of overpaint.

scientific investigation of materials used on works of art because it combines high sensitivity with non-destructiveness, and in addition it can be performed *in situ*, thus obviating the need for sampling and consequently any damage to the object under examination. The use of a microscope is essential to the strength of this technique as it permits probing of isolated grains of pigment (down to 1 μm across) by the laser beam (from a continuous wave (cw) laser). Moreover the very high spatial resolution thus achieved gives rise to clean spectra, which are reasonably free of interference from surrounding materials. The identification of pigments is made on the basis of spectral bands in the Raman spectra recorded, which correspond to specific vibrational transitions characteristic of the pigment probed.

The combined use of Raman microscopy and LIBS in the analysis of objects of cultural heritage has been reported in the literature [54–56, 61, 114]. For example the LIBS analysis of the yellow paint on a Byzantine icon (The Annunciation, eighteenth century, Greece) revealed the presence of lead and chromium suggesting the use of chrome yellow (Figure 9.13a), which was indeed verified by the analysis of the paint with Raman

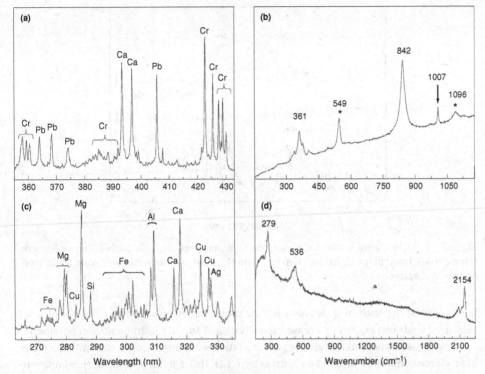

Figure 9.13. (a) LIBS spectrum from yellow paint and (b) Raman spectrum from green paint (asterisks: ultramarine blue; arrow: gypsum) on a Byzantine icon (The Annunciation, eighteenth century, Greece). (c) LIBS and (d) Raman spectra from blue paint on a French miniature (Figure 9.6a).

microscopy [55]. The Raman spectrum also showed the green paint to be a mixture of chrome yellow and ultramarine blue (Figure 9.13b). In another case, the blue paint on the French miniature shown in Figure 9.6a was analyzed and found to contain iron on the basis of the LIBS analysis (Figure 9.13c). Raman analysis proved the presence of Prussian blue based on characteristic vibrational frequencies in the Raman spectrum (Figure 9.13d) [55].

The combined application of LIBS and Raman microscopy, as described above, involves the use of two separate instrumentation units. However, the simultaneous application of the two techniques on a unified instrument has been demonstrated in a different case [115] and this result is indeed quite promising for achieving parallel LIBS and Raman analysis optimizing the output of the combined analytical approach.

9.8.2 LIBS and fluorescence spectroscopy

Fluorescence, or more generally photoluminescence, spectroscopy is a sensitive, non-destructive technique, shown in several cases to be capable of identifying pigments used

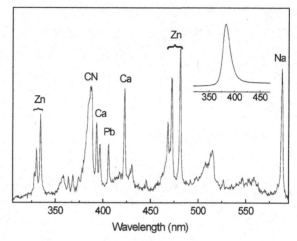

Figure 9.14. LIBS spectrum from white overpaint layer, analyzed on the St. Nicholas Byzantine icon (Figure 9.6d). Inset: fluorescence emission spectrum (λ_{exc}: 248 nm) from the same point on the icon, proving the presence of ZnO.

in painting on the basis of their characteristic molecular emission bands. The use of laser excitation leads obviously to increased sensitivity and simpler instrumentation because no monochromator is needed for selecting the excitation wavelength from a conventional white light source commonly used. The common term LIF (laser-induced fluorescence) refers to fluorescence or photoluminescence emission spectroscopy employing pulsed or cw lasers as excitation source.

Several examples of the application of LIF spectroscopy to the analysis of selected pigments that are photoluminescent have been reported in the literature [10–13, 116]. In this respect LIF analysis can be complementary to LIBS as it can provide molecular information on materials. In fact the combined use of LIBS and LIF was critical in fully characterizing the white retouching of the Byzantine icon shown in Figure 9.6d [54]. The LIBS spectrum showed that zinc was the main component of the retouching while the LIF spectrum confirmed the presence of zinc white, based on the characteristic intense narrow-band photoluminescence of ZnO at 383 nm (Figure 9.14).

A particularly interesting example of a practically simultaneous application of LIBS and LIF, which makes use of a single optical setup, has been described [51]. It involves the analysis of a model sample consisting of a cadmium yellow ($CdS \cdot ZnS \cdot BaSO_4$) paint layer applied on a white background made of gypsum ($CaSO_4 \cdot 2H_2O$) containing small amounts of zinc white (ZnO), a highly luminescent material. The application exploits the depth profile and elemental analysis capabilites of LIBS in combination with the spectroscopic analytical features of LIF. Employing the optical setup shown in Figure 9.1, the sample surface is irradiated with 355 nm pulses from the third harmonic of a Nd:YAG laser. At first, the time-integrated fluorescence/luminescence emission of the sample is collected. It is important to stress that for properly making the fluorescence measurement, the laser

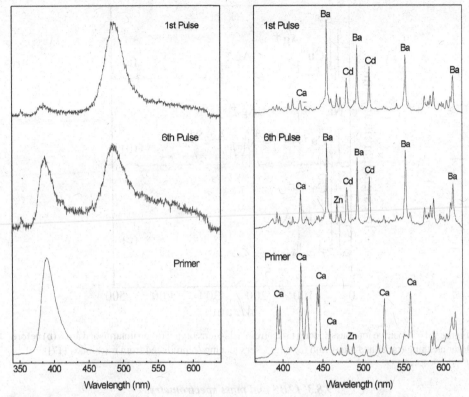

Figure 9.15. LIF (left) and LIBS (right) spectra obtained in the depth profile analysis of a model sample composed of a yellow paint layer (cadmium yellow: $CdS \cdot ZnS \cdot BaSO_4$) on top of a white background ($CaSO_4 \cdot 2H_2O$ containing low amounts of ZnO). For both LIF and LIBS spectra, $\lambda_{exc} = 355$ nm.

energy per pulse employed is much lower than that corresponding to the ablation threshold of the material examined. Following the fluorescence experiment, a single "high" energy laser pulse is delivered on the same spot and a LIBS experiment is done, where the time-resolved emission spectrum is recorded. This completes the first cycle of the measurement where two spectroscopic tools are used in the analysis of the sample. Additional cycles are repeated, in which an additional LIF measurement probes each time a newly exposed layer of the sample as a result of the ablation of material that took place in the LIBS measurement done in the previous cycle. In the example described, indicative LIF and LIBS spectra are shown (Figure 9.15); these were collected while performing a depth profiling study of the sample. A transition from the characteristic luminescence emission of CdS at *c.* 488 nm from cadmium yellow to that of ZnO at *c.* 383 nm from the background layer is seen in the LIF spectra. Similarly, atomic emission lines due to Cd, Zn, and Ba are observed in the LIBS spectra corresponding to the upper paint layer while mainly Ca and Zn emission (with only minor contribution from Cd and Ba) is recorded from the ground layer.

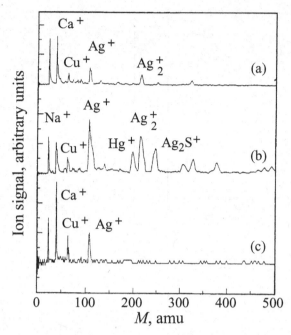

Figure 9.16. Positive ion mass spectra taken from a daguerreotype: (a) on untarnished area, (b) before laser cleaning (tarnished area), and (c) after laser cleaning (reproduced from reference [117]).

9.8.3 *LIBS and mass spectrometry*

Laser ablation–mass spectrometry relies on measuring the masses of species ejected from solid surfaces as a result of the laser ablation process. Ionized species or even neutral species (post-ionized) can be detected and their masses can be related to the material analyzed. The complementarity of LIBS, which records the optical emission resulting from laser ablation, and laser mass spectrometry has been known to scientists and was recently shown in the field of artwork analysis with a recent study on daguerreotypes [117]. Mass spectrometric analysis has identified the tarnish layer (thin dark corrosion layer appearing in the daguerreotype's surface) as consisting of silver sulfide (Ag_2S) and has shown it to be efficiently removed upon laser cleaning of the daguerreotype using the second harmonic (532 nm) of a picosecond Nd:YAG laser (Figure 9.16). It was also sensitive in detecting mercury used in the development process and gold used in the gilding process of the developed silver plate. LIBS, as already shown, was used in mapping the depth profile of the silver plate and for detecting pigment used in several colored daguerreotypes. The complementarity of the two techniques is obvious upon examination of the spectra [101, 117]. Both approaches can detect Ag, Cu, and Au quite well. However, mass spectrometry can detect Hg, S, and some organic contaminants, which LIBS can detect only with difficulty. That is, the expected S lines are in the ultraviolet and Hg is difficult to detect. In addition, molecular or organic species are usually not seen in LIBS mode.

9.9 Concluding remarks

As discussed in this chapter and illustrated with the examples presented, we can see that LIBS features several unique advantages that make its use in archaeological and artwork analysis attractive. First of all, it requires no removal of sample from the object. The analysis can be performed *in situ* and it requires only optical contact with the object. Material consumption in a typical LIBS experiment is minimal and any damage to the sample's surface is practically invisible to the naked eye. Thus LIBS can be considered as a nearly non-destructive technique. The absence of sampling and sample preparation in combination with the fact that a single laser pulse measurement is complete in less than a second offer unparalleled speed to the technique. The spatial resolution achieved by LIBS across a surface is nearly microscopic. In addition, the technique has the capability of providing depth profile information if spectra from successive laser pulses delivered at the same point are recorded individually. Finally, the equipment used is more compact and can potentially be put together in a portable or transportable unit.

The authors are grateful to Ms. K. Melessanaki for her valuable assistance in the preparation of this chapter and essential contribution in much of the work described and to Professor C. Fotakis and Dr. V. Zafiropulos for stimulating discussions. Dr. M. Doulgeridis (National Gallery of Athens) is gratefully acknowledged for making available the paintings examined. D. A. is indebted to Professor R. J. H. Clark (University College London; Raman microscopy) and Professor P. P. Betancourt (Temple University, analysis of archaeological objects). Financial support of the work at FORTH has been provided in part by the Ultraviolet Laser Facility (ULF-FORTH), through the TMR (ERB-FMGE-CT-950021) and IHP (HPRI-1999-CT-00074) programs of the EU, by the General Secretariat for Research and Technology (Greece) through projects ΕΠΕΤ II/640 and ΠΕΝΕΔ/99ΕΔ6 and by the Institute of Aegean Prehistory. J. C. M. acknowledges the use of a NATO International Collaboration Grant (CRG 950660) during the course of this work and also helpful discussions with Dr. C. Fotakis and Dr. V. Golovlev. Dennis Waters is gratefully acknowledged for the loan of several colored daguerreotypes examined. Research at ORNL is sponsored by the Office of Biological and Environmental Research, US Department of Energy (DOE) under contract number DE-AC05-96OR22464 with UT-Battelle, LLC.

9.10 References

[1] E. Ciliberto and G. Spoto (editors), Modern analytical methods in art and archaeology, *Chemical Analysis, A Series of Monographs on Analytical Chemistry and its Applications*, vol. 155, J. D. Winefordner (editor) (New York: Wiley, 2000).
[2] M. Ferreti, *Scientific Investigations of Works of Art* (Rome: ICCROM-International Centre for the Study of Preservation and the Restoration of Cultural Property, 1993).
[3] A. M. Pollard and C. Heron, *Archaeological Chemistry* (Cambridge: Royal Society of Chemistry, 1996).
[4] P. Mirti, *Ann. Chim.*, **79** (1989), 455.
[5] A. L. Beilby, *J. Chem. Educ.*, **69** (1992), 437.

[6] M. Bacci, F. Baldini, R. Carla and R. Linari, *Appl. Spectrosc.*, **45** (1991), 26.

[7] S. P. Best, R. J. H. Clark, M. A. M. Daniels, C. A. Porter and R. Withnall, *Stud. Conserv.*, **40** (1995), 31.

[8] A. Casini, F. Lotti, M. Picollo, L. Stefani and E. Buzzegoli, *Stud. Conserv.*, **44** (1999), 39.

[9] M. Bacci, UV-VIS-NIR, FT-IR and FORS spectroscopies, chapter 12 in reference [1], pp. 321–361.

[10] E. R. de la Rie, *Stud. Conserv.*, **27** (1982), 1.

[11] (a) T. Miyoshi, M. Ikeya, S. Kinoshita and T. Kushida, *Jpn. J. Appl. Phys.*, **21** (1982), 1032; (b) T. Miyoshi, *Jpn. J. Appl. Phys.*, **26** (1987), 780.

[12] D. Anglos, M. Solomidou, I. Zergioti *et al.*, *Appl. Spectrosc.*, **50** (1996), 1331.

[13] I. Borgia, R. Fantoni, C. Flamini *et al.*, *Appl. Surf. Sci.*, **127–129** (1998), 95.

[14] T. Learner, *Spectrosc. Europe*, **8** (4) (1996), 14.

[15] J. C. Shearer, D. C. Peters, G. Hoepfner and T. Newton, *Anal. Chem.*, **55** (1985), 874A.

[16] R. J. Meilunas, J. G. Bentsen and A. Steinberg, *Stud. Conserv.*, **35** (1990), 33.

[17] B. Guineau, *Stud. Conserv.*, **29** (1984), 35.

[18] R. J. H. Clark, An arts/science interface: medieval manuscripts, pigments and spectroscopy, in *Proceedings of the Royal Institution of Great Britain*, Vol. 69, P. Day (editor) (Oxford: Oxford University Press, 1998), pp. 151–167.

[19] R. J. H. Clark, *Chem. Soc. Rev.*, **24** (1995), 187.

[20] L. Burgio, R. J. H. Clark and H. Toftlund, *Acta Chem. Scand.*, **53** (1999), 181.

[21] P. Vandenabeele, B. Wheling, L. Moens *et al.*, *Anal. Chim. Acta*, **407** (2000), 261.

[22] H. G. M. Edwards and M. J. Falk, *Appl. Spectrosc*, **51** (1997), 1134.

[23] F. Cariati and S. Bruni, Raman Spectroscopy, chapter 10 in reference [1], pp. 255–278.

[24] A. Casoli and P. Mirti, *Fresenius J. Anal. Chem.*, **334** (1992), 104.

[25] J. I. Garcia Alonso, J. Ruiz Enhinar, J. A. Martinez and A. J. Criado, *Spectrosc. Eur.*, **11** (1999), 10.

[26] B. Gratuze, *J. Archaeol. Sci.*, **26** (1999), 869.

[27] W. Noll, R. Holm and L. Born, *Angew. Chem. Int. Ed.*, **14** (1975), 602.

[28] P. Mirti, *X-Ray Spectrom.*, **29** (2000), 63.

[29] V. F. Hanson, *Appl. Spectrosc.*, **27** (1973), 309.

[30] M. Mantler and M. Schreiner, *X-Ray Spectrom.*, **29** (2000), 3.

[31] R. Klockenkamper, A. van Bohlen and L. Moens, *X-Ray Spectrom.*, **29** (2000), 119.

[32] L. Moens, A. von Bohlen and P. Vandenabeele, X-ray fluorescence, chapter 4 in reference [1], pp. 55–79.

[33] S. E. Philippakis, B. Perdikatsis and T. Paradelis, *Stud. Conserv.*, **21** (1976), 143.

[34] S. E. Philippakis, B. Perdikatsis and K. Assimenos, *Stud. Conserv.*, **24** (1979), 54.

[35] J.-C. Dran, T. Calligardo and J. Salomon, Particle-induced X-ray emission, chapter 6 in reference [1], pp. 135–166.

[36] C. P. Swann, S. Ferrence and P. P. Betancourt, *Nucl. Instrum. Meth. Phys. Res. B*, **161–163** (2000), 714.

[37] N. Kallithrakas-Kontos, A. A. Katsanos, C. Potiriadis, M. Oeconomidou and J. Touratsoglou, *Nucl. Instrum. Meth. Phys. Res. B*, **75** (1993), 440.

[38] H. Neff, Neutron activation analysis for provenance determination in archaeology, chapter 5 in reference [1], pp. 81–134.

[39] V. Kilikoglou, Y. Bassiakos, A. P. Grimanis *et al.*, *J. Archaeol. Sci.*, **23** (1996), 343.

[40] G. Spoto, *Thermochim. Acta*, **365** (2000), 157.

[41] J. S. Mills and R. White, *Stud. Conserv.*, **22** (1977), 87.

[42] H. H. Hairfield Jr. and E. M. Hairfield, *Anal. Chem.*, **62** (1990), 41A.

[43] G. A. van der Doelen, K. J. van der Berg and J. J. Boon, *Stud. Conserv.*, **43** (1998), 249.

[44] S. O Troja and R. G. Roberts, Luminescence dating, chapter 18 in reference [1], pp. 585–640.

[45] R. Grun, Electron spin resonance dating, chapter 19 in reference [1], pp. 641–679.

[46] R. E. M. Hedges, Radiocarbon dating, chapter 16 in reference [1], pp. 465–502.

[47] G. A. Wagner, Isotope analysis, dating, and provenance methods, chapter 15 in reference [1], pp. 445–464.

[48] N. H. Gale and Z. Stos-Gale, Lead isotope analysis applied to provenance studies, chapter 17 in reference [1], pp. 503–584.

[49] L. J. Radziemski and D. A. Cremers (editors), Spectrochemical analysis using laser plasma excitation, in *Laser-Induced Plasmas and Applications*, chapter 7 (New York: Marcel Dekker, 1989).

[50] R. S. Adrain and J. Watson, *J. Phys. D: Appl. Phys.*, **17** (1984), 1915.

[51] D. Anglos, *Appl. Spectrosc.*, **55** (2001), 186A.

[52] D. Anglos, S. Couris and C. Fotakis, *Appl. Spectrosc.*, **51** (1997), 1025.

[53] D. Anglos, C. Balas and C. Fotakis, *Am. Lab.*, **31** (1999), 60.

[54] L. Burgio, R. J. H. Clark, T. Stratoudaki, M. Doulgeridis and D. Anglos, *Appl. Spectrosc.*, **54** (2000), 463.

[55] L. Burgio, K. Melessanaki, M. Doulgeridis, R. J. H. Clark and D. Anglos, *Spectrochim. Acta*, **B56** (2001), 905.

[56] M. Castillejo, M. Martin, D. Silva *et al.*, *J. Molec. Struct.*, **550–551** (2000), 191.

[57] K. Melessanaki, V. Papadakis, C. Balas and D. Anglos, *Spectrochim. Acta*, **B56** (2001), 2337.

[58] I. Borgia, L. M. F. Burgio, M. Corsi *et al.*, *J. Cult. Herit.*, **1** (2000), S281.

[59] M. Corsi, V. Palleschi, A. Salvetti and E. Tognoni, *Res. Adv. Appl. Spectrosc.*, **1** (2000), 41, and references therein.

[60] A. Ciucci, V. Palleschi, S. Rastelli *et al.*, in *Proceedings of the Fifth International Conference on Optics within the Life Sciences – OWLS-V*, C. Fotakis, T. G. Papazoglou and C. Calpouzos (editors) (Berlin: Springer-Verlag, 2000), pp. 163–169.

[61] M. Bicchieri, M. Nardone, P. A. Russo *et al.*, *Spectrochim. Acta*, **B56** (2001), 915.

[62] L. Bolognesi, M. Corsi, V. Palleschi *et al.*, in *Proceedings of the Second International Congress on Science and Technology for the Safeguard of Cultural Heritage in the Mediterranean Basin* (Paris: Elsevier, 1999) pp. 431–436.

[63] F. Colao, R. Fantoni, V. Lazic and V. Spizzichino, *Spectrochim. Acta*, **B57** (2002), 1219.

[64] I. Gobernado-Mitre, A. C. Prieto, V. Zafiropulos, Y. Spetsidou and C. Fotakis, *Appl. Spectrosc.*, **51** (1997), 1125.

[65] P. V. Maravelaki, V. Zafiropulos, V. Kylikoglou, M. Kalaitzaki and C. Fotakis, *Spectrochim. Acta*, **B52** (1997), 41.

[66] S. Klein, T. Stratoudaki, V. Zafiropulos *et al.*, *Appl. Phys.*, **A69** (1999), 441.

[67] L. Moenke-Blackenburg, *Prog. Anal. Spectrosc.*, **9** (1986), 335, and references therein.

[68] A. Roy, *Nat. Gall. Tech. Bull.*, **3** (1979), 43.

[69] F. Brech and L. Cross, *Appl. Spectrosc.*, **16** (1962), 59.

[70] V. Majidi and M. R. Joseph, *Crit. Rev. Anal. Chem.*, **23** (1992), 143.

[71] D. A. Rusak, B. C. Castle, B. W. Smith and J. D. Winefordner, *Crit. Rev. Anal. Chem.*, **27** (1997), 257.

[72] I. Schechter, *Rev. Anal. Chem.*, **16** (1997), 173.

[73] D. A. Rusak, B. C. Castle, B. W. Smith and J. D. Winefordner, *Trends Anal. Chem.*, **17** (1998), 453.

[74] S. Palanco, J. M. Baena and J. J. Laserna, *Spectrochim. Acta*, **B57** (2002), 591.

[75] I. B. Gornushkin, B. W. Smith, H. Nasajpour and J. D. Winefordner, *Anal. Chem.*, **71** (1999), 5157.

[76] D. Body and B. L. Chadwick, *Rev. Sci. Instr.*, **72** (2000), 1625.

[77] D. Body and B. L. Chadwick, *Spectrochim. Acta*, **B56** (2001), 725.

[78] H. E. Bauer, F. Leis and K. Niemax, *Spectrochim. Acta*, **B53** (1998), 1815.

[79] C. Haisch, U. Panne and R. Niessner, *Spectrochim. Acta*, **B53** (1998), 1657.

[80] M. Corsi, G. Crisoforetti, V. Palleschi, A. Salvetti and E. Tognoni, *Eur. Phys. J., D* **13** (2001), 373.

[81] A. Uhl, K. Loebe and L. Kreuchwig, *Spectrochim. Acta Part B*, **56** (2001), 795.

[82] K. Y. Yamamoto, D. A. Cremers, M. J. Ferris and L. E. Foster, *Appl. Spectrosc.*, **50** (1996), 222.

[83] B. C. Castle, A. K. Night, K. Visser, B. W. Smith and J. D. Winefordner *J. Anal. Atom. Spectrom.*, **13** (1998), 589.

[84] S. Rosenwasser, G. Assimielis, B. Bromley *et al.*, *Spectrochim. Acta Part B*, **56** (2001), 707.

[85] J. C. Miller (editor), *Laser Ablation: Principles and Applications*, Springer Series in Material Science, Vol. 28 (Berlin: Springer-Verlag, 1994).

[86] J. C. Miller and R. F. Haglund Jr. (editors), *Laser Ablation and Desorption*, Experimental Methods in the Physical Sciences, Vol. 30 (New York: Academic Press, 1998).

[87] K. Niemax, *Fresenius J. Anal. Chem.*, **370** (2001), 332.

[88] T. Stratoudaki, A. Manousaki, K. Melesanaki, G. Orial and V. Zafiropulos, *Surf. Eng. J.*, **17** (2001), 249.

[89] P. Pouli, D. C. Emmony, C. E. Madden and I. Sutherland, *Appl. Surf. Sci.*, **173** (2001), 252.

[90] T. Kim, C. T. Lin and Y. Yoon, *J. Phys. Chem. B*, **102** (1998), 4284.

[91] D. Romero and J. J. Laserna, *J. Anal. Atom. Spectrom.*, **13** (1998), 557.

[92] D. R. Anderson, C. W. McLeod, T. English and A. T. Smith, *Appl. Spectrosc.*, **49** (1995), 691.

[93] J. M. Vadillo, C. G. Garcia, S. Palanco and J. J. Laserna, *J. Anal. Atom. Spectrom.*, **13** (1998), 793.

[94] R. Wishburn, I. Schechter, R. Niessner, H. Schroder and K. Lompa, *Anal. Chem.*, **66** (1994), 2964.

[95] V. Bulatov, R. Krasniker and I. Schechter, *Anal. Chem.*, **70** (1998), 5302.

[96] A. Ciucci, M. Corsi, V. Palleschi *et al.*, *Appl. Spectrosc.*, **53** (1999), 960.

[97] (a) R. J. Gettens and G. L. Stout, *Painting Materials* (New York: Dover, 1956); (b) H. G. Friedstein, *J. Chem. Educ.*, **58** (1981), 291.

[98] A. Roy (editor), *Artists' Pigments Vol. 2* (Washington, DC: National Gallery of Art, 1993).

[99] E. West Fitzhugh (editor), *Artists' Pigments Vol. 3* (Washington, DC: National Gallery of Art, 1997).

[100] M. S. Barger and W. B. White, *The Daguerreotype* (Baltimore: The Johns Hopkins University Press, 1991).

[101] D. Anglos, V. Zafiropulos, K. Melessanaki, M. J. Gresalfi and J. C. Miller, *Appl. Spectrosc.*, **56** (2002), 423.

[102] P. Maravelaki-Kalaitzaki, D. Anglos, V. Kilikoglou and V. Zafiropulos, *Spectrochim. Actà*, **B56** (2001), 887.

[103] S. Morel, M. Durand, P. Adam and J. Amouroux, Analysis of stained-glass windows by time-resolved laser induced breakdown spectroscopy, in *Book of Abstracts of 1st International Conference on Laser Induced Plasma Spectroscopy and Applications*, Pisa, Italy (2000), p. 77.

[104] (a) J. M. Vadillo and J. J. Laserna, *Talanta*, **43** (1996), 1149; (b) J. M. Vadillo, I. Vadillo, F. Carrasco and J. J. Laserna, *Fresenius J. Anal. Chem.*, **361** (1998), 119.

[105] M. K. Sanford, D. B. Repke and A. L. Earle, Elemental analysis of human bone from Carthage: a pilot study, in *The Circus and a Byzantine Cemetery at Carthage*,. *Vol. I*, J. H. Humphrey (editor), chapter 9 (Ann Arbor: University of Michigan Press, 1988).

[106] J. H. Kyle, *J. Archaeol. Sci.*, **13** (1984), 403.

[107] L. L. Klepinger, J. K. Kuhn and W. S. Williams, *Amer. J. Phys. Anthropol.*, **70** (1986), 325.

[108] O. Samek, D. C. S. Beddows, H. H. Telle *et al.*, *Spectrochim. Acta*, **B56** (2001), 865.

[109] M. Cooper, *Laser Cleaning* (Oxford: Butterworth-Heinemann, 1998).

[110] R. Salimbeni, R. Pini and S. Siano, *Spectrochim. Acta*, **B56** (2001), 877.

[111] A. Ceccarini, M. P. Colombini, M. Rosato *et al.*, Characterization of old lakes in old paintings, in *Book of abstracts of 1st International Conference on Laser Induced Plasma Spectroscopy and Applications*, Pisa, Italy (2000), p. 72.

[112] I. M. Bell, R. J. H. Clark and P. J. Gibbs, *Spectrochim. Acta*, **A53** (1997), 2159.

[113] L. Burgio and R. J. H. Clark, *Spectrochim. Acta*, **A57** (2001), 1491.

[114] R. J. H. Clark, *C. R. Chimie*, **5** (2002), 7.

[115] B. J. Marquardt, D. N. Stratis, D. A. Cremers and S. Michael Angel, *Appl. Spectrosc.*, **52** (1998), 1148.

[116] G. Pozza, D. Ajo, G, Ciari, F. De Zuane and M. Favaro, *J. Cult. Herit.*, **1** (2000), 393.

[117] D. L. Hogan, V. V. Golovlev, M. J. Gresalfi *et al.*, *Appl. Spectrosc.*, **53** (1999), 1161.

10

Civilian and military environmental contamination studies using LIBS

J. P. Singh, F. Y. Yueh and V. N. Rai

Diagnostics Instrumentation and Analysis Laboratory (DIAL), Mississippi State University, USA

R. Harmon

US Army Research Laboratory, Research Triangle Park, USA

S. Beaton and P. French

ADA Technologies, Inc., USA

F. C. DeLucia, Jr., B. Peterson, K. L. McNesby and A. W. Miziolek

US Army Research Laboratory, Aberdeen Proving Ground, USA

10.1 Introduction

Laser-induced breakdown spectroscopy (LIBS) has demonstrated its capability in quantitative determination of elemental composition in various samples in laboratories for decades. Recently, the interest in applying LIBS to detect various hazardous materials in the environment has grown rapidly. This chapter reviews some recent work related to environmental contamination studies using LIBS. Two portable LIBS instruments developed for field application are described in this chapter. The first instrument has been used by the US Army Research Laboratory to analyze the contaminated soil from Army sites and also for other geological applications. The other system was used by Mississippi State University to detect the resources conservation and recovery act (RCRA) metals in the off-gas of industrial plants and in liquids. The practical problems with LIBS application in environmental application such as calibration and sensitivity are also discussed in this chapter. The concept of using a new generation of broadband spectrometers to improve LIBS' capability in monitoring multiple emission lines of the same element to improve the detection limit and other signal enhancement techniques is addressed. The initial result of LIBS application in environmental problems is encouraging. We believe that the performance of LIBS in environmental application will continuously improve with the commercial development of various components in LIBS detection system.

Laser-Induced Breakdown Spectroscopy: Fundamentals and Applications, ed. Andrzej W. Miziolek, Vincenzo Palleschi and Israel Schechter. Published by Cambridge University Press. © A. W. Miziolek, V. Palleschi and I. Schechter 2006. A. W. Miziolek's contributions are a work of the United States Government and are not protected by copyright in the United States.

LIBS is a chemical sensor technology with a capability for real-time *in situ* analysis that was first demonstrated in the laboratory over two decades ago and is now at the threshold of widespread commercialization. Over the past ten years, the LIBS technique has proven capable of detecting many hazardous substances of environmental quality interest and, more recently, military concern. There has been a significant growth in the number of LIBS applications to environmental problems, with current interest particularly focused on the analysis of liquids and aerosols [1–6], spatially resolved microanalysis [7], and toxic chemical surveying [8–12].

Early environmental applications of LIBS were directed at trace pollutants in air [13]. Soils and related natural materials have been the focus of more recent study. For example, Wisbrun *et al.* used time-resolved LIBS in a laboratory setting to study physical factors such as aerosol production, crater formation, size effects, timing effects, laser intensity, and humidity, which affect the ability to undertake quantitative elemental analysis and developed a rapid screening method for heavy metals (Cd, Cr, Cu, Ni, Pb, and Zn) based upon their findings [8]. Good calibration plots were achieved, with detection limits in the 10 p.p.m. range. Subsequently, Ciucci *et al.* undertook a detailed laboratory study of 19 heavy metals in two certified soil samples in which the elemental concentrations were well established [9]. The results of this study demonstrated the identification of most elements in both the ultraviolet and visible spectral regions, with detection limits for some elements extending as low as 500 p.p.b. The potential for LIBS as a field analytical tool has been realized based upon the pioneering work of Cremers *et al.* [14] and Yamamoto *et al.* [15]. A LIBS system has been developed for deployment with a cone penetrometer for the *in situ*, subsurface detection of metals in contaminated soils [16, 17]. The analysis of Pb in paint has also been demonstrated by using laboratory and field LIBS systems [12, 18].

In this chapter, we describe the environmental applications of LIBS based on the work done by Diagnostics Instrumentation and Analysis Laboratory (DIAL) at Mississippi State University and US Army Research Laboratory (ARL). LIBS groups at DIAL and ARL are both very active in environmental applications. ARL has been investigating a variety of environmental, geological, forensic, and military applications. The LIBS system used in ARL's field surveys is described in Section 10.2.1. The results of the field survey are given in Section 10.2.2. The method to calibrate the field system is discussed in Section 10.2.3. An exploration of other geological applications with this portable LIBS system is described in Section 10.2.4. DIAL has developed a LIBS system for off-gas application, which had been evaluated as a continuous emission monitor at an EPA test facility. The description of DIAL LIBS system is given in Section 10.3.1. The result of the CEM test is summarized in Section 10.3.2. The sensitivity of the DIAL LIBS system was studied under bulk and liquid jet configuration. Different techniques to enhance LIBS signal for liquid application were also explored. These study results are also presented in Section 10.3.3.

10.2 Applications of the ADA portable LIBS unit

10.2.1 The field-portable LIBS system

ADA Technologies, Inc., of Littleton, Colorado, has developed and manufactured a field-portable LIBS instrument aimed specifically at the detection of Pb in paint and soil under an SBIR program award from the ARL Army Research Office. The prototype ADA field LIBS system was validated under laboratory conditions through the analysis of samples with differing Pb contents and then extensively tested and compared with a laboratory bench-top system at the Army Research Laboratory. Continuing developmental work by ADA has significantly enhanced the performance of this field-portable LIBS system.

The ADA field-portable LIBS instrument consists of two parts: (i) a laser-bearing sample probe with a safety lock that also contains an optical fiber for signal detection, and (ii) a central detector/analyzer unit that houses the spectrometer/detector, timing, power, and data acquisition and analysis equipment (see Figure 10.1). The laser power cables and fiber optics connect the two units. The complete LIBS instrument is contained within a 23 cm × 51 cm × 38 cm aluminum case.

The laser utilized in the LIBS unit is passively Q-switched and provides a nominal 17 mJ energy per pulse at 1064 nm. The nominal single pulse lamp energy threshold for lasing is approximately 5 J and the pulse width is 4 ns. The beam diameter is 3 mm, with a beam divergence of 1 mr. Upon focusing by a 45 mm focal length lens, the laser spot size is 60 μm. The laser is fired once every 4 s. The instrument uses a 0.5 mm round bundle to linear array fiber optic to transport the light from the plasma to the spectrograph. No focusing lens is used in front of the fiber optic to collect the light and the width of the individual fibers (100 μm, N.A. 0.22) defines the aperture for the light input to the spectrograph. Additional information about this portable LIBS system is provided by Wainner *et al.* [12].

Two portable units were built by ADA during the SBIR technology development program. In one configuration, the fiber optic cable collects and transmits the light to a single, high-resolution spectrograph (1/6 m f.l., 2400 g mm^{-1}, 250 nm blaze) with a thermoelectrically cooled, 250 ×12 element CCD (24 μm pixel) or with a 512 element PDA. This configuration had a 20–30 nm tunable spectral range and is ideal for spectral region of interest analysis. In the second configuration, which was designed for full-spectrum analysis, the fiber optic cable collects and transmits the light to an assembly of four small spectrometers (125 nm band pass each with about 1.5 nm resolution) with either LPDA or CCD detectors, for a total spectral coverage of 500 nm. Both designs have certain advantages. The LIBS instrument with a fixed-range spectrograph is of lower cost. It is ideal for the detection and quantitative analysis of a single element, such as a toxic metal contaminant in soil or collected on air filters, because it can capture, at high resolution, just the small wavelength region of the LIBS spectrum where the metal emits. The multi-spectrometer LIBS instrument is more versatile, but is more expensive. This design can capture a large portion of the LIBS spectrum from each individual microplasma event, albeit at a lower resolution. It can be used for a variety of applications, such as the detection and identification of multiple contaminant species, or trace amounts of explosive residues and chemical warfare agents, because the

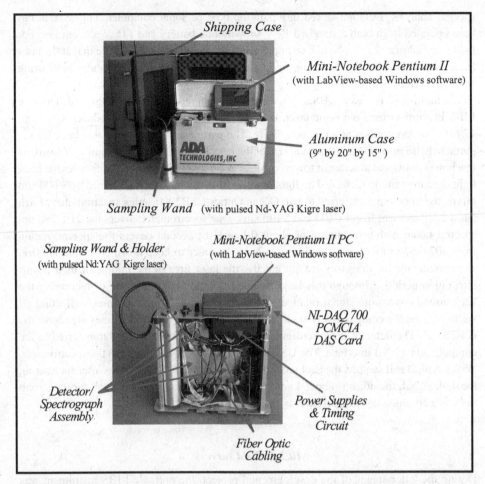

Shipping Case

Mini-Notebook Pentium II
(with LabView-based Windows software)

Aluminum Case
(9" by 20" by 15")

Sampling Wand (with pulsed Nd-YAG Kigre laser)

Sampling Wand & Holder
(with pulsed Nd:YAG Kigre laser)

Mini-Notebook Pentium II PC
(with LabView-based Windows software)

NI-DAQ 700
PCMCIA
DAS Card

Detector/
Spectrograph
Assembly

Power Supplies
& Timing
Circuit

Fiber Optic
Cabling

Figure 10.1. Upper illustration: rear, cutaway view of the ADA portable LIBS analyzer with major components indicated. This system comes in two configurations, a single spectograph model for high-resolution spectral region of interest analysis and a four spectrometer model for full spectrum analysis. Lower illustration: front view of the LIBS instrument and shipping case.

full spectrum contains information from most constituent elements within the material sampled.

In both configurations, the LIBS analyzer is run off an independent clock circuit, which triggers the laser, the detector, and the data collection system. This design separates the power systems for the detector and the laser in order to avoid false triggers of the data acquisition system owing to the high voltage power supply. The detectors are read, but not recorded, immediately prior to the plasma event to clear any readings from ambient light since the previous plasma event. The detectors are then read and recorded immediately after the plasma event (i.e. within a few milliseconds) and these readings form the basis for the spectral analysis. Data are stored in a palmtop-type personal computer. This system has been operated from both a standard 12 V snowmobile battery and 115 V a.c. current. For field use the laser/fiber probe unit terminates in a flat plate with a small hole in it at the laser focal point. This plate is also connected to a safety pressure switch to inhibit laser firing when no surface is present.

The multi-spectrometer LIBS unit recently has been redesigned and upgraded. The new LIBS instrument uses the same laser, a Kigre model 367. This laser produces 1064 nm, 17 mJ, 4 ns long pulses at a rate of 1/3 Hz. The particular advantage of this laser for the portable LIBS instrument is the small size of the laser head ($100\,\text{mm} \times 30\,\text{mm} \times 20\,\text{mm}$), its low power draw, and absence of forced cooling. A 50 mm focal length lens focuses the laser light onto the sample surface. The light emitted from the plasma is collected by a 600 μm bifurcated fiber and transferred to two Ocean Optics USB2000 series spectrometers. Each has a 2048 element linear CCD and 5 μm slits. One spectrometer covers the 210–280 nm spectral range with a resolution of about 0.1 nm, the second covers the spectral region from 307–457 nm with a resolution of 0.2 nm. In order to keep down cost, weight, size, and complexity, the detectors are not gated – the laser fires in the middle of their 50 ms integration period. Although this lack of time resolution has the effect of increasing the background continuum signal relative to that for elemental emission lines, collecting all the light from the comparatively small plasma ensures that the atomic lines are above the noise level. The detector data are digitized within the spectrometer and transferred to the computer via a USB interface. The USB interface also supplies power to the spectrometer. With a typical soil sample, the background signal is around 50–100 counts after subtracting the dark signal, the noise is about 4 counts, and the atomic emission signal, for the strong lines of a common constituent such as Al, can exceed 1500 counts.

10.2.2 Field surveys

During the latter stages of the developmental project, the portable LIBS instrument was field tested at the Sierra Army Depot in California and at Fort Carson Military Reservation (FCMR) in Colorado for the detection of Pb in contaminated soils and Pb in paint on painted surfaces [12]. The LIBS instrument was also able to detect Pb collected on PM-10 air filters from local air monitoring stations in Panama City, Panama. Subsequently, additional non-Pb applications have been explored at Yuma Proving Ground in Arizona and the Idaho Springs–Central City mining district of Colorado.

Sierra Army Depot, CA

The Sierra Army Depot (SIAD) is located in an arid setting in northeastern California east of the Sierra Nevada mountains. The soil at SIAD consists of a thin surficial cover of highly permeable, windblown sand that overlies low-permeability unconsolidated carbonate-rich lacustrine and fluvial lake bed sediments and alluvial fan deposits of sand, silt, and clay. The location at SIAD selected for the detailed LIBS field test is an area known as the

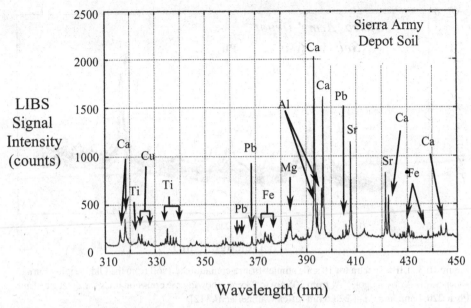

Figure 10.2. Cumulative 10-shot LIBS spectrum from 310 nm to 450 nm for a Pb-contaminated soil from the Old Popping Furnace site at Sierra Army Depot, CA.

"Old Popping Furnace" site. Here, a furnace was used from the 1940s to mid 1950s for the demilitarization of small arms ammunition by burning. Metal casings and Pb were recovered from the furnace, which was operated without air pollution controls, and the ash and solid furnace residues were buried locally in the soil at shallow depths. The furnace was dismantled and removed some time after operations ceased, but the concrete pad on which the furnace was situated remains at the site. There is also physical evidence that small amounts of small arms ammunition were burned on the surface in the area immediately surrounding the furnace. Thus, it is expected that the Pb contamination in the soils of the Old Popping Furnace site is most likely particulate materials in the form of PbO from furnace ash and solid furnace residues.

The lacustrine soils at SIAD are a mixture of carbonate and aluminosilicate clays. A LIBS spectrum in the range 310–450 nm for a typical surface soil is presented in Figure 10.2. In addition to Pb, which is a major contaminant at SIAD, the emission lines identified are Ca, Mg, Sr, Al, Fe, and Ti, as expected for such a soil. Oxygen, which typically comprises the bulk by weight of any silicate or carbonate rock, or derivative soil, is not present in the LIBS spectrum because its major peaks at 130–131 nm and 777–778 nm are outside the 310–450 nm portion of the spectrum illustrated in Figure 10.2. Similarly, Si, which is a major component of the clay component of the SIAD soil, is not seen in Figure 10.2 because its major lines are at 250–253, 288, and 299 nm.

Bulk surface soil and samples from the Old Popping Furnace site have been analyzed for concentrations of 23 trace metals including Pb during an environmental survey conducted by

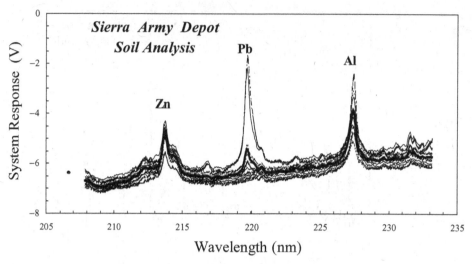

Figure 10.3. LIBS spectra for 10 soil samples from a contaminated soil from the Old Popping Furnace site at Sierra Army Depot, CA, from 205 nm to 235 nm showing the emission lines for Zn at 213.8 nm, Pb at 220.4 nm, and Al at 228.6 nm (after Wainner *et al.* [12].

Harding Lawson Associates under contract to SIAD [19]. Also, 11 subsurface samples from five boreholes were analyzed for their Pb content. This survey recognized distinctly elevated and highly variable concentrations of several metals such as Pb, Cu, and Zn across the Old Popping Furnace site compared with background soil levels. Pb is the most prominent contaminant at the Old Popping Furnace site soils, extending up to 180 000 p.p.m. Pb compared with background soil Pb levels, which typically are in the range of 2–3 p.p.m. In general, Pb concentrations were highest to the east of the furnace site in the prevailing downwind direction and decreased with distance from the furnace site, an observation that is consistent with the deposition of Pb on the soil surface from furnace stack emissions. However, the highest soil lead levels were observed in an area some 250 m southeast of the concrete pad, suggesting that this location may have been a site of furnace ash and residue disposal. The subsurface soil samples exhibited much lower Pb values, with concentrations declining to local ambient soil concentrations below 2 m depth.

The portable LIBS system was employed to conduct a Pb survey at the SIAD Old Popping Furnace site. Five locations were sampled. Individual soil samples were collected over an area ~5 cm and to a depth of ~1 cm, homogenized, and sieved to remove large particles. A split of each sample was formed into a firm pellet in an aluminum dish by using a small hydraulic press. The head of the LIBS analyzer was placed on the soil pellet for analysis, the laser fired, and the laser head then moved incrementally by hand to a new spot on the pellet for the next shot. Two sets of tests were made: (i) one which involved the collection of a LIBS spectrum for 10 localities across the entire Old Popping Furnace area based upon the Pb emission line at 220.4 nm (Figure 10.3) and (ii) a second, Pb-only survey along a southeast traverse using the Pb emission line at 405.8 nm.

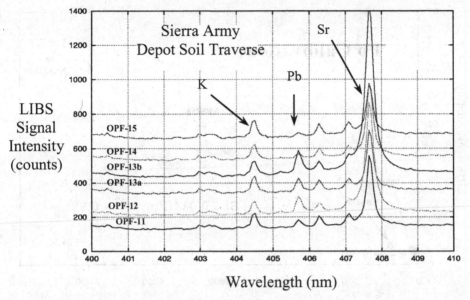

Figure 10.4. Ten-shot average LIBS spectra from 310 nm to 450 nm for Pb-contaminated soils along a 120 m traverse southeast from the Old Popping Furnace concrete pad. Note the variable intensity of the 405.8 nm Pb emission line along the traverse compared with the constant intensity of the 407.7 nm Sr emission line peak.

Ten single-shot LIBS emission spectra for a soil from the Old Popping Furnace area are plotted in Figure 10.3 as functions of wavelength for the 205–235 nm portion of the spectrum. The thick line in the figure is the average of the 10 analyses. Three distinct peaks are present: Zn at 213.8 nm, Pb at 220.4 nm, and Al at 228.2 nm. Al is a primary component of the aluminosilicate component of the lacustrine silts and clay soils at SIAD. The other two metal elements are anthropogenic contaminants derived from the burning of small arms munitions; Pb from the bullets and Zn from the brass shell casings.

The results of the LIBS analysis along the Old Popping Furnace southeast traverse are presented in Figure 10.4, where emission intensity for the 405.8 nm wavelength Pb line is shown as a function of shot number for the five sample sites. The strong Pb enrichments that characterize samples OPF12 and OPF13 are clearly detected by the portable LIBS system. As illustrated by the two samples from the OPF13 location (labeled 13a and 13b), which were collected <2 cm apart, the strong signal variation seen in the LIBS spectra for these two samples is a result of the heterogeneous distribution of Pb in the soil. This heterogeneity is most likely the result of the presence of Pb as small but discrete particles at this location, where the disposal of furnace ash is likely to have occurred.

10.2.3 Calibration

Following the developmental field tests described above, an initial calibration of the portable LIBS analyzer was undertaken for Pb using the spectral line at 405.8 nm. Twelve soil

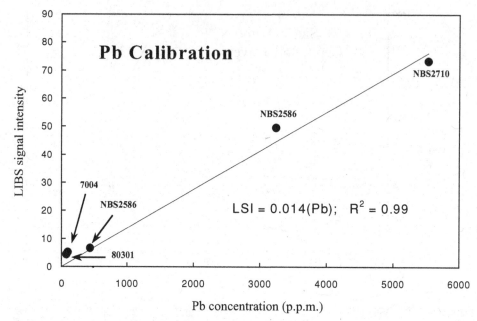

Figure 10.5. Calibration of Pb in soil standards. The Pb concentrations in the 12 standards ranged from 12 p.p.m. to 5532 p.p.m. Only the 5 samples with Pb contents in excess of 70 p.p.m. produced a signal significantly different from the background continuum emission to permit measurement.

standards were used in the calibration: five from the US National Institute of Technology and Standards (SRM-1646, SRM-2586, SRM-2587, SRM-2709, and SRM2710), two from the Canadian Certified Reference Materials Project (Till-3 and Till-4), four from the Czech Metrological Institute (CRM-7001, CRM-7002, CRM-7003, and CRM-7004), and one from the China National Analysis Center for Iron and Steel (NCS-DC-80301). Calibrations were performed on pressed soil pellets that had been dried at 105 °C for 3 h prior to analysis with 10 laser shots taken at different spots on each standard and accumulated. Net line intensities were calculated by subtracting the background intensity. The calibration curve in Figure 10.5 was constructed by plotting the net intensity for the 405.8 nm Pb line versus the known Pb concentration in the standard.

The calibration shown in Figure 10.5 was used to estimate the Pb contents of nine soil samples from the Old Popping Furnace area at SIAD, a paint standard from Los Alamos National Laboratory, and a mine dump soil from a Pb-mineralized area in Central City, Colorado (see discussion below). We found that the difference between the expected and estimated Pb concentrations for the LANL paint standard is <20%. It was also observed that the samples from the Old Popping Furnace traverse exhibit the same relative pattern of Pb variation, although a large difference in Pb content is observed for the samples of highest Pb content. Several factors may contribute to this discrepancy, given the reasonable results for the LANL paint standard. First, the soils at the Old Popping Furnace site are

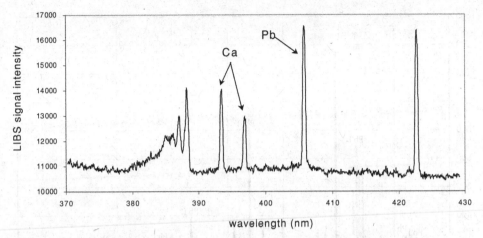

Figure 10.6. Single-shot spectrum of 1.04 p.p.m. Pb^{2+} (using $Pb(NO_3)_2$) dissolved in deionized water after adsorption onto ion-exchange resin pellets – Purolite S-930. Pellets were filtered and dried before LIBS analysis.

highly heterogeneous with respect to Pb as a result of the way that the furnace ash and solid residues were disposed of randomly across the area. Second, the quality of the Old Popping Furnace soil analyses is not known, but values in excess of 100 000 p.p.m. Pb reported for some Old Popping Furnace site samples seem unusually high given that no distinct difference in density is observed for soils across the area. Finally, small differences in pellet thickness and surface roughness may have contributed to an uneven plasma development, which can affect analytical performance [9]. Thus, it may be necessary to account for optical and thermal properties of the substrate and the influence of laser performance on measured Pb peak intensities in attempting quantitative LIBS analysis.

Current research is addressing the important issues of sensitivity and calibration. In order to detect Pb at sub-p.p.m. concentration levels such as might occur in contaminated groundwater, the use of concentrating agents has been considered. In particular, the commercially available (Purolite S-930) ion exchange resin was added to a low concentration (1.04 p.p.m.) aqueous solution of Pb $(NO_3)_2$. Figure 10.6 shows the LIBS spectrum taken of the ion-exchange pellet surface after filtering and drying the pellets. Clearly, the Pb signal at 406 nm is significantly intense and indicates successful binding and concentration of soluble Pb on the pellet surface. The use of ion-exchange resins, of course, adds an additional step in the analysis. However, this step is quite simple, straightforward, and effective. Recently there has been a significant advance in LIBS instrumentation with the commercial development of a multiple-spectrometer system (LIBS 2000+ broadband spectrometer by Ocean Optics, Inc.) that allows for broadband detection of microplasma emission in the spectral range 200–980 nm. The new generation of echelle spectrometers also offer broad spectral coverage with high resolution (0.1 nm range). Figure 10.7 shows the broadband LIBS spectrum of solid Pb. Clearly the use of such broadband spectrometers will greatly

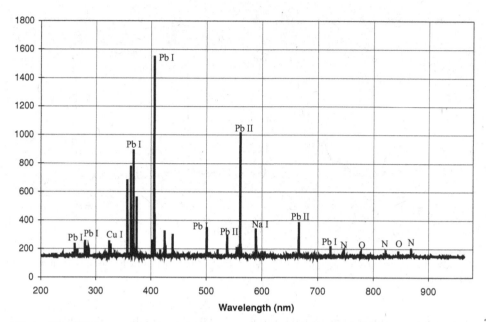

Figure 10.7. LIBS spectrum of a lead fishing weight. Spectrum was obtained using a broadband spectrometer. (LIBS 2000+ Ocean Optics, Inc.). The N and O peaks are from laboratory air.

aid in analyses of metal contamination in the future by: (1) providing the flexibility of monitoring multiple emission lines for a given element, and (2) allowing the monitoring of multiple metals simultaneously. This new technology will improve both the speed and accuracy of future LIBS analyses of contamination.

10.2.4 Other geological applications

Desert varnish

Rock varnish is a dark shiny patina, typically a few tens to a hundred micrometers in thickness, that accumulates in discrete laminae continuously and slowly on rock surfaces in semi-arid and arid regions [20]. It is a common feature on rock ledges and rock faces over which water has flowed and is a ubiquitous characteristic of desert pavements, stony surfaces composed of a thin layer of tightly packed gravel that rest upon a fine, clast-free soil layer. Typically, desert varnish consists of iron and manganese oxides, oxides of other trace elements, silica, clay minerals [21], with Mn–Fe oxide layers alternating with layers of detrital silicate and carbonate material. It is an important geological material because variations in the composition, thickness, and rate of deposition of the micro-laminations of desert varnish can provide important geomorphologic and paleoclimatic information [22].

We have conducted an experiment to determine the applicability of LIBS analysis to the study of desert varnish, which is widespread on military installations in the western

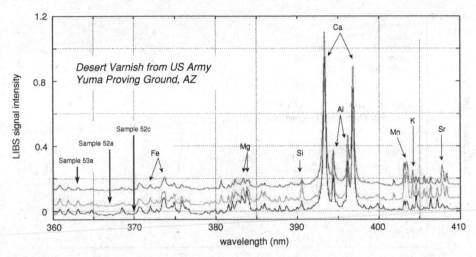

Figure 10.8. Single-shot LIBS spectra from 360 nm to 410 nm for the top surfaces of desert varnish coating on cobbles from US Army Yuma Proving Ground, AZ.

USA. The concept behind our experiment was that the LIBS system could be used to successively ablate the coating on a desert pavement pebble on a laser shot by shot as a means of determining both the chemical character and thickness of the varnish. Figure 10.8 shows LIBS spectra in the range 360–410 nm for the top surface of three varnished cobbles (samples 52a, 52c, and 53a) from desert pavement at US Army Yuma Proving Ground in the desert of southwestern Arizona. Figure 10.9 shows cumulative five-shot spectra for sample 52c as the laser was used to successively ablate through the varnish coating on the cobble. In both diagrams, the spectra for the different samples were normalized to the largest Ca line at 393.4 nm, then offset vertically for visibility.

Elemental emission lines for Mn and Fe from the oxide component and Mg, Si, Al, Ca, K, and Sr from the mixed silicate-carbonate detrital component are clearly resolved in both figures. The three spectra in Figure 10.8 are similar, attesting to the broad compositional similarity of the varnish layer presently forming on desert pavement surfaces at Yuma Proving Ground. The four spectra in Figure 10.9 were each cumulates of five laser shots taken on the top surface of the cobble (shots 1–5) and at depths of 21–25, 201–205, and 501–505 shots as the laser was repetitively fired on a single spot to ablate through the varnish coating. Although the same set of elemental emission lines are present, there are some important differences observed among the four spectra. The intensity of the Mn peak is greatest in the spectra from the surface layer that has formed under present-day climate conditions. Si, Al, and Mg are strongly enriched in layer 2 and in the lowest layer defined by shots 501–505, indicating the presence of a detrital silicate layer at these depths. The result of our very preliminary work is highly encouraging and, once we have calibrated the ablation characteristics of the Kigre laser, it should be possible to determine the thickness

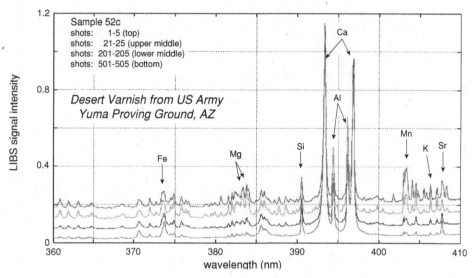

Figure 10.9. Five-shot averaged LIBS spectra from 360 nm to 410 nm acquired during successive laser ablation of the varnish coating on cobble sample 52c from US Army Yuma Proving Ground, AZ.

of varnish layers, which will open the possibility of using the portable LIBS unit as a stratigraphic and paleoclimatic tool in desert areas.

Hydrothermal alteration

Central City is one of the most economically important mining districts in the Colorado Mineral Belt. Located only 50 km west of Denver in the Colorado Front Range, Central City has been worked for Au, Ag, Cu, Pb, Zn, and U ores over the century since the discovery of gold in Gregory Gulch in 1859 [23]. Mineralization at Central City is in the form of extensive veins and minor stockworks that formed along joints and faults in Precambrian gneisses and granites [24] as a result of hydrothermal alteration [23] associated with the emplacement of Tertiary dikes and porphyritic plutons [25, 26]. A distinct pattern of concentric hydrothermal alteration and concomitant mineralization has been documented at Central City [24, 27]. An elliptical, inner area of pyrite (FeS_2) veins (central zone) is surrounded by an outer region of galena (PbS) and sphalerite (ZnS) veins (peripheral zone). A transitional zone containing both types of mineralization (intermediate zone) separates the central pyrite zone from the peripheral galena–sphalerite zone. Gold and silver mineralization are found throughout the district within these three zones [23]. A barren zone containing some well-developed veins containing galena extends outward several kilometers from Central City. Hydrothermal alteration of the country rocks is present throughout the Central City district, with every fissure fracture, veins and stockwork exhibiting some degree of wallrock alteration [28].

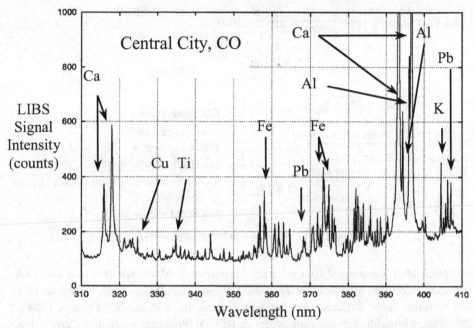

Figure 10.10. Cumulative 10-shot sprectrum from 310 nm to 450 nm for hydrothermally altered country rock from a mine dump within the intermediate zone of mixed pyrite + galena and sphalerite mineralization and mixed sericitic/argillic hydrothermal alteration.

The alteration products at Central City are K–Al silicates, carbonates, and Fe–Ti oxides. The main ore is Fe sulfide in the central zone and Pb–Zn sulfide in the intermediate zone. The 310–450 nm LIBS spectrum in Figure 10.10 for a hydrothermal clay from a mine dump in the intermediate zone at Central City exhibits emission lines for the elements expected – Ca, Fe, K, Al, Ti, Pb, and Cu.

10.3 Applications of DIAL's portable LIBS system

10.3.1 DIAL LIBS system

The schematic diagram of the experimental setup for recording the laser-induced breakdown emission from off-gas or liquid samples is shown in Figure 10.11. A Q-switched frequency doubled Nd:YAG laser (Continuum Surelite III) that delivers energy of ~400 mJ in a 5 ns duration was used in this experiment. The laser was operated at 10 Hz during this experiment and was focused on the target by using an ultraviolet (UV)-grade quartz lens with a focal length of 20 cm. The same focusing lens was used to collect the optical emission from the laser-induced plasma. Two UV-grade quartz lenses with focal lengths of 100 mm and 50 mm were used to couple the LIBS signal to an optical fiber bundle. The fiber bundle consisted of a collection of 80 single fibers with a core diameter of 0.1 mm. The rectangular exit end of

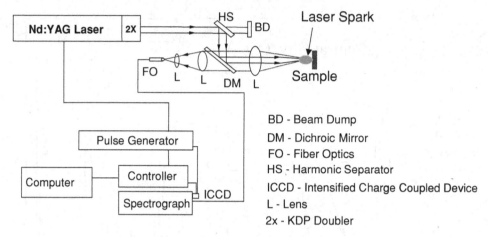

Figure 10.11. Schematic of DIAL LIBS system.

the optical fiber was coupled with an optical spectrograph (Model HR 460, Instrument SA Inc., Edison, NJ) and used as an entrance slit. The spectrograph was equipped with an 1800 and 3600 line mm^{-1} diffraction grating, with dimensions of 75 mm × 75 mm. A 1024 × 256 element intensified charge coupled device (ICCD) (Princeton Instrument Corporation, Princeton, NJ) with a pixel width of 0.022 mm was attached to the exit focal plane of the spectrograph and used to detect the dispersed light from the laser-produced plasma. The detector was operated in gated mode with the control of a high-voltage pulse generator (PG-10, Princeton Instruments Corporation, Princeton, NJ) and was synchronized to the laser output. Data acquisition and analysis were performed with a personal computer. The gate delay time and gate width were adjusted to maximize the signal-to-background (S/B) and signal-to-noise (S/N) ratios, which are dependent on the emission characteristics of the elements as well as the target matrix. LIBS data were recorded using either an 1800 or a 3600 line mm^{-1} grating, depending on the requirement of spectral resolution that provides the best signal-to-noise ratio. In order to increase the sensitivity of the system, a 100-spectra laser pulse was accumulated to obtain one spectrum.

10.3.2 Continuous emission monitor

The detection of trace metals in the off-gas of various industrial plants such as coal-fired power plants, cement kilns, incinerators, etc., is a great public-health and environmental concern. Conventional analytical techniques involved getting samples, sent to laboratories for analysis. This is because of sampling problems in off-gas systems. LIBS can perform off-gas measurements by focusing the laser beam on the gas stream through a window and collecting the signal through an optical fiber. It offers a technique to perform remote and *in situ* measurements. Radziemski and Cremers have applied LIBS to analyze effluent gases from a prototype fixed-bed coal-gasifier at the DOE Morgantown Energy Technology

Center [29]. They demonstrated that LIBS has the capability for near real-time monitoring of the concentrations of major and minor species in the off-gas emission. Neuhauser *et al.* tested their online Pb aerosol detection system with aerosol diameters between 10 and 800 nm. A detection limit of 155 μg m^{-3} was found [30].

The amount of toxic metal added to the atmosphere is restricted and controlled by various US Environmental Protection Agency (EPA) rules and permits. The EPA is modifying regulations to further reduce metal emissions. Thus, the measurement of toxic metals is very important for compliance with the existing EPA rules and also for the proposed Maximum Achievable Control Technology (MACT) rule in the future. LIBS is one of the techniques currently being evaluated by the EPA as a multi-metal continuous emissions monitor (CEM).

DIAL's LIBS system has been tested in two US Department of Energy (DOE)/EPA CEM tests [31, 32]. The CEM test was designed to measure the performance of multi-metal CEMs for regulatory compliance applications. It was conducted at the EPA's Rotary Kiln Incinerator Simulator (RKIS) facility, which consists of a primary combustion chamber, a transition section, and an afterburner in the secondary combustion chamber [32]. The kiln and secondary combustion chamber were operated with natural gas during the tests. Metals were introduced into the flue gas by injecting an aqueous metal solution directly into the secondary flame of the incinerator to achieve the target flue gas concentrations. To simulate actual flue gas conditions, fly ash particles were also injected into the incinerator. The LIBS system used a port located 5.7 m downstream from the air dilution damper. The EPA RM sample port was 1.4 m upstream from the LIBS port.

The first test demonstrated LIBS' rapid sampling rate and potential for metal CEM. It also helped us pinpoint various problems associated with LIBS field measurements. These problems include higher limits of detection (LOD), a need for online calibration, degradation of optical components, and the need for simultaneous monitoring of all the RCRA metals. These problems have been extensively studied in the laboratory since the first CEM test. Calibration techniques have been tested in the laboratory. The LOD has been reduced by a factor of eight or more for most of the metals. A method to correct the signal loss due to the degradation of optical components during the field test was developed. The various improvements made to the DIAL/LIBS system were evaluated in the second DOE/EPA CEM test [32]. The results of the LIBS calibration study, and the results of LIBS measurements during the second DOE/EPA CEM test, are given here.

The CEM test focused on As, Be, Cd, Cr, Pb, and Hg, which are the RCRA metals regulated in the EPA's MACT rules. The test program consisted of a high-metal test and a low-metal test. The target concentrations were 75 μg dscm^{-1} in high-metal tests and 15 μg dscm^{-1} in low-metal tests. The EPA's Reference Method (RM) and CEM measurements were performed concurrently for each test condition. The number of RM measurements performed for each test depended on the target metal concentration. The RM sampling time was one hour for the high-target-metal test and two hours for the low-metal tests. There were in total 20 RM samplings during the entire test, 10 for low-metal tests, and 10 for high-metal tests. Because of the difficulty in injecting a known amount of sample into a practical gas stream, LIBS calibrations were all performed in the laboratory before the

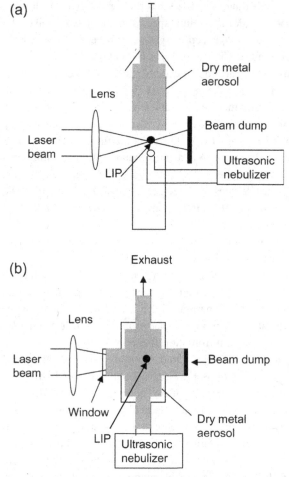

Figure 10.12. DIAL/LIBS calibration set up for gas sample: (a) open and (b) closed system.

field test. The two LIBS detection systems used in the field test were calibrated for all the RCRA metals. The peak area of an analyte line from the calibration LIBS data was used to construct the calibration curves. Linear regression was used to obtain the calibration factor.

On-site calibrations for Cr, Pb, Cd, and Be were performed at RKIS during the shakedown test with a calibration setup similar to that shown in Figure 10.12a. The on-site calibrations were done by injecting metal aerosol into the RKIS gas stream with a probe. The sample injection probe was mounted across the gas stream on the opposite port. As the gas flow quickly diluted the injected sample in the gas stream, the metal concentration near the focal volume could not be accurately estimated. Therefore, the on-site calibrations were mainly used to check system response. The temperature, flue gas flow rate, and particle loading in the test environment were $\sim$232 °C, 3.4 scm min^{-1}, and 25–50 mg dscm^{-1}, respectively.

The effects of these gas-stream parameters on LIBS calibration had not been systematically studied before.

The concentrations of Be, Cd, Cr, and Pb were monitored simultaneously in near real-time during the four-day test. Analyte lines of Cd and Be were monitored in the 220–260 nm spectral region with a 1200 line mm^{-1} grating, while analyte lines of Pb 405.8 nm and Cr 425.44 nm were monitored simultaneously in the 400–429 nm spectral region with an 1800 line mm^{-1} grating. During the test, we found that the high-quality optics used in the LIBS system degraded quickly, causing the LIBS signal to drop significantly. The dichroic mirrors used in LIBS all have high-damage thresholds ($>$GW cm^{-2}) under normal operating conditions. However, the properties of the optical coating changed in the humid and hot test environment, resulting in a lower damage threshold than the specification and damaging rapidly in the field test. Figure 10.13a shows the LIBS background recorded during one RM sampling period. It clearly shows the LIBS signal falling as the optics gradually degraded. We used the background normalization technique mentioned above to correct the LIBS raw data. Figures 10.13b and 10.13c show the raw CEM data and background corrected CEM data during the RM sampling period, respectively. A more steady inferred metal concentration was obtained after correction. This indicates that this technique is effective for the problems of optics damage. This technique was then used to correct all the LIBS data collected during this field test. The effects of fly ash and temperature were taken into account in recalling the laboratory calibration factor for CEM test.

Our LIBS system was successfully used to simultaneously monitor concentrations of Cd, Be, Cr, and Pb in near real-time during both the high- and low-metal tests. The system response time mainly depends on the sampling rate of our system. In this CEM test, the LIBS system response time was 10–20 s. The measured metal concentrations have been compared with the results from EPA's RM. Figure 10.14 shows the time-averaged LIBS data (over the RM sampling period) along with the data obtained with RM. The relative accuracy of LIBS based on the RM results was found to be in the range 19%–78%. They are in reasonable agreement. The expected accuracy in these measurements was 20% or 50%, which is much higher than expected in an analytical laboratory measurement. The LIBS data during the four test days roughly followed the trend of the RM data. LIBS data were more consistent with RM data for the last test day. This is because the experimental setup was more stable on that test day owing to a cooler probe and a new dichroic mirror. During the first three test days, we encountered more technical problems such as optics damage and laser power dropping due to the sensitivity of the frequency doubler affected by the environmental temperature. The rough correction with the background used in this test has shown promising results. However, a more refined correction taking into account the effects of gas-stream parameters should improve the accuracy of LIBS.

LIBS has shown its capability as a multi-metal CEM for Cd, Be, Cr, and Pb. Background normalization has proved to be a useful method to correct the signal variation due to optics damage. Currently LIBS' sensitivity, precision, and accuracy for certain toxic metals still need to be further improved before it can be accepted by the EPA as a metal CEM. Future development in this area includes improving the detection sensitivity of all the RCRA

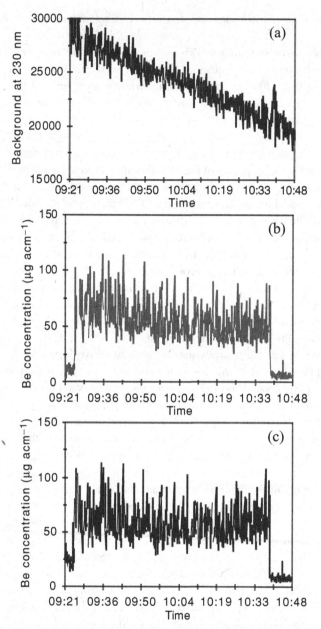

Figure 10.13. (a) LIBS background, (b) raw CEM data, and (c) corrected CEM data during RM sampling period.

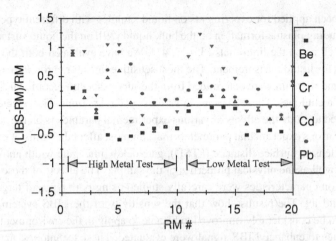

Figure 10.14. Comparison of CEM and RM data.

metals. A calibration routine for automatically compensating plasma-condition changes due to variations of gas-stream conditions or laser pulse-to-pulse fluctuations will be added in future tests.

10.3.3 Containment in liquid

The detection and quantification of light and heavy elements within liquid samples are also important from application points of view, particularly in environmental monitoring and waste treatment. Initially, liquid samples were studied by focusing the laser on the surface of the bulk liquid, which creates heavy splashing as well as shock waves [1, 2]. The major disadvantage of bulk analysis is the severely reduced plasma emission intensity in comparison with the results obtained at the liquid's surface. Watcher and Cremers overcame this problem by using a surface excitation scheme in which the sample solutions were placed in cylindrical glass vials to minimize handling of the toxic liquid [33]. The plasma was then bounded on one side by the rigid glass body of the vial. The light emission from the plasma displayed much longer durations – of the order of several microseconds with an enhanced emission intensity. LIBS analysis of liquid samples has also been realized using the laminar flows of liquid jets [34–38]. This approach was first used by Ito *et al.* for the detection of colloidal iron in a turbid solution of FeO(OH) in water [35]. The authors observed iron emission lines up to about 3.5 μs after the laser pulse. The limit of detection (LOD) for iron was estimated at 0.6 p.p.m. The sensitivity of LIBS for quantitative analysis of liquid samples is often poorer than that of other analytical techniques, such as atomic emission spectroscopy ICP-AES. However, the importance of LIBS comes into prominence if a remote online analysis is required. Remote online analysis is preferred when the measurements are to be carried out under hazardous or difficult environmental conditions. It is not possible by any analysis technique other than LIBS.

LIBS has been applied for experiments on liquid samples with different types of configurations, including plasma formation on the bulk liquid [39], on the liquid surfaces [40], on droplets as well as on the liquid jets [35–37, 41]. We have evaluated both the bulk liquid and liquid jet for liquid measurement. The most sensitive lines for chromium, magnesium, manganese, and rhenium were determined from the survey spectra recorded using different experimental techniques. These lines lay in various spectral regions. The selected emission lines were used to study the effects of various experimental parameters on the sensitivity of the system. Various experimental parameters that can most affect the limits of detection are laser energy, lens-to-surface distance (LTSD), gate delay time, gate width and its ambient condition, as well as the physical properties of the sample. The effects of these parameters on the emission characteristics were carefully studied using both types of targets, such as bulk and liquid jet. The results show that the sensitivity of the LIBS system in a liquid solution has to be considerably improved to be able to apply to the environmental samples. Two techniques to enhance LIBS signal were evaluated. These techniques, magnetic field confinement and double laser pulse excitation [35, 42], are also discussed.

Liquid configuration for plasma formation

Trace elements in different types of liquid matrices without any sample preparation are likely to be detected in the nuclear fuel cycle. Matrices such as colloids, turbid, liquids, sludge, oils, etc., require the production of plasma at the liquid surface [43]. Bulk-liquid analysis is not possible in many liquid samples having a turbid nature, which would prevent the laser beam from reaching the bulk liquid.

The general configuration used for LIBS on solid samples with the laser beam perpendicular to the surface leads to splashing in the case of liquids. Splashing results in covering the focusing optics with droplets and therefore prevents further use of this technique. This is because the plasma expansion at atmospheric pressure is directed perpendicularly to the surface. Thus a titled configuration can minimize this phenomenon [43]. Using the low laser frequency of 1 Hz can minimize the perturbation that takes place at the liquid surface following the laser pulse. It was shown that measurements with a 1 Hz laser repetition rate were more reproducible. Although the droplet and jet configurations of the liquid sample demand a sample preparation, the use of a pump-backed jet has several advantages. (1) The volume evaporated in the plasma formation process is extremely well defined, being equal to the laser spot diameter multiplied by the thickness of the jet; little or no interaction with the residual material takes place, since nearly all of the sample volume is vaporized. (2) A suitable flexible tube system in the entrance side of the pump can be used, in a principal arbitrary location within a large volume of a liquid, and can be probed, both in lateral direction and in depth; this, therefore, provides the possibility of probing in real time the spatial distribution of concentrations in a specimen. (3) One can add known amounts of elements to the flow of the sample in order to get a standard element for normalization, which will help in providing a better calibration curve. The comparison of bulk- and liquid-jet experimental results from the spectroscopic point of view is discussed in this section.

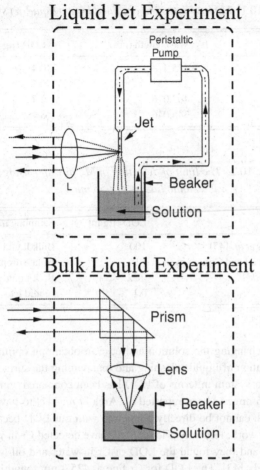

Liquid Jet Experiment

Peristaltic Pump

Jet

L

Beaker

Solution

Bulk Liquid Experiment

Prism

Lens

Beaker

Solution

Figure 10.15. Experimental setup for recording laser-induced breakdown emission from liquid samples.

The schematic diagram of the experimental setup for recording the laser-induced break-down emission on the bulk liquid surface as well as in the case of a laminar jet is shown in Figure 10.15 [44]. A Teflon nozzle of ≤ 1 mm was used with a peristaltic pump (Cole-Parmer Instrument Co.) to form a laminar liquid jet. The laser was focused on the jet such that the direction of laser propagation was perpendicular to the direction of the liquid jet. The laser was focused ~15 mm below the jet exit, where the liquid flow was laminar. However, the extent of laminar flow depended on the speed of the pump. The liquid jet was aligned in a vertically downward direction.

The LODs for Mn, Mg, Cr and Re in jet measurement were found to be better than or comparable to those of the bulk liquid measurement. Table 10.1 lists the LODs obtained from a liquid jet. The jet system is also preferable over the bulk measurement because

Table 10.1. *Limit of detection obtained from a liquid-jet system*

Element	Wavelength (nm)	LOD (μg ml^{-1})
Cr	425.4	0.4
Mg	279.55	0.1
Mn	403.076	0.7
Re	346.046	8

Table 10.2. *The limit of detection of Mg in water using the ionic line at 279.55 nm*

	LOD (μg ml^{-1})	Sampling method
Cremers *et al.* [45]	100	Bulk liquid
Archontaki and Crouch [42]	1.9	Isolate droplet
This work	0.22	Bulk liquid
	0.1	Liquid jet

of the flexibility in changing the solution and ease in obtaining optimized LIBS signals without the complication of liquid splashing and maintaining the surface-to-lens distance. The sensitivity of our system in terms of LOD has been compared with others. The LOD for Cr line at 283.56 nm has been reported by Arca *et al.* [41] to have a detection limit of 100 μg l^{-1}, which cannot be directly compared with our LOD because a different Cr line was used in our work. Recently, Fichet *et al.* have detected Cr in water and oil using Cr 425.435 nm line and they found the LOD of Cr in water and oil to be 20 p.p.m. and 30 p.p.m., respectively [43]. The LOD for Cr line at 425.4 nm found in this work with a liquid jet system is 0.4 p.p.m., which is lower by a factor of 50 than that reported by Fichet *et al.* The LOD of Mg 279.55 nm found from the literature and determined in this work has been compared. The results are shown in Table 10.2. It is clear that we have achieved much better LOD using the liquid jet than the previous works [42, 45].

LIBS signal enhancement in liquid application

The limit of detection reported for various elements in this experiment as well as in the literature has proved the LIBS technique suitable for finding pollutant (trace) elements at high and moderate concentrations. However, it is not possible to detect them at very low concentrations. A serious effort is needed to make the system versatile for very low concentration measurements. Recently, two techniques to enhance the sensitivity of the LIBS system in liquid sample measurement have been developed and are described in this section [46, 47].

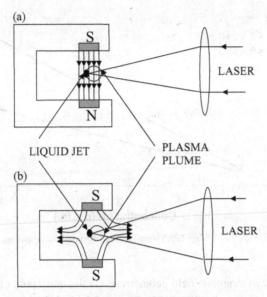

Figure 10.16. Schematic diagram (top view) of the experimental setup in the presence of a magnet. (a) Opposite poles of magnets facing each other (linear field); (b) same poles of magnets facing each other (cusp field).

(a) LIBS signal enhancement with a steady magnetic field   The main aim here is to evaluate the effect of a steady magnetic field on the emission and analytical characteristics of laser-induced plasma from Mg, Mn, Cr, and Ti in aqueous solution. To minimize the splashing and surface distortion, we used the liquid-jet analysis for the liquid sample. The limits of detection (LOD) of Mg and Mn were estimated in the presence of the magnetic field to determine the improvement in LIBS sensitivity.

The use of a steady magnetic field with the LIBS system was found to be a simple technique, which enhanced the system. The main advantage of using a steady magnetic field with the LIBS system is its simplicity in comparison with the generation of a pulsed magnetic field and its synchronization with the laser and detection systems. In fact, the enhancement in the sensitivity of LIBS in the presence of the magnetic field is small, but the combination of a steady magnetic field with other signal-enhancing techniques, such as the use of purge gas and double laser pulse excitation, may be useful for further improving the sensitivity of the LIBS system.

Two rare earth (neodymium and samarium cobalt) permanent magnets of size $0.5 \times 0.5 \times 0.125$ inches were used for generating a steady magnetic field during this experiment. Both magnets were held in a mild-steel (MS) structural arrangement separated by $\sim$5 mm. This arrangement provides a horseshoe-type magnet with a $\sim$5 kG magnetic field between the poles. The magnet system was held in such a way that the liquid jet passed vertically between and at an equal distance from the two poles (Figure 10.16). The laser was focused on the jet such that the plasma plume expanded in a nearly uniform magnetic field. These experiments

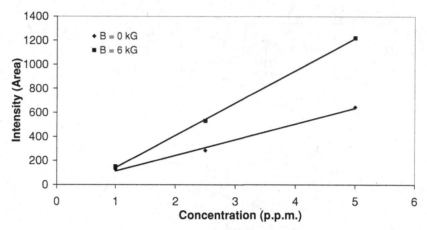

Figure 10.17. Calibration curve of Mg (279.55 nm) in the presence and the absence of the magnetic field.

were performed for two magnetic field geometries: (1) the north pole of one magnet was facing the south pole of the other (Figure 10.16a) and the field lines were passing straight from one pole to the other (N–S configuration), and (2) similar poles from both the magnets were facing each other (Figure 10.16b) and making a cusp magnetic field geometry (N–N configuration). In the first case, the plasma expanded across a nearly uniform magnetic field present between the two poles, whereas in the second case, the plasma was confined in a nearly zero magnetic field produced by field lines forming a cusp geometry. The MS bars holding the magnets were wrapped in adhesive insulating tape so as to avoid a direct interaction of the laser or plasma particles with the MS bar.

The calibration curve and the limit of detection (LOD) are important parameters for getting analytical information about the elements present in the sample matrix. The LIBS spectra of magnesium and manganese present in the liquid solution were recorded for different concentrations ranging from 1 p.p.m. to 5 p.p.m. in the absence, as well as in the presence, of the magnetic field. These experiments were performed at the laser energy of $\sim$140 mJ. Figure 10.17 shows the calibration curve, of Mg in the absence and in the presence of the magnetic field for the most sensitive ion line at 279.55 nm. The calibration curves show that the value of the slope increases in the presence of the magnetic field. The limit of detection is defined here as the concentration that produces net line intensity equivalent to three times the standard deviation of the background signal (σ_B). The LOD was calculated considering $CL = 3\sigma_B/S$, where σ_B is the standard deviation of the background signal and S is the slope of the calibration curve. Table 10.3 shows the LODs obtained for various line emissions from magnesium and manganese in the absence and presence of the magnetic field. The LOD of magnesium in the presence of the magnetic field was obtained as 0.23 p.p.m. in comparison with 0.43 p.p.m. in the absence of the magnetic field. Similarly, the LOD of manganese improved to 0.83 p.p.m. in the presence of the magnetic field from 1.74 p.p.m. in the absence of the magnetic field. An improvement of a factor of $\sim$2 was

Table 10.3. *Limits of detection obtained for Mg and Mn lines for laser energy of 140 mJ in the absence and the presence of magnetic field*

Elements	Wavelength (nm)	LOD	
		No magnetic field	With magnetic field[a]
Mg[a]	279.55	0.43	0.23
	280.27	1.93	1.10
	285.20	12.20	7.15
Mn[b]	403.08	1.74	0.83
	403.31	2.47	1.53
	403.45	6.23	2.21

Notes: [a]Gate delay of 4 μs; gate width of 2 μs. [b]Gate delay of 10 μs; gate width of 10 μs. [c]$B = 6$ kG.

noted in the LOD in the presence of the magnetic field for both the elements. The experimental results presented here show that even the steady magnet with comparatively low magnetic field intensity ($\sim$5 kG) may also be useful in improving the LOD by a factor of $\sim$2.

(b) Double-pulse excitation LIBS   The schematic diagram of the experimental set-up for making the two laser beams collinear for recording the laser-induced breakdown emission from the liquid sample under double pulse excitation is shown in Figure 10.18. It consisted of two Q-switched, frequency-doubled Nd:YAG laser (Continuum Surelite III and Quanta-Ray DCR-2A-10) that delivers energy of $\sim$300 mJ at 532 nm in 5 ns time duration. Both the lasers were operated at 10 Hz during this experiment and were focused on the target (in the center of the liquid jet). The first laser provided a p-polarized laser beam, whereas the second laser beam was s-polarized. Both the lasers were made collinear using a thin-film polarizer (CVI Lasers), which transmitted the p-polarized light, but reflected s-polarized light. For an optimum performance of the thin-film polarizer (TFP) it was necessary that p- and s-polarized light kept an angle of incidence nearly 57°. Both the beams became collinear after TFP, which was focused on the target with the help of a dichroic mirror and a spherical, ultraviolet, quartz focusing lens of 20 cm focal length. The combination of dichroic mirror and the focusing lens was used to collect the optical emission from the laser-induced plasma. The laser operation was synchronized using a programmable trigger pulse generator (Stanford Research System Inc. Model DG 535) which makes available the arrival of both the lasers at a certain time delay. The delay between lasers can be changed from nanosecond to microsecond range. The detector was operated in gated mode with the control of a high voltage pulse generator (PG-10, Princeton Instruments Corporation, Princeton, NJ) and was synchronized generally to the output of the second laser pulse. The remaining detection system was similar to the system reported in Section 10.3.1. The gate delay time and gate width were adjusted to maximize the signal-to-background (S/B) and

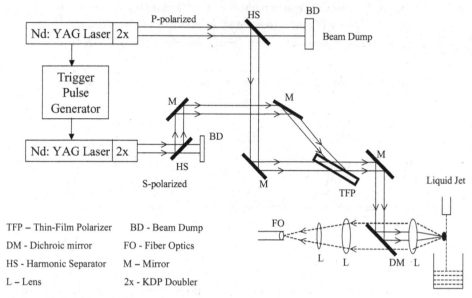

Figure 10.18. Schematic diagram for recording LIBS under double laser pulse excitation as well as the ray diagram for making two synchronized lasers collinear.

signal-to-noise (S/N) ratios, which are dependent on the emission characteristics of the elements as well as the target matrix. Emission spectra during the present experiment were recorded mainly using 2400 line mm^{-1} grating for a better spectral resolution.

The aqueous solution of Mg and Cr (5 p.p.m.) in 2% HNO_3 was used in this experiment for producing the jet. The concentrated solution of Mg and Cr ($\sim$1000 p.p.m.) present in 2% HNO_3 (E M Science, New Jersey) was used to make a dilute solution ($\sim$5 p.p.m.) for this experiment. The variation in the emission intensity with concentration was recorded for magnesium solution in the single and double pulse operation in order to find the LOD. Figure 10.19 shows the calibration curve for magnesium ion emission (279.55 nm) recorded using single and double pulse excitation. The emission intensity from ions showed a linear variation between 0.1 p.p.m. and 5 p.p.m. in single pulse operation mode, whereas emission increased in the double pulse excitation mode. Two slopes were observed in double pulse excitation mode in comparison with single pulse excitation for different concentration ranges. The second slope (1–5 p.p.m.) seemed to be the result of saturation of emission as a result of self-absorption. The limit of detection was defined here as the ratio of three times standard deviation to the slope of calibration curve. The limit of detection was calculated as 69 p.p.b. in double pulse excitation in the 0–1 p.p.m. concentration range, whereas it was 230 p.p.b. in single pulse excitation mode. Figure 10.20 shows the calibration curve for neutral magnesium emission recorded under single and double pulse excitation. Magnesium neutral line emission showed only one slope between 0.1 p.p.m. and 5 p.p.m. concentration range for single as well as double pulse excitation modes. The limit of detection for neutral

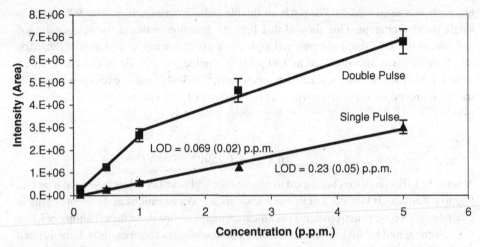

Figure 10.19. Variation in emission (279.55 nm) from magnesium ions with concentration under single pulse excitation (laser = 140 mJ, gate delay/gate width = 5 μs/1 μs) and double pulse excitation (laser (1) = 140 mJ, laser (2) = 140 mJ, gate delay (from first laser)/gate width = 5 μs /1 μs, delay between lasers = 2.5 μs). Confidence levels in the estimation of LOD are given in the brackets with the value of LOD.

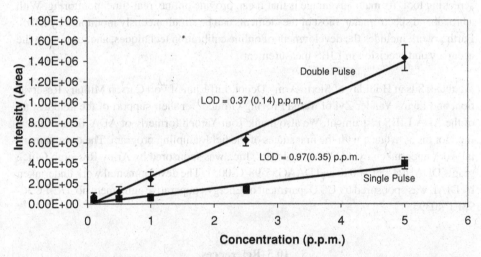

Figure 10.20. Variation in emission intensity (285.20 nm) from neutral magnesium with concentration under single pulse excitation (laser = 130 mJ, gate delay/gate width = 5 s/1s) and double pulse excitation (laser (1) = 140 mJ, laser (2) = 140 mJ, gate delay (from first laser) /gate width = 5 s/1 s, delay between lasers = 2.5 s). Confidence levels in the estimation of LOD are given in the brackets with the value of LOD.

emission was estimated as 370 p.p.b. in double pulse in comparison with 970 p.p.b. in single pulse excitation. This showed that limit of detection obtained for magnesium ion and neutral line emission improved in double pulse excitation mode. The limit of detection for chromium was also obtained as 120 p.p.b. in double pulse mode in comparison with 1300 p.p.b. in single pulse excitation mode. Thus, the double pulse excitation technique has been proved useful in improving the sensitivity of LIBS system; see also References [48–51].

10.4 Conclusion

A portable LIBS probe has been used to monitor the soil near US Army bases and near Old Popping Furnace. It was used to monitor toxic metal concentrations at different depths in ground using a cone penetrometer. These measurements demonstrate the capability of LIBS as a screening tool for toxic elements in soil. LIBS has also been successfully demonstrated as a multi-metal continuous emission monitor in off-gas emission at EPA's RKIS facility in Raleigh, NC. It shows that LIBS can be used as a CEM for incinerator, coal power plant, waste processing facilities, and other chemical plants. Monitoring the toxic elements in liquid sample has also been demonstrated. Practical application in this area is relayed on the development of the various techniques to further improve the system sensitivity for trace elements. In most of the environmental applications, LIBS has been well demonstrated as a screening tool. Its main advantage is that it can provide online, real-time monitoring. With a broadband spectrometer, most of the elements can be simultaneously monitored by LIBS. Further work includes the development of online calibration techniques, and improving the accuracy and precision in LIBS measurement.

We thank Susan Holliday of Sierra Army Depot, Jeff Linn of Fort Carson Military Reservation, and Lance Vander Zyl of Yuma Proving Ground, for their support of the field testing of the ADA LIBS instrument. We also thank Tom Yaroch formerly of ADA Technologies, Inc., for his assistance with the first stages of the field sampling program. The developmental work undertaken by ADA Technologies, Inc., was sponsored by Army Research Office grants DDAH04-96C-0030 and DAAG55-98-C-0031. The developmental work undertaken by DIAL was sponsored by US Department of Energy cooperative agreement no. DE-FC26-98FT 40395.

10.5 References

[1] D. A. Cremers, L. J. Radziemski and T. R. Loree, Spectrochemical analysis of liquids using the laser spark. *Appl. Spectrosc.*, **38** (1984), 721–729.
[2] R. Knopp, F. J. Scherbaum and J. I. Kim, Laser-induced breakdown sectroscopy (LIBS) as an analytical toool for the detection of metal ions in aqueous solutions. *Fres. J. Anal. Chem.*, **355** (1996), 16–20.

[3] R. E. Neuhauser, U. Panne and R. Niessner, Laser-induced plasma spectroscopy (LIPS): a versatile tool for monitoring heavy metal aerosols. *Anal. Chim. Acta*, **392** (1999), 47–54.

[4] J. O. Caceres, J. Tornero Lopez, H. H. Telle and A. Gonzalez Urena, Quantitative analysis of trace metal ions in ice using laser-induced breakdown spectroscopy. *Spectrochim. Acta B*, **56** (2001), 831–838.

[5] J. E. Carranza, B. T. Fisher, G. D. Yoder and D. W. Hahn, On-line analysis of ambient air aerosols using laser-induced breakdown spectroscopy. *Spectrochim. Acta B.*, **56** (2001), 851–864.

[6] U. Panne, R. E. Neuhauser, M. Theisen, H. Fink and R. Niessner, Analysis of heavy metal aerosols on filters by laser-induced plasma spectroscopy. *Spectrochim. Acta B*, **56** (2001), 839–850.

[7] A. E. Pichahchy and D. A. Cremers, Elemental analysis of metals under water using laser-induced breakdown spectroscopy. *Spectrochim. Acta*, **B52** (1997), 25–39.

[8] D. Kossakovski and J. L. Beauchamp, Topographical and chemical microanalysis of surfaces with a scanning probe microscope and laser-induced breakdown spectroscopy. *Anal. Chem.*, **72** (2001), 4731–4736.

[9] R. Wisbrun, I. Schechter, R. Niessner, H. Schroeder and K. L. Kompa, Detector for trace elemental analysis of solid environmental samples by laser plasma spectroscopy. *Anal. Chem.*, **66** (1994), 2964–2975.

[10] A. Ciucci, V. Palleschi, S. Rastelli *et al.*, Trace pollutants analysis in a soil by a time-resolved laser-induced breakdown spectroscopy technique. *Appl. Phys. B – Las. Opt.*, **63** (1996), 185–190.

[11] I. B. Gornushikin, J. I. Kim, B. W. Smith, S. A. Baker and J. D. Winefordner, Determination of cobalt in soil, steel, and graphite using excited-state laser fluoresence induced in a laser spark. *Appl. Spectrosc.*, **51** (1997), 1055–1059.

[12] C. Haisch, J. Liermann, U. Panne and R. Niessner, *Anal. Chem. Acta*, **346** (1997), 23.

[13] F. Hilbk-Kortenbruck, R. Noll, P. Wintjens, H. Falk and C. Becker, Analysis of heavy metals in soils using laser-induced breakdown spectrometry combined with laser-inducred fluorescence. *Spectrochim. Acta B*, **56** (2001), 933–945.

[14] R. Wainner, R. S. Harmon, P. D. French, A. W. Miziolek and K. L. McNesby, Analysis of environmental lead contamination: comparison of LIBS field and laboratory instruments. *Spectrochim. Acta B*, **56** (2001), 777–794.

[15] L. J. Radziemski, T. R. Loree, D. A. Cremers and N. M. Hoffman, Time resolved laser induced spectrometry of aerosols. *Anal. Chem.*, **55** (1983), 1246–1252.

[16] D. A. Cremers, J. E. Barefield and A. C. Koskelo, Remote elemental analysis by laser-induced breakdown spectroscopy using a fiber optic cable. *Anal. Chem.*, **68** (1987), 997–981.

[17] K. Y. Yamamoto, D. A. Cremers, M. J. Ferris and L. E. Foster, Detection of metals in the environment using a portable laser-induced breakdown spectroscopy instrument. *Appl. Spectrosc.*, **50** (1996), 1175–1181.

[18] B. Miles and J. Cortes, Subsurface heavy metal detection with the use of laser-induced breakdown spedtroscopy (LIBS) penetrometer system. *Field Anal. Chem. Tech.*, **2** (1998), 75–87.

[19] G. A. Theriault, S. Bodensteiner and S. H. Liberman, A real-time fiber optic LIBS probe for the in-situ delineation of metals in soils. *Field Anal. Chem. Tech.*, **2** (1998), 117–125.

[20] B. J. Marquardt, S. R. Goode and S. M. Angel, In situ determination of lead in paint by laser-induced breakdown spectroscopy using a fiber-optic probe. *Anal. Chem.*, **68** (1996), 977–981.

[21] Harding Lawson Associates, *Remedial Investigation and Feasibility Study Report, Sierra Army Depot, Lassen County, California.* Unpublished internal report for Sierra Army Depot Contract # DACA31-94-D-0069 (2000).

[22] T. Liu, Visual microlaminations in rock varnish: a new paleoenvironmental and geomorphic tool in drylands. Ph.D. thesis, Arizona State University (1994).

[23] R. M. Potter and G. R. Rossman, The manganese- and iron-oxide mineralogy of desert varnish. *Chem. Geol.*, **25** (1979), 79–94.

[24] T. M. Oberlander, Rock varnish in deserts, in *Geomorphology of Desert Environments* (London: Chapman & Hall, 1994).

[25] P. K. Sims, A. A. Drake and E. W. Tooker, *Economic Geology of the Central City district, Gilpin County, Colorado*, USGS Professional Paper 359, 231 pages (1963).

[26] T. S. Lovering and E. N. Goddard, *Geology and ore deposits of the Front Range, Colorado*, USGS Professional Paper 223, 319 pages (1950).

[27] J. D. Wells, *Petrography of radioactive Tertiary igneous rocks, Front Range mineral belt, Colorado*, USGS Bulletin 1032-E, pp. 223–272 (1960).

[28] C. M. Rice, D. R. Lux and R. M. MacIntyre, Timing of mineralization and related intrusive activity near Central City, Colorado: *Econ. Geol.*, **77** (1982), 1655–1666.

[29] E. S. Bastin, and J. M. Hill, *Economic geology of Gilpin County and adjacent parts of Clear Creek and Boulder Counties, Colorado*, USGS Professional Paper 94, 379 pages (1917).

[30] E. W. Tooker, *Altered wall rocks in the central part of the Front Range mineral belt, Gilpin and Clear Creek Counties, Colorado*, USGS Professional Paper 439, 102 pages (1963).

[31] L. J. Radziemski and D. A. Cremers, *Laser Induced Plasma and Applications* (New York: Marcel Dekker, 1989).

[32] R. E. Neuhauser, U. Panne, R. Niessner *et al.*, On-line and in-situ detection of lead aerosols by plasma-spectroscopy and laser-excited atomic fluorescence spectroscopy *Anal. Chim. Acta*, **346** (1997), 37–48.

[33] H. Zhang, F. Y. Yueh and J. P. Singh, 1999, Laser-induced breakdown spectrometry as a multimetal continuous-emission monitor. *Appl. Opt.*, **38** (1999), 1459–1466.

[34] J. P. Singh, H. Zhang and F. Y. Yueh, Technical report for continuous emission monitor (CEM) test at the Rotary Kiln Incinerator Simulator (RKIS) at the EPA Environmental Research Center, Research Triangle Park, Raleigh, NC, September 1997.

[35] J. R. Watcher and D. A. Cremers, Determination of uranium in solution using laser-induced breakdown spectroscopy. *Appl. Spectrosc.*, **41** (1987), 1042–1048.

[36] O. Samek, D. C. S. Beddows, J. Kaiser *et al.*, Application of laser-induced breakdown spectroscopy to in situ analysis of liquid samples. *Opt. Eng.*, **38** (2000), 2248–2262.

[37] Y. Ito, O. Ueki and S. Nakamura, Determination of colloidal iron in water by laser-induced breakdown spectroscopy. *Anal. Chim. Acta*, **299** (1995), 401–405.

[38] C. W. Ng, W. F. Ho and N. H. Cheung, Spectrochemical analysis of liquids using laser-induced plasma emissions: effects of laser wavelength on plasma properties. *Appl. Spectrosc.*, **51** (1997), 976–983.

[39] S. Nakamura, Y. Ito, K. Sone, H. Hiraga and K. Kaneko, Determination of an iron suspension in water by laser-induced breakdown spectroscopy with two sequential laser pulses. *Anal. Chem.*, **68** (1996), 2981–2986.

[40] W. F. Ho, C. W. Ng and N. H. Cheung, Spectrochemical analysis of liquids using laser-induced plasma emissions: effects of laser wavelength. *Appl. Spectrosc.*, **51** (1997), 87–91.

[41] J. P. Singh, H. Zhang and F. Y. Yueh, *Plasma Arc Centrifugal Treatment PACT 6 Slip Stream Test Bed (SSTB) 100-hour duration Controlled Emission Demonstration CED) test*, DIAL Trip Report 96–3 (1996).

[42] J. R. Wachter and D. A. Cremers, Determination of uranium in solution using laser-induced breakdown spectroscopy. *Appl. Spectrosc.*, **41** (1987), 1042–1048.

[43] G. Arca, A. Ciucci, V. Palleschi, S. Rastelli and E. Tognoni, Trace element analysis in water by the laser-induced breakdown spectroscopy technique. *Appl. Spectrosc.*, **51** (1997), 1102–1105.

[44] H. A. Archontaki and S. R. Crouch, Evaluation of an isolated droplet sample introduction system for laser-induced breakdown spectroscopy. *Appl. Spectrosc.*, **42** (1988), 741–746.

[45] P. Fichet, P. Mauchien, J. F. Wagner and C. Maulin, Quantitative elemental determination in water and oil by laser induced breakdown spectroscopy. *Anal. Chim.*, **429** (2001), 269–278.

[46] F. Y. Yueh, R. C. Sharma, J. P. Singh and H. Zhang, Evaluation of the potential of laser-induced breakdown spectroscopy for detection of trace element in liquid. *J. Air Waste Manage. Assoc.*, **52** (2002), 174–185.

[47] F. Y. Yueh, J. P. Singh and H. Zhang 2000. Laser-induced breakdown spectroscopy, elemental analysis. In *Encyclopedia of Analytical Chemistry* (Chichester: John Wiley and Sons, 2000), pp. 2066–2087.

[48] D. A. Cremers, L. J. Radziemski and T. R. Loree, Spectrochemical analysis of liquids using the laser spark. *Appl. Spectrosc.*, **38** (1984), 721–729.

[49] C. G. Engel and R. S. Sharp, Chemical data on desert varnish. *Geol. Soc. Amer. Bull.*, **69** (1958), 487–518.

[50] V. N. Rai, A. K. Rai, F. Y. Yueh and J. P. Singh, Optical emission from laser-induced breakdown plasma of solid and liquid samples in the presence of a magnetic field. *Appl. Opt.*, **42** (2003), 2085–2093.

[51] V. N. Rai, F. Y. Yueh and J. P. Singh, Study of laser-induced breakdown emission from liquid under double pulse excitation. *Appl. Opt.*, **42** (2003), 2094–2101.

11

Industrial applications of LIBS

Reinhard Noll, Volker Sturm and Michael Stepputat

Fraunhofer-Institut für Lasertechnik (ILT), Germany

Andrew Whitehouse, James Young and Philip Evans

Applied Photonics Ltd, Skipton, North Yorkshire, UK

11.1 Introduction

The availability of compact and reliable laser sources, sensitive optical detectors, and powerful computers has helped to stimulate significant growth in industrial applications of LIBS during the past decade. This, together with a better understanding of the physical processes involved when intense laser radiation interacts with a material, has helped researchers to exploit the LIBS technique for various industrial applications ranging from process control of materials during manufacturing to rapid sorting of scrap materials during recycling and remote characterization of highly radioactive nuclear waste. LIBS is still regarded as an emerging technology and there remain many technological barriers that must be overcome before widespread industrial use becomes a reality.

This chapter aims to provide the reader with a general overview of industrial applications of LIBS and is not meant to provide an exhaustive review of the field. The scope has been restricted to applications of LIBS in an industrial rather than laboratory environment. Accordingly, the various laboratory-based LIBS instruments that are now available from a number of manufacturers are not discussed here. The chapter has been written in four sections relating to the following general areas of industry: (i) metals and alloys processing, (ii) scrap sorting and recycling, (iii) nuclear power generation and spent fuel reprocessing, and (iv) miscellaneous industrial applications.

11.2 Metals and alloys processing

11.2.1 Background

The continuously increasing requirements for productivity and product quality in the metal producing and processing industries initiate the demand for measuring methods having the potential to analyze the chemical composition of the processed materials at high speed and – if possible – on-line. LIBS is predestined for this task. Sample taking can be simplified or even avoided by directly measuring on the specimen in the production line. Even

Laser-Induced Breakdown Spectroscopy: Fundamentals and Applications, ed. Andrzej W. Miziolek, Vincenzo Palleschi and Israel Schechter. Published by Cambridge University Press. © A. W. Miziolek, V. Palleschi and I. Schechter 2006. A. W. Miziolek's contributions are a work of the United States Government and are not protected by copyright in the United States.

the sample preparation can be performed by the laser itself. In a first step surface layers such as oxides, scale, oil residues, and coatings that are not characteristic for the bulk composition are ablated by the laser used as a non-contacting drilling tool prior to the start of the spectrochemical analysis with the laser used as excitation source. Today, the analytical performance for a multi-element analysis achieves a level which equals or is even better than that of classical methods. The possibility of performing non-contact measurements over large distances opens up new fields of applications that are not accessible for conventional physical methods such as X-ray fluorescence analysis or spark emission spectrometry. Such fields are identification testing of products and online analysis of liquid steel, for example, which will be presented together with others in this chapter. Quality assurance close to the process on the basis of an automated online measurement is the first step to fast and efficient feedback actions, which result in cost savings and improved competitiveness.

11.2.2 Identification of pipe fittings

Especially in the oil and gas industry, pipe-lines and process piping are increasingly exposed to corrosion as the fields become more sour, and wells are deeper with higher pressures and hotter products. In order to meet the requirements with regard to corrosion resistance and mechanical stability, different steel grades are used for the production of pipe fittings. The range of processed materials extends from high alloy steel grades to nickel base alloys. The use of incorrect steel grades can lead to corrosion with severe consequential damage. Increasing quality requirements, particularly in the nuclear industry, and environmental responsibilities demand a material identification of each produced pipe fitting. An automatic inspection machine called LIFT (laser identification of fittings and tubes) based on LIBS was developed to prevent any mix of material grades [1]. The machine is able to inspect more than 35 different material grades.

The conventional inspection method for pipe fittings is spark optical emission spectrometry (spark OES). The fittings had to be cleaned and a considerable percentage of the fittings had to be measured two or three times before they had been correctly identified. The duration of one inspection was 4 s and the electrode of the spark OES had to be cleaned every three measurements. The throughput was 60 fittings per hour. Considering only the overhead of the spark OES, the costs for inspection were €0.5 per fitting. Moreover, the surface of the fittings had to be abraded after the spark discharge. In a third job step the fittings had to be marked with an ink jet printer or by electrolytic etching. In view of a material inspection of each produced pipe fitting, the economic efficiency had to be improved.

With the laser-based inspection only one job step is necessary for the inspection and marking of the pipe fittings. Inspection and marking are performed by one machine, and cleaning and abrading are no longer necessary. The schematic set-up is shown in Figure 11.1.

An operator puts the workpieces on a table type circular conveyor. The table rotates by 90° and positions the pipe fitting in front of the measuring window of the inspection

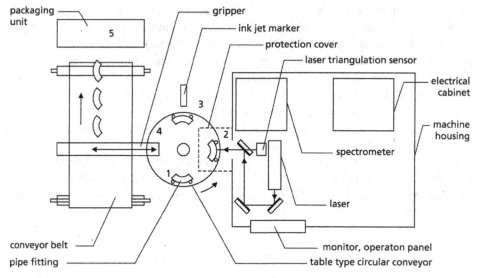

Figure 11.1. Schematic set-up of the inspection machine based on LIBS to test pipe fittings. The numbers 1 to 5 show the sequence of steps: 1, the pipe fitting is put onto the table type circular conveyor; 2, inspection with LIBS; 3, marking of the workpiece with an ink jet marker; 4, transport of pipe fitting to the conveyor belt; 5, packaging of the inspected pieces.

machine. A triangulation sensor, whose measuring beam propagates collinear to the pulsed laser beam for LIBS, measures the distance to the specimen. If the surface of the specimen is in a set tolerance range of distance the pulsed laser is activated to generate the plasma. The inspection time is 2 s. Within this time 100 spectra are generated and evaluated. If the identified material grade corresponds to the expected one, the workpiece will be marked by an ink jet marker at the next cycle of the handling system. Afterwards the fitting is carried to a conveyor belt by a gripper and transported to the packaging station. When material mixing is identified, the fitting will not be marked. Instead, it will be sorted out by the conveyor belt. The handling system can be used for sizes of the pipe fittings from 0.5″ up to 8″ and the theoretical throughput is 450 fittings per hour. Figure 11.2 shows a view of the LIFT system installed in the production facility of a pipe manufacturer.

In order to operate LIFT in dusty surroundings at a humidity of up to 90% and at an ambient temperature of up to 40 °C, all components of the laser inspection machine are installed in a closed air-conditioned cabinet; see Figure 11.2. For maintenance purposes the cabinet can be entered through a door, which is usually locked.

The light emitted from the laser-induced plasma is collected by a fiber optic cable and transferred to a Paschen–Runge spectrometer equipped with 12 photomultipliers for the elements Fe, Ni, Cr, Mo, Ti, Cu, Nb, Al, and W. For time-resolved spectrometry, the photomultiplier signals are processed by a multi-channel electronic device with fast gateable integrators and analog-to-digital converters.

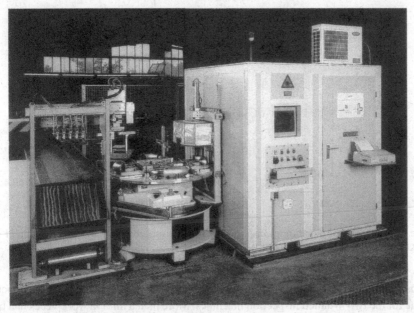

Figure 11.2. View of LIFT (on the right-hand side) and the handling system for the pipe fittings. The table-type circular conveyor is located on the left-hand side of LIFT and feeds the inspection machine with the pipe fittings for the material identification.

The use of a laser for the plasma generation is favorable to a flexible automation, for the following reasons.

- The contactless measurement with working distances of up to 0.5 m simplifies the integration into a production line.
- Laser light is an energy source that can be applied to all kinds of materials. It can be used for material processing as well as for measurement purposes. Therefore sample preparation is normally not required. Rust, varnish or dust can be ablated before the measurement is started.
- The energy transfer to the sample is limited to a small area (< 1 mm^2) and a very short time interval (< 10 ns). Only a small amount (of the order of micrograms) of material is ablated and the properties of the fitting with regard to corrosion resistance and mechanical stability are not changed.

The inspection machine is able to identify more than 30 different steel grades, in particular alloy steels, stainless steels, Duplex, Super Duplex, 6MO grades, high nickel alloys, titanium and clad steels. The concentration range of Fe and Ni extends to 100%, and the range of Cr, Mo, and Cu extends to 30%. For the classification of the measured data set concerning the steel grade an expert system was developed. Key features of the expert system are high decision reliability, minimization of the consultation time, and flexibility.

Two LIFT machines have been in routine use since October 1998. The possible throughput is about 500 000 fittings per year for one-shift operation. The percentage of falsely rejected

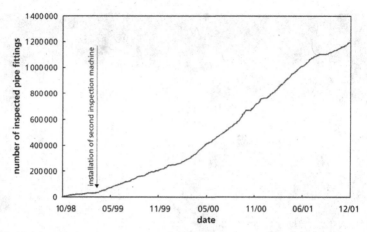

Figure 11.3. Number of pipe fittings inspected with two inspection machines as a function of time from October 1998 until the end of 2001.

fittings is below 0.1%. Figure 11.3 shows the number of inspected pipe fittings as a function of time for the two installed inspection machines. More than a million workpieces had been inspected by the end of the year 2001.

11.2.3 Slag analysis

Improved steel qualities are a demanding task for the steel producing industry. For example, concentrations below 10 p.p.m. of sulfur and phosphorus are required to achieve the desired material properties of the pre-products. Hence the goal is continuously to improve the metallurgical processing steps from the pre-product to the final product to achieve well-defined material qualities within narrow tolerances for the specified element concentrations. The composition of a steel melt is influenced by chemical reactions of the melt with slag components. Therefore the chemical analysis of the slag provides essential information for an efficient metallurgical process control. One important issue concerning steel production is the analysis of slag from the converter and the ladle. For the future an increasing demand for slag analysis is expected further to optimize the metallurgical process guiding. For an efficient process control the total time needed for slag analysis has to be reduced to less than 3 min.

X-ray fluorescence (XRF) spectroscopy is a state of the art measuring technique for slag analysis. The slag sample taken is transported to a laboratory. The total time necessary for this procedure in an off-line laboratory including the steps sampling, sample transportation, sample preparation, and analysis amounts to 6.5 min. Nearly 40% of the total time is necessary for sample preparation and sample transportation. However, the analytical results during steel production are, in most cases, required within a time scale of a few minutes, since the next processing step must be controlled on the basis of this analytical information. Hence,

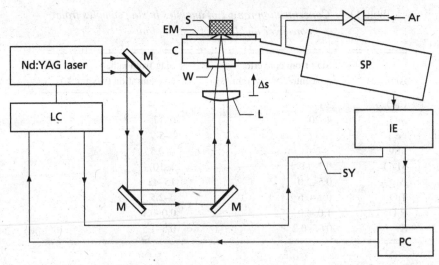

Figure 11.4. Schematic set-up for LIBS of slag samples. M, mirror; L, lens; W, window; C, measurement chamber; S, sample; EM, eccentric mechanism; LC, laser control; SP, Paschen–Runge spectrometer; IE, integrator electronics; SY, signal cable for synchronization; PC, personal computer.

there is an increasing need to simplify or even avoid those steps of sample transportation or sample preparation.

LIBS is a method where the laser-based analysis can be easily combined with a sample preparation or conditioning by the laser beam itself. Before a measurement starts, the sample surface is irradiated by a number of pre-pulses to ablate a small amount of material from the surface that may be not representative of the bulk composition. The following laser bursts and the laser-induced plasma emission produced thereby are used for analysis. The necessary time for the surface preparation by the pre-pulses and the subsequent analysis is in the range of 0.5–2 min. Thus more than 2.5 min can be saved in comparison with the XRF standard analysis method.

An inspection machine based on LIBS was developed for the quantitative analysis of slag samples from steel production [2]. Figure 11.4 shows the schematic set-up. A Q-switched Nd:YAG laser operating at 1064 nm was used to excite the plasma. The Q-switch electronics was equipped with a double-pulse option to generate up to two separate laser pulses within a single flashlamp pulse instead of a single laser pulse. The influence of multiple laser pulses on the dynamics and physical state of the induced plasma was described elsewhere [3]. The laser emits within a single flashlamp discharge two laser pulses of approximately equal energy. The laser beam is guided via mirrors to a plano-convex lens with a focal length of 200 mm. The converging laser radiation propagates via an optical window into a measurement chamber. The cover plate of the measurement chamber can be moved by an eccentric mechanism in order to shift the sample relative to the laser beam on a circular trajectory. Thus, a spatial averaging can be performed.

Table 11.1. *Concentration ranges of analytes in slag samples from converter and ladle of a steel plant*

Analyte	Slag from converter concentration (weight %)	Slag from ladle concentration (weight %)
CaO	30–64	43–64
SiO_2	4–20	0–17
Fe_{tot}	6–34	0–5
Mn	2–6	0–0.5
MgO	0.7–11	1–10
Al_2O_3	0.5–2.0	15–42
TiO_2	0.1–0.6	0–2.5
P_2O_5	1.0–4.0	0.0–0.1
S	0.05–0.2	0.5–1.5

The spectrometer has a Rowland circle diameter of 500 mm. The dispersed radiation passing the exit slits is detected by photomultiplier tubes (PMT). An integrator electronics developed for the simultaneous processing of up to 64 PMT signals in a pulse-by-pulse way is triggered by the Pockels cell synchronization monitor of the laser (see connection SY in Figure 11.4).

Table 11.1 shows typical concentrations of the analytes of slag samples from the converter and the ladle. All elements vary in a range of at least 10% (relative) and there is no single element whose concentration remains approximately constant. These variations of analyte concentrations lead to matrix effects. Detailed parameter studies were performed to minimize the influence of these matrix variations. Furthermore multivariate calibration models have been applied to reduce the relative standard deviation $s(c)$ (average for all elements) – describing the uncertainty in the determination of the concentration of an analyte – down to 1.5%.

Figure 11.5 shows a view of the analytical system that enables the measurement of slag samples taken directly from the converter using LIBS. The integrated Paschen–Runge spectrometer simultaneously detects emission lines of the relevant elements Ca, Fe, Si, Mn, Mg, Al, and Ti. Most of the lines belong to atomic transitions of the respective elements.

The duration of the quantitative analysis is about 2 min. The achieved precision given as the empirical standard deviation (in weight %) for a repetitive measurement under repeatability conditions lies between 0.05% for low concentrations and 0.8% for high concentrations.

11.2.4 Liquid steel analysis

In steel works, the goal of direct analysis of liquid steel is to receive a fast elemental analysis of the steel composition, for example during the secondary metallurgical processes. Direct analysis helps to achieve the target composition more closely in shorter process times with

Figure 11.5. Analytical system for slag analysis. The hood of the sample stand is open.

less energy and material consumption, and hence the steel quality and the productivity can be improved. A future perspective is the online analysis of the melt composition.

Today's conventional analysis methods are off-line and discontinuous. The required steps for analysis are taking a sample from the liquid steel, solidification of the sample, transport of the sample to a container laboratory (installed close to the aggregates) or to a central laboratory, sample preparation, and finally analysis, for example by spark optical emission spectrometry (spark OES) or combustion methods. Recent progress in container laboratory installations and the degree of automation have reduced the analysis times to about 3–5 min under best conditions. With direct laser analysis using LIBS it is feasible to reduce these times by about 50%.

Analyzing liquid steel by LIBS started as early as 1965. Runge *et al.* [4] used a ruby laser to analyze nickel and chromium of molten stainless steel samples in the laboratory. Several experiments have been conducted over the years demonstrating the feasibility of a LIBS-based liquid steel analysis in a steel works [5–14]. Applications at different aggregates, such as the converter or the ladle in secondary metallurgy, have been under investigation [6, 7, 8, 12]. The major challenge is still to realize reliable access to the melt that additionally allows the transmission especially of the favored analytical emission

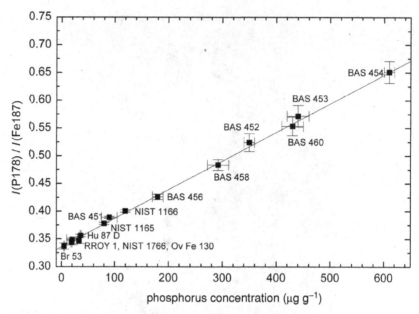

Figure 11.6. Calibration curve of phosphorus for the emission line 178.28 nm for solid steel samples. Limit of detection (3s) is 9 μg g^{-1}. Sample denominations: NIST = National Institute of Standards & Technology, BAS = Bureau of Analysed Samples Ltd, Hu = Hanau, Br = Breitländer, Ov = Ovako Steel.

lines of carbon (wavelength 193.09 nm), sulfur (180.73 nm), and phosphorus (178.28 nm) in the short ultraviolet wavelength range. The configurations use (1) the bottom-blow nozzles of a converter, (2) a bore in the side-wall of a converter, or (3) a lance immersed from the top of the melt. The analytical performance of LIBS with detection limits of 250 μg g^{-1} for C even with solid samples has been sufficient for applications at the converter but not for secondary metallurgy [9]. Recent improvements demonstrated the feasibility of limits of detection (LOD$_{3s}$) near and below 10 μg g^{-1} for the most relevant elements [15–17]. For example, calibration curves of phosphorus for solid samples and for liquid steel are given in Figures 11.6 and 11.7.

Figure 11.8 shows a view of a prototype LIBS analyzer during operation at an induction furnace with a 100 kg steel melt [17]. To analyze the important elements like carbon, sulfur, and phosphorus, the radiation of the laser-induced plasma is guided through a protection window, a 200 mm collimating lens (both made of MgF$_2$) and two beam steering mirrors into a vacuum Paschen–Runge spectrometer with a Rowland circle diameter of 750 mm. The Nd:YAG laser emits pulse bursts at 10 Hz repetition rate. More details of the laser and spectrometer equipment are given in references [15] and [17]. The refractory lance for the immersion into the steel melt of about 1630 °C consists of a water-cooled copper shielding and a refractory lance tip. The lance was immersed over several hours in the melt.

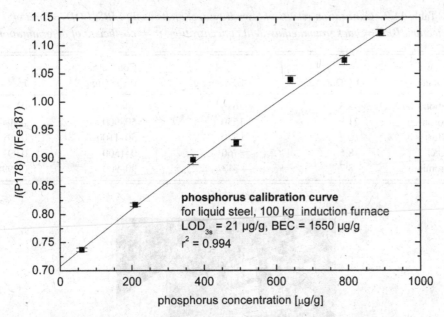

Figure 11.7. Calibration curve of phosphorus for the emission line 178.28 nm for liquid steel. Limit of detection (3s) is 21 μg g^{-1}. For calibration, P was added successively to the melt and samples are taken for conventional solid sample analysis (spark OES and combustion).

The LIBS analyzer can determine, within 60 s, up to 20 elements and even more – depending on the lines installed in the spectrometer. The calibration is performed by adding a defined amount of alloying material to the melt and taking samples that are analyzed conventionally by spark OES and combustion methods. Table 11.2 lists the results of the measurements with a 100 kg steel melt. These values are nearly the same as measured for solid samples [15]. With that the preconditions for the application in secondary metallurgy at larger aggregates in steel works are given.

11.3 Scrap material sorting and recycling

11.3.1 Background

A significant increase in the commercial value of scrap materials may be obtained through reliable and rapid sorting of the material prior to recycling. Examples include the segregation of brominated plastics from non-brominated plastics, the sorting of metal alloys by grade type or group, and the sorting of technical glasses according to their chemical composition. In order for the recycling of these materials to be commercially viable, however, high material throughput is required and hence the sorting must be performed at very high speed and with appropriate reliability of identification. The non-contact and rapid analysis capabilities of LIBS lend themselves to high-speed sorting of scrap materials and a considerable amount

Table 11.2. *Calibration results of low alloy steel melts using LIBS, LOD = limit of detection, BEC = background equivalent concentration, r^2 = coefficient of determination*

Element	LOD$_{3s}$ (μg g^{-1})	BEC (μg g^{-1})	Concentration range (μg g^{-1})	r^2
carbon	3.3	680	50–300	0.999
phosphorus	21	1550	50–900	0.994
sulfur	10.7	870	50–1100	0.993
nickel	8.5	760	0–1500	0.997
chromium	8.7	313	90–900	0.9998

Figure 11.8. View of the LIBS analyzer for the direct analysis of liquid steel. The laser beam and the measuring radiation are guided inside a refractory lance, which is immersed from the top into the steel melt. The laser, spectrometer, and evaluation equipment are mounted in a shielding box on a platform of a lifting device to dip the lance into the melt.

of work has been done during the past ten years to develop full-scale scrap material sorting systems based on LIBS. Some of the development work is ongoing and is subject to commercial confidentiality and so cannot be disclosed here; however, two examples of industrial scrap material sorting systems based on LIBS are discussed in the following sections.

11.3.2 Identification of technical polymers for material specific recycling

Technical polymers are used to manufacture components and housings for electrical appliances such as telephones, computer keyboards or household appliances. Various polymer matrices, for example ABS (acrylonitrile-butadiene-styrene), PC (poly-carbonate), SB (styrene butadiene), and SAN (styrene acrylonitrile), are doped with additives to improve their mechanical, electrical and chemical properties. Common additives are flame retardants, antioxidants, light stabilizers, fillers, dyes and pigments. Their concentration varies from traces to several percent [18].

During the recycling process of end-of-life waste electric and electronic equipment (EOL-WEEE), downgrading of valuable technical polymers has to be avoided by separating the collected material in fractions of high purity. Further, waste pieces containing brominated flame retardants (BFR) and heavy metals have to be automatically identified and removed from the waste stream to be recycled due to the significant environmental problems during the waste management phase caused by these substances. To establish an economically feasible recycling process that meets these requirements, high-speed automatic sorting systems performing the identification of both the polymer and the critical additives for several parts per second are required. A prototype automatic identification and sorting line has been set up for material specific sorting of EOL-WEEE pieces [19]. For the detection unit of this automatic sorting line, a multi-sensor system for a rapid identification of the polymer matrix and the contained additives has been developed comprising three spectroscopic modules based on NIR (= near infrared spectroscopy), MIR (= mid-infrared spectroscopy) and LIBS. The goal is to combine the spectroscopic information measured in the infrared region with the element specific LIBS information to identify the polymer type, the heavy metals and the flame retardants. The task of the LIBS module is the rapid quantification of heavy metal and halogen concentrations in EOL-WEEE pieces moving at a speed of 0.5 m s^{-1} to 1.0 m s^{-1} on a conveyor belt. The LIBS detection of heavy metals is focused on Pb, Cr, Cd, and Hg found in pigments, nucleating agents and heat stabilizers. Brominated flame retardants are detected via Br and via Sb as a component of their widely used synergist Sb_2O_3. Frequently applied methods for the determination of metals and flame retardants are atomic absorption spectrometry (AAS), inductively coupled plasma atomic spectrometry (ICP-AES, ICP-MS) or electrochemical techniques [20]. However, these methods require the time-consuming step of sample dissolution prior to the analysis and are therefore not applicable for high-speed applications. Direct sampling analytical methods such as laser ablation atomic emission spectrometry (LA-AES) [21], laser ablation mass spectrometry (LA-MS) [22], glow discharge spectrometry (GD-OES) [23], sliding spark discharge optical emission spectrometry (sliding SD-OES) [24], and X-ray fluorescence spectrometry (XRF) [25] have been applied for polymer additive analysis. However, these methods are not feasible for on-line application with several parts per second. On the other hand, LIBS as a rapid measuring method without sample contact has the potential for on-line high-speed applications. The identification of the type of polymer based on LIBS spectra has been demonstrated for several polymers [26, 27]. The LIBS detection of heavy metals in

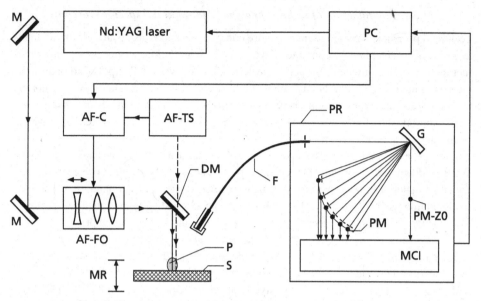

Figure 11.9. Set-up of the LIBS analyzer. PC = personal computer, PR = Paschen–Runge polychro-mator, PM = photomultiplier, PM-ZO = photomultiplier for zeroth order, G = concave grating, MCI = multi-channel integration electronics, F = fiber optics, P = plasma, S = sample, MR = measuring range, AF = autofocusing unit, AF-TS = triangulation sensor, AF-C = controller, AF-FO = focusing optics.

polymers with LODs of about 100 $\mu g\, g^{-1}$ has been published [28]. Because of the require-ment to provide a LIBS analysis for on-line measurements under industrial conditions, all measurements are constrained to be conducted in air under atmospheric pressure. Signal-to-noise enhancing detection in low pressure or noble gas atmosphere [29] is not applicable in this case.

The major problem is the large variation in size and shape of the real waste EOL-WEEE pieces such as PC monitors, telephones and keyboards. To provide a reduced dependency of the measuring distance on the sample shape, the samples are presented to the sensor on a tilted conveyor belt sliding on the side panel. The measurement is conducted through a side-on identification gap in the conveyor panel. To compensate for the remaining dis-tance variations due to surface topology of the sample, an autofocusing unit for LIBS was developed which provides a measuring range of 50 mm.

The set-up of the LIBS module is shown in Figure 11.9. The Nd:YAG laser delivers laser bursts consisting of up to two laser pulses with a total energy of 360 mJ at a repetition rate of 30 Hz. The laser pulses are focused onto the sample surface by the autofocusing optics and the emission of the laser-induced plasma is picked up by a fiber optics at a fixed position. The photomultiplier signals of the Paschen–Runge polychromator are processed by the MCI multi-channel integration electronics developed at ILT where the signals are integrated during defined gating intervals [30]. The integrated values are digitized and transmitted

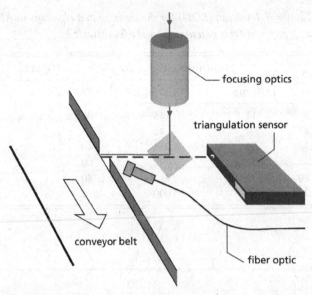

Figure 11.10. Schematic illustration of the LIBS analyzer at the conveyor belt of the pilot sorting plant.

to the control PC for further data processing. The configuration of the Paschen–Runge polychromator and the MCI multi-channel integration electronics provides the simultaneous detection of a wide spectral range along with measuring frequencies of up to 1000 Hz with single pulse evaluation.

In the LIBS autofocusing unit, a laser triangulation sensor measures the distance to the sample surface, see Figures 11.9 and 11.10. The distance information is used to control the focusing optics of the autofocusing unit. The focusing optics is a variable focal length optical system with an average focal length of about 220 mm. Moving the dispersion lens along the optical axis by means of a linear motor allows a 50 mm variation of the focus position. The autofocus control converts the distance provided by the triangulation sensor with a fixed frequency of 50 Hz into the control signal for lens positioning. Using static focusing, the strong variation of the detected LIBS signal due to a varying distance to the sample surface of greater than a few millimeters cannot be compensated by referencing the LIBS signals picked up by the fixed fiber optics to the zeroth-order signal. However, using dynamic focusing with the autofocusing unit, LIBS signals referenced to the zeroth order are constant within 5% over a measuring range of 50 mm. For different distances between the focusing optics and the sample, the referenced Pb (I) 405.78 nm LIBS signals are compared for static and dynamic focusing in Figure 11.11.

The limits of detection (LOD) determined for the investigated heavy metals and halogens with a laboratory set-up using static focusing as well as LODs determined for the on-line LIBS module with dynamic focusing are shown in Table 11.3.

Table 11.3. *Limits of detection (LOD) for the investigated elements in ABS samples.*
LODs calculated by the 3s-criterion

Element	λ(nm)	Static focusing		Dynamic focusing	
		LOD ($\mu g\ g^{-1}$)	BEC ($\mu g\ g^{-1}$)	LOD ($\mu g\ g^{-1}$)	BEC ($\mu g\ g^{-1}$)
Cd	228.80	11	45	96	272
Cr	425.43	4	42	73	226
Pb	405.78	24	356	140	448
Hg	253.65	22	198	60	73
Sb	259.80	75	213	80	93
Br	827.24	11 000	78 000	n. d.	n. d.

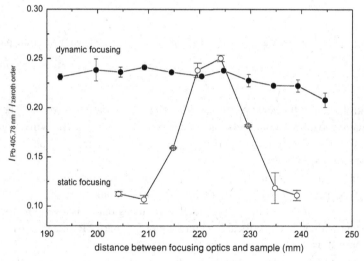

Figure 11.11. Comparison of Pb (I) 405.78 nm LIBS signal referenced to zeroth order with static and dynamic focusing.

In the on-line test, the concentrations of each detected heavy metal or halogen were classified by the LIBS module into four different categories: (a) 0–100 $\mu g\ g^{-1}$, (b) 100–1000 $\mu g\ g^{-1}$, (c) 1000–2000 $\mu g\ g^{-1}$, and (d) greater than 2000 $\mu g\ g^{-1}$. Lead was evaluated despite the fact that the LOD of 140 $\mu g\ g^{-1}$ was higher than the upper concentration limit of category zero. This was possible since the test samples did not contain lead with concentrations in the range of 110–140 $\mu g\ g^{-1}$. The classification results obtained for samples without surface preparation moving on the conveyor belt with a velocity of about 0.5 m s^{-1} are summarized in Table 11.4.

Table 11.4. *Results of the on-line LIBS module at the pilot sorting plant with real waste EOL-WEEE monitor samples moving at a velocity of* 0.5 m s^{-1}. *Classification of the measured concentration is performed into the categories* 0–100 μg g^{-1}, 100–1000 μg g^{-1}, 1000–2000 μg g^{-1}, > 2000 μg g^{-1}

	Sb	Cd	Pb	Cr	Hg[a]
No. of correct classifications	152	135	152	152	(160)
No. of total classifications	160	160	160	160	160
Ratio of correct classification (%)	95	84	95	95	(100)

[a]No mercury concentration >1 μg g^{-1} was found in real waste EOL-WEEE pieces with the ICP-OES reference analysis.

Figure 11.12. Commercial glass recycling machine integrated with the LIBS instrument at WEREC GmbH.

For Sb, Cd, Pb and Cr, high classification ratios of up to 95% are achieved. The on-line LIBS analyzer has proved the potential to characterize the heavy metal content in real WEEE monitors moving with 0.5 m s^{-1} on a conveyor belt with high accuracy. No specific surface treatment of the EOL-WEEE monitors is needed and height variations of up to 50 mm of the specimen surface owing to the shape were compensated by the presented autofocusing unit.

11.3.3 Classification of technical glasses during the recycling of electronic waste

The increasing yearly volume of electronic waste in Germany requires the development of methods and devices for the recycling of valuable materials. Technical glass, especially as used in cathode ray tubes, fluorescent lamps and discharge lamps, is of particular interest for recycling. Normally, the waste materials are collected as mixed fractions of glass featuring

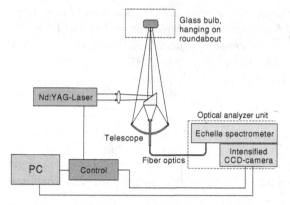

Figure 11.13. Principle of function of glass sorting LIBS instrument.

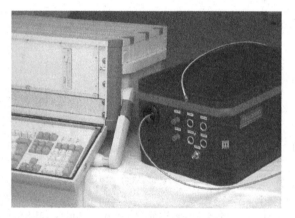

Figure 11.14. Echelle spectrometer of the glass sorting LIBS instrument.

an inhomogeneous chemical composition and a partial high content of toxic substances such as lead, strontium and barium. Only a small percentage is reusable, the main portion has to be sent to special landfill sites. Additionally, the production costs for discharge lamps are high.

A commercial laser–plasma spectrometer for classification of technical glass has been developed and manufactured by LLA Instruments GmbH [31]. The LIBS instrument was designed in such a way as to facilitate its integration with a commercial glass recycling machine, as illustrated in Figures 11.12 and 11.13. A Q-switched Nd:YAG laser with a pulse repetition rate of 10 Hz was used to generate a plasma on the surface of the glass and an echelle spectrograph was used to record the plasma emissions. The echelle spectrograph, illustrated in Figure 11.14, allows simultaneous multi-element analysis over the spectral range 200–780 nm and the 16-bit detector provides sufficient dynamic range to allow measurements to be taken over the full concentration range of the elements of interest. For each analysis, an accumulated spectrum produced from ten laser pulses was analyzed by purpose-designed software; the graphical user interface of the computer software is

Table 11.5. *Classification of analyzed glass and corresponding quantitatively determined elements*

Group of glass	B	Ba	Pb	Na
Hard glass 1	$B_2O_3 > 8\%$	$BaO < 0.1\%$	$PbO < 1\%$	
Hard glass 2	$B_2O_3 > 8\%$	$BaO > 1\%$	$PbO < 1\%$	
Soft glass 7	$B_2O_3 < 1\%$	$BaO < 1\%$	$PbO < 1\%$	$Na_2O > 14\%$

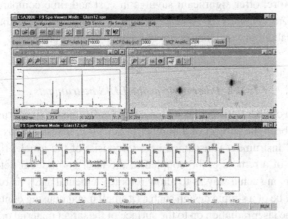

Figure 11.15. Definition of measuring task, monitoring of measurement process and quantitative analysis using customized software.

illustrated in Figure 11.15. The readout time of the CCD is approximately two seconds and hence the total analysis time per sample is approximately three seconds.

The LIBS instrument is used for determination of main glass components and trace elements. Firstly, a rough sorting is performed according to the main important glass types: borosilicate, aluminosilicate, alumoborosilicate, etc. Then a fine sorting is performed on the basis of different percentages of single glass components, e.g. Pb-content or Ba-content. The criteria for the recycling of high-pressure discharge lamps at WEREC GmbH are given in Table 11.5. Three different glass fractions are sorted on the basis of the quantitative analysis of Si, B, Pb, Ba and Na.

Owing to the universal measuring principle of LIBS, the technology is applicable to other industrial applications, including recycling of contaminated wood [32], analysis of mineral nutrients in wheat grain and wheaten flour [33], and classification of material defects in glass and tool steel [34].

11.4 Nuclear power generation and spent fuel reprocessing

11.4.1 Background

The remote, non-contact capabilities of LIBS make it particularly attractive for situations where information is required on the elemental composition of materials within radioactive environments. For example, there are many cases throughout the nuclear industry where compositional analysis of a plant component is required during routine inspection campaigns. Options for carrying out these inspections are often limited to removal of a physical sample for laboratory analysis; however, this is sometimes not possible as the removal of a sample may compromise the mechanical integrity of the component being examined. Compositional analysis of highly radioactive process materials during spent fuel reprocessing poses major problems since the removal, transport and chemical analysis of samples require highly specialized equipment and facilities. Analysis of components and process materials *in situ* can, therefore, offer significant savings in cost and time compared with physical sampling and laboratory analysis. Another benefit of *in situ* analysis is that, unlike the use of wet-chemical laboratory techniques, no secondary waste is produced.

11.4.2 Fiber optic probe LIBS instruments

The incorporation of fiber optic cables within a LIBS instrument to perform remote analysis of a material has been the subject of a number of scientific investigations [35–56]. A fiber optic probe LIBS instrument is assumed here to consist of the usual LIBS hardware but with the addition of a remote probe coupled to the laser and optical detector via one or more optical fibers in such a way as to allow remote elemental analysis of a material. The remote probe may be of a simple design such as a small metal housing containing the lenses needed to focus the laser radiation on to the surface of the target material or a more complex design incorporating additional components such as a miniature camera, a lighting unit and a clamping mechanism for attaching the probe to the component being examined.

In situ *batch identification of Magnox reactor control rods*

The first reported industrial use of a fiber optic probe LIBS instrument was undertaken only relatively recently for a unique application within the UK nuclear power generation industry [37]. In August 1993, crack-like indications were discovered in a secondary shut-down control rod at two of the UK's Magnox nuclear power stations. The ensuing investigation concluded that the outer casings of a batch of control rods were manufactured from steel containing abnormally low concentrations of silicon and were thus at risk of increased corrosion and possible premature failure. The location of this batch of control rods within the two reactors was, unfortunately, not known and hence a method of identification was required. The control rods in question were in service and were therefore highly radioactive. Physical sampling followed by laboratory analysis, besides being a time-consuming and costly process, was ruled out because of the risk of compromising the mechanical integrity of the control rods. The remote and essentially non-destructive analysis capabilities of LIBS were viewed as having the potential to provide a cost-effective solution to this problem and a fast-track development program was initiated to design and build a fiber optic probe LIBS instrument capable of identifying the suspect components. It was realized early on in the development program that measurement of the low levels of silicon by LIBS would be

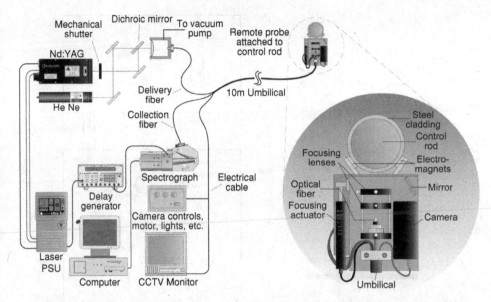

Figure 11.16. Schematic of fiber optic probe LIBS instrument developed by Duckworth in 1995.

difficult and the necessary level of accuracy was unlikely to be achieved. The specification of the control rod casing steel, however, showed that the low-silicon batch could be uniquely identified if the concentration of manganese and copper could be measured with sufficient accuracy. Accordingly, the LIBS instrument was designed and calibrated to measure the manganese and copper content of the steel. The general design of the LIBS instrument developed for this application is illustrated schematically in Figure 11.16.

The suspect control rods were removed from the reactor and lowered into a hot-cell facility equipped with remote handling equipment. The remote head of the LIBS instrument was designed to pass through a 180 mm diameter access port into the hot-cell and attach to the control rod under examination using a Master–Slave Manipulator (MSM). The remote head was connected to the main control unit of the LIBS instrument via a 10 m length of umbilical that provided mechanical protection for the optical fibers and electrical cables. A low-power He–Ne laser was included in the design of the instrument, the output of which was launched into the same optical fiber used to transmit the high-power Nd:YAG laser radiation to the measurement site. A time trace of the reflected He–Ne laser light was recorded to establish when the surface oxide layer had been removed by laser ablation. In this way, the necessary sample preparation was monitored so that LIBS measurements were carried out only when the underlying steel was adequately exposed. In practice, several hundred laser shots were used to prepare the surface of the steel followed by a further one hundred laser shots for the LIBS analysis. The deployment of the LIBS instrument proved to be a great success with each of the suspect control rods being readily identified.

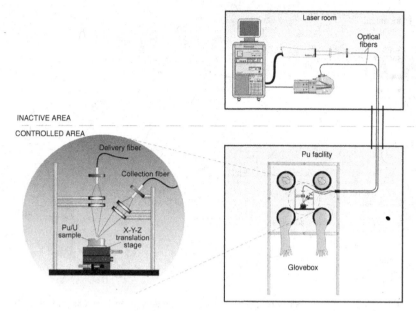

Figure 11.17. Schematic of glovebox fiber optic probe LIBS instrument used remotely to analyze the composition of uranium and plutonium nuclear fuel.

Trace metal analysis of uranium metal fuel

During the mid to late 1990s, fiber optic probe LIBS instruments found application in the nuclear fuel manufacturing and spent fuel reprocessing industries. One of these applications [47] involved a specially adapted glovebox facility suitable for performing LIBS analysis of actinide materials including plutonium metal, uranium metal and mixed-oxide (MOX) nuclear fuel. As illustrated schematically in Figure 11.17, the remote probe was situated within the glovebox with the remainder of the LIBS hardware situated in another room outside of the controlled (radiation) area. Two 30 m lengths of optical fiber were used to connect the remote probe to the LIBS hardware, one to deliver the Nd:YAG laser radiation and the other to collect the plasma light. Although primarily intended for research and development work, the system was used as a full-scale prototype of an on-line system for measurement of the iron and aluminum content of uranium metal fuel.

In situ *compositional analysis of Advanced Gas-Cooled Reactor (AGR) superheater tubes*

A novel-design fiber optic probe LIBS instrument incorporating a 75 m armored umbilical has been developed for use within the pressure vessels of Advanced Gas-Cooled Reactor (AGR) nuclear power stations [56]. The instrument was designed to remotely determine the copper content of 316H austenitic stainless-steel superheater bifurcation tubing with the aim of identifying components manufactured from a certain batch of steel. It was suspected that this batch of steel suffered from low-creep ductility, a condition that could result in premature failure of the bifurcation welds with the consequent risk of steam leaks. This represented a significant commercial threat to the operation of the reactors and hence identification of the suspect components was necessary. It was believed that the suspect batch of steel exhibited higher than normal concentrations of copper and that this may have helped in the process of identification of suspect bifurcations. Chemical analysis of the steel could be used to measure the copper concentration but the severely restricted physical access to the bifurcations and the hostile environment of the reactor pressure vessel were such that removal of physical samples of the components for laboratory analysis was not feasible. The problem was exacerbated by the number of bifurcations requiring inspection, 528 in each of four reactors, and the limited time available to conduct a survey of each reactor. Inspections could be carried out only during a routine reactor outage, during which time the reactor is shut down and the pressure vessel opened up to allow man-access to the steam generators. Reactor outages are typically six weeks in duration but, since the reactor pressure vessel must be de-pressurized and allowed to cool before man-access can go ahead, only two to three weeks are available for in-vessel inspection work. When allowing for other in-vessel work, approximately only ten days were available for identification of the suspect bifurcations. The rapid analysis and remote capabilities of the LIBS technique offered an attractive solution to this problem. The environment within the pressure vessel during a routine reactor outage is particularly demanding for the in-vessel inspection personnel. Ambient temperatures of up to 60 °C, high noise levels, buffeting from the flow of cooling air, severely restricted physical access, and confined working space placed severe physical demands on the in-vessel inspection team. The environment of the pressure vessel is classed in radiological terms as an R4/C3 area, indicating that radiation levels are relatively high and airborne radioactive contamination is possible. Specially designed air-fed "hot-suits" are used to provide protection for the vessel entrants against heat, noise and radioactive contamination. An umbilical is used to supply cooling and breathing air to the suit and to provide a communications link with the vessel entry controller and other vessel entrants. The descent into the vessel from the pile cap is made via a 15 m vertical access ladder which leads to a circular walkway situated between the dome of the reactor and the twelve steam generators, as illustrated in Figure 11.18. Between each steam generator is another vertical ladder leading down to the sub-boiler annulus at the lowest part of the pressure vessel. After descending this ladder for about 4 m, access to a superheater is made via a small hatch. Each superheater is approximately 4 m × 2.5 m in cross-section with a central baffle plate situated along the centre line; see Figure 11.19. Either side of the central baffle plate are located three rows of bifurcations. During inspection, a section of the baffle plate is removed to allow access to all of the 44 bifurcations. There is approximately 1 m of headroom inside the superheaters, meaning that the inspection and repair work must be carried out with the vessel entrant in a semi-prone position on top of the banks of bifurcations. Each vessel entry period is limited to a maximum of 2 h; after allowing for the time taken to enter and exit the vessel, this leaves little more than 1 h to carry out the inspection/repair work. A three-shift

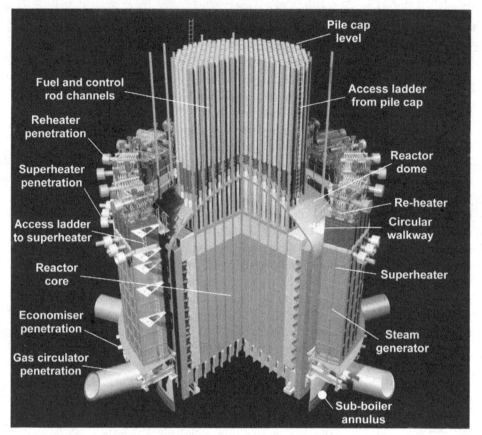

Figure 11.18. CAD view of an Advanced Gas-Cooled Reactor illustrating the general layout of the major components of the reactor pressure vessel interior (reproduced with the permission of British Energy Generation Ltd).

working pattern was adopted with the aim of completing a survey of a superheater within two to three shifts.

The design of the fiber probe LIBS instrument used during this work is illustrated schematically in Figure 11.20 and is described in detail elsewhere [56]. The remote probe was integrated with a purpose-designed pneumatic clamping mechanism suitable for attachment to a bifurcation. Laser safety was a significant factor in the design of the system because it was not practicable for the vessel entrant with the task of deploying the probe to wear safety goggles. A safety interlock system was designed which monitored the air pressure and flow-rate to the umbilical and probe. A laser shutter was interlocked to the pneumatic system in such a way that the shutter could be opened only when the air pressure reached a pre-set value and the flow-rate reduced to zero. Incorrect attachment of the probe or damage to the

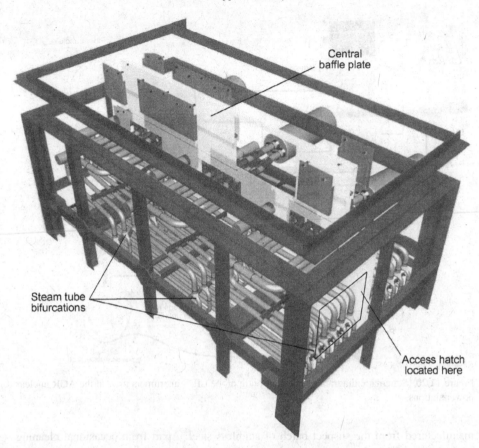

Figure 11.19. CAD view of an AGR superheater illustrating the location of the access hatch, the central baffle plate and the 44 bifurcations (reproduced with the permission of British Energy Generation Ltd).

umbilical would result in a loss of air pressure and an increase in flow-rate, the detection of either by the interlock system would cause the laser shutter to close.

The control module of the LIBS instrument was located in a cabin above the reactor pile cap. The probe and umbilical were fed into the vessel via the man-access route that leads down from the pile cap to the circular walkway surrounding the dome of the reactor. During deployment (see Figure 11.21), the vessel entrant would attach the probe to a bifurcation, indicating via the communication link to the laser operator that a measurement could commence. Each measurement would take approximately 2 min with the results being recorded to the hard disk of the computer and cross-referenced with the identification number of the bifurcation.

The accuracy of the copper concentration measurements from the LIBS instrument was approximately ±25%, which was adequate to allow identification of bifurcations

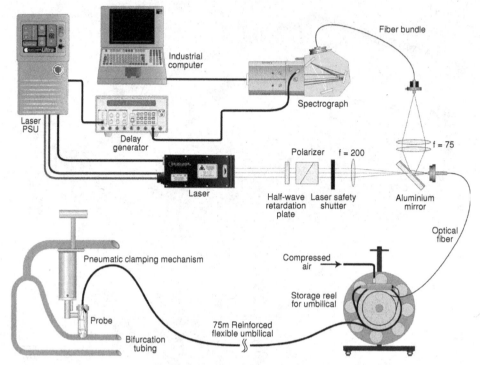

Figure 11.20. Schematic diagram of the fiber optic probe LIBS instrument used at the AGR nuclear power stations.

manufactured from the suspect batch of stainless steel. Apart from occasional cleaning of the lenses within the remote probe and periodic in-vessel calibration checks, the LIBS instrument operated without failure of any of the components. Two spare optical fibers were included in the design of the instrument; however, only one fiber was used throughout all measurements taken initially at Hunterston "B" station in Scotland and subsequently at Hinkley Point "B" station in the south of England.

In situ *compositional analysis of economizer tubes within the sub-boiler annulus of an AGR pressure vessel*

In early 2001, a modified version of the fiber optic LIBS instrument described above was used to remotely determine the chromium content of mild-steel economizer tubing within the sub-boiler annulus (SBA) of reactor 4 at Hunterston B station (see Figure 11.18 for location of the SBA). The measurements were required as part of an inspection program undertaken to identify tubes that had suffered damage through the effects of a process known as erosion-corrosion. Erosion-corrosion is a significant problem in the nuclear industry since it essentially leads to the thinning of tube walls at rates of up to several millimeters per year. Erosion-corrosion is a mechanical process affecting tubes whereby oxide crystals are

Figure 11.21. Deployment of the LIBS probe within an AGR superheater. The upper sections of the first bank of bifurcations can be seen in the lower foreground of the image.

removed from the surface of the tube wall by either high shear-stress or particles introduced via the flow. Erosion-corrosion has been found to be particularly associated with mild-steel tubing at elevated temperatures of 90–300 °C [57]. The rate of erosion-corrosion depends on various factors including temperature, flow-rate, water chemistry and the composition of the tube material. Tubes manufactured from mild steel containing less than approximately 0.1% Cr are far more likely to be affected by erosion-corrosion. Accordingly, identification of "at risk" tubes could be achieved by measuring the chromium content of the tube material.

Access to the SBA was achieved by removing one of the gas-circulators and installing a temporary air-lock and change-room structure within the circulator port. As shown in Figure 11.22, the vessel entrants are required to wear protective clothing and equipment due to the nature of the environment. The fiber optic umbilical of the LIBS instrument was fed into the SBA via a ventilation duct situated adjacent to the air-lock. Temporary scaffold towers were erected within the SBA to facilitate access to the banks of economizer tubes located directly beneath the steam generators and approximately 6 m above the floor level, as shown in Figure 11.23.

The LIBS instrument included a 30 m umbilical and a purpose-designed probe suitable for attaching to the 19.1 mm diameter tubes. Deployment of the instrument required two people; one person to operate the main control unit (the laser operator) and a vessel entrant (the probe operator). The main control unit of the instrument was located outside of the SBA adjacent to the vessel-entry change-room area. A safety electrical switch was fitted to the probe in such a way that the probe operator needed to hold the switch in the on position before the laser operator could activate the Nd:YAG laser.

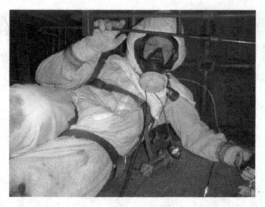

Figure 11.22. Protective clothing required for working within the sub-boiler annulus.

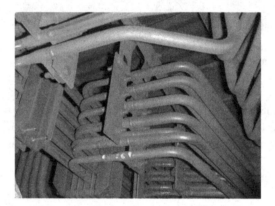

Figure 11.23. View of a bank of economizer tubes situated approximately 6 m above the floor of the SBA.

Calibration of the instrument was achieved by using the reference materials given in Table 11.6. All steel samples were certified reference materials (CRMs) with the exception of MGNX 001 which was analyzed by an independent laboratory.

Calibration measurements were taken using 1600 pre-conditioning, "burn in" laser pulses (80 s) followed by two measurements each consisting of 100 laser pulses (10 s). These measurements were repeated for five different locations on each of the reference materials. The resulting calibration curve is given in Figure 11.24. Small sampling statistics were used to calculate the mean values for each set of measurements. The error bars indicate the 95% confidence limits of a given measurement.

The limit of detection for chromium was approximately 100 p.p.m. and the measurement precision was typically better than 6%. By performing multiple repeat measurements on a selection of the reference materials over a period of several days, the measurement accuracy was determined to be better than 15%.

Table 11.6. *Reference materials used to calibrate the LIBS instrument*

Reference material	Cr (% by mass)	Fe (% by mass)
ECRM 097-1	0.0016	99.9
MGNX 001	0.027	99.0
SS 452	0.042	98.8
SS 434/1	0.055	97.5
SS 452/1	0.067	97.6
BS 13B	0.081	98.6
SS 451/1	0.104	98.3
BS CA 5A	0.128	97.3
SS 435/1	0.14	98.2
CENIM 301	0.165	98.2
SS 455/1	0.21	97.6
SS 433/1	0.26	98.5
SS 432/1	0.31	98.0
SS 404/1	0.48	96.4

The LIBS instrument was deployed successfully at Hunterston "B" AGR power station in February 2001. Measurements of a number of tubes were carried out, the results of which indicated that the tubes examined were manufactured from steel having a chromium content of higher than 0.1% and hence were unlikely to be affected by erosion-corrosion. These findings were consistent with the results of previous video probe inspections of the internal bore of the tubes which showed that the tubes had not suffered from excessive erosion.

11.4.3 Telescope LIBS instruments

A telescope LIBS instrument is assumed here to consist of the usual LIBS hardware configured in such a way as to direct the laser beam to the sample by a system of mirrors and lenses. Sample stand-offs of centimeters to tens of meters are possible with this configuration, and the instrument may be used remotely to analyze materials where line-of-sight access exists. The optical path between the LIBS instrument and the sample may include transparent materials such as a lead-glass shield window and/or additional mirrors used to redirect the laser beam where necessary.

In situ *measurements of oxide thickness inside a Magnox nuclear reactor*

The first reported industrial use of a telescope LIBS instrument within the nuclear industry was for an application within Magnox nuclear power reactors [35]. The instrument was developed at the Marchwood Engineering Laboratories of the UK's Central Electricity Generating Board during 1974 and was used to measure the thickness of magnetite (Fe_3O_4)

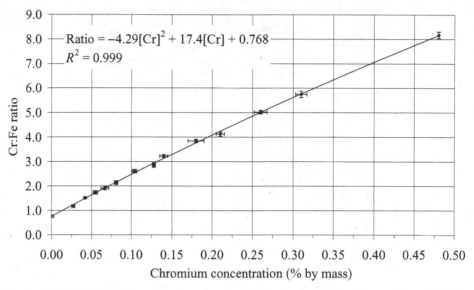

Figure 11.24. Calibration curve for chromium in mild steel using a 30 m fiber optic probe LIBS instrument.

layers on mild-steel surfaces within the hot and highly radioactive environment of the reactor core. The instrument, illustrated schematically in Figure 11.25, consisted of a He–Ne laser and a pulsed ruby laser enclosed in a cylindrical canister approximately 490 mm long with 150 mm diameter. The canister employed a water-cooled jacket to protect the components of the instrument from the ambient temperature of the reactor core (~200 °C). The magnetite layer was ablated in a controlled manner by focusing the output of the ruby laser onto the surface of the steel component under examination. The output of the He–Ne laser was focused at precisely the same point as the ruby laser beam and the reflected intensity monitored. Once the magnetite layer had been breached by the ruby laser, the intensity of the reflected He–Ne laser beam increased sharply since the surface reflectivity of the mild-steel substrate is significantly higher than that of the magnetite. With appropriate calibration, the thickness of the magnetite layer could be obtained by counting the number of ruby laser pulses required to cause a sharp increase in the reflected He–Ne laser beam. In order to allow remote measurements of the composition of the reactor components, the instrument was later modified to include plasma light collection optics and fiber optic delivery of the light to a remote spectrometer.

Identification of cooling-water tubes and reinforcing bars within Magnox reactor pressure vessels

During a maintenance and repair program at a Magnox nuclear power station in the UK, there was a requirement to drill a large number of bore-holes into the steel-reinforced concrete pressure vessels of the two reactors. The bore-holes, which were of 20 mm diameter and up to 1 m in depth, were to be used for anchor points for a reinforcing structure being added to

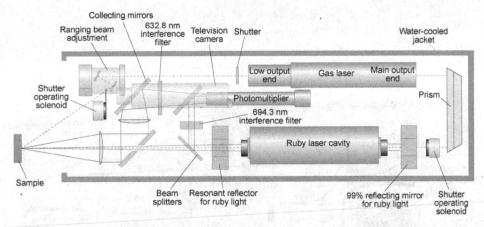

Figure 11.25. Schematic diagram of the telescope LIBS instrument used to measure the oxidation thickness for steel components within a Magnox reactor.

the steam penetration tubes. Within the vicinity of the steam tubes, cooling-water tubes were embedded in the concrete structure to maintain the temperature of the concrete within certain limits. It was imperative not to damage these cooling-water tubes when drilling the bore-holes and hence a method was devised to stop the drilling process automatically in the event of contact with a metallic object. Because of the large number of reinforcing bars also present within the concrete structure, a method was required to distinguish between a cooling-water tube and a reinforcing bar. Physical access to the component was severely restricted owing to the diameter and length of the bore-hole, making the task of identification particularly difficult. As there was line-of-sight access to the steel component, it was proposed to use a LIBS instrument to identify the component through differences in material composition.

It was known that the reinforcing bars contained approximately 1.5% manganese, whereas the cooling-water tubes contained less than 0.8% manganese. The measurement precision of the LIBS instrument was sufficient to resolve this difference and hence identification of the component could be achieved. A purpose-designed LIBS instrument, illustrated schematically in Figures 11.26 and 11.27, was subsequently developed for this application and consisted of a control unit with umbilical connection to a laser module.

The laser module, which weighs approximately 25 kg, was attached to the pressure vessel wall by means of either (i) mechanical attachment to bolt anchor points within the wall or (ii) heavy-duty vacuum cups designed for use with rough surfaces. The latter method of attachment incorporated a multiply redundant design and other safety features to minimize the risk of the instrument becoming detached from the wall through, for example, failure of the vacuum pumps. The instrument incorporated a remote viewing camera and an array of high-intensity LEDs used to illuminate the inside of the bore-hole. A view of the inside of the bore-hole was displayed on a monitor and the operator could adjust the optical alignment of the instrument using remote-controlled steering and focusing of the laser beam. Identification of the metallic component was achieved by activating the high-power laser for a period of several seconds. Results of the measurement were displayed on the

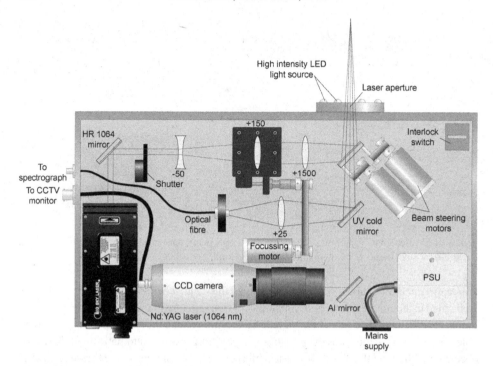

Figure 11.26. Schematic diagram of the laser module of the telescope LIBS instrument used to identify reinforcing bars and cooling-water tubes within the concrete pressure vessel of a Magnox reactor.

computer monitor and stored to the hard-drive together with the bore-hole identification number. Identification of a reinforcing bar/cooling-water tube could be performed within 15 m, including the time needed to attach the laser module to the reactor pressure wall and align the laser beam with the metallic component. A view of the laser module attached to the pressure vessel wall is given in Figure 11.28.

Non-invasive compositional analysis of radioactive contamination within a hot-cell

The first use of a telescope LIBS instrument remotely to analyze material within a hot-cell by directing the laser beam through a lead-glass radiation shield window was carried out in September 2001 at a nuclear reprocessing plant in the UK. A brief description of the background to this work is now given.

Routine inspections of some steel components in a hot-cell had identified an accumulation of a significant quantity of surface contamination. Characterization of the contamination was required prior to decontamination and waste sentencing of the components. Radiometric measurements were taken to identify the radionuclide inventory of the contamination but, as these provided no information on the non-active components, full characterization was not possible. The material was highly radioactive, making sample removal and laboratory analysis very difficult.

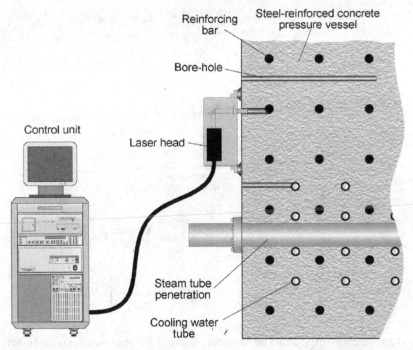

Figure 11.27. Schematic diagram illustrating the method of deployment of the LIBS instrument.

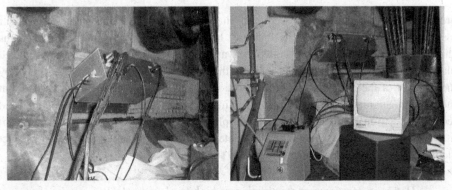

Figure 11.28. Images of the LIBS instrument being deployed at the reinforced concrete pressure vessel wall of a Magnox reactor.

Optical access to the material was possible via a 1 m thick lead-glass radiation shield window. The steel component could be positioned approximately 3 m beyond the window and raised/lowered by means of a hoist within the hot-cell. The laser beam of the LIBS instrument could then be transmitted into the hot-cell via the radiation shield window and focused onto the surface of the contamination, as illustrated in Figure 11.29. The optical

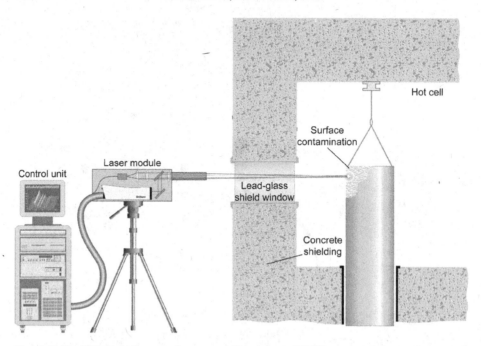

Figure 11.29. Schematic diagram of a transportable telescope LIBS instrument being used to remotely analyze the composition of a highly radioactive material by directing the laser beam through the shield window of the hot-cell.

properties of the shield window were such that it was reasonably transparent to the laser radiation (>75% transmission at 1064 nm) but was effectively opaque to wavelengths shorter than about 500 nm. This limited the wavelength range over which the laser-induced plasma could be monitored to 500–800 nm, the long-wavelength limit being governed by the detector of the optical spectrograph. The results of the measurements suggested that the contaminant material was rich in zirconium and molybdenum, as illustrated in a typical recorded spectrum given in Figure 11.30. The plant operators suspected that the contaminant material was likely to be zirconium molybdate with traces of uranium and fission-product elements. The LIBS measurements helped to support this view.

Detection of surface contamination on high-level waste (HLW) containers

The preferred method of processing various types of radioactive waste including high-level waste (HLW) is vitrification whereby the waste is blended with a glass or ceramic matrix to produce a chemically stable material suitable for long-term storage or disposal in an underground depository. The vitrification process is currently employed by a number of countries including France, the UK and the USA. After the containers have been filled with waste and sealed by welding a lid in place, the containers are then decontaminated by using a water spray system. The external surfaces are subsequently monitored for residual

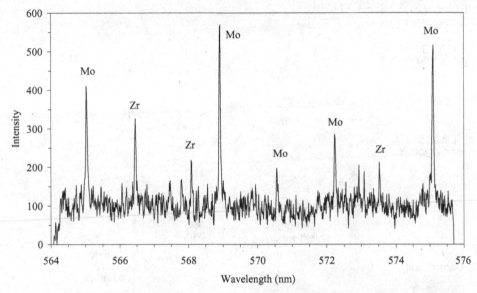

Figure 11.30. Example of a (single laser shot) spectrum obtained during "through-window" LIBS compositional analysis of the contaminant material. The detection of zirconium and molybdenum in relatively high abundance was consistent with the material being composed mainly of zirconium molybdate.

contamination prior to being transported to an interim storage facility. The monitoring process is based on the use of a robotic swabbing technique whereby the external surfaces of the container are swabbed to pick up a fraction of any loose contamination present. The swab is then posted to a radiation monitoring instrument that is shielded from the gamma radiation emanating from the filled container, thus allowing highly sensitive measurements of loose surface contamination to be made. An undesired aspect of the vitrification process is that small quantities of vitrified product arising from splashing during the filling process may become strongly adhered to the outside walls of the container. This so-called "fixed" contamination may not be detected by the robotic swab monitoring system and could go unnoticed. The contamination may later become detached from the container, either during transport or while in storage, and could constitute a significant radiological hazard.

A new type of monitoring instrument based on LIBS has been developed by Applied Photonics Ltd [58]; this instrument provides a complementary monitoring technique to robotic swabbing by allowing the direct detection of "fixed" surface contamination. Because LIBS, in general, lacks the sensitivity required to detect the radionuclides directly, the LIBS instrument operates by detecting the matrix material (glass or ceramic) used to encapsulate the radioactive materials. The glass matrix contains high concentrations of sodium and lithium, both of which exhibit very intense emission lines and hence are readily detected by the LIBS instrument. The mode of operation is illustrated schematically in Figure 11.31.

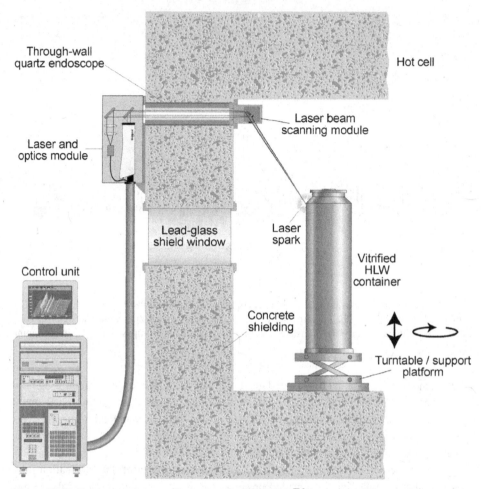

Figure 11.31. Schematic diagram illustrating the LIBSCAN™ instrument being used to remotely detect "fixed" contamination on the surface of an HLW container.

The LIBS instrument incorporates a "through-wall" optical endoscope similar in design to camera viewing endoscopes commonly used throughout the nuclear industry. These endoscopes consist essentially of an optical-quality quartz rod inside a shielding plug that fits into a port within the shield wall of the hot-cell. The majority of the instrument hardware is located on the "cold" side of the hot-cell with only the beam steering mirror and focusing lens exposed to the nuclear radiation, both of which may be manufactured from radiation-tolerant materials.

A typical spectrum obtained from the instrument during a full-scale inactive trial using highly representative simulated contamination is shown in Figure 11.32. The intense atomic emission lines of Na and Li are readily detected by the instrument, allowing reliable and

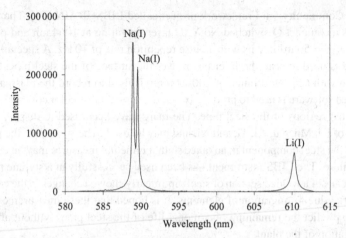

Figure 11.32. Spectra obtained from simulated vitrified HLW showing very intense Na and Li atomic emission lines.

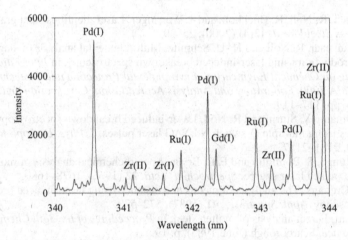

Figure 11.33. Spectra obtained from simulated vitrified HLW showing atomic and ionic emission lines characteristic of various fission-product elements.

sensitive detection of "fixed" contamination. Owing to the relatively high abundance of fission-products within the HLW, the instrument is able also to detect certain of these species, as illustrated in Figure 11.33.

11.5 Miscellaneous industrial applications of LIBS

11.5.1 In situ *compositional analysis of steel pipes at high temperature*

A transportable LIBS instrument designed remotely to measure the composition of steel pipes whilst in service and at a temperature of around 700 °C has been developed as part of a European Community 5th Framework project called LIBSGRAIN [59]. The instrument incorporates a compact Q-switched Nd:YAG laser operating at 1064 nm and producing a pulsed output of ~50 mJ in 7 ns with a pulse repetition rate of 10 Hz. A specially designed optical head is used to scan the laser beam over the surface of the steel pipe. An echelle spectrograph with a spectral range of 200–800 nm is used to record the plasma emissions and dedicated software is used to produce two-dimensional false-color compositional maps of the selected regions of the steel pipe. The maps have been used to display the spatial distribution of Fe, Mn, Cu, Al, Ti, and Ni, and may be used to help estimate the probability of failure of the steel component in advance so that remedial measures may be taken before an actual failure. The LIBS instrument has been used successfully at a styrene plant in the UK where scans of a welded section of steel pipe clearly showed changes in the composition of the material due to migration of elements. It is hoped that these measurements may be used to help predict the remaining operational life of the steel pipes without affecting the normal operation of the plant.

11.6 References

[1] I. Mönch, R. Noll, R. Buchholz and J. Worringer, Laser identifies steel grades, *Stainless Steel World*, **12**(4) (2000), 25–29.
[2] M. Kraushaar, R. Noll and H.-U. Schmitz, Multi-elemental analysis of slag from steel production using laser-induced breakdown spectroscopy, in *International Meeting on Chemical Engineering, Environmental Protection and Biotechnology, ACHEMA 2000, Laboratory and Analysis Accreditation, Certification and QM* (2000), pp. 117–119.
[3] R. Sattmann, V. Sturm and R. Noll, Laser-induced breakdown spectroscopy of steel samples using multiple Q-switch Nd:YAG laser pulses. *J. Phys. D: Appl. Phys.*, **28** (1995), 2181–2187.
[4] E. F. Runge, S. Bonfiglio and F. R. Bryan, Spectrochemical analysis of molten metal using a pulsed laser source. *Spectrochim. Acta*, **22** (1966), 1678–1680.
[5] D. A. Cremers, The analysis of metals at a distance using laser-induced breakdown spectroscopy. *Appl. Spectrosc.*, **41** (1987), 572–579.
[6] R. Jowitt, Direct analysis of molten steel, in *Proceedings of the 38th Chemists' Conference*, Scarborough (June 1985), p. 19.
[7] R. Jowitt, Laser analysis of liquid steel, in *Proceedings of the International Conference on Progress Analysis in the Chemical Iron and Steel Industry*, Luxembourg (May 1991).
[8] C. Carlhoff and S. Kirchhoff, Direct analysis in steelmaking converters using laser-induced emission spectrometry, in *Proceedings of the 3rd International Conference on Progress Analysis in the Chemical Iron and Steel Industry*, R. Nauche (editor), ECSC-EEC-EAEC, Brussels, Luxembourg (1992), pp. 150–153.
[9] C. Aragon, A. Aguilera and J. Campos, Determination of carbon content in molten steel using laser-induced breakdown spectroscopy. *Appl. Spectrosc.*, **47** (1993), 606–608.
[10] L. Moencke-Blankenburg, Möglichkeiten und Grenzen der Analytik flüssigen Stahls mittels laserinduzierter Atomemissionsspektrometrie, in *CANAS '93 Colloquium Analytische Atomspektroskopie*, Klaus Dittrich and Bernhard Walz (editors), Universität Leipzig, Germany (1993), pp. 165–180 (in German).

[11] R. Noll, V. Sturm, L. Peter and I. R. C. Whiteside, Analysis using lasers, in *49th Chemists' Conference*, Scarborough (Oct. 14–16, 1997), pp. 22–27.

[12] R. Noll, R. Sattmann, V. Sturm, S. Lüngen and H.-J. von Wachtendonk, Schnelle Multielement-Analyse in der Stahlschmelze mit laserinduzierter Emissionsspektrometrie. *Stahl und Eisen*, **117** (1) (1997), pp. 57–62 (in German).

[13] L. M. Cabalín, D. Romero, P. Lucena, J. M. Baena and J. J. Laserna. Laser induced breakdown spectrometry for stainless steel characterization, in *Proceedings of the 5th International Conference on Progress Analysis in the Chemical Steel and Metals Industries*, European Commission, Luxembourg (1999), p. 336.

[14] J. Gruber, J. Heitz, H. Strasser, D. Bäuerle and N. Ramaseder, Rapid in-situ analysis of liquid steel by laser-induced breakdown spectroscopy. *Spectrochim. Acta Part B*, **56** (2001), 685–693.

[15] V. Sturm, L. Peter and R. Noll, Steel analysis with laser-induced breakdown spectrometry in the vacuum ultraviolet. *Appl. Spectrosc.*, **54** (2000), 1275–1278.

[16] M. Hemmerlin, R. Meilland, H. Falk, P. Wintjens and L. Paulard, Application of vacuum ultraviolet laser-induced breakdown spectrometry for steel analysis. *Spectrochim. Acta, Part B*, **56** (2001), pp. 661–669.

[17] V. Sturm, L. Peter, R. Noll *et al.*, Elemental analysis of liquid steel by means of laser technology, in *International Meeting on Chemical Engineering, Environmental Protection and Biotechnology*, ACHEMA 2000, Materials Technology and Testing (May 2000), pp. 9–11.

[18] *Ullmann's Encyclopedia of Industrial Chemistry*, **A20** (VCH Publishers Inc., 1992).

[19] Brite Euram III project, *Development of multipurpose industrial units for recycling of plastic wastes by on-line pattern recognition of polymer feature (Sure-Plast)*, contract number BRPR-CT98-0783.

[20] J. Marshall, J. Carrol, J. S. Crighton and C. Barnard, Industrial analysis: metals, chemicals and advanced materials. *J. Anal. Atom. Spectrom.*, **9** (1994), 319–356.

[21] K. J. Grant, G. L. Paul and J. A. O'Neill, Quantitative elemental analysis of iron ore by laser-induced breakdown spectroscopy. *Appl. Spectrosc.*, **45** (1991), 701–705.

[22] R. Zenobi, Modern laser mass spectrometry. *Fresenius J. Anal. Chem.*, **348** (1994), 506–509.

[23] Z. Weiss, New method of calibration for glow-discharge optical emission spectrometry, *J. Anal. Atom. Spectrom.*, **9** (1994), 351–354.

[24] A. Golloch and D. Siegmung, Sliding spark spectroscopy – rapid survey analysis of flame retardants and other additives in polymers. *Fresenius J. Anal. Chem.*, **358** (1997), 804–811.

[25] R. Gijbels, Elemental analysis of high-purity solids by mass spectrometry. *Tantala*, **37** (4) (1990), 363.

[26] R. Sattmann, I. Mönch, H. Krause *et al.*, Laser-induced breakdown spectroscopy for polymer identification. *Appl. Spectrosc.*, **52** (1998), 456.

[27] J. M. Anzano, I. B. Gornushkin, B. W. Smith and J. D. Winefordner, Laser-induced plasma spectroscopy for plastic identification. *Polymer Eng. Sci.*, **40** (2000), 2423–2429.

[28] H. Fink, U. Panne and R. Niessner, Analysis of recycled thermoplastics from consumer electronics by laser-induced plasma spectroscopy. *Anal. Chim. Acta*, **440** (2001), 17–25.

[29] M. Tran, Q. Sun, B. W. Smith and J. D. Winefordner, Determination of F, Cl, and Br in solid organic compounds by laser-induced plasma spectroscopy. *Appl. Spectrosc.*, **55** (2001), 739–744.

[30] R. Noll, H. Bette, A. Brysch *et al.*, Laser-induced breakdown spectrometry – applications for production control and quality assurance in the steel industry, *Spectrochim. Acta, Part B – Atom. Spectrosc.*, **56** (2001), 637–649.

[31] LLA Instruments GmbH, *Verfahren und Vorrichtung zur Bestimmung der Materialzusammensetzung von Stoffen*, patent DE-4341462.1.

[32] A. Uhl, K. Loebe and L. Kreuchwig, Fast analysis of wood preservers using laser breakdown spectroscopy. *Spectrochim. Acta, Part B*, **56**, (2001), 795–806.

[33] K. Löbe, H. Lucht, B. Handreck and J. Dörfer, On-line Mineralstoffanalyse von Weizenkörnern und -mehlen mittels laserinduzierter Plasmaspektroskopie. *Getreide Mehl und Brot.*, **51** (1997), 131–136.

[34] K. Loebe, H. Lucht and A. Uhl, Laser micro analysis of glass and tool steel, *LIBS 2002 Technical Digest, Conference on Laser-Induced Plasma Spectroscopy and Applications*, Orlando, Florida, September 25–28 (2002), pp. 206–208.

[35] B. A. Tozer, Remote measurement of oxide thickness and steel composition using laser techniques. *Opt. Las. Technol.*, (1976), pp. 57–64.

[36] R. Nyga and W. Neu, Double-pulse technique for optical emission spectroscopy of ablation plasmas of samples in liquids. *Opt. Lett.*, **18** (9) (1993), 747–749.

[37] A. Duckworth, Remote metal analysis by laser-induced breakdown spectroscopy, *British Nuclear Engineering Society (BNES) Conference Proceedings, Remote Techniques for Hazardous Environments* (1995), pp. 259–263.

[38] C. M. Davies, H. H. Telle, D. J. Montgomery and R. E. Corbett, Quantitative analysis using remote laser-induced breakdown spectroscopy (LIBS). *Spectrochim. Acta, Part B*, **50** (1995), 1059–1075.

[39] G. A. Theriault and S. H. Lieberman, Remote in-situ detection of heavy metal contamination in soils using a Fiber Optic Laser Induced Breakdown Spectroscopy (FOLIBS) System. *SPIE*, **2504** (1995), 75–83.

[40] D. A. Cremers, J. E. Barefield II and A. C. Koskelo, Remote elemental analysis by laser-induced breakdown spectroscopy using a fibre-optic cable. *Appl. Spectrosc.*, **49** (1995), 857–860.

[41] W. E. Ernst, D. F. Farson and D. J. Sames, Determination of copper in A533b steel for the assessment of radiation embrittlement using laser-induced breakdown spectroscopy. *Appl. Spectrosc.*, **50** (1996), 306–309.

[42] J. Young, I. M. Botheroyd and A. I. Whitehouse, Remote analysis of nuclear and non-nuclear materials using laser-induced breakdown spectroscopy, in *Conference on Electro-optics and Lasers (CLEO/Europe)*, Hamburg, Germany (1996), p. 152.

[43] S. Saggese and R. Greenwell, LIBS fiber optic sensor for subsurface heavy metals detection. *SPIE*, **2836** (1996), 195–205.

[44] G. A. Theriault and S. H. Lieberman, Field deployment of a LIBS probe for rapid delineation of metals in soils. *SPIE*, **2835** (1996), 83–88.

[45] B. J. Marquardt, S. R. Goode and S. M. Angel, In-situ determination of lead in paint by laser-induced breakdown spectroscopy using a fibre-optic probe. *Anal. Chem.*, **68** (1996), 977–981.

[46] C. M. Davies, H. H. Telle and A. W. Williams, Remote in situ analytical spectroscopy and its applications in the nuclear industry. *J. Anal. Chem.*, **355** (1996), 895–899.

[47] J. Young, I. M. Botheroyd and A. I. Whitehouse, Trace element analysis of uranium metal fuel using laser-induced breakdown spectroscopy, in *Conference on Electro-optics and Lasers (CLEO/Europe)*, Glasgow, UK (1998), p. 201.

[48] I. M. Botheroyd, J. Young, A. I. Whitehouse and A. Duckworth, Remote analysis of steels and other solid materials using laser-induced breakdown spectroscopy (LIBS),

in *Conference on Lasers and Electro-optics-Europe (CLEO/Europe)*, Glasgow, UK (1998), p. 198.

[49] B. J. Marquardt, D. N. Stratis, D. A. Cremers and S. M. Angel, Novel probe for laser-induced breakdown spectroscopy and Raman measurements using an imaging optical fibre. *Appl. Spectrosc.*, **52** (9) (1998), 1148–1153.

[50] G. A. Theriault, S. Bodensteiner and S. H. Lieberman, A real-time fibre-optic LIBS probe for the in situ delineation of metals in soils. *Field Anal. Chem. Technol.*, **2** (2) (1998), 117–125.

[51] R. E. Neuhauser, U. Panne and R. Niessner, Laser-induced plasma spectroscopy: a versatile tool for monitoring heavy metal aerosols. *Anal. Chem. Acta*, **392** (1999), pp. 47–54.

[52] D. N. Stratis, K. L. Eland and S. M. Angel, Characterisation of laser-induced plasmas for fiber optic probes. *Proc. SPIE*, **3534** (1999), 592–600.

[53] K. L. Eland, D. N. Stratis, J. C. Carter and S. M. Angel, Development of a dual-pulse fiber optic probe for in-situ elemental analyses. *Proc. SPIE*, **3853** (1999), pp. 288–294.

[54] R. E. Neuhauser, U. Panne and R. Niessner, Utilization of fibre optics for remote sensing by laser-induced plasma spectroscopy (LIPS). *Appl. Spectrosc.*, **54** (6) (2000), p. 923.

[55] P. A. Mosier-Boss and S. H. Lieberman, Direct push fiber-optic laser induced breakdown spectroscopy (FO-LIBS) sensor probe for real-time, in-situ measurements of metals in soils. *National Meeting (Chicago, Aug. 26–30, 2001) – American Chemical Society, Division of Environmental Chemistry*, **41** (No. 2, Part 3) (2001), 603–609.

[56] A. I. Whitehouse, J. Young, I. M. Botheroyd *et al.*, Remote material analysis of nuclear power station steam generator tubes by laser-induced breakdown spectroscopy. *Spectrochim. Acta, Part B*, **56** (2001), 821–830.

[57] B. Poulson, Complexities in predicting erosion-corrosion. *Proc. Of Wear – Lausanne*, **233–235** (1999), pp. 497–504.

[58] Applied Photonics Ltd, *Laser spectroscopic remote detection of surface contamination*, International Patent Application No. PCT/GB01/00866.

[59] G. Cristoforetti *et al.*, In-situ LIBS analysis of steel pipes at high temperature, *LIBS 2002 Technical Digest, Conference on Laser Induced Plasma Spectroscopy and Applications*, Orlando, Florida, September 25–28 (2002), pp. 172–174.

12

Resonance-enhanced LIBS

N. H. Cheung

Department of Physics, Hong Kong Baptist University,
People's Republic of China

12.1 Introduction to resonance-enhanced LIBS

An analytical technique based on resonance-enhanced laser-induced plasma spectroscopy is reviewed in this chapter. The technique differs from conventional LIBS in that the plasmas are formed and heated by photoresonant excitation of the host species in the plume. The chaotic explosion and intense continuum emissions associated with thermal breakdown are therefore avoided. Pilot cases of solid and liquid analysis are discussed. The key experimental parameters are identified, the plasma dynamics are explained, and the improvements over non-resonant LIBS are reported. For solid samples, the ratio of the analyte line signal to the background continuum noise increased by an order of magnitude typically. For aqueous samples, the relative limits of detection (LODs) improved by $20\times$ to $1000\times$ for the range of elements tested. The absolute LODs were low enough to enable the measurement of Na and K in single human erythrocytes.

Laser-induced breakdown spectroscopy (LIBS) is a versatile technique for elemental analysis. While it is finding niche applications where alternative technology is inferior or simply does not exist, it is also challenged with ever more difficult analytical tasks (numerous examples of which can be found elsewhere in this book). Among the many problems encountered, some are specific and require solutions tailored for the particular application. But there are also general problems, such as the reproducibility and sensitivity issues associated with LIBS analysis.

The universality of the problem arises from the very nature of laser-induced breakdown. When a powerful laser pulse ablates the sample, an intrinsically violent and chaotic explosion occurs at the target site. This produces a plume so intensely white hot that all analyte emissions are overwhelmed. The plume then follows its idiosyncratic course, challenging the experimenter to catch a glimpse of the analyte signal. Quite obviously, chaotic ablation compromises reproducibility, and huge background reduces signal-to-noise ratios. Detection sensitivity is therefore adversely affected.

In this chapter, it will be argued that the violent ablation and the bright background can be avoided if the laser energy is coupled to the sample vapor in more efficient and predictable

Laser-Induced Breakdown Spectroscopy: Fundamentals and Applications, ed. Andrzej W. Miziolek, Vincenzo Palleschi and Israel Schechter. Published by Cambridge University Press. © A. W. Miziolek, V. Palleschi and I. Schechter 2006. A. W. Miziolek's contributions are a work of the United States Government and are not protected by copyright in the United States.

ways than thermal breakdown. It will be shown that one such method is photoresonant pumping, when the laser wavelength is tuned to match the resonance absorption of the host species in the vapor plume.

Resonance-enhanced laser-induced plasma spectroscopy, or RELIPS for short, has been investigated since the early 1990s. In this chapter, the conceptual basis of RELIPS will be outlined in Section 12.2. The experimental findings will then be summarized in Sections 12.3, 12.4, and 12.5 for the respective cases of solid, liquid, and gas analysis. Section 12.6 concludes this chapter.

12.2 Basic principles of spectrochemical excitation in laser-induced plasmas

The conceptual basis of RELIPS will be discussed in this section. In Section 12.2.1 the principles of plasma emissions and the conditions favorable for analyte excitations will be reviewed. This will allow us to assess whether LIBS plasmas are ideal for spectrochemical excitation in Section 12.2.2. In Sections 12.2.3 and 12.2.4 it will be shown that plasmas produced and sustained by laser resonance excitation/ionization will perform better analytically. Section 12.2.5 summarizes the foregoing.

12.2.1 Plasma conditions for maximum line-to-continuum ratios

The optimal plasma for spectrochemical excitation has been a subject of intensive study [1, 2]. While real spectrochemical plasmas are complicated [3], a simplified computer model can still be illustrative [4]. This simple model assumes a known amount of analyte (guest) atoms, such as sodium, in a homogeneous plasma of host atoms. The plasma is assumed to be in local thermal equilibrium (LTE). The quantities to be computed are the plasma continuum background intensity, and the analyte emission intensity. The line-to-continuum ratio, which is the main concern in spectrochemical analysis, could then be computed.

Continuum emissions

The continuum background arises from two processes. One is Bremsstrahlung radiation emitted by a moving electron that slows down. The emission coefficient (energy per unit volume, time, solid angle, and angular frequency interval) of this free–free transition, $j_{ff}(\omega, T, n_e)$, as a function of the angular frequency, ω, of the emitted photons of energy $\hbar\omega$ (in eV), plasma temperature T (in eV), and electron density n_e (in cm^{-3}), can be shown [5] to be

$$j_{ff}(\omega, T, n_e) = 8.0 \times 10^{-43} n_e n_i Z^2 \frac{\bar{g}}{\sqrt{T}} e^{-\hbar\omega/T} \text{ W m}^{-3} \text{ sr}^{-1} \text{ s}. \qquad (12.1)$$

The ion density n_i is assumed to be equal to n_e and the ion charge Z is set to 1 in the simplified model. That means a neutral plasma with atoms singly ionized is assumed. The Maxwell-averaged Gaunt factor $\bar{g}$ can be shown to be about 1 if T is between fractions of electronvolts and a few electronvolts, and if the emissions are in the visible spectral range [5].

The other continuum emission process is recombination radiation when a free electron is captured by an ion of the host species. The emission coefficient, $j_{fb}(\omega, T, n_e)$, of this free–bound transition can be shown [5] to be

$$
j_{fb}(\omega, T, n_e) = 8.0 \times 10^{-43} n_e n_i Z^2 \frac{1}{\sqrt{T}} e^{-\hbar\omega/T}
$$

$$
\times \left(G_n \frac{\xi}{n^3} \frac{\chi_i}{T} e^{\chi_i/T} + \sum_{v=n+1}^{\infty} \frac{2Z^2}{v^3} \frac{R_y}{T} e^{Z^2 R_y/v^2 T} \right) \text{W m}^{-3} \text{ sr}^{-1} \text{ s.}
$$

$$(12.2)$$

The first term comes from free electrons dropping to the lowest unfilled shell (n) of the ion; ξ is the number of holes in that shell, and χ_i (in eV) is the ionization potential of the recombined ion. This transition is allowed energetically only if the photon energy $\hbar\omega$ is larger than χ_i. The Gaunt factor G_n is then about 1, or else G_n is zero. The second term comes from free electrons dropping to all the higher unfilled levels. For these high lying levels, the nuclear charge is assumed to be screened and hydrogenic energy levels are used. The Rydberg constant R_y (in eV) is 13.6. An ionic charge Z of unity is again assumed in the model.

Line emissions

The analyte emission coefficient, $j_b(\omega)$, corresponding to a bound–bound transition, is given by

$$
j_b(\omega) = \frac{1}{4\pi} N^* A \, \hbar\omega L(\omega) \text{ W m}^{-3} \text{ sr}^{-1} \text{ s,} \qquad (12.3)
$$

where N^* is the analyte population density in the upper level of that transition, A is the Einstein coefficient, and $L(\omega)$ is the lineshape function. Note that, for high enough temperatures, N^* depends on T in a complicated way because the energy is now partitioned among numerous possible states, including ionized states. As such, N^* actually depends on both T and n_e.

The functional form of $N^*(T, n_e)$ may be derived from the Saha equation [5]

$$
\frac{N_i}{N_0} = \frac{g_i}{g_0} \left[\frac{1}{n_e} \frac{2m^3}{h^3} \left(\frac{2\pi T}{m} \right)^{3/2} \right] e^{-\varphi_i/T}, \qquad (12.4)
$$

which gives the ratio of the number density of ground state ions to ground state neutrals, N_i/N_0, at a plasma temperature T and an electron density n_e; g_i and g_0 are the degeneracy factors of the ion and neutral states respectively, m is the electron mass, and h is the Planck constant. Note that φ_i is the ionization potential of the analyte atom, which is not to be confused with the ionization potential χ_i of the host atoms in equation (12.2).[1]

---

[1] Of course, the Saha equation can also be used to compute the extent of ionization of the host species in the plasma. In that case, φ_i will be the ionization potential of the dominant species.

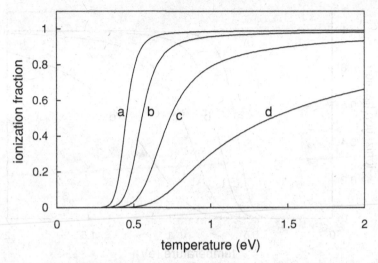

Figure 12.1. Based on the Saha equation, the ionization fraction of sodium, $f(T, n_e)$, as a function of the plasma temperature T and electron density n_e, can be computed. Shown are the computed $f(T, n_e)$ as functions of T at four different values of n_e: (a) 10^{16}, (b) 10^{17}, (c) 10^{18}, and (d) 10^{19} cm^{-3}.

The ionization fraction, $f(T, n_e)$, defined as the total ion population divided by the total (ion + neutral) population, can be shown to be

$$f(T, n_e) = 1 \bigg/ \left(1 + \frac{N_0}{N_i}\frac{Z_0}{Z_i}\right), \tag{12.5}$$

where Z_0 and Z_i are the partition functions of the neutrals and ions, respectively. They are functions of T. The expression N_i/N_0 is given by the Saha equation (equation (12.4)) and is a function of T and n_e. The ionization fractions at four different electron densities, as functions of T, are shown in Figure 12.1 for the case of sodium. As can be seen, for $T >$ 1 eV, ionization is nearly complete if the electron density is below 10^{19} cm^{-3}. Ionization is equally extensive for elements with higher ionization potentials. Figure 12.2 shows the case for lead. Although the respective φ_i values of sodium and lead are 5.14 and 7.42 eV, the difference in T to achieve the same degree of ionization for both elements is only a small fraction of an electronvolt. In fact, at high enough temperatures, lead will ionize more because lead ions have a higher density of state than sodium ions.

The density of analyte atoms in the upper level, N^*, that appears in equation (12.3), can be shown to be equal to

$$N^* = N_{\text{total}}(1 - f)\frac{g^* e^{-E^*/T}}{Z_0}, \tag{12.6}$$

where N_{total} is the total (neutral + ion) analyte concentration, and g^* and E^* are the respective degeneracy and energy of the excited level. Neutral instead of ionic N^* has been assumed. Equation (12.6) can be easily modified to accommodate the opposite case.

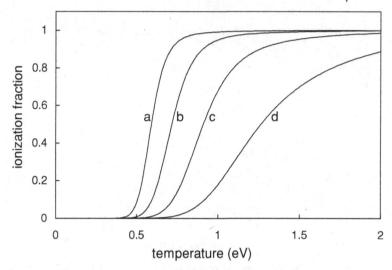

Figure 12.2. Identical to Figure 12.1, but for the case of lead.

Line-to-continuum ratio

The line-to-continuum ratio, R, defined as $j_b(\omega_o)/[j_{ff}(\omega_o) + j_{fb}(\omega_o)]$ where ω_o corresponds to the peak of the analyte line emission, can be computed as a function of T, with n_e as a constant parameter. Figure 12.3 shows the case of the 589.0 nm line of sodium in hydrogen-like plasma, with n_e of 10^{16}, 10^{17}, 10^{18}, and 10^{19} cm^{-3}. The total sodium number density was assumed to be 10^{14} cm^{-3} and an instrumental linewidth of 0.2 nm was convoluted with the Stark- and Doppler-broadened lineshape function $L(\omega)$. As can be seen, maximum R occurs at $T = 0.44$ eV and $n_e = 10^{16}$ cm^{-3}. The temperature dependence of R may be understood this way. At low T, Na(3p) is hardly populated, whereas, at high T, ionization depletion of Na becomes serious. The dependence of R on electron density can be easily explained as well. As n_e increases, three things happen. First, the continuum emissions increase as n_e^2. This explains the huge drop in R. Second, the ionization fraction decreases as n_e increases (see Figure 12.1). So the Na(3p) population increases and peaks at a slightly higher temperature. For a 10-fold increase in n_e, [Na(3p)] increases by a factor of about 2 (see Figure 12.1). This explains the drop by a factor of 45 instead of a factor of 100 in R as n_e increases from 10^{16} to 10^{17} cm^{-3}. Third, the Stark width $\Delta\lambda$ increases approximately linearly with n_e [6]. It is about 0.03 nm when n_e is 10^{17} cm^{-3}. So the overall linewidth remains dominated by the instrumental linewidth of 0.2 nm until n_e reaches 10^{18} cm^{-3}. Beyond that, the peak intensity of the observed 589.0 nm line drops as $1/n_e$. This explains why R drops by a factor of 4×10^4 as n_e increases from 10^{17} to 10^{19} cm^{-3}. The best R, occurring at T of 0.44 eV and n_e of 10^{16} cm^{-3}, is about 10^4. It should be pointed out that, although the line-to-continuum ratio improves significantly as electron density drops, n_e much lower than 10^{16} cm^{-3} might not be meaningful, for two reasons. First, when the plasma continuum emissions become exceedingly weak, other background signal such as stray light or wings of neighboring

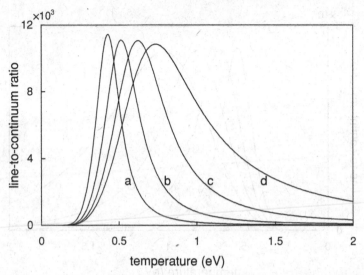

Figure 12.3. By assuming a homogeneous LTE plasma, the line-to-continuum ratio $R(T, n_e)$ of the sodium 589.0 nm line, as a function of the plasma temperature T and electron density n_e, can be computed. Shown here are the computed $R(T, n_e)$ as functions of T at four different values of n_e: (a) 10^{16}, (b) 10^{17}, (c) 10^{18}, and (d) 10^{19} cm^{-3}. Note that in order to plot all four curves on the same graph, they have to be magnified by: 1 for (a), 45 for (b), 4000 for (c), and 1.8×10^6 for (d). A density of 10^{14} cm^{-3} sodium atoms and an instrumental line width of 0.2 nm were also assumed.

lines will be dominant [7]. Second and more importantly, when the electron density is much lower than 10^{16} cm^{-3}, the LTE assumption may not be valid [8].

Analogous to Figure 12.3, Figure 12.4 shows the line-to-continuum ratio R for the 405.8 nm line of lead, which corresponds to the Pb(I) (6p7s $^3P_1°$) to (6p^2 $^3P_2$) transition. Despite the higher excitation energy of 4.4 eV for the Pb(I) upper level, relative to the 2.1 eV of the Na(3p) level, the optimum temperature can be seen to be still between 0.58 and 1 eV. A maximum R of about 2700 occurs at T of 0.58 eV and n_e of 10^{16} cm^{-3}.

We may now conclude that, for the purpose of spectrochemical analysis, a plasma temperature between 0.4–0.6 eV and an electron density of about 10^{16} cm^{-3} are optimum. These conditions will maximize the intensity ratio of the analyte line to the background continuum for the cases of Na 589 nm and Pb 405.8 nm lines. At lower temperatures, the number of analyte atoms excited to the upper level of the desired transition will be too few. At higher temperatures, the analyte population will be seriously depleted by thermal ionization. If these sodium and lead analytical lines, with their very different ionization potential and excitation energy, can be optimized under nearly identical plasma conditions, analytical lines of numerous other elements will be equally well served.[2]

---

[2] We have modeled the Na 589.0 nm and Pb 405.8 nm lines. Na is a typical "soft" line, where as Pb is "hard," although not representative. A more typical "hard" line is the Cd(I) 228.8 nm line (excitation energy of 5.42 eV and IP of 8.991 eV), the modeling of which was carried out recently. At n_e of 10^{16} cm^{-3}, the line-to-continuum ratio peaked at T of 0.62 eV. It should be pointed out, however, that extra "hard" lines, such as the 281.62 nm Al(II) line (excitation energy of 11.82 eV and IP of 5.99 eV), would not be optimized under the present scheme.

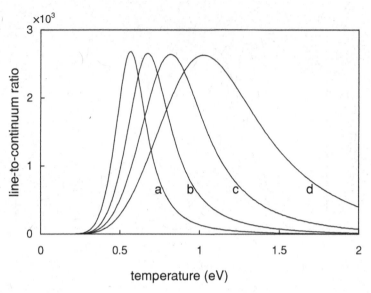

Figure 12.4. Identical to Figure 12.3 but for the case of the Pb 405.8 nm line. The magnification factors for the four curves are: (a) 1, (b) 45, (c) 6200, and (d) 3.6×10^6.

12.2.2 Are LIBS plasmas ideal for spectrochemical analysis?

It is well known that typical LIBS plasmas are a few electronvolts in temperature and 10^{18}–10^{19} cm^{-3} in electron density near the beginning [9]. Only continuum emissions are visible at that time. As the plasma cools below 1 eV and electron density drops to about 10^{17} cm^{-3}, analyte line emissions emerge briefly. With further cooling, the plume turns dark. These observations correlate well with the predictions of Figures 12.3 and 12.4.

Instead of passively waiting for the plasma conditions to turn optimal for just a fleeting moment, is it possible to produce and sustain LIBS plasmas at the desired temperature and electron density? To answer that question, the nature of laser-induced thermal breakdown has to be considered. The physics of laser ignition of plasmas is complicated and not well understood [10]. A useful empirical model showed that, for laser wavelengths from near-ultraviolet through infrared, the threshold irradiance I (in W cm^{-2}) needed to induce breakdown is fairly material independent and can be estimated from the following inequality [11]:

$$I\sqrt{\tau} \geq 4 \times 10^4 \text{ W cm}^{-2} \text{ s}^{1/2},$$ (12.7)

where the pulse duration τ (in s) can range from nanoseconds to milliseconds. For τ of 5 ns, I is some fractions of gigawatts per square centimeter, and the threshold fluence is accordingly a couple of joules per square centimeter. The corresponding plasma temperature T (in eV), which is dependent on the laser wavelength λ (in cm), can be estimated [11] by using

$$T = 3\sqrt{I \lambda \sqrt{\tau}}.$$ (12.8)

For ultraviolet through near-infrared laser pulses at the breakdown threshold, T is already of the order of a few electronvolts.

In other words, the observed high temperature (>1 eV) of the LIBS plasma is essential for its breakdown. At that temperature, extensive thermal ionization will occur while neutral analytes will be depleted. The high electron density will give rise to an intense continuum background while the analyte signal will be minuscule. Yet, without breakdown, there will be no plasma; nor will there be electron impact excitation of the analytes to give spectral signals. This dilemma is illustrated in Figure 12.5 [12]. Figure 12.5a shows time-resolved LIBS spectra of an aqueous sample containing 6 mM of sodium as analyte, ablated by Nd:YAG ($2\times\nu$, 532 nm, 5 ns) laser pulses at a fluence of 3.2 J cm^{-2}. The sodium 589 nm doublet was visible, though weak relative to the huge continuum background. When the laser fluence was reduced, as shown in Figures 12.5b and 12.5c, the continuum background dropped but the analyte emissions now totally disappeared.

12.2.3 Laser-induced cool plasmas

The breakdown condition of elevated temperatures ($T > 1$ eV) and high electron densities ($n_e > 10^{18}$ cm^{-3}) is certainly not compatible with the mild T ($\sim$0.5 eV) and low n_e (10^{16} cm^{-3}) preferred for spectrochemical excitation. The dilemma can be avoided if the plasma is laser-generated yet without thermal breakdown in the first place. An early experiment is particularly interesting in this regard [13]. It was demonstrated in that experiment that a sodium vapor of about 10^{16} cm^{-3} density was nearly completely ionized when irradiated by a dye laser tuned to 589.6 nm with an energy of about 300 mJ in a 500 ns pulse. Subsequent experiments showed that a broad range of elements could be similarly ionized to yield dense ($n_e \sim 10^{16}$ cm^{-3}) and cool ($T \sim 0.3$ eV) plasmas [14–17]. The extremely efficient channeling of excitation energy into ionization is believed to be the result of multiple processes occurring in four stages [15]. Using the 589.6 nm pumping of sodium vapors as an example, in stage one, the laser rapidly saturates the upper Na ($3p^2P_{1/2}$) level in much less than 1 ns if the laser irradiance is higher than of the order of megawatts per square centimeter. In stage two, primary electrons are produced by energy pooling, associative ionization, stimulated Raman scattering, and other less probable ionization processes. Electron density grows linearly during this stage. In stage three, super-elastic collisions generate hot electrons:

$$\text{Na (3p)} + e \rightarrow \text{Na (3s)} + e^*. \tag{12.9}$$

These hot electrons then impact ionize Na (3p) atoms in multiple steps, with or without laser assistance. This leads to exponential growth of free electrons and full plasma formation. In the fourth and final stage, runaway collisional ionization of Na quickly "burns out" the population of neutral Na atoms. Resonant absorption of laser energy ceases and the electron temperature drops. This implies that plasma heating is self-terminated and the electron temperature is capped. It was shown that the time τ_B required to reach the

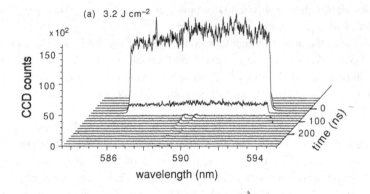

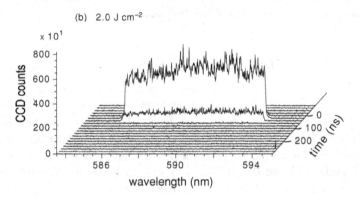

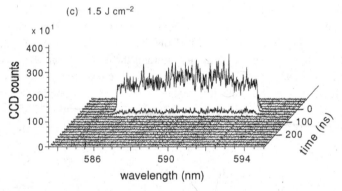

Figure 12.5. A 6 mM Na solution was ablated by 532 nm laser pulses. Shown are time sequences of the plume spectra generated by laser pulses of different energy densities. At 3.2 J cm^{-2} (a), an intense continuum could be seen during the first 60 ns while the weak sodium 589 nm doublets were just barely observable afterwards. With decreasing fluence, the continuum background dropped but the analyte signal disappeared altogether. Adapted from reference [12].

burn out phase is very short. For dense (10^{17} cm^{-3}) vapor irradiated with powerful (≥ 10 MW cm^{-2})) lasers, τ_B was only a few nanoseconds [15].

The production of seed electrons during stage two may be augmented by laser resonance ionization. Resonance photoionization schemes for practically all elements have been devised [18]. Take potassium as an example. The K atom can be resonantly photoionized by absorbing two 404.4 nm photons:

$$K(4\,^2S_{1/2}) + h\nu(404.4\text{ nm}) \rightarrow K^*(5\,^2P_{3/2}{}^\circ), \tag{12.10}$$

$$K^*(5\,^2P_{3/2}{}^\circ) + h\nu(404.4\text{ nm}) \rightarrow K^+ + e. \tag{12.11}$$

A potassium vapor pumped by a 404.4 nm laser pulse will therefore have an additional, direct pathway for electron seeding. Subsequent full plasma formation and burn out would still depend on super-elastic collisions of electrons with $K^*(5\,^2P_{3/2}{}^\circ)$.

Both laser resonance excitation and laser resonance ionization have long been exploited by analytical scientists, in the form of laser-enhanced ionization [19], laser ionization mass spectrometry [20], and the more recent resonant laser ablation [21], and photoresonant LIBS [22, 23]. In all these schemes, the resonantly excited species are the trace amounts of analytes. In contrast, for the present purpose of plasma generation, the pumped species are the major, abundant species in the vapor plume. Prior knowledge of the sample matrix (not the analyte) is therefore necessary. However, for real-world analysis, the matrix chemicals are generally known, such as water in aqueous samples, silicon in microcircuits, and iron in steel specimens.

12.2.4 Sustaining non-thermal plasmas

Cool, photoresonant plasmas are not Saha equilibrated. This can be illustrated by considering the following course of events, beginning with a neutral vapor of sodium about 10^{18} cm^{-3} in density. When irradiated at 10^8 W cm^{-2} with 589.6 nm light, ionization will be complete in less than a couple of nanoseconds. The free electron temperature is probably around 0.4 eV [22]. Once the laser pulse is over, the extent of thermal ionization, as predicted by the Saha equation (equation (12.4)), is only about 0.2% (see Figure 12.1) at T of 0.4 eV and n_e of 10^{18} cm^{-3}. The Saha-predicted electron density is therefore only 2×10^{15} cm^{-3}. The electron density has to drop.

The plasma relaxation rate depends on the recombination rate of the electron and ion in the presence of a third body (usually an electron). This three-body recombination time τ_{3BR} is given by [24]

$$\tau_{3BR} \approx \frac{10^{26}}{5} \frac{T^{4.5}}{n_e^2} \text{ s}, \tag{12.12}$$

where the temperature T is in electronvolts and the electron density n_e is given per cubic centimeter. For $T \sim 0.4$ eV and $n_e \sim 10^{18}$ cm^{-3}, τ_{3BR} is ~ 400 fs. As n_e drops to 10^{17} cm^{-3}, and assuming T is still 0.4 eV, τ_{3BR} becomes 40 ps, and the Saha ionization fraction is 2%

while the corresponding electron density is 2×10^{16} cm^{-3}. It can be shown that the plasma will eventually relax to an equilibrium state of $T \sim 0.4$ eV and $n_e \sim 4 \times 10^{16}$ cm^{-3} in less than 1 ns.

At that point, the plasma becomes relatively stable, and dissipative effects other than electron–ion recombination, such as vapor cooling and dispersion, will have more impact. These heat and mass transports are slow. For laser-ablated plumes, they occur on a time scale of tens of nanoseconds to microseconds, and are affected by ambient gases in terms of cooling and plume confinement [25]. With suitably chosen ambient gases, laser-induced cool plasmas could be made to last for more than hundreds of nanoseconds [26].

It was also demonstrated that, except for rarefied atmosphere, cooling is the usual cause of plasma extinction [26]. Plasma re-heating by a second laser pulse is therefore logical, and has been carefully investigated [27–29]. Usually, the plasmas are laser-heated by inverse Bremsstrahlung absorption. The cross-section σ_{IB} (in cm^2) per electron, as a function of electron density n_e (in cm^{-3}), photon energy $\hbar\omega$ (in eV), and plasma temperature T (in eV), is given by [11]

$$\sigma_{IB} \approx 3 \times 10^{-37} n_e \left[\frac{1 - e^{-\hbar\omega/T}}{(\hbar\omega)^3 \sqrt{T}} \right]. \tag{12.13}$$

For visible lasers irradiating an extinguishing plume of $T \sim 0.3$ eV and $n_e \sim 10^{16}$ cm^{-3}, σ_{IB} is only about 10^{-22} cm^2. With 10.6 µm lasers, it is still only about 10^{-18} cm^2. The absorption coefficient, which is the product of the cross-section and the electron density, is no more than 10^{-2} cm^{-1}. Very high fluence (more than a few joules per square centimeter) is therefore necessary to produce heating effects, resulting inevitably in another cycle of thermal breakdown.

A more effective way to rekindle the plasma is to pump it photoresonantly, i.e. to irradiate at 589.6 nm if it is a sodium vapor, or 404.4 nm if it is potassium. The absorption cross-section is $\sim 10^{-16}$ cm^2. The absorbent density is about 10^{18}–10^{19} cm^{-3} for typical LIBS plumes [11]. Accordingly, the absorption coefficient is about 10^2–10^3 cm^{-1}, which is many orders of magnitude larger than the inverse Bremsstrahlung case. Subsequent conversion of the excitation energy to other forms of energy, including free electron kinetic energy and ionization, will be rapid. As a result, the optimal plasma can be revived without the complications of thermal breakdown.

12.2.5 A summary of the conceptual basis of RELIPS

Numerical modeling showed that a cooler ($T \sim 0.5$ eV) plasma at $\sim 10^{16}$ cm^{-3} electron density should give the highest line-to-background ratio for the spectrochemical analysis of a broad range of elements. In comparison, LIBS plasmas are created too hot and electrons too dense. These unfavorable conditions are inevitable if the plasmas are ignited by thermal breakdown. Fortunately, non-breakdown pathways exist. For example, full plasma can be generated from the neutral vapor by tuning the laser wavelength to match a strong absorption line of the vapor atom. Excitation energy is converted efficiently and promptly to ionization.

This kind of photoresonant plasma was demonstrated for vapors of alkali metals as well as other elements. Alternatively, free electrons can be produced by laser resonance ionization.

These non-thermal plasmas are generally over-ionized. Once the laser beam is off, the electron density will relax at sub-nanosecond rates and soon stabilize to conditions favorable for spectrochemical excitations. That may last for more than hundreds of nanoseconds, until the luminous plasmas turn cold and dark. The extinguishing plumes may be rekindled resonantly by a second laser pulse. Plasma reheating by photoresonant pumping should be orders of magnitude more effective than non-resonant inverse Bremsstrahlung absorption.

12.3 RELIPS analysis of solids

LIBS analysis has been successfully applied to gaseous, liquid, and solid samples [30, 31]. Whether RELIPS can be equally versatile is a question of practical importance. RELIPS analysis of solids was reported recently [32]. Its feasibility, instrumentation, and underlying mechanisms have since been investigated [26, 32, 33]. They form the subject matter of the present section. The analyses of liquid and gaseous samples will be discussed in the following sections. As is well known, laser-induced plasmas are extremely rich and complex, and each decade of laser power features distinctly different dynamics [34]. For analytical applications, fluence (power density) ranging from 10^0 to 10^1 J cm^{-2} (from 10^8 to 10^9 W cm^{-2}) is typical. In the case of RELIPS analysis of solids, the less ablative regime of $\sim$J cm^{-2} was investigated.

The experiments are first described in Section 12.3.1. Evidence of resonance enhancement is then presented in Section 12.3.2. The strategy for signal optimization is outlined in Section 12.3.3, and results of multielement analysis are given in Section 12.3.4. The dynamics of RELIPS plasma is presented in Section 12.3.5, followed by a summary in Section 12.3.6.

12.3.1 Experiments

Two general aspects of RELIPS analysis of solids are noteworthy. First, the generation of plasmas non-thermally to avoid the high (>1 eV) temperature and the continuum emissions proves to be challenging. The minimum fluence needed for solid ablation may already induce thermal breakdown. This is especially true for refractory targets such as silicon wafers and geological samples. A double-pulse scheme was thus necessary [32]. The first pulse ablates. When the plume has cooled, a second pulse resonantly photoexcites and reheats the sample vapor.

The second aspect concerns the choice of sample matrix. At this relatively early phase of the study, demonstration of feasibility and identification of key parameters are the goals. Analysis of real-world samples is for the next phase. For that reason, a well-characterized test matrix that calls for the least-complicated instrumentation should be used. It was shown that pellets made from pressing crystalline potassium iodate (KIO$_3$) were a suitable matrix [32].

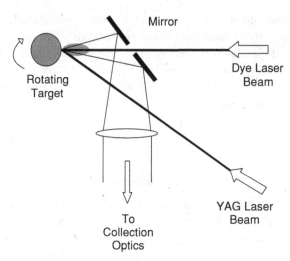

Figure 12.6. Experimental schematics for the RELIPS analysis of solid samples are shown. A rotating target of cylindrical pellets of KIO_3 housed in a sealed chamber was ablated by the second harmonic (532 nm) output of a Nd:YAG laser pulse of 10 ns width. The expanding plume was intercepted 30 ns later by a dye laser pulse of 9 ns duration and 0.2 nm line width centering on 404.4 nm. The plume emissions were directed onto the entrance slit of a spectrograph equipped with an intensified array detector. Adapted from reference [32].

The experimental setup for doing RELIPS is similar to a typical LIBS setup. Instrumentation details have been reported [32, 33, 35]. The schematic is reproduced in Figure 12.6. Briefly, frequency-doubled Nd:YAG laser pulses (532 nm, 10 ns, 10 Hz) were apertured and delivered onto the side of a rotating cylindrical pellet of KIO_3 which was housed in a sealed chamber.[3] The ambient atmosphere inside the sample chamber could be controlled. Frequency-tripled laser pulses (355 nm, 10 ns, 10 Hz) from the same Nd:YAG laser were optically delayed by 30 ns before pumping a pulsed dye laser running on Ex-404 (Exciton) dye. The dye laser beam, with linewidth of about 0.2 nm, was focused normally onto the same target site. The beam spot was oval in shape, with major and minor diameters (full width at half maximum, FWHM) of 700 and 250 μm, respectively. Because all ablations were to be minimally invasive, the 532 nm ablation fluence used was low. The peak energy density, defined as the fluence of the central hottest (~90% maximum) region, was never more than about 1.3 J cm^{-2}. The dye laser fluence was even lower because it was meant to be non-ablative. The peak fluence was no more than about 700 mJ cm^{-2}.

Light emissions from the plume were collected axially and directed onto the entrance slit of a spectrograph. An intensified charge-coupled device (ICCD) array detector was mounted at the exit plane of the spectrograph. The intensifier was gated on after a delay

---

[3] The 532 nm beam was first apertured with a 5 mm diameter iris. The aperture was then imaged onto the target surface through a 50 mm focal length lens at 20× demagnification. The spot size at the target was about 800 μm (FWHM). By using such a delivery scheme, the beam profile at the target was shown to be more Gaussian than that produced by simply focusing the laser beam onto the target.

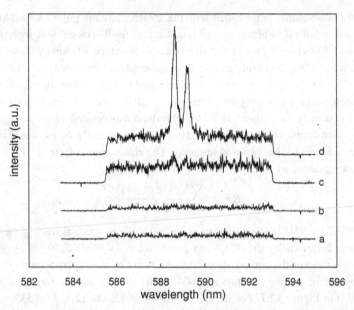

Figure 12.7. Plume emission spectra generated by double-pulse RELIPS. The KIO_3 pellets containing 35 p.p.m. Na were ablated by a 532 nm laser pulse at a fluence of about 800 mJ cm^{-2}; 30 ns later, the expanding plume was intercepted by a marginally ablative dye laser pulse of about 460 mJ cm^{-2}. The detector gate delay was 10 ns from the firing of the dye laser; the gate width was 5 μs. Emission spectra produced by (a) the dye laser alone, (b) the 532 nm laser alone, (c) a 532 nm pulse followed by an off-resonance 407 nm pulse, and (d) a 532 nm pulse followed by an on-resonance 404.4 nm pulse. All four traces were averages of 300 double-pulse events. They were offset vertically for clarity. Because the edge pixels of the CCD were not intensified, the spectral trace at the edges served conveniently as a baseline in all cases.

of t_d from the firing of the Nd:YAG laser, and remained on for a duration of t_b. A few hundred ablation events were averaged before the spectral signal was stored and processed on a personal computer. Typically, for better signal-to-noise ratios, t_d was set to 40 ns to avoid the initial continuum background and t_b was set to 5 μs to envelop the signal pulse. Alternatively, for plume dynamics studies, t_b was set to 50 ns while t_d was stepped.

12.3.2 Resonance enhancements

Resonance enhancement is shown in Figure 12.7. The target was an iodate pellet containing 35 p.p.m. sodium. Ablation was done in open air. The peak fluence (90% maximum) of the ablative 532 nm pulse was about 800 mJ cm^{-2}. The fluence of the dye laser pulse was about 460 mJ cm^{-2}, which was marginally ablative. The time delay between the two pulses was 30 ns. The spectrograph slit-width was 300 μm, yielding a 0.2 nm resolution. The ICCD time delay t_d was 40 ns from the firing of the 532 nm pulse. Gate width t_b was 5 μs. All spectral traces shown were averages of 300 events. Spectral trace a in Figure 12.7 was with the 404.4 nm pulse alone. Trace b was with the ablative 532 nm pulse alone. Note that the first 40 ns was gated off to eliminate the bright initial flash. Trace c was with the 532 nm pulse followed 30 ns later by a non-ablative and off-resonance 407 nm dye laser pulse. The Na doublet at 589.0 and 589.6 nm can be seen, although the line intensities were not much enhanced relative to the superposition of traces a and b. Trace d was similar to trace c but with an on-resonance 404.4 nm pulse. Resonance enhancement is clearly demonstrated.

In order to quantify the enhancement, the Na signal was defined as the average intensity under the sodium doublet minus the average background intensity. Noise was defined as the standard deviation in the background intensity. The enhancement factor (EF) was defined by the following equation;

$$EF = \frac{S/N(532 \otimes 444)}{S/N(532 + 404)} \tag{12.14}$$

where $S/N(532 \otimes 404)$ is the signal-to-noise ratio of the spectral trace gotten with the 532 nm pulse followed by the 404.4 nm pulse (trace d), while $S/N(532 + 404)$ is the signal-to-noise ratio of the composite spectrum formed by the superposition of the spectra produced by the respective 532 nm and 404.4 nm lasers alone, i.e. the superposition of traces a and b in Figure 12.7. For the results shown in Figure 12.7, $S/N(532 + 404)$ was about 4 while $S/N(532 \otimes 404)$ was about 53. An EF of 13 was therefore demonstrated.

12.3.3 Important experimental parameters

Four parameters that critically affect the RELIPS performance are listed below. First, as expected, the signal depended sensitively on the wavelength of the dye laser. This is shown in Figure 12.8 [32]. The persistence of the enhancement even at 1 nm detuning indicates the excessive linewidth of the K absorption, the exact cause of which remains to be identified.

Second, the enhancement was extremely sensitive to the beam profile of the Nd:YAG laser [33]. Even small asymmetries in an otherwise near-Gaussian profile would lower the signal intensity. This is shown in Figure 12.9 where trace b was captured using a beam profile more Gaussian than that used to generate trace a. The two corresponding beam profiles are shown in Figure 12.10. Asymmetries in the beam profile probably gave rise to instabilities in the flow of the ablated vapor, which in turn caused accelerated dispersion of the plasma plume.

Third, the accuracy of the spatial overlap of the two beams was demanding [33]. Offsets of more than about one-tenth of the FWHM size of the Nd:YAG beam spot would halve the signal. This is shown in Figure 12.11. This indicates that only the central hottest region of the 532 nm beam was ablative, which is not unreasonable since the fluence of the Nd:YAG laser was intentionally kept low to minimize target destruction.

Fourth, the ambient atmosphere affected the RELIPS signal significantly [26]. This is illustrated in Figure 12.12 when the spectral traces of Na emissions captured in various ambient gases are plotted. For example, going from vacuum to 13 mbar xenon, the signal-to-noise ratio increased from about 39 to over 100. The outcome should not be surprising.

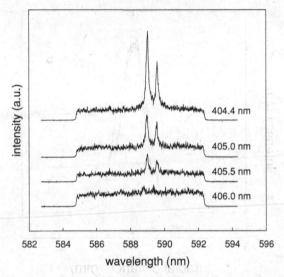

Figure 12.8. Detuning effect of the dye laser in RELIPS analysis is shown. Spectral traces were obtained under conditions similar to those of trace d in Figure 12.7, except with progressive detuning of the dye laser away from the resonance wavelength of 404.4 nm and 404.7 nm. Spectral traces were offset vertically for clarity. Adapted from reference [32].

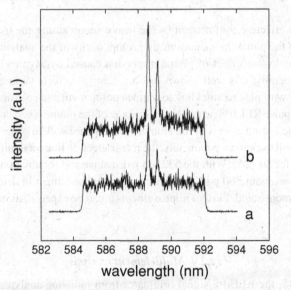

Figure 12.9. Beam profile effect in RELIPS analysis is shown. Spectral traces were obtained under conditions similar to those of trace d in Figure 12.7, except that two different beam profiles of the ablation (532 nm) laser were used. Trace a was generated with a slightly asymmetric beam profile, as shown in curve a of Figure 12.10. Trace b was generated with a more symmetric beam profile as shown in curve b of Figure 12.10. Adapted from reference [33].

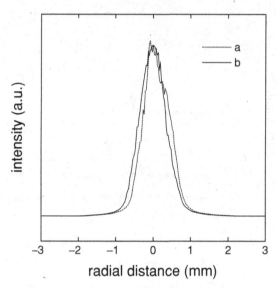

Figure 12.10. Intensity profiles of the ablation (532 nm) laser used to generate the spectra shown in Figure 12.9. Profile a was obtained by focusing the laser beam onto the target with a 200 mm focal length plano-convex lens. Profile b was obtained by delivering the laser beam through a 5 mm diameter aperture and imaging the aperture onto the target at 20-fold demagnification. Both profiles were measured along a diameter of the circular beam spot at the target. Adapted from reference [33].

In contrast to the effective confinement by the heavy xenon atoms, the free expansion and rapid thinning of the plume in vacuum mean prompt decay of the analyte signal. We have mentioned the analogous effect of plume dispersion caused by asymmetries in the beam profile. More generally, it is well known that ambient gas affects various stages of LIBS plasmas, starting with plasma shielding to plume confinement and collisional cooling [25]. Because double-pulse RELIPS probes the later stage of the plume when the thermal plasma has cooled, all the lasting effects of the ambient atmosphere will be observed.

By optimizing all the known parameters, the mass detection limit of double-pulse RELIPS was about 14 pg for Na [26]. With the 532 nm pulse alone and for ablation in air, the corresponding limit was about 360 pg [26]. Twenty times enhancement in detection sensitivity was therefore demonstrated. Further improvements are to be expected as our understanding of RELIPS evolves.

12.3.4 Multielement analysis

Similarly to LIBS, the RELIPS signal originates from radiating analytes that are excited by non-selective electron impact. Multielement analysis should therefore be possible. That was demonstrated recently when iodate pellets containing 1.25 p.p.m. of Na, 6 p.p.m. of Li, and 50 p.p.m. of Al were analyzed [33]. The 670.8 nm Li(I) line, the 589 nm Na(I) doublet, and the 308.2 nm and 309.3 nm Al(I) doublet were chosen as the analytical lines. They

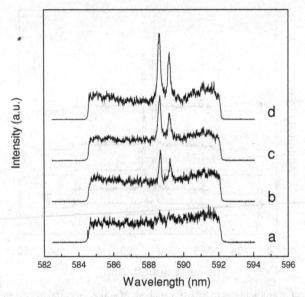

Figure 12.11. Beam offset effect. The emission spectra of the sodium 589 nm doublet were obtained under conditions similar to those of trace d of Figure 12.7, except with the 532 nm and 404 nm laser beams offset spatially by (a) 320 μm, (b) 220 μm, (c) 110 μm, and (d) 0 μm. The full width at half maximum of the 532 nm beam was 800 μm, while that of the 404 nm beam was 250 μm. Offsets by as little as one-tenth of the convoluted spot size would halve the signal. This indicates that the dye laser beam had to precisely bracket the expanding plume in order to give a RELIPS signal.

correspond to excitation energies ranging from 1.8 eV to over 4 eV. All three analytes were detected under the same experimental conditions.

The Na and Li spectra are shown respectively in Figures 12.13 and 12.14. Spectral traces a were taken with the dye laser alone, while traces b were with the 532 nm laser alone. Traces c were generated by 532 nm laser ablation followed by an off-resonance 406 nm dye laser pulse. Traces d were similar to traces c except with an on-resonance 404.4 nm dye laser pulse. Resonance enhancement is clearly evident for both Na and Li. The enhancement factor EF, as defined in equation (12.14), is about 8 for the Na spectra shown in Figure 12.13. It is less than the EF of 13 shown in Figure 12.7 because of a narrower spectrograph slit and lower dye laser fluence. The value of the EF is about 4.5 for the case of Li shown in Figure 12.14.

Resonance enhancement of the Al spectral signal is shown in Figure 12.15. Traces a and b were produced respectively by the dye and 532 nm laser alone. Trace c was produced by a 532 nm pulse followed by a 404.4 nm pulse. The Al 308 nm doublet can be clearly seen, although it is not as pronounced as the Na and Li spectral lines (Figures 12.13 and 12.14). The enhancement factor is about 3.

Based on the enhancement results shown in Figures 12.13, 12.14 and 12.15, multielement analysis using RELIPS was demonstrated for Al, Li and Na. The smaller enhancement for

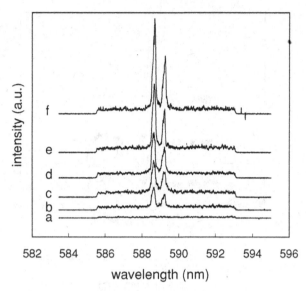

Figure 12.12. Effect of ambient gas in RELIPS analysis of solids. Spectral trace a was obtained under conditions similar to those of trace b of Figure 12.7, that is, with the 532 nm beam alone and in open air. The signal-to-noise ratio, *S/N*, is about 4. The other traces were obtained under conditions similar to those of trace d of Figure 12.7, that is, with the 532 nm pulse followed by a 404 nm pulse, except the ambient conditions were varied. The exact ambient conditions and the corresponding *S/N* are: (b) vacuum: 39; (c) open air: 53; (d) 350 mbar air: 70; (e) 50 mbar argon: 75; and (f) 13 mbar xenon: 110.

Al could be explained by the energy (4.022 eV) of the radiative state that was higher than the 2.104 eV of the Na excited level and was therefore less populated. As pointed out later, hotter RELIPS plasma will alleviate the problem.

12.3.5 Time-resolved characterization of the plasma plume

In order to understand the RELIPS dynamics, a series of time-resolved spectra with incremental delay t_d were captured [26]. The time-resolved Na signal, $s(t)$, which was defined earlier as the spectral area of the 589 nm doublet minus the background, was plotted as a function of time t. This is shown in Figure 12.16a (open circles). A slightly rarefied atmosphere of 350 mbar was used to boost the analyte signal. The dye laser was also tuned off-resonance and the corresponding $s(t)$ was also plotted in Figure 12.16a (crosses). As can be seen, the on-resonance signal was about four times stronger while its half-life was about 10 ns longer than the off-resonance one. This implies a marked increase in the population of excited sodium Na (3p) atoms when resonantly pumped. Because the etch rates and ambient conditions were similar, the vapor density of the plume and therefore the total sodium population should all be comparable. It thus indicated that the sodium atoms were much more likely to be excited to the Na (3p) level in the resonance case.

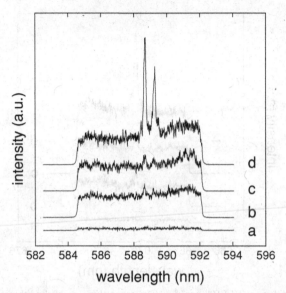

Figure 12.13. Multielement RELIPS analysis. The spectral region of the sodium 589 nm doublet is shown. KIO_3 pellets containing 1.25 p.p.m. Na, 6 p.p.m. Li, and 50 p.p.m. of Al were ablated by a 532 nm laser pulse at a fluence of about 1.3 J cm^{-2}. The expanding plume was intercepted 30 ns later by a non-ablative dye laser pulse of about 300 mJ cm^{-2}. The plume emissions were time-integrated for 1 μs starting 10 ns after the firing of the dye laser. Emission spectra produced by (a) the dye laser alone; (b) the 532 nm laser alone; (c) the 532 nm pulse followed by an off-resonance 406 nm dye laser pulse; and (d) the 532 nm pulse followed by an on-resonance 404.4 nm dye laser pulse. All traces were offset vertically for clarity. The resonance enhancement of the Na doublet is clearly seen. Adapted from reference [33].

The electron density $n_e(t)$ and temperature $T(t)$, as functions of time, were also measured [26]. The time-resolved electron density $n_e(t)$, derived from the Stark widths of the Li 670.8 nm line (target doped with Li), is shown in Figure 12.16b for both on (open circles) and off (crosses) resonance cases. Resonance-enhanced ionization is obvious. Upon 404 nm irradiation at $t = 30$ ns, the initial decay trend of $n_e(t)$ was promptly reversed. It shot to 10^{17} cm^{-3} and subsequently decayed again with a half-life of about 30 ns. It dropped below the detectable threshold of 3×10^{16} cm^{-3} at $t \sim 200$ ns. In contrast, the off-resonance $n_e(t)$ decayed monotonically and soon became unmeasurable at $t \sim 120$ ns. During the first 120 ns, n_e was higher than the LTE threshold even for the off-resonance case. A plasma temperature T should therefore be well defined in both cases.

The plasma temperature $T(t)$, deduced from the intensity ratio of the Li 610.4 nm and 670.8 nm lines, is plotted in Figure 12.16c for the on (open circles) and off (crosses) resonance cases. The intensity of the Li 610 nm emissions is also plotted (lighter symbols). It serves as a measure of the heated mass. The off-resonance T was only briefly measurable between 60 and 65 ns, when it was found to be about 0.42 eV. At earlier times, the background masked the weak Li 610 nm line. At later times, T became too low. The lowest measurable

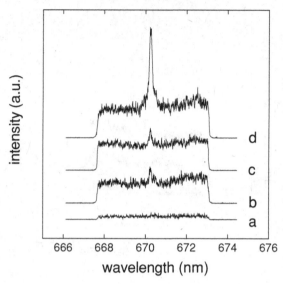

Figure 12.14. Multielement RELIPS analysis. The spectral region of the lithium Li(I) 671 nm line is shown. Experimental conditions were identical to those of Figure 12.13. Spectral traces were generated by (a) the dye laser alone; (b) the 532 nm laser alone; (c) the 532 nm pulse followed by an off-resonance 406 nm dye laser pulse; and (d) the 532 nm pulse followed by an on-resonance 404.4 nm dye laser pulse. All traces were offset vertically for clarity. The resonance enhancement of the Li line is clearly seen. Adapted from reference [33].

T was about 0.35 eV. In comparison, the on-resonance plasma was heated up earlier and remained hot longer; T reached 0.4 eV beginning around 45 ns. It then went through two cycles, each time attaining maximum temperatures of 0.45 eV and 0.53 eV, respectively; T remained measurable out to 140 ns. It is also evident from Figure 12.16c that the resonant Li signal was ten times stronger even though the temperatures were comparable at $t = 65$ ns. This suggests that a much larger volume of the plume was heated to that temperature in the resonance case. Quite clearly, an extended hot plasma is the key to brighter sodium emissions.

We may now understand resonance enhancement in terms of efficient reheating of the plasma plume. The resonant 404 nm heating was probably the result of three possible mechanisms. The first is inverse Bremsstrahlung absorption. Because the heating efficiency per sample volume is proportional to n_e^2, the doubling of n_e upon 404 nm irradiation will give rise to a four times increase in absorption. The absorption cross-section per electron is about 3×10^{-19} cm^2 at the peak n_e of 10^{17} cm^{-3} (equation (12.13)). The overall absorption coefficient is about 0.03 cm^{-1}.

The second heating mechanism is the production of hot electrons by resonance $1 + 1$ photoionization (equations (12.10) and (12.11)). The total energy of two 404.4 nm photons is more than the ionization potential of K by 1.8 eV. The absorption cross-section for the $1 + 1$ photoionization is about 3×10^{-18} cm^{-2}. The density of K (4s) atoms should be

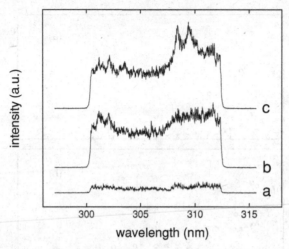

Figure 12.15. Multielement RELIPS analysis. The spectral region of the aluminum Al(I) 308 nm doublet is shown. Experimental conditions were identical to those of Figure 12.13. Spectral traces were generated by: (a) the dye laser alone; (b) the 532 nm laser alone; (c) the 532 nm pulse followed by an on-resonance 404.4 nm dye laser pulse. All traces were offset vertically for clarity. The resonance enhancement of the Al doublet is evident. Adapted from reference [33].

about 10^{18} cm^{-3}. The absorption coefficient is therefore about units of cm^{-1} which is about $30\times$ higher than that of inverse Bremsstrahlung absorption.

The third heating mechanism is the super-elastic collision of electrons with excited potassium atoms K (5p) that were formed by resonance absorption of 404.4 nm photons (equation (12.9)). This kind of resonant pumping is known to be extremely efficient in converting photon energy to electron thermal energy if the sample vapor density is higher than about 10^{16} cm^{-3} and when the laser irradiance is above units of MW cm^{-2} [15]. The resonance absorption cross-section for the transition (12.10) is 6×10^{-16} cm^2. Assuming the density of ground state K is $\sim 10^{18}$ cm^{-3}, the absorption coefficient is $\sim 10^3$ cm^{-1}, or four orders of magnitude bigger than that of inverse Bremsstrahlung absorption.

12.3.6 RELIPS analysis of solids: a summary

We have reviewed the RELIPS analysis of solids in the low fluence ($\sim$ J cm^{-2}) regime. Potassium iodate pellets doped with trace amounts of sodium, lithium, and aluminum were used as test samples. The target pellet was ablated by a 532 nm laser pulse. After 30 ns, the plasma plume was reheated by a 404.4 nm laser pulse. The analyte emissions were shown to be much enhanced. By optimizing the experimental conditions, detection sensitivity could be improved by an order of magnitude relative to the more conventional non-resonance case. Multielement RELIPS analysis was also demonstrated. It was now clear that cooled potassium plasma, with electron density less than 3×10^{16} cm^{-3} and temperature less than 0.35 eV, could be successfully rekindled by 404.4 nm light. The electron density could be

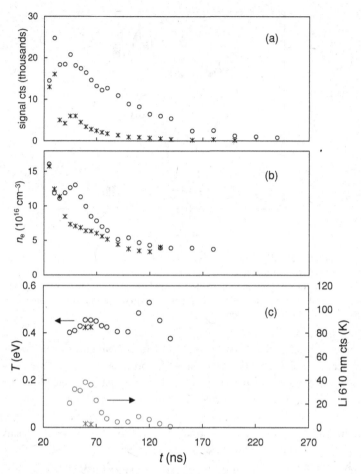

Figure 12.16. Time-resolved plume emissions generated under conditions identical to those of trace d in Figure 12.7, except the ambient gas was 350 mbar air and the ICCD gate width was 50 ns. The effective optical gate width was about 37 ns. The time axis t was the ICCD gate delay t_d plus half of the gate width t_b. Time t was therefore the center of the time window. Both on-resonance (circles) and off-resonance (crosses) behaviors are shown for easy comparison. The three panels are: (a) sodium signal; (b) electron density determined from Stark widths of the Li 610.3 nm and 670.8 nm lines; and (c) plasma temperature determined from the intensity ratio of the same pair of Li lines, together with the Li 610.3 nm line intensity (lighter symbols).

doubled while the plasma temperature could reach 0.53 eV. It was believed that the light energy was efficiently converted to plasma thermal energy through resonant absorption and super-elastic collisions. Still-higher plasma temperature favorable for the excitation of "hard" transitions may be attainable. It is important to note that the successful rekindling of the plasma was carried out in a gentle and controlled manner without thermal ignition or chaotic explosion. This naturally provides the fascinating opportunity of customizing the electron density and plasma temperature for optimal spectrochemical analysis. There are still open questions. For instance, the exact mechanism of photoresonant heating and the role of resonance photoionization have to be clarified. Extension to other sample matrices and attempts at real-world analysis should be taken up as well.

12.4 Liquid samples

Numerous works on LIBS analysis of liquids ranging from small droplets and aerosols to organic solutions and molten steel have been reported [30, 31]. Among them, aqueous samples are particularly interesting. This is not only because water is omnipresent in industrial, pharmaceutical, and environmental specimens, but also because water is the common denominator of practically all biological samples. For that reason, all resonance-enhanced analyses of liquids were done on aqueous samples [12, 36–39].

In contrast to refractory targets mentioned in the previous section, water is relatively volatile. The minimum laser fluence needed to vaporize liquid water is only tens of millijoules per square centimeter [36], while that needed to induce thermal breakdown is a few joules per square centimeter [12]. The huge fluence "window" in between implies that the double-pulse approach designed for solids is no longer necessary. A single laser pulse at the right wavelength can readily do both vaporization and resonant pumping without inducing thermal breakdown. The remaining challenge is to find ways to excite the water molecule photoresonantly. We will first theorize about viable schemes in Section 12.4.1. Experimental studies and analytical performance will be reported in Sections 12.4.2 and 12.4.3. An example of ultra-micro analysis will be described in Section 12.4.4. The section ends with a summary (Section 12.4.5).

12.4.1 A model of photoresonant pumping of water molecules

The photoabsorption and photoionization of water vapor have been extensively studied [40–44]. Two broad absorption bands peaking at 165.5 nm and 128.0 nm in the visible–ultraviolet region were known [45, 46]. They correspond to transitions from the ground X^1A_1 electronic state to the respective $\tilde{A}^1B_1$ and B^1A_1 excited electronic states. At wavelengths of common lasers, absorption is extremely weak.

Interestingly, for hot (>1000 K) water vapor, strong absorption at 193 nm was recently reported [47]. The enhanced absorption was attributed to favorable Franck–Condon overlap between the vibrationally excited (0,0,1) state of X^1A_1 and the purely repulsive $\tilde{A}^1A''$ electronic state. The absorption cross-section for this transition was estimated to be about 2×10^{-18} cm^2 at 193 nm. This development is encouraging for two reasons. First, powerful pulsed lasers operating on 193 nm are readily available in the form of ArF excimer lasers. Second, even at sub-breakdown fluences, the laser-induced plumes are easily ~1000 K in temperature. Given that the (0,0,1) state is 3755.7 cm^{-1} (5400 K) above the ground (0,0,0) vibrational state [46], the vibrationally excited population can be as high as 0.5%. For a

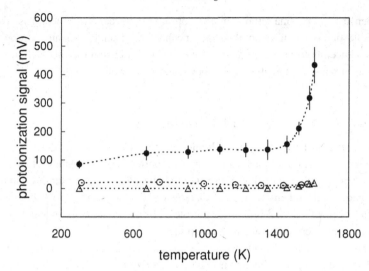

Figure 12.17. The 193 nm photoionization of water vapor at elevated temperature. When the ionization chamber was filled with 5×10^{15} cm^{-3} of H_2O molecules, ionization signal (solid circles) rapidly increased with temperature T for $T > 1500$ K. When the laser wavelength λ was changed to 248 nm (KrF, open triangles), or when the chamber was evacuated while λ was still 193 nm (open circles), the ionization signal became negligibly small. Adapted from reference [49].

typical plume density of 10^{19} cm^{-3}, the absorption coefficient will be about 0.1 cm^{-1}. The process feeds back positively, with the absorption coefficient increasing with plume temperature. This leads to very efficient pumping of water vapor at 193 nm.

However, the 12.6 eV ionization potential of H_2O is much higher than typical thermal energies. Stepwise photoionization by absorbing two 193 nm photons is also prohibited by the dissociative nature of the intermediate $\tilde{A}^1A''$ state. The production of free electrons for plasma initiation is therefore highly improbable through this channel.

A more likely channel was suggested by a quantum *ab initio* calculation [48]. It was shown that the potential energy curve of the $\tilde{A}^1A''$ state will change from purely repulsive to weakly binding if the molecular symmetry is changed from C_s to C_{2v}. That corresponds to converting from an antisymmetric O–H stretch to a symmetric one. The weakly bound state, labeled $\tilde{A}^1B_1$, is 7.08 eV above the ground X^1A_1 state and can be reached from the (2,0,0) vibrational state by absorbing a 193 nm photon. Subsequent absorption of a second 193 nm photon will ionize the molecule. The (2,0,0) state is 2×3657 cm^{-1} (10 520 K) above the (0,0,0) state [46]. So, if the temperature is high enough, a good fraction of the water molecules can be photoionized by absorbing two 193 nm photons.

The 193 nm photoionization of water vapor at elevated temperature has been investigated experimentally [49]. The results are shown in Figure 12.17. The steam density was 5×10^{15} cm^{-3}. For temperature above 1500 K, efficient ionization induced by ArF laser is clearly evident (solid circles). When the sample cell was evacuated (open circles), or when ArF was changed to KrF (248 nm, open triangles), the ionization signal became negligible. Because the steam density was too low for efficient heating via super-elastic collisions, the ionization should be directly photoinduced. These results are consistent with the spectroscopic model.

By measuring the photoionization signal as a function of ArF laser energy at various steam temperatures, bounds on the cross-sections for the two-photon absorption steps could be established [49]. The cross-section for the second step, σ_2, corresponding to the ionization from the intermediate $\tilde{A}^1B_1$ state, should be much bigger than 10^{-18} cm^2. It is likely to be between 10^{-17} and 10^{-16} cm^2. The cross-section for the first step, σ_1, corresponding to the excitation from the X^1A_1 (2,0,0) state to the $\tilde{A}^1B_1$ state, was shown to be much less than σ_2. A reasonable value for σ_1 is 10^{-18} cm^2.

Summing up, ArF laser pulses at 193 nm can heat up as well as photoionize water molecules that are vibrationally excited. Free electrons at the desired temperatures may be produced without thermal breakdown.

12.4.2 ArF laser ablation of aqueous samples

ArF (193 nm) pulsed laser ablation of aqueous samples was investigated [12, 36, 38]. The samples were also ablated at 248 nm and 532 nm for comparison. The setup is shown schematically in Figure 12.18. Experimental details have been reported [50]. Briefly, an aqueous solution containing 6 mM of sodium was ejected through a tubing to form a stable vertical stream. The solution was seeded with 12 mM methyl violet in order strongly to absorb at all three wavelengths. The absorption coefficient was about 10^3 cm^{-1}. The liquid jet was intercepted by the ablation laser beam that was focused normally onto the liquid surface. The luminous plume produced was imaged along a transverse direction onto a 30 μm slit of a spectrograph. The overall spectral resolution was 0.04 nm. Mounted at the exit port was an intensified charge-coupled device (ICCD) that was synchronously gated with the laser pulse. The gate width was 20 ns and the gate delay was decreased at 20 ns steps. By scanning the plume image across the entrance slit of the spectrograph, emissions from various position of the plume could be detected. About 50–100 shots were averaged before each spectral trace was captured and processed on a personal computer.

For the purpose of measuring the electron density n_e and temperature T of the plasma plume, lithium (15 mM LiCl) was used as the reporter element. Hydrogen α and β lines and the oxygen 777 nm line were also used for crosschecking.

The time-evolution of the sodium doublet emissions is shown in Figure 12.19 for the case of 193 nm ablation. The corresponding case of 532 nm ablation was shown earlier, in Figure 12.5a. The laser fluence was about 3 J cm^{-2} and the hottest core of the plume near the base was imaged in both cases. As can be seen from Figure 12.5a, with 532 nm ablation, the intense continuum background masked the weak sodium emissions during the first 60 ns. In sharp contrast, the continuum background was eliminated in the 193 nm case (Figure 12.19) and the sodium signal was about ten times stronger. This is obviously ideal for spectrochemical analysis. Results of 248 nm ablation were similar to the 532 nm case,

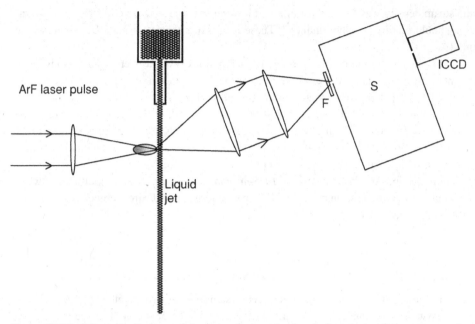

Figure 12.18. Schematic diagram of the setup for spectrochemical analysis of aqueous samples using ArF laser ablation. A 0.4 mm diameter sample jet was ablated by an ArF laser pulse. The plume emissions were imaged transversely through UV blocking filters (F) onto the entrance slit of a 0.5 m spectrograph (S). An intensified charge-coupled device (ICCD) array detector was mounted at the exit port of the spectrograph. Adapted from reference [12].

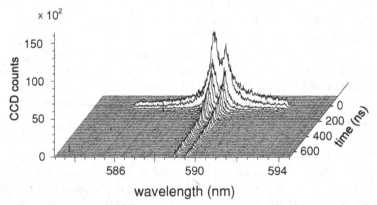

Figure 12.19. Effect of 193 nm ablation. A time sequence is shown of spectra captured under conditions identical to those of Figure 12.5a except an ArF (193 nm) laser was used. The fluence was about 3 J cm^{-2}. Relative to the case of 532 nm ablation shown in Figure 12.5a, the suppression of the continuum background and the enhancement of the sodium doublet are very apparent for the 193 nm case. Adapted from reference [12].

Table 12.1. *Ablation at 532 nm: electron temperature T and electron density n_e determined from line intensity ratios and linewidths, at plume position z (normal to target surface) and time t*

		T (eV)[b]			n_e (10^{17} cm^{-3})[c]		
$z(\mu m)$	t^a(ns)	H	H/O	Li	H$_\alpha$	H$_\beta$	Li (610)
0	60	8	4	—	10	5	—
110	100	1	—	0.6	5	2	4
230	120	0.2	—	0.4	3	2	2

[a]$t = 0$ was defined as the rising edge (20%) of the incident laser pulse.
[b]Based on intensity ratios of H$_\alpha$/H$_\beta$, H$_\alpha$/O$_I$ (777 nm), and Li (671 nm)/Li (610 nm). All intensities were integrated over line profiles.
[c]Based on linewidths of H$_\alpha$, H$_\beta$, and Li (610 nm) transitions.

Table 12.2. *Ablation at 193 nm: electron temperature T and electron density n_e determined from line intensity ratios and linewidths, at plume position z (normal to target surface) and time t*

z (μm)	t (ns)[a]	T (eV)[b]	n_e (10^{17} cm^{-3})[c]
0	20	0.5	6
0	60	0.5	3
100	100	0.4	2
200	120	0.4	2

[a]$t = 0$ was defined as the rising edge (20%) of the incident laser pulse.
[b]Based on intensity ratio of Li 671:610. Intensities were integrated over line profiles.
[c]Based on line widths of the Li (610 nm) transition.

except that the background was less intense by a factor of around five, while the sodium signal was not even detectable.

Results of n_e and T measurements are tabulated in Tables 12.1 and 12.2 for the 532 nm and 193 nm cases, respectively. The 532 nm ablation produced electron densities ($>10^{18}$ cm^{-3} decaying to 10^{17} cm^{-3} in 100 ns) and temperatures (10 eV dropping to 0.3 eV in 100 ns) that were typical of LIBS plasmas created under comparable conditions. Interestingly, the electron densities produced by 193 nm ablation were not much lower but the initial temperatures were only 0.5 eV.

All these observations may now be interpreted in terms of resonant versus non-resonant pumping. With 532 nm ablation, the laser radiation was non-resonantly coupled to the plasma plume through inverse Bremsstrahlung absorption. Once the temperature reached 1 eV or so, the fraction of plume species ionized thermally could be as high as tens of percent. As a result, the nascent plasma was dense and n_e could reach 10^{19} cm^{-3}. Under these conditions, the absorption cross-section for inverse Bremsstrahlung could be estimated (equation (12.13)) to be about 3×10^{-19} cm^2. The absorption coefficient is therefore about 3 cm^{-1}.

With 193 nm ablation, inverse Bremsstrahlung heating was much less efficient. For example, assuming n_e of 10^{18} cm^{-3}, T of 0.5 eV and λ of 193 nm (see Table 12.2), the absorption coefficient was only about 10^{-3} cm^{-1}. As mentioned above, if the temperature of the vapor plume was about 1000 K at birth, the resonant pumping of the (0,0,1) state of the water molecules would lead to an effective absorption coefficient of about 0.1 cm^{-1}. The vapor was therefore only mildly heated up relative to the 532 nm case. At such low temperature, thermal ionization was unlikely. The seed electrons were more probably generated by $1 + 1$ resonance photoionization of vibrationally excited H_2O when two 193 nm photons were absorbed. As T increased, the indirect ionization channel through resonant pumping of the (0,0,1) and (2,0,0) states would start to contribute [51].[4] At T of 0.5 eV, the overall ionized fraction could reach 10% [49]. This is consistent with the measured n_e of 10^{18} cm^{-3}.

With 248 nm ablation, unlike the 193 nm case, no resonant absorption channels were known to be available; and non-resonant inverse Bremsstrahlung heating was less effective than the 532 nm case because of the shorter wavelength. As a result, only weak continuum emissions were observed initially. Sodium emissions were not detectable because there were too few electron impacts to sustain analyte excitations.

12.4.3 Analytical performance

The analytical performance of 193 nm ablation in terms of both *relative* and *absolute* limits of detection (LODs) has been investigated [36, 37, 49]. Relative LOD is the primary concern in trace analysis when the sample supply is abundant while the analyte concentration is low. These limits for Pb, Ba, Ca, and Na were reported [49, 52]. The experimental setup was similar to that shown in Figure 12.18, except the plume emissions were imaged (1:2 magnified) onto a 250 μm slit to optimize the signal. Aqueous specimens containing graded amounts of the test analytes were analyzed. Because only minute amounts of the analytes were present, a higher ArF laser fluence of about 10 J cm^{-2} was needed to remove a detectable number of analyte atoms from the target. Thermal breakdown became inevitable. Nevertheless, the plasmas were shown to be still cooler and the time-windows for useful spectral detection were significantly longer than typical LIBS conditions. The ICCD gate delay was 1–2 μs while the gate width was 3–4 μs. For each sample, at least 20 spectra were taken, each being the average of 500 shots. The mean analyte signal $\langle S \rangle$ and the standard

---

[4] Laser heating of molecular vapors by resonance excitation may not be as efficient as atomic vapors because the energy losses to vibrational and rotational degrees of freedom are too large.

Table 12.3. *Wavelength* λ *and excited-state energy* E_k *of the analytical lines, the gate delay* t_d *and width* t_b *and laser pulse energy* E *employed, and the limit of detection for various elements*

| element | transition | | gate-setting | | | limit of detection | | |
	λ (nm)	E_k (eV)	t_d (μs)	t_b (μs)	E (mJ)	193 nm	LIBS[a]	ICP-AES[b]
Na (I)	589.0	2.10	2.1	3.0	16.0	0.4 p.p.b.	7.5 p.p.b.	0.5 p.p.b.
Ca (I)	422.7	2.93	1.6	3.5	16.0	3 p.p.b.	130 p.p.b.	0.04 p.p.b.
Ba (II)	455.4	2.72	0.9	4.0	13.0	7 p.p.b.	6.8 p.p.m.	0.03 p.p.b.
Pb (I)	405.8	4.38	1.3	3.5	14.4	300 p.p.b.	13 p.p.m.	1 p.p.b.

[a] See reference [52].
[b] See reference [53].

deviation σ of the 20 or so spectra were computed. The detection limits, at $\langle S \rangle / \sigma = 3$, for the four elements are listed in Table 12.3, together with the best non-193-nm LIBS results for easy comparison [53]. As can be seen, the detection limits are 20–1000 times better. For the case of Na, the detection limit is better than that of ICP-AES.[5]

The plasmas were also characterized by measuring the electron density n_e and temperature T as functions of time [12, 49]. Stark widths of the H_β 486.1 nm line and Li(I) 610.4 nm line were used for n_e determinations, and intensity ratios of the Fe lines and the Li(I) 610.4 nm line and 670.8 nm line were used for T evaluations. The ICCD gate width was set to 1 μs while the delay was stepped at 1 μs intervals. The results are shown in Figure 12.20. As can be seen, n_e was in the 10^{16} cm^{-3} range during the time interval of 1–5 μs when analyte emissions were detected. This density was low enough to minimize continuum background, yet high enough to guarantee LTE. The plasma temperature decayed from about 0.45 eV to about 0.35 eV in that same time interval. This temperature range was near optimal for easily excited analytes such as Na, but too low for species with higher excited levels such as Pb (see Figures 12.3 and 12.4). This could explain why some of the LODs achieved with 193 nm ablation (Table 12.3) are inferior to that of ICP-AES, and would point to plasma re-heating as a remedy.

For the experiments just described, if the *absolute* instead of the *relative* LODs were considered, the results would not be impressive. The heavy averaging and the draining of unablated liquids were wasteful if the sample supply was limited. Different strategies were adopted when the sample mass available was minuscule [37]. First, the mass ablated per pulse was minimized. While the jet was still used, the sample stream was shrunk to 17 μm in diameter by using a sheath flow arrangement. Second, single-shot sensitivity was optimized. That required sub-breakdown laser fluences to maximize the target stability and

[5] Perkin Elmer Optima 4300, introduced in 2000.

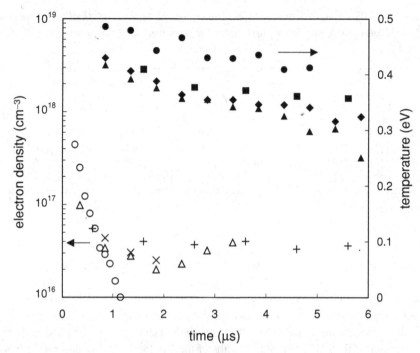

Figure 12.20. Electron density (left axis) and temperature (right axis) of plasmas produced by ArF laser ablation of aqueous solutions of Fe or Li salts, as functions of time. The laser energy density was about 10 J cm^{-2}. Time zero corresponded to the firing of the ArF laser pulse. The electron density was determined spectroscopically by measuring the Stark-broadened linewidths of the H$_\beta$ 486.1 nm (O) and Li(I) 610.4 nm emissions. Li concentrations of 0.1 M (+), 98 μM (Δ), and 20 μM (×) were used. The plasma temperature was determined from the intensity ratios of the Fe lines (●) and the Li(I) 610.4 nm line and 670.8 nm line. Fe concentration of 480 μM, and Li concentrations of 0.1 M (■), 98 μM (▲), and 20 μM (♦) were used. Adapted from reference [52].

suppress the background [36, 37]. Normalization of the optical signal by the acoustic signal was also found to aid overall data reproducibility [36, 37]. Using sodium as the test analyte, and counting only the liquid mass ablated, the absolute LOD was estimated to be 8 fg [37]. This is to be compared with the picogram sensitivity of typical atomic emission spectrometry [36]. The draining of unused sample is still an issue. Synchronized sample delivery such as ink jets may be a suitable alternative.

12.4.4 Detection of sodium and potassium in single red blood cells

Given the femtogram sensitivity of 193 nm ablative sampling, the 16 fg of Na and 340 fg of K found in a typical human erythrocyte could be detected [37]. That was recently demonstrated [54]. The details of the instrumentation can be found elsewhere [37, 50], Briefly, the sample feed assembly was modeled after a flow cytometer. Erythrocytes suspended in either 8%

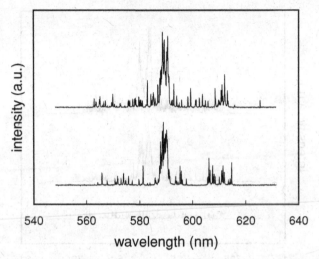

Figure 12.21. Typical plasma plume emission spectra produced by 193 nm laser ablative sampling of human red blood cells that had been centrifuge-washed and resuspended in 8% glucose. The laser fluence was about 4 J cm^{-2}. The initial 70 ns of each event was gated off, and the subsequent 1 μs was integrated. The instrumental resolution was about 3 nm, and therefore the sodium 589.0 nm and 589.6 nm doublets are not resolved. Both traces correspond to single-shot events when single blood cells were believed to have been sampled. Adapted from reference [54].

glucose or phosphate-buffered saline were dripped through a 50 μm internal diameter capillary and dragged into an 8 μm stream by a fast-flowing sheath of deionized water. The narrow sample stream, with blood cells lined up in single file about 400 μm apart, was forced to flow on the *outside* of the sheath to face the ablating laser. The ArF beam was focused to an 80 μm × 700 μm ($h \times w$) spot on the sample stream. The laser fluence was about 4 J cm^{-2}. A tightly focused He–Ne laser beam intercepted the sample stream at a point 200 μm above the ablation region. The 633 nm Mie signal scattered off falling blood cells triggered the ArF laser after a suitable delay. Because of the huge gap between adjacent cells, no more than one erythrocyte was ablated per shot. The plume emissions were again collected transversely as shown in Figure 12.18. The spectrograph slit was widened to 1 mm. The ICCD gate delay was either 35 ns or 70 ns since the firing of the ArF laser. The gate width was 1 μs.

Typical single-cell spectra are shown in Figure 12.21 for the sodium spectral region. The ICCD gate delay was 70 ns. At an instrumental resolution of 3 nm, the sodium doublet was not resolved. Erythrocytes were suspended in 8% glucose that contained negligible amounts of sodium. The 590 nm signal was therefore attributed to single blood cells. The signal intensity, which was defined as the net spectral area with the background subtracted, was about 18 times that of the root-mean-square fluctuation of the background for the two traces shown. The *absolute* sensitivity should therefore be adequate for quantitative analysis of sodium in single blood cells.

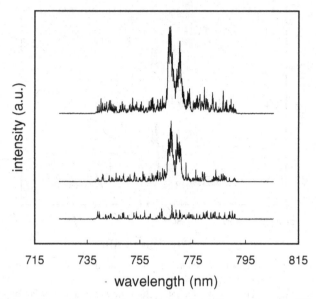

Figure 12.22. Typical plasma plume emission spectra produced by synchronous 193 nm laser ablative sampling of human red blood cells that had been centrifuge-washed and resuspended in phosphate-buffered saline. The laser fluence was about 4 J cm^{-2}. The initial 35 ns of each event were gated off, and the subsequent 1 μs was integrated. The ArF laser was timed to fire at the blood cells for the top two traces, while it was timed to miss the blood cells for the bottom trace. Spectral resolution was about 3 nm, and the potassium 766.5 nm and 769.9 nm doublets are clearly resolved in the top two single-cell spectra. Adapted from reference [54].

Single-cell spectra for the potassium spectral region are shown in Figure 12.22. The ICCD gate delay was 35 ns. The potassium 766.5 nm and 769.9 nm doublets, now resolved, are clearly visible in the top two traces when the ArF laser was timed to fire at the blood cells. They are no longer observable in the bottom trace when the ArF laser was timed to purposely miss the blood cells. This shows that the phosphate-buffered saline used to suspend the blood cells did not generate any significant interference. The signal-to-background fluctuation, as defined previously, was about 30 for the top two traces. This again illustrates the adequate sensitivity for quantitative analysis of potassium at the single cell level.

12.4.5 Liquid analysis: a summary

The spectrochemical analysis of aqueous samples using 193 nm laser-induced plasma emissions has been reviewed. A model of resonant pumping of water molecules at 193 nm was described. It predicted a radiation-plume coupling that is far stronger than inverse Bremsstrahlung absorption. That prediction was consistent with the experimental observations. In particular, ArF (193 nm) laser ablation of aqueous samples at sub-breakdown fluences yielded enhanced analyte spectra that were background free. Impressive detection limits, both relative as well as absolute, were demonstrated. The feasibility of quantitative analysis of sodium and potassium in single blood cells was demonstrated.

Future efforts may be directed to the following areas. Definitive experimental verification of the proposed pathways of H_2O excitation at 193 nm should be pursued. An understanding of the basic mechanisms is essential for guiding the design of ablative sampling and plasma rekindling schemes. Real-world problems that would benefit from the 193 nm probe should be identified. The required instrumentation developments, be they synchronized specimen feeding for pulsed sampling, or cytometric modifications for ultra-micro analysis, can be tailored to meet the specific needs. More general issues such as calibration standards or extension to other liquids will need to be addressed as well.

12.5 Gaseous samples

Although RELIPS analysis of gaseous samples has not been performed, resonance-enhanced production of laser-induced plasmas in gases has been demonstrated [55, 56]. In that work, it was shown that, for carbon-containing gases such as CO, CO_2, CH_3OH, and $CHCl_3$, the threshold irradiance for breakdown was an order of magnitude lower if ArF instead of Nd:YAG (fundamental through fourth harmonics) laser was used. The easier breakdown at 193 nm was attributed to resonance-enhanced two-photon ionization of metastable carbon atoms. However, units of gigawatts per square centimeter were still required even with 193 nm radiation. This is because breaking down gases is more difficult than condensed matter. As a result, the initial plasma temperatures ($>$eV) and electron densities (10^{18}–10^{19} cm^{-3}) were higher than optimal for spectrochemical analysis. It was shown that a stronger coupling of the laser light and the target gas, such as the efficient two-photon dissociation of CO at 193 nm, would further reduce the breakdown threshold [56]. Future attempts at RELIPS analysis of gases should consider these factors.

12.6 Conclusion: resonance-enhanced LIBS as an analytical tool

We have seen how resonance can be exploited in the laser generation and rekindling of plasmas for analytical applications. Suppression of background and improvements in sensitivity have been demonstrated. It may still be early to critically evaluate the analytical performance of RELIPS at this time. A few general and descriptive remarks are probably more appropriate. The first remark concerns universality. In order to resonantly excite a major species in the vapor plume, prior knowledge of the sample matrix (though not the analytes) is necessary. This is certainly an additional requirement over non-resonance techniques such as LIBS. Nonetheless, as mentioned earlier, the matrix chemicals are generally known in practical applications and this should not be a major handicap.

The second remark concerns selectivity. Quite unlike most ultra-sensitive techniques such as laser ionization mass spectrometry where both the excitation and detection steps are selective, electron impact excitation in plasmas is *nonselective*. The spectral dispersion step provides all the discrimination. Therefore, RELIPS will probably fare better in high throughput, simultaneous multielement analysis than in high-resolution isotope separation.

The third remark concerns sensitivity. It is generally assumed that the sensitivity of optical spectrometry will never approach that of mass spectrometry, for the simple reason that the collection efficiency of ions is orders of magnitude better than that of photons. After all, single atom detection was first demonstrated by using laser resonance ionization and ion detection [57]. This explains why the successful detection of a single dye molecule in solution by laser-induced fluorescence came as a pleasant surprise [58]. One now realizes that the repetitive laser pumping of the fluorophore through numerous excitation/radiation cycles can generate enough photons to compensate for the poor collection efficiency. An interesting analogy may be drawn between repetitive photon pumping and electron collision excitation. The need for laser-sustained plasmas is again evident.

The final remark concerns controllability. We have shown how cooler plasmas can be generated photoresonantly in a mild and tractable manner, which is very different from the violent explosions associated with thermal breakdown. The predictability and repro-ducibility of the process mean not only superior detection limits, but also better-defined plasma properties, such as temperature, to aid spectrochemical analysis. A well-defined plasma temperature will facilitate calibration-free LIBS [59]. This can have far-reaching implications. While we are working with laser-*induced* plasmas today, we may be toying with laser-*customized* plasmas tomorrow.

Nevertheless, our present understanding of RELIPS is still limited and evolving. Hope-fully this review will stimulate more and deeper studies of the fundamental as well as the applied aspects of this promising tool.

Note added in proof. More recent studies of the mechanisms [60] and applications [61–63] of RELIPS have been reported.

The author wishes to thank the following for their contributions: E. S. Yeung of Iowa State University, USA; J. D. Wu of Fudan University, China; and S. Y. Chan, W. F. Ho, K. M. Lo, S. L. Lui, C. W. Ng, and S. F. Wong of Hong Kong Baptist University. Special thanks to L. St-Onge of NRC, Canada, for sharing preprints and relevant references. Supports from the Faculty Research Grants of Hong Kong Baptist University, the Research Grants Council of the University Grants Committee of Hong Kong, and the U.S. Department of Energy are also gratefully acknowledged.

12.7 References

[1] P. W. J. M. Boumans (editor), *Inductively Coupled Plasma Emission Spectroscopy* (New York: Wiley-Interscience, 1987).
[2] C. Th. J. Alkemade, Tj. Hollander, W. Snelleman and P. J. Th. Zeegers, *Metal Vapours in Flames* (Oxford: Pergamon, 1982).
[3] M. I. Boulos and R. M. Barnes, Plasma modeling and computer simulation; in reference [1].

[4] S. F. Wong, Plasma emission spectroscopy: modeling line-to-continuum ratios. Honours Thesis, Department of Physics, Hong Kong Baptist University (1998).

[5] I. H. Hutchinson, *Principles of Plasma Diagnostics* (Cambridge: Cambridge University Press, 1987).

[6] H. R. Griem, *Plasma Spectroscopy* (New York: McGraw-Hill, 1964).

[7] P. W. J. M. Boumans, Basic concepts and characteristics of *ICP-AES;* in reference [1].

[8] R. W. P. McWhirter, Spectral intensities, in *Plasma Diagnostic Techniques*, R. H. Huddlestone and S. L. Leonard (editors) (New York: Academic Press, 1965).

[9] T. L. Thiem, Y.-L. Lee and J. Sneddon, *Microchem. J.*, **45** (1992), 1–35. (See especially section 3.5.)

[10] R. J. Harrach, *Theory of Laser-Induced Breakdown over a Vaporizing Target Surface*, Lawrence Livermore Laboratory Report, No. UCRL-52389, 1977.

[11] C. R. Phipps and R. W. Dreyfus, Laser ablation and plasma formation, in *Laser Ionization Mass Analysis*, A. Vertes, R. Gijbels and F. Adams (editors) (New York: Wiley, 1993).

[12] C. W. Ng, W. F. Ho and N. H. Cheung, *Appl. Spectrosc.*, **51** (1997), 976–983.

[13] T. B. Lucatorto and T. J. McIlrath, *Phys. Rev. Lett.*, **37** (1976), 428–431.

[14] T. B. Lucatorto and T. J. McIlrath, *Appl. Opt.*, **19** (1980), 3948–3956.

[15] R. M. Measures and P. G. Cardinal, *Phys. Rev. A*, **23** (1981), 804–815.

[16] A. Kopystynska and L. Moi, *Phys. Rep.*, **92** (1982), 135–181.

[17] O. L. Landen, R. J. Winfield, D. D. Burgess, J. D. Kilkenny and R. W. Lee, *Phys. Rev. A*, **32** (1985), 2963–2971.

[18] G. S. Hurst and M. G. Payne, *Principles and Applications of Resonance Ionization Spectroscopy* (Bristol: Adam Hilger, 1988).

[19] J. C. Travis and G. C. Turk, *Laser Enhanced Ionization Spectrometry* (New York: Wiley-Interscience, 1996).

[20] A. Vertes, R. Gijbels and F. Adams (editors), *Laser Ionization Mass Analysis* (New York: Wiley, 1993).

[21] T. M. Allen, C. H. Smith, P. B. Kelly *et al.*, *SPIE*, **2385** (1995), 39–50.

[22] M. Capitelli, F. Capitelli and A. V. Eletskii, *Las. Phys.*, **10** (2000), 1244–1250.

[23] D. Y. Tsipenyuk, and M. A. Davydov, *Las. Phys.*, **6** (1996), 806–810.

[24] Ya. B. Zel'dovich and Yu. P. Raizer, *Physics of Shock Waves and High-Temperature Hydrodynamic Phenomena* (New York: Academic Press, 1966).

[25] Y. Iida, *Spectrochim. Acta*, **45**B (1990), 1353–1367.

[26] S. L. Lui and N. H. Cheung, *Spectrochim. Acta B*, **58** (2003), 1613–1623.

[27] J. Uebbing, J. Brust, W. Sdorra, F. Leis and K. Niemax, *Appl. Spectrosc.*, **45** (1991), 1419–1423.

[28] L. St-Onge, V. Detalle and M. Sabsabi, *Spectrochim. Acta B*, **57** (2002), 121–135.

[29] See also Chapter 15 of the present book for an extensive review of plasma reheating.

[30] D. A. Rusak, B. C. Castle, B. W. Smith and J. D. Winefordner, *Crit. Rev. Anal. Chem.*, **27** (1997), 257–290.

[31] Y.-I. Lee, K. Song and J. Sneddon, *Laser-Induced Breakdown Spectrometry* (Huntington: Nova Science Publisher, 2000).

[32] S. Y. Chan and N. H. Cheung, *Anal. Chem.*, **72** (2000), 2087–2092.

[33] J. D. Wu and N. H. Cheung, *Appl. Spectrosc.*, **55** (2001), 366–370.

[34] R. G. Root, Modeling of post-breakdown phenomena, in *Laser-Induced Plasmas and Applications*, L. J. Radziemski and D. A. Cremers (editors) (New York: Marcel Dekker, 1989).

[35] S. Y. Chan, Resonance-enhanced laser-induced plasma spectroscopy for elemental analysis. M.Phil. Thesis, Hong Kong Baptist University (1999).

[36] N. H. Cheung and E. S. Yeung, *Appl. Spectrosc.*, **47** (1993), 882–886.

[37] N. H. Cheung and E. S. Yeung, *Anal. Chem.*, **66** (1994), 929–936.

[38] W. F. Ho, C. W. Ng and N. H. Cheung, *Appl. Spectrosc.*, **51** (1997), 87–91.

[39] N. H. Cheung, C. W. Ng, W. F. Ho and E. S. Yeung, *Appl. Surf. Sci.*, **127–129** (1998), 274–277.

[40] D. H. Katayama, R. E. Huffman and C. L. O'Bryan, *J. Chem. Phys.*, **59** (1973), 4309–4319.

[41] H.-T. Wang, W. S. Felps and S. P. McGlynn, *J. Chem. Phys.*, **67** (1977), 2614–2628.

[42] G. N. Haddad and J. A. R. Samson, *J. Chem. Phys.*, **84** (1986), 6623–6626.

[43] R. H. Page, R. J. Larkin, Y. R. Shen and Y. T. Lee, *J. Chem. Phys.*, **88** (1988), 2249–2263.

[44] M. J. J. Vrakking, Y. T. Lee, R. D. Gilbert and M. S. Child, *J. Chem. Phys.*, **98** (1993), 1902–1915.

[45] K. Watanabe and M. Zelikoff, *J. Opt. Soc. Amer.*, **43** (1953), 753–755.

[46] G. Herzberg, *Molecular Spectra and Molecular Structure III. Electronic Spectra and Electronic Structure of Polyatomic Molecules* (Princeton: Van Nostrand, 1967).

[47] W. J. Kessler, K. L. Carleton and W. J. Marinelli, *J. Quant. Spectrosc. Radiat. Transfer*, **50** (1993), 39–46.

[48] V. Staemmler and A. Palma, *Chem. Phys.*, **93** (1985), 63–69.

[49] K. M. Lo, Laser ablation of aqueous samples at 193 nm: mechanism and applications. M.Phil. Thesis, Hong Kong Baptist University (2000).

[50] C. W. Ng, Detection of sodium and potassium in single human erythrocytes by laser-induced plasma spectroscopy: instrumentation and feasibility demonstration. M.Phil. Thesis, Hong Kong Baptist University (1999).

[51] G. M. Weyl, Physics of laser-induced breakdown: an update, in *Laser-Induced Plasmas and Applications,* L. J. Radziemski, and D. A. Cremers (editors) (New York: Marcel Dekker, 1989).

[52] K. M. Lo and N. H. Cheung, *Appl. Spectrosc.*, **56** (2002), 682–688.

[53] R. Knopp, F. J. Scherbaum and J. I. Kim, *Fresenius J. Anal. Chem.*, **355** (1996), 16–20.

[54] C. W. Ng, and N. H. Cheung, *Anal. Chem.*, **72** (2000), 247–250.

[55] J. B. Simeonsson and A. W. Miziolek, *Appl. Opt.*, **32** (1993), 939–947.

[56] J. B. Simeonsson and A. W. Miziolek, *Appl. Phys. B*, **59** (1994), 1–9.

[57] G. S. Hurst, M. H. Nayfeh and J. P. Young, *Appl. Phys. Lett.*, **30** (1977), 229–231.

[58] S. Nie, D. T. Chiu and R. N. Zare, *Science*, **266** (1994), 1018–1021.

[59] See also Chapter 3 of the present book.

[60] S. L. Lui and N. H. Cheung, *Appl. Phys. Lett.*, **81** (2002), 5114–5116.

[61] X. Y. Pu and N. H. Cheung, *Appl. Spectrosc.*, **57** (2003), 588–590.

[62] X. Y. Pu, W. Y. Ma and N. H. Cheung, *App. Phys. Lett.*, **83** (2003), 3416–3418.

[63] S. L. Lui and N. H. Cheung, *Anal. Chem.*, **77** (2005), 2617–2623.

13

Short-pulse LIBS: fundamentals and applications

R. E. Russo

Lawrence Berkeley National Laboratory, USA

13.1 Introduction

The basis of laser-induced breakdown spectroscopy (LIBS) is well established – real time, simultaneous multielement chemical analysis without sample preparation. Although the advantages are well recognized, understanding the fundamental laser ablation processes underlying this technology remains a quest. The heart of LIBS is the luminous plasma; its initiation and history are critical to applications. The plasmas' properties depend on experimental parameters, including the laser pulse (energy, duration, repetition rate, and wavelength), the sample (physical and optical), and ambient atmosphere (gas, pressure). Except for a few research papers in the analytical literature, most LIBS applications are based on nanosecond pulsed lasers. There are two compelling reasons to delve into the short-pulse regime for LIBS – expected differences in the laser–material and the laser–plasma interactions. Stated differently, there is reason to believe that improved analytical accuracy, precision, and sensitivity can be achieved by the use of short-pulsed lasers for LIBS applications.

In the LIBS plasma, spectral emission intensity of elemental lines is based on the amount of mass ablated as well as on the concentration of that mass in the sample (ignoring plasma temperature and electron number density for the moment). For accurate chemical analysis, the chemistry of the ablated mass must be representative of the sample (no fractionation). Ideally, there also would be matrix independence; the same amount of mass would be ablated independent of sample properties. These two goals can be realized by better understanding and controlling the laser–material interaction. Finally, the plasma itself can influence the ablation process. Less available energy for ablation, plasma heating, and increased broadband-background emission can result when the plasma interacts with the laser pulse. This chapter discusses the influence of short (picosecond and femtosecond) pulses on the laser–material and laser–plasma interactions, and the latest research using short pulses for LIBS applications.

Laser-Induced Breakdown Spectroscopy: Fundamentals and Applications, ed. Andrzej W. Miziolek, Vincenzo Palleschi and Israel Schechter. Published by Cambridge University Press. © A. W. Miziolek, V. Palleschi and I. Schechter 2006. A. W. Miziolek's contributions are a work of the United States Government and are not protected by copyright in the United States.

13.2 Effect of pulse duration on ablation

Without fully understanding the complexity of ablation processes, it is believed that the pulse duration is the most important parameter influencing the laser–material interaction. The physical response of a material system subjected to an intense laser pulse is based on the energy deposited per unit time and volume. From a classical point of view, a material system can accommodate an energy impulse through thermodynamic pathways: absorption, diffusion, melting, boiling, and vaporization. Phonon relaxation (heat dissipation) in a solid occurs on the order of approximately 100 fs. For nanosecond pulsed lasers, there will be ample time *during* the laser pulse for the system to equilibrate. This is not to say that nanosecond ablation can be described completely by classical thermodynamics. For nanosecond laser pulses with high peak power ($> 10^8$ W cm^{-2}), ablation mechanisms involve nonlinear processes such as multiphoton absorption/ionization and inverse Bremsstrahlung [1–5]. For picosecond and femtosecond lasers, the pulse duration can be comparable to or shorter than the phonon relaxation time. From a simple point of view, all of the laser energy can be deposited into the material before it can *thermally* respond. A phenomenological theory to describe short-pulse laser ablation mechanisms is a two-temperature model in which the electrons are heated by absorption of photons due to inverse Bremsstrahlung. The energy of the hot electron gas is subsequently transferred to the lattice and thermalized in the bulk [6]. Before the material undergoes any change in its thermodynamic state, the laser pulse is over. Material removal (ablation) occurs *after* the laser pulse.

Thermal diffusion and the heat-affected zone *during* the laser pulse will be negligible. The temperature-affected volume may be that defined by the optical penetration depth, although intensities are so high that absorption mechanisms are not well established. The thermal diffusion length L is proportional to $(t)^{1/2}$, where t is the pulse duration [7]. For metals, L will be on the order of micrometers and nanometers with nanosecond and femtosecond pulses, respectively. Using femtosecond laser pulses, the thermal diffusion length becomes comparable to the optical penetration depth.

Exact mechanisms to describe ablation are not well established. Some of the proposed mechanisms include photo-ionization, photo-dissociation, melting, boiling, vaporization, phase explosion, spallation, and others [1–5, 8]. Such a list clearly demonstrates the complexity and misunderstanding of this laser–material interaction. If the entire irradiated/heated volume is not ablated, thermal diffusion can take place *after* the laser pulse, contributing to elemental migration in the heated sample. Also, the influence of the plasma on the "temperature" of the sample has not been investigated; radiation from the plasma can heat the sample for a time period much longer than the laser pulse. Short-pulse LIBS of metal samples shows significant melting around the craters, possibly a result of these two post-pulse processes.

Fundamentals aside, significant benefits have been discovered in short-pulse LIBS studies that can be correlated to the laser material interaction. Lower ablation threshold (energy and fluence) and more efficient ablation have been achieved by using short-pulsed lasers [7, 9, 10]. The ablated depth per pulse is less, providing better depth resolution for analyzing

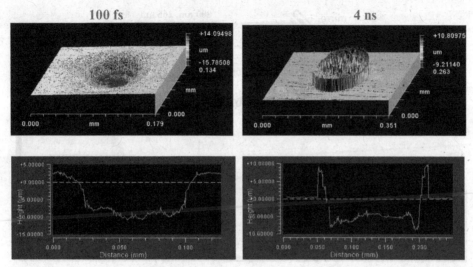

Figure 13.1. Crater profiles in brass samples after 40 pulses using a 100 fs and 4 ns laser. The images were obtained using a Zygo interferometric microscope.

multilayers [11, 12]. A clear sign that the laser–material interaction is different can be seen in the craters (Figure 13.1). Craters are cleaner (less debris) and there is no raised rim with femtosecond versus nanosecond pulses [9, 13, 14]. However, this last effect can be caused by changes in the laser–plasma interaction or expansion dynamics of the femtosecond-induced plasma. Shorter pulses reduce material deposition around the crater and increase the reproducibility of ablation – because of reduced thermal damage and lower fluence threshold.

13.3 Effect of pulse duration on plasma

The second primary reason for considering shorter pulses is to eliminate the interaction of the laser beam with the expanding mass plasma. For nanosecond ablation, the efficiency of coupling laser-beam energy into a sample target drops off as the energy density is increased [8]. The smaller amount of mass removed per pulse is the result of plasma shielding, a process in which the plasma electron number density is high enough to absorb or scatter incident photons. Laser-plasma absorption by inverse Bremsstrahlung processes can further increase temperature and electron number density. The extent of this process on LIBS has not been established. On the time-scale of a femtosecond pulse, to the best of our knowledge, mass (including fast electrons) cannot move away from a sample surface into the ambient. Therefore, the laser pulse cannot interact with the plasma. For picosecond laser pulses, a high density (10^{20} cm^{-3}) of fast electrons (10^9 cm s^{-1}) can form a few micrometers above the sample surface. A picosecond laser pulse can interact with this electron-plasma (discussed below).

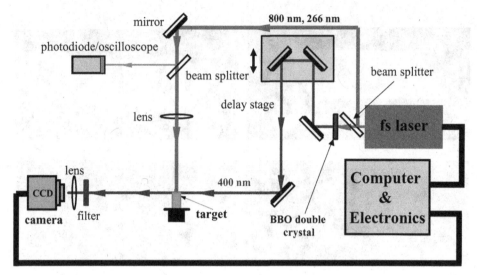

Figure 13.2. Experimental system for measuring time-resolved images of laser-induced plasmas. As shown, the system provides femtosecond time resolution with femtosecond pulsed ablation. In this case, the first harmonic of the laser is used for ablation with the second harmonic traversing an optical delay line and used as the probe beam.

Empirically, picosecond and femtosecond LIBS plasmas have been shown to be spatially smaller (than nanosecond induced plasmas) because of the lower laser-beam energy required for ablation, and the elimination of plasma heating. Short laser pulses produce a shorter lifetime plasma; there is a more rapid decrease in intensity of both background and spectral line emission [15, 16]. In some cases, gating the detector is not required to achieve good sensitivity owing to lower broadband background emission [16]. Because femtosecond-induced plasmas decay so rapidly and there is no laser–plasma interaction, higher repetition rate pulsed lasers can be used to get approximately the same signal to noise (S/N) and signal to background (S/B) in a shorter experimental time [13]. Alternatively, sensitivity can be increased for a given measurement time using higher repetition rate lasers.

The resolution for spatially analyzing a sample material is associated with the ablated-crater dimensions. The diameter of the crater is determined not only by the laser-beam waist, but by the laser-beam interaction with the sample and the plasma [14]. Laser–plasma interaction can change the laser-beam intensity distribution on the surface.

13.4 Picosecond-induced electron plasma

Time-resolved imaging can be used to study the initiation, propagation, and properties of plasmas induced by short-pulsed lasers. Figure 13.2 shows an experimental system for measuring shadowgrams and interferograms with picosecond and femtosecond time resolution [17]. Two beams from the same laser, traversing an optical delay path, allow

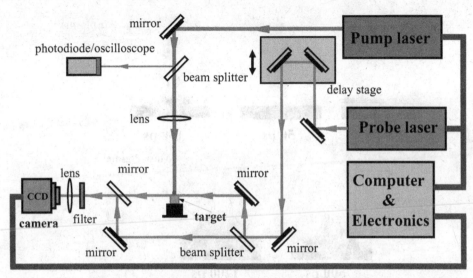

Figure 13.3. Experimental system for measuring time-resolved images of laser-induced plasmas. As shown, the system provides nanosecond and longer time resolution by using two lasers, one for ablation and the other for probe. The system also provides the capability to measure number density of species in the plume by configuring the probe-beam path as an interferometer.

precise time resolution. For example, the fundamental 1064 nm wavelength of a Nd:YAG laser can be used as the ablation or pump beam. The 532 nm second harmonic from the same laser provides the probe beam; it passes an optical delay stage and is directed to a CCD camera. The direction of the probe beam is perpendicular to the pump beam. A narrowband filter placed in front of the CCD camera eliminates emission from the plasma. Femtosecond and picosecond time resolution are achieved by adjusting the delay time between the pump and probe beams; zero time delay is defined when the peaks of the two beams overlap. The experimental system in Figure 13.3 shows how two separate lasers would be utilized for nanosecond and longer time-resolved studies. For either experimental system, the two beams can be configured as a Mach–Zehnder interferometer or to provide shadowgraph images. Shadowgraphs provide visual information on plasma propagation. Phase-shift images, obtained from Fourier analysis of the interferograms, provide electron and mass number density analyses.

Time-resolved shadowgraph images for a plasma induced by a 35-ps-pulsed Nd:YAG laser focused onto a Cu sample with an irradiance of 10^{12} W cm^{-2} are shown in Figure 13.4 [18]. A plasma is observed immediately in the air above the sample surface. The expansion velocity of this plasma was measured to approximately 10^9 cm s^{-1} initially, decreasing to zero after 100 ps. Its longitudinal extent remains approximately constant after about 100 ps, and the plasma expands principally in the radial direction. The velocity of this plasma was much greater than the velocity of atom and ions (10^6 cm s^{-1}). The air plasma was induced by electron emission from the target. After 100 ps, the electron number density was about

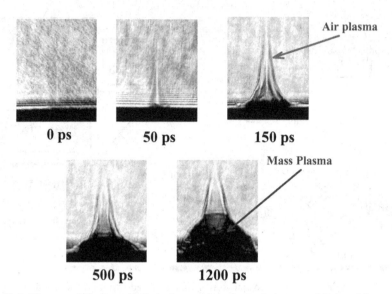

Figure 13.4. Picosecond time-resolved shadowgraph images of picosecond-induced laser plasma. The pulse duration was 35 ps, energy was 7.5 mJ and spot size was 100 μm in diameter. The thin line plasma observed almost immediately is an electron-induced air plasma, which occurs before the main mass plume ejects from the surface. The laser beam is incident from the top and the sample at the bottom in these images. The spatial scale is approximately 150 μm × 100 μm.

10^{20} cm^{-3}. The longitudinal expansion of the plasma was suppressed owing to the development of a strong space charge (4×10^6 V m^{-1}) just above the target surface [19].

Ablated ions or atoms from the target did not cause the air plasma. After approximately 400 ps, atomic and ionic mass from the target can be seen in the shadowgrams, which correspond to the white area in the phase-shift maps (Figure 13.5). While the Cu plasma continues to expand, the shape of the air plasma remains the same. No plasma was observed for delay times smaller than 400 ps when the same laser conditions were used with lower air pressure (300 Torr) or without the Cu sample. Therefore, the electrons had to originate from the material surface. Surface electron emission due to thermionic and photoelectric effects was calculated using a two-temperature model for the picosecond laser–metal interactions [18, 19]. This emission from the sample provides the initial seed electrons for the breakdown of the ambient gas. The electron number density near the sample surface was calculated to be approximately 10^{20} cm^{-3} within 30 ps, in excellent agreement with the measurements. The expansion velocity was calculated to be 10^9 cm s^{-1}, also in support of the experimental data. The electrons ejected from the Cu sample interact with air and absorb laser energy by inverse Bremsstrahlung processes.

13.5 Femtosecond plasma

Analytical and physical-property measurements related to femtosecond LIBS will be described in the following section. Initial fundamental research imaging the femtosecond

Laser Beam

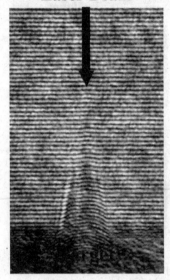

Interferometry Image

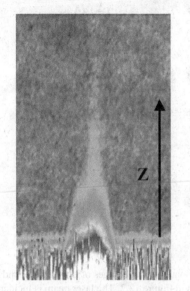

Phase Shift Image

Figure 13.5. Interferometric and phase-shift images measured using the pump-probe configuration as in Figure 13.2, with the probe-beam path configured as an interferometer. Electron number density is calculated from the phase shifting of the fringes. Time is approximately 100 ps after the 35 ps pulse. Spatial dimension are approximately 100×50 µm.

laser induced plasma demonstrated that expansion dynamics are different than that measured using nanosecond pulsed lasers. Figure 13.6 shows picosecond time-resolved images of the femtosecond (100 fs, 800 nm) induced plasma from ablation of a steel sample [20]. The fluence of the laser beam was approximately 5.5 J cm^{-2}. The spot size was approximately 80 µm in diameter. Visual observation up to about 1 ns shows that the plasma expands principally in one dimension, away from the sample surface. These results are different than those measured for nanosecond-induced plasmas during the same time period, in which case plasma expansion is three dimensional [21]. The difference in plasma expansion dynamics could be related to difference in plasma absorption during the laser pulse.

13.6 Short-pulse LIBS

The analytical community has only recently begun investigating the use of short-pulsed laser for LIBS, and much of this initial research has primarily emphasized the differences in plasma properties, not so much analytical performance. This section reviews the results from the latest research using short pulses in LIBS studies.

LeDrogoff *et al.* [22] used a Ti:Sapphire chirped pulsed amplification laser system to produce pulses with 50 mJ at 800 nm and 100 fs duration. By focusing the laser beam to a

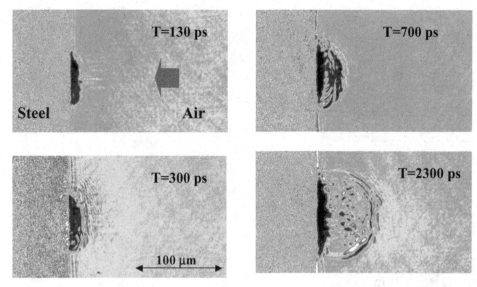

Figure 13.6. Shadowgraph images of the femtosecond expanding plume measured using the experimental system in Figure 13.2. The laser beam is incident from the right with the sample (steel) target on the left. The femtosecond induced plasma exhibits one-dimensional expansion up to approximately 1ps after the laser pulse.

60 μm spot diameter, the fluence at the sample surface was 20 J cm^{-2}. Although this fluence is comparable to that used in many nanosecond LIBS studies, the irradiance is six orders greater (10^{15} W cm^{-2} compared with 10^9 W cm^{-2}). Aluminum alloys were used as samples. Spatially integrated, time-resolved continuum and spectral emission lines were measured to decrease in intensity much more rapidly than respective emissions measured from nanosecond-induced plasmas; femtosecond-induced plasmas had a shorter lifetime. Most emission lines decayed in <10 μs. Plasma properties (temperature T and electron number density N_e) were calculated from the time-resolved emission measurements. Because of strong inverse Bremsstrahlung background, data were reported for measurements 50 ns after the laser pulse. The N_e at 50 ns was 2×10^{18} cm^{-3}, decaying by 2 orders in 5 μs. Surprisingly, the initial value for N_e was similar to that measured for nanosecond-induced plasmas at the same fluence [21]. The primary differences were in temperature and decay kinetics of the plasmas. The nanosecond-induced plasma exhibited a higher excitation temperature that sustained a longer duration. The higher temperature supports laser–plasma heating with nanosecond pulses.

This work also compared analytical performance of nanosecond and femtosecond LIBS. Calibration curves exhibited good linearity for several non-resonant spectral lines. RSD for measurements were 2%–10%. LODs were measured to be a few parts per million (ppm) (Cu, Mn, Mg) to tens of parts per million (Fe, Si) depending on the spectral line and concentration of the element in the sample. This analytical performance (RSD and LOD) of femtosecond LIBS was similar to that achieved using nanosecond pulses. Research is under way to better utilize the rapid decay characteristics of the femtosecond-induced plasma to improve analytical performance.

Angel and co-workers compared nanosecond, picosecond, and femtosecond LIBS in a series of papers [13, 16, 23]. A sync-pumped dye laser Nd:YAG system with 1.3 ps pulses, 1 mJ energy, and 570 nm wavelength was used for these experiments. A Ti:Sapphire laser system was also used; it provided 1mJ energy, 810 nm wavelength, 140 fs duration, with repetition rates variable from 1 Hz–1 kHz. These papers demonstrated general differences in short-pulse plasmas compared with nanosecond-induced plasmas. For example, plasma temperature decayed at least 1 order faster using femtosecond/picosecond versus nanosecond pulses. Atomic line intensity and broadband emission background also decayed very quickly in the short-pulsed plasmas. The spatial extent of short-pulsed plasmas was smaller than nanosecond-induced plasma using a similar fluence, probably because of the lower energy threshold for ablation and no plasma heating. The energy threshold required for ablation was less with short pulses, and the threshold power at which ablation occurred was more reproducible.

In these papers, LIBS intensity was correlated with crater volume as laser energy was increased. Even though the plasma temperature rolled off with increasing energy, the LIBS/volume ratio was linear. The best ablation efficiency was obtained by using femtosecond pulses. Crater diameters were found to be larger than the laser-beam diameter for nanosecond and picosecond pulses, except at very low fluence. Unique opportunities were proposed and demonstrated. Non-gated detection was used in some cases, simplifying the experimental detection system, and higher repetition rates were used to get approximately the same S/N and S/B in a shorter experimental time.

Hergenroder and co-workers [12] investigated the use of femtosecond-pulsed LIBS for depth profiling of Cu–Ag multilayers on Si substrates. Each layer was about 600 nm thick. They were able to discern the interfaces of Cu–Ag but the Gaussian beam caused mixing as deeper layers were accessed. The ablation rate was estimated to be about 15–20 nm per pulse at the center of the Gaussian beam. SEM showed that no significant mixing or melting of the layers occurred, the mix in signal was due to the Gaussian beam non-uniformly penetrating layers. A flat top beam profile would significantly improve these data. Nanosecond-pulsed LIBS is not generally suited for depth profiling because the large thermal influence can cause mixing of the layers by thermal diffusion in the solid or in the melt. This same group [24] compared LIBS plasmas from brass using 170 fs and 6 ns pulses. Nonlinear calibration curves for Zn/Cu ratios were measured and correlated with a two-temperature model. They also calculated the ablated mass versus Zn and Cu concentrations and correlated the measured signals to these calculated values. Linear calibration curves were established for signal intensity versus ablated mass for both Zn and Cu.

In another paper by Hergenroder and co-workers [15], the time evolution of Cu and Mg was reported for nanosecond and femtosecond LIBS. The femtosecond-induced plasma showed strong ion lines near zero delay that decreased quickly. The intensity of neutral lines

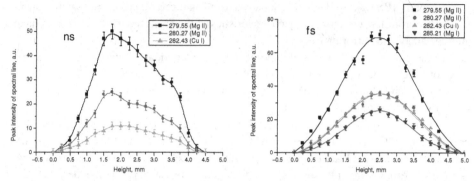

Figure 13.7. Spectral intensity versus distance from target surface for nanosecond (left) and femto-second (right) pulsed LIBS.

slightly increased with time and then slowly decayed. For nanosecond-induced plasmas, the maximum intensity of ionic lines was shifted to longer times. The Cu(II) to Cu(I) ratio demonstrates the heating effect of the nanosecond-induced laser plasma. The spatial distribution of the emission was also different (Figure 13.7). The nanosecond-induced plasma showed an asymmetry in the spatial distribution; enhanced emission skewed towards later times represents the laser plasma interaction. The femtosecond-induced plasma looked like it came from a symmetrically expanding cloud and its spatial extent fit nicely to a Gaussian profile. As discussed above, the expansion dynamics of the femtosecond plasma are initially one-dimensional versus the three-dimensional expansion of the nanosecond-induced plasma. The time dependence (plasma decay time) of the Cu line for nanosecond- and femtosecond-induced plasmas in 40 and 140 mbar pressure Ar also was studied. Results were described in terms of the influence of the laser–plasma interaction on the decay times. The femtosecond plasma without laser heating was influenced by pressure at early times (<4 μs) whereas the ns plasma was not. After 4 μs, both plasmas experienced about the same change in decay versus pressure.

In this work, fractionation of Zn or Cu was not measured for either femtosecond or nanosecond ablation. However, only 250 pulses with micro-joule energies were used in each case. The ablation rate was about 0.7–1.3 μm per pulse. The mass ablation rate was Zn/Cu ratio dependent for all laser pulse durations. For nanosecond ablation, fractionation becomes significant when the crater aspect ratio reaches approximately 6 [25]. This is not the case for femtosecond ablation in that much larger aspect ratios can be achieved without changes in fractionation [26]. Good calibration was achieved for both nanosecond and femtosecond LIBS when the Zn/Cu ratio was reported versus Zn/Cu ratio of the standards. However, the data were better for the femtosecond laser case; better correlation coefficients. For precision of analysis, femtosecond LIBS was about a factor of two better, which could be correlated to the better ablation precision using femtosecond pulses.

Semerok and colleagues compared crater profiles and plasma expansion using a 6 ns pulsed Nd:YAG and a variable 70 fs–10 ps pulsed Ti:Sapphire laser [14, 27, 28]. For all studies, the wavelength and fluence were similar, and only the pulse duration was changed for comparison. Although LIBS was not addressed in these studies, the crater properties themselves demonstrated the advantages of using shorter pulses for ablation. The crater depth per pulse was measured to be similar for pulses from 70 fs to 800 fs, but it decreased by further increasing the pulse duration (> 1 ps). These measurements confirm that plasma shielding plays a role in laser ablation process. The femtosecond-induced craters did not have a raised rim (called a convexity). Laser ablation efficiency was found to be independent of the wavelength for 70 fs laser pulses. These characteristics point to the possibility of using femtosecond pulses for matrix-independent sampling – a Holy Grail for chemical analysis.

There are numerous papers in the literature on the properties of short-pulsed plasmas for applications other than LIBS. Only an example is mentioned here to show additional capabilities and benefits of short laser pulses. Femtosecond plasmas initiated in air have molecular lines; they dominate initially [29]. Molecular emission, continuum emission, and line emission were separated in time. The availability of molecular emission lines allows new applications for LIBS. Focusing a high-intensity short laser pulse onto the surface of a material can efficiently generate X-rays. This bright X-ray source can be used for X-ray lithography, time-resolved X-ray spectroscopy, and micro-crystallography. Hard X-ray emission from high-density plasmas produced by femtosecond laser pulses also has been suggested for elemental analysis [30]. A note of caution – we should all remember that the femtosecond-induced plasma is producing X-rays!

13.7 Conclusion

The jury is still out on the use of short pulses for LIBS, mainly because the use of these lasers in analytical applications is relatively new with only a few initial studies. However, these initial studies do show promising opportunities. Smaller heat affected zone, improved ablation efficiency, lower ablation threshold, finer depth resolution, no laser–plasma interaction, faster broadband-background decay, no gating for LIBS, matrix-independent sampling, are but a few of the already discovered opportunities.

It is surprising that the initial analytical performance (sensitivity, accuracy, fractionation) is not improved compared with nanosecond LIBS. There are several possibilities that need to be investigated. Short-pulsed plasmas are generally smaller, especially when produced near the ablation threshold. Therefore, most researchers have used higher energy than threshold to get good signal statistics using LIBS. The configuration of the detection system should be addressed in order to benefit from the different plasma properties. A characteristic of using short pulses is reduced plasma heating. However, plasma heating may be better for LIBS. Most nanosecond LIBS applications utilize IR wavelengths, at which plasma absorption is strong. Maybe a more robust plasma is necessary to excite ablated species. Finally, it is important to remember that the plasma effects the ablation process; even though it is not present during the laser pulse, the plasma is established. The plasma will interact with the sample surface and affect final results.

There always will be cases in which inexpensive, portable nanosecond lasers will be appropriate for LIBS applications. However, femtosecond laser technology is advancing rapidly and these lasers are expected to be more economical in the near future. As the benefits of short-pulsed ablation are borne out in new research activities, the use of these lasers will surely increase in LIBS applications.

Thanks are due to all the authors whose work was referenced in this chapter. Special thanks to Xianglei Mao and Sam Mao for their work on the picosecond and femtosecond laser properties. This work was supported by the US Department of Energy, Office of Basic Energy Sciences, Division of Chemical Sciences, and the Office of Nonproliferation and National Security, at the Lawrence Berkeley National Laboratory, under Contract No. DE-AC03-76SF00098.

13.8 References

[1] *Laser Ablation and Desorption*, J. C. Miller and R. F. Haglund (editors) (New York: Academic Press, 1998).

[2] D. A. Cremers and L. J. Radziemski, in *Laser Spectroscopy and Its Applications*, L. J. Radziemski, R. W. Solarz and J. A. Paisner (editors) (New York: Marcel Dekker, Inc., 1987).

[3] M. von Allmen, *Laser-Beam Interactions with Materials – Physical Principles and Applications* (New York: Springer-Verlag, 1987).

[4] N. Bloemberger, in *Laser Solid Interaction and Laser Processing*, S. D. Ferris, H. J. Leamy and J. M. Poate (editors) (New York: American Institute of Physics, 1979).

[5] J. F. Ready, *Effect of High-Power Laser Radiation* (New York: Academic Press, 1971).

[6] S. I. Anisimov and V. A. Khokhlov (editors), *Instabilities in Laser–Matter Interactions* (Boca Raton, FL: CRC Press, 1995).

[7] I. Zergioti, M. Stuke, Short pulse UV laser ablation of solid and liquid gallium. *Appl. Phys. A*, **67** (4) (1998), 391–395.

[8] R. E. Russo, X. L. Mao and S. S. Mao, The physics of laser ablation in micro chemical analysis. *Anal. Chem.* **74**, (2002), 70A–77A.

[9] B. N. Chichkov, C. Momma, S. Nolte, F. Vonalvensleben and A. Tunnermann, Femtosecond, picosecond and nanosecond laser ablation of solids. *Appl. Phys. A*, **63**(2) (1996), 109–115.

[10] S. Preuss, E. Matthias and M. Stuke, Sub-picosecond UV laser ablation of Ni-films: strong fluence reduction and thickness independent removal. *Appl. Phys. A*, **59**(1) (1994), 79–82.

[11] R. E. Russo, X. L. Mao, J. J. Gonzalez and S. S. Mao, Femtosecond laser ablation ICP-MS. *J. Anal. Atom. Spectrom.*, **17**(9) (2002), 1072–1075.

[12] V. Margetic, M. Bolshov, A. Stockhaus, K. Niemax and R. Hergenroder, Depth profiling of multi-layer samples using femtosecond laser ablation. *J. Anal. Atom. Spectrom.*, **16**(6) (2001), 616–621.

[13] S. M. Angel, D. N. Stratis, K. L. Eland *et al.*, LIBS using dual- and ultra-short laser pulses. *Fresenius J. Anal. Chem.*, **369**(3–4) (2001), 320–327.

[14] A. Semerok, B. Salle, J. F. Wagner and G. Petite, Femtosecond, picosecond, and nanosecond laser microablation: laser plasma and crater investigation. *Las. Part. Beams*, **20**(1) (2002), 67–72.

[15] V. Margetic, A. Pakulev, A. Stockhaus *et al.*, A comparison of nanosecond and femtosecond laser-induced plasma spectroscopy of brass samples. *Spectrochim. Acta B*, **55**(11) (2000), 1771–1785.

[16] K. L. Eland, D. N. Stratis, T. S. Lai *et al.*, Some comparisons of LIBS measurements using nanosecond and picosecond laser pulses. *Appl. Spectrosc.*, **55**(3) (2001), 279–285.

[17] X. L. Mao, S. S. Mao and R. E. Russo, Imaging femtosecond laser-induced electronic excitation in glass. *Appl. Phys. Lett.*, **82**(5) (2003), 697–699.

[18] S. S. Mao, X. L. Mao, R. Greif and R. E. Russo, Initiation of an early-stage plasma during picosecond laser ablation of solid. *Appl. Phys. Lett.*, **77**(16) (2000), 2464–2466.

[19] S. S. Mao, X. L. Mao, R. Greif and R. E. Russo, Simulation of a picosecond laser ablation plasma. *Appl. Phys. Lett.*, **76**(23) (2000), 3370–3372.

[20] X. L. Mao and R. E. Russo, unpublished data.

[21] H. C. Liu, X. L. Mao, J. H. Yoo and R. E. Russo, Early phase laser induced plasma diagnostics and mass removal during single-pulse laser ablation of silicon. *Spectrochim. Acta B*, **54**(11) (1999), 1607–1624.

[22] B. Le Drogoff, J. Margot, M. Chaker *et al.*, Temporal characterization of femtosecond laser pulses induced plasma for spectrochemical analysis of aluminum alloys. *Spectrochim. Acta B.*, **56**(6) (2001), 987–1002.

[23] K. L. Eland, D. N. Stratis, D. M. Gold, S. R. Goode and S. M. Angel, Energy dependence of emission intensity and temperature in a LIBS plasma using femtosecond excitation. *Appl. Spectrosc.*, **55**(3) (2001), 286–291.

[24] V. Margetic, K. Niemax and R. Hergenroder, A study of non-linear calibration graphs for brass with femtosecond laser-induced breakdown spectroscopy. *Spectrochim. Acta B*, **56**(6) (2001), 1003–1010.

[25] P. R. D. Mason and A. J. G. Mank, Depth-resolved analysis in multi-layered glass and metal materials using laser ablation inductively coupled plasma mass spectrometry (LA-ICP-MS). *J. Anal. Atom. Spectrom.*, **16**(12) (2001), 1381–1388.

[26] R. E. Russo and X. L. Mao, unpublished data.

[27] V. Detalle, J. L. Lacour, P. Mauchien and A. Semerok, Investigation of laser plasma for solid element composition microanalysis. *Appl. Surf. Sci.*, **139** (1999), 299–301.

[28] A. Semerok, C. Chaleard, V. Detalle *et al.*, Experimental investigations of laser ablation efficiency of pure metals with femto, pico and nanosecond pulses. *Appl. Surf. Sci.*, **139** (1999), 311–314.

[29] F. Martin, R. Mawassi, F. Vidal *et al.*, Spectroscopic study of ultrashort pulse laser-breakdown plasmas in air. *Appl. Spectrosc.*, **56**(11) (2002), 1444–1452.

[30] N. Takeyasu, Y. Hirakawa and T. Imasaka, Elemental analysis of stainless steel using hard X-ray emission arising from a high-density plasma produced by an intense femtosecond KrF laser pulse. *Appl. Spectrosc.*, **56**(9) (2002), 1161–1164.

14

High-speed, high-resolution LIBS using diode-pumped solid-state lasers

Holger Bette

Lehrstuhl für Lasertechnik (LLT), Germany

Reinhard Noll

Fraunhofer-Institut für Lasertechnik (ILT), Germany

14.1 Introduction

Most of the investigations and applications of LIBS use laser sources operating at repetition frequencies in the range of 10–20 Hz, yielding laser pulse energies of typically 50–300 mJ. This is a consequence of the broad commercial availability of Q-switched flashlamp-pumped solid-state lasers. The most important laser type is the Nd:YAG laser with an emission wavelength in the near infrared at 1064 nm. Such lasers are reliable radiation sources for a variety of research topics and industrial applications. The dominant industrial application field is laser material processing, for example marking, drilling, cutting and cleaning. The first industrial applications of LIBS have also been realized on the basis of flashlamp-pumped Nd:YAG lasers [1–4].

The prevailing task for LIBS with repetition frequencies between 10 and 20 Hz is bulk analysis of materials. A few papers deal with spatially resolved measurements [5–8]. A relative motion between the focused laser beam and the sample enables a scanning across the surface. The spectral information gained is linked to the location of irradiation of the laser pulse onto the sample surface. The result is presented, for example, as an element line intensity of the laser-induced plasma versus one or two spatial coordinates. The latter situation yields maps of element-specific intensities of a sample surface. However, a mapping with a step size of 20 μm by scanning of an area of 1 cm^2 of a sample requires 250 000 single measurements. A laser operating at 10 Hz would require a measuring time of nearly 7 h, which is far too much for any practical application. Moreover the beam quality of flashlamp-pumped Nd:YAG lasers is in many cases not sufficient to achieve small focal diameters in the range <10 μm.

During the past few years the availability of high-power semiconductor laser diodes has stimulated the development of all-solid-state laser systems [9–12]. Instead of the spectrally broad emitting flashlamp, laser diodes pump the laser crystal in a narrow spectral band selected to increase the efficiency of the pumping by minimizing energy dissipation. These

Laser-Induced Breakdown Spectroscopy: Fundamentals and Applications, ed. Andrzej W. Miziolek, Vincenzo Palleschi and Israel Schechter. Published by Cambridge University Press. © A. W. Miziolek, V. Palleschi and I. Schechter 2006. A. W. Miziolek's contributions are a work of the United States Government and are not protected by copyright in the United States.

Table 14.1. *Typical diode-pumpable laser crystals, their strongest laser emission wavelengths and suitable pumping wavelengths to be generated by laser diodes*

Laser crystal	Laser wavelength (nm)	Pumping wavelength (nm)
Nd:YAG	1064, 1319	808
Nd:YVO$_4$	1064, 1340	809
Nd:YLF	1047, 1053	792, 798
Yb:YAG	1023–1052	941
Cr:LiSAF	800–920	670, 760

diode-pumped solid-state lasers (DPSSL) in Q-switch operation offer high pulse repetition rates of up to 1 kHz and more [12, 13]. Although the pulse energy is still limited to a few millijoules, high intensities in the focal spot can be achieved owing to the excellent beam quality of DPSSL systems. The potential benefits of DPSSL compared with flashlamp-pumped lasers for LIBS are:

- faster acquisition of spectral information,
- higher spatial resolution,
- improved precision on the basis of significantly enlarged data ensembles (mean values, pulse discrimination analysis),
- increased lifetime,
- lower energy consumption.

This chapter describes the use of DPSSL to perform high-speed LIBS with repetition rates of up to 1 kHz. The application presented is a multi-element mapping system for metallic samples achieving a spatial resolution of <20 μm. Element lines with upper energy levels of up to 10 eV and more are excited with single laser pulses, thus enabling the detection of microscopic amounts of light elements such as oxygen and nitrogen in solid samples. As a direct consequence of the high pulse repetition rate and high spatial resolution the experimental configuration and especially the acquisition of the spectral information have to be matched to this new parameter range for LIBS.

14.2 Diode-pumped solid-state lasers

Diode-pumped solid-state lasers were developed in recent years and are increasingly replacing the flashlamp-pumped laser systems [9, 10]. One important advantage of DPSSL systems is the lifetime of the pumping source. Laser diodes have typical lifetimes of 10 000 h in continuous wave mode and 10^9–10^{10} pulses in pulsed mode. In contrast to this, typical flashlamp lifetimes amount to about 5×10^7 pulses.

Typical diode-pumped laser materials and their important pumping wavelengths are listed in Table 14.1 [9, 10]. The main advantage of diode-pumped solid state laser systems is the high beam quality in combination with the achievable pumping efficiency. This means that a

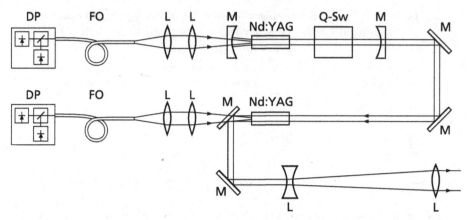

Figure 14.1. Typical layout of a diode-pumped solid-state laser (DPSSL) in an end-pumped configuration. DP = diode pump module, FO = fiber optic, L = lens, M = mirror, Q-Sw = Pockels cell for Q-switch operation. Because of the high degree of overlap of pumped volume and laser mode, and the improved matching of the pumping wavelength to the absorption band of the laser medium, the efficiency of DPSSL systems is a factor of 3–20 higher than for flashlamp-pumped lasers.

small laser focus diameter is easier to achieve with a DPSSL than with a flashlamp-pumped system. To get a small focus with a flashlamp-pumped system requires the use of apertures within the oscillator for mode selection, which in turn limits the efficiency factor (optical-to-optical) down to 1%–3%. Diode-pumped systems achieve typical efficiencies of 10%–20%. Therefore DPSSL systems save energy costs and secondary costs like chillers. The reduced heat load of the laser crystal leads to smaller internal stresses and changes of the optical properties, improving the mode quality of the laser beam.

DPSSL systems can be pumped in a transversal configuration as it is the common arrangement for flashlamp-pumped systems, as well as in a longitudinal (called "end-pumped") configuration, see Figure 14.1. High beam quality and efficiency in an end-pumped configuration are achieved by gain guiding. In this case the pumped volume and laser mode have a high degree of overlap. Figure 14.2 shows a photograph of a DPSSL based on this principle developed at ILT.

Q-switched DPSSL systems have tunable repetition rates from below 1 Hz to several kilohertz, favoring them for the generation of high repetitive laser pulses in the millijoule range. Pre-flashing of the laser to achieve thermal stability of the system and to generate reproducible laser pulses is strongly reduced or even not necessary, thus eliminating dead times.

The intensity within a laser focus is determined by the laser pulse energy, the pulse duration and the beam quality. A sufficiently high intensity in the beam focus is necessary to excite analytes having excitation energies of several electronvolts. The average intensity in the focus of a laser beam – assuming a collimated laser beam focused by an aberration free lens – is given by

$$I_f = \frac{\pi E_L D^2}{4\tau_L \cdot f^2 \cdot \lambda^2 \cdot M^4}$$

Table 14.2. *Data of the Nd:YAG diode-pumped laser developed at ILT*

Wavelength	1064 nm
Pulse energy	2 mJ
Pulse duration	5.5 ns
Beam quality M^2	≤ 1.3

Figure 14.2. Photograph of the Nd:YAG DPSSL developed at ILT; data are given in Table 14.2.

where E_L = laser pulse energy, D = diameter of illuminated aperture of focusing lens (intensity at edge of aperture: $1/e^2$ of intensity peak on axis), τ_L = pulse duration (FWHM), f = focal length of lens, λ = wavelength of laser, and M^2 = times-diffraction-limit factor.

The minimal achievable radius of the laser focus for a fixed optical setup depends therefore only on the beam quality of the laser beam denoted by M^2. The M^2 value is a number describing the beam quality of a laser beam in relation to a laser beam yielding the smallest possible focus diameter achievable with a TEM_{00} laser beam having a Gaussian intensity distribution. Note that M^2 is called the times-diffraction-limit factor and is defined in an international standard [14]. A typical value for conventional flashlamp-pumped Nd:YAG lasers is $M^2 = 2.8$.

With this formula and the typical data of the Nd:YAG diode-pumped laser given in Table 14.2, the average intensity in the focus of a laser pulse can be calculated. With $D = 17$ mm and $f = 60$ mm, the average intensity amounts to 1.2×10^{12} W cm^{-2}. As will be shown later in this chapter, this intensity is sufficient for LIBS to excite spectral lines with excitation energies beyond 10 eV, such as the nitrogen 149.26 nm line.

For the numbers given in the previous paragraph the diameter of the focus spot on the target surface amounts to about 6 μm. So in principle a DPSSL allows a high spatial resolution for LIBS to be achieved. The spatial resolution of the chemical analysis by LIBS

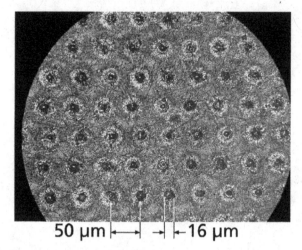

Figure 14.3. Photograph of craters produced by single laser pulses of the DPSSL system on a steel surface; the laser data are given in Table 14.2.

depends on the size of the laser-produced crater in a solid sample. The crater diameter is not only a function of the laser focus diameter but also depends on the heat affected zone and therewith on the thermal diffusivity of the material to be analyzed and on the laser pulse duration. Figure 14.3 shows a photograph of a series of laser craters in a steel sample each produced by a single laser pulse of the DPSSL with the data given in Table 14.2. The crater diameter amounts to about 16 μm, which is nearly a factor of 3 greater than the beam diameter. In Section 14.5 the influence of the laser pulse duration on the size of the crater will be discussed in more detail.

14.3 State of the art

A LIBS set-up can be described by the following frequencies mainly determining the speed of a LIBS measurement: (a) the repetition rate of the laser source defines the number of laser pulses per second irradiating the sample; and (b) the measuring frequency given by the read-out rate of the detectors – such as charged coupled devices (CCD, one-dimensional and two-dimensional), photodiode arrays (PDA), photomultipliers (PMT) – used to acquire the spectral emission from the laser-induced plasma. In many cases the measuring frequency is chosen to be equal to the repetition rate of the laser. In this case the spectral information of each event, that is of each transient laser-induced plasma, is acquired separately and processed for quantitative or qualitative analysis. Pulse discrimination methods can be applied to improve the precision of the measuring results or to extract specific signals. In some cases the measuring frequency is less than the laser pulse repetition rate, for example in echelle spectrometers equipped with megapixel CCD-cameras. Here several events are accumulated on the detector before a read-out cycle is activated. Only a sum of a series of single spectra is obtained.

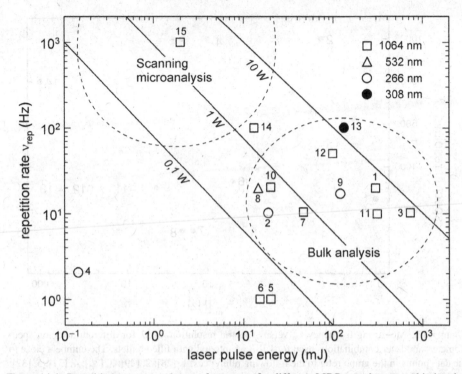

Figure 14.4. Repetition rate versus laser pulse energy for different LIBS experiments and industrial applications. The numbers close to the data points in the graph refer to the following references: 1, [15]; 2, [16]; 3, [17]; 4, [18]; 5, [19]; 6, [20]; 7, [21]; 8, [22]; 9, [23]; 10, [24]; 11, [25]; 12, [3]; 13, [26]; and 14, [27]. Data point 15 is the high-speed scanning LIBS system described in this chapter, see Figure 14.9 and Table 14.4.

Figure 14.4 shows the parameter ranges of LIBS in terms of energy of the applied laser pulses and laser pulse repetition rates for different applications related to the analysis of metals, steel, slags and oxides. Most of the research and development (R&D) activities reported so far are clustered around pulse energies of 200 mJ and repetition rates of 20 Hz. The prevailing task in this area is bulk analysis. The plotted diagonals in Figure 14.4 depict constant laser power. Only a few papers have reported the use of lasers with repetition rates of 50 Hz and 100 Hz [3, 26, 27]. The respective laser systems are an excimer laser emitting at 308 nm and conventional lamp-pumped Nd:YAG lasers.

Several authors have studied the possibilities of LIBS for spatially resolved element analysis [5–8, 18]. However, the lasers used were all running in the range of pulse repetition frequencies between 2 Hz and 20 Hz, thus strongly limiting the area to be scanned with small step sizes.

The restriction to repetition frequencies for LIBS measurements to ≤ 100 Hz is mainly determined by the use of conventional lamp-pumped solid-state lasers. DPSSL systems have opened up the region of high-speed LIBS measurements with repetition frequencies an order of magnitude larger than is common today.

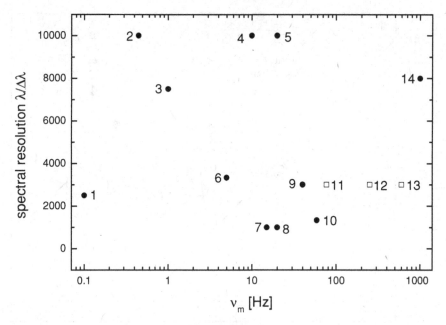

Figure 14.5. Measuring frequency ν_m versus spectral resolution $\lambda/\Delta\lambda$ for different detector, spectrometer, and laser combinations used for LIBS are plotted with filled bullets. The numbers close to the data points in the graph refer to the following references: 1, [28]; 2, [29]; 3, [30]; 4, [31]; 5, [32]; 6, [33]; 7, [34]; 8, [35]; 9, [36]; and 10, [37]. Data point 14 is the high-speed LIBS system presented in Section 14.4.1. The open squares for data points 11–13 represent commercially available detectors which can be used for LIBS [38], the assumed resolution of 3000 can be achieved with standard spectrometers and monochromators.

Figure 14.5 shows an overview of the LIBS parameters measuring frequency and spectral resolution. The measuring frequency is strongly linked to the type of detector used for the acquisition of the spectral information. Common configurations of spectrometers and detector types are as follows: (a) Czerny–Turner spectrometer and intensified PDA or CCD; (b) echelle spectrometer and intensified CCD-camera; and (c) Paschen–Runge spectrometer and PMT. The requirement concerning the spectral resolution $\lambda/\Delta\lambda$ depends on the expected density of emission lines in the spectrum of the laser-induced plasma to clearly separate analyte emissions from interfering lines. A quantitative analysis of iron-based samples generally requires a spectral resolution of about 5×10^3 and more. Most of the existing LIBS set-ups are presented by the data points 1–10, where the measuring frequency is mostly limited by the detector read-out time or the laser repetition frequency to less than 60 Hz. The data points 11–14 illustrate the potential to push the measuring frequency into the range of 100 Hz to 1 kHz.

The combination of a Paschen–Runge spectrometer with PMT detectors achieves the highest measuring frequencies for LIBS to date; see data point 14 in Figure 14.5. This is at present the configuration best suited for high-speed LIBS. The type of signal gained with such a set-up for a plasma induced by 2 mJ pulses of a DPSSL on a silicon nitride and

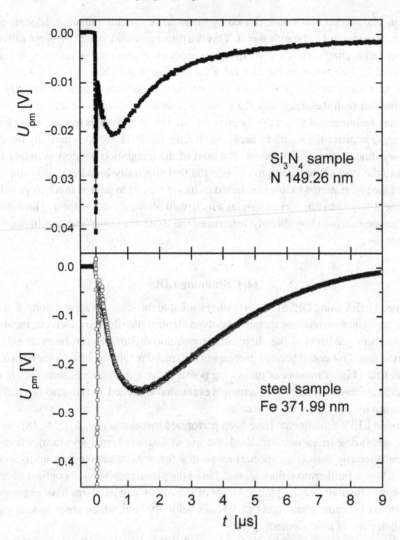

Figure 14.6. Photomultiplier signals as functions of time. Top: nitrogen channel at 149.26 nm of a Paschen–Runge spectrometer, Si_3N_4 sample. Bottom: iron channel at 371.99 nm, steel sample. In both cases the plasma was induced by a DPSSL pulse of 2 mJ under argon atmosphere. Each displayed signal is an accumulated average over 10^4 laser pulses, where each laser pulse is irradiated on a new part of the sample surface.

steel sample is shown in Figure 14.6. The photomultiplier signal of a nitrogen channel at 149.26 nm and an iron channel at 371.99 nm is measured with a high-bandwidth oscilloscope. The first signal peak corresponds to the spectrally broad emission just after the ignition of the plasma. The rising edge of this peak coincides temporally with the laser pulse. The temporal width (FWHM) is less than 25 ns. The subsequent signal contains the element-specific information, in Figure 14.6 this is shown for the matrix elements nitrogen and iron. The second maximum, caused by the element-specific radiation, appears earlier for the nitrogen than for the iron signal. This is a direct consequence of the large difference in the excitation energies of the corresponding upper energy levels. For nitrogen this energy level lies at 10.7 eV, whereas the iron value for this line is 3.3 eV.

If the area under the first peak, that is the spectrally broad emission, is sufficiently small in comparison with the area under the subsequent element-specific signal part, the whole signal may be integrated to get the spectral information of the respective element channel. However, to improve the signal-to-background ratio, the PMT signal is partially integrated during a defined integration window. The start of the integration window is delayed with respect to the start of the signal to exclude the first spectrally broad emission spike. The width of the integration window is selected in such a way as to gain as much as possible of the element-specific emission as long as this is well above the noise level. The definition of the integration window directly influences the dynamic range and sensitivity of the measurement.

14.4 Scanning LIBS

High-speed LIBS using DPSSL is particularly suitable for scanning applications. Scanning LIBS allows one to determine spatially resolved element distributions. Defects, inclusions, segregations or gradients of the chemical composition of surfaces can be analyzed. High measuring speed is needed because increasing the spatial resolution by a factor of two leads to a four times higher number of measuring points for two-dimensional scanning. In order to detect single events a single measurement evaluation is needed, thus requiring a sufficient high sensitivity and signal-to-noise ratio.

Scanning LIBS experiments have been performed by several groups [5–8, 18]. Besides the low measuring frequencies realized, the use of Czerny–Turner spectrometers limited the simultaneously detectable spectral range to a few tens of nanometers up to approximately 100 nm. Furthermore the transmission of the plasma emission via optical fibers and the spectral sensitivity of the photocathode of the microchannel plates have excluded the observation of vacuum ultraviolet (VUV) lines below 190 nm, which are of special interest for the detection of light elements [39].

An alternative way to perform fast scanning LIBS is to move a line focus of the laser beam across the sample [8]. The spatial resolution in the direction of the line focus amounts to about 20 μm (calculated) and in the perpendicular direction to 25 μm. However, along the plasma line light scattering and reflection will cause a crosstalk effect, which will reduce the spatial resolution. The detection of VUV lines below 200 nm like oxygen and nitrogen has not been reported.

14.4.1 Set-up for high-speed, high-resolution LIBS

Fraunhofer ILT has set up a scanning LIBS system with a Paschen–Runge-type spectrometer equipped with 41 detectors to measure line emissions in the spectral range 130–777 nm

Table 14.3. *Data of the integrator electronics developed at ILT*

Number of channels	up to 64
Delay time for the integration window	240 ns–300 μs
Width of the integration window	50 ns–200 μs
Digitalization resolution	15 bit
Maximum measuring frequency	1000 Hz
Charge detection limit	20×10^{-15} C
Sensitivity, delay time, width of integration	programmable per pair of channels

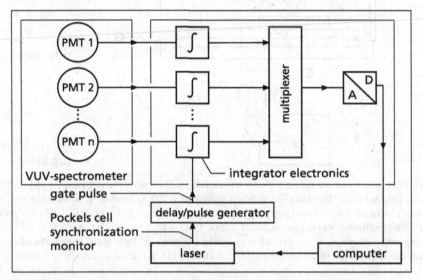

Figure 14.7. Block diagram of the multi-channel integrator electronics. PMT = photomultiplier, VUV = vacuum ultraviolet, A/D = analog-to-digital converter.

including elements with high excitation energies like nitrogen and oxygen. The Nd:YAG DPSSL system developed at ILT was used as the excitation source; its set-up is shown in Figure 14.1. The specifications are given in Table 14.2. For the detection and processing of the photomultiplier signals a multi-channel integrator electronics was developed. Figure 14.7 shows the block diagram; the data are given in Table 14.3. The electronics is capable of processing up to 64 PMT signals in a pulse-by-pulse manner. The anode currents of the PMT are transmitted via coaxial cables to a gated multi-channel integrator electronics.

Figure 14.8 shows schematically the set-up for high-speed scanning microanalysis with LIBS. The surface of a flat sample is scanned by a laser beam of the DPSSL as described in Section 14.2. The plasma light is guided into a vacuum spectrometer via a mirror. The intensities of 41 spectral lines are detected simultaneously and evaluated by a computer. With respect to the spectral range below 200 nm, the measurements are performed in a

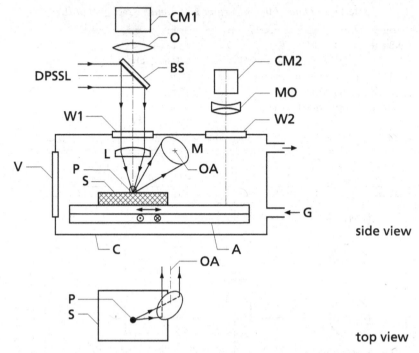

Figure 14.8. Schematic set-up for high-speed scanning microanalysis with LIBS. DPSSL = beam of diode-pumped solid-state laser, BS = beam splitter, W1, W2 = window, L = focusing lens, S = sample, A = translation stages in two perpendicular directions (x, y), O = objective, CM1 = camera for process monitoring, MO = microscope objective, CM2 = high-resolution camera, G = gas supply, C = measuring chamber, V = valve, M = mirror to collimate the light of the laser-induced plasma towards the entrance slit of the spectrometer, P = plasma, OA = optical axis of the spectrometer.

vacuum-tight measuring chamber filled with an argon atmosphere. The achieved spatial resolution with regard to the spectroscopic analysis of metallic samples and the diameter of the craters (see Figure 14.3) is smaller than 20 μm. The measuring frequency amounts to 1 kHz.

Figure 14.9 shows a photograph of the high-speed scanning LIBS system SML 1 (= scanning microanalysis with laser spectrometry) developed at ILT. The technical data of SML 1 are given in Table 14.4.

The scanning principle of the system shown in Figures 14.8 and 14.9 is performed by serially scanning the measuring area point by point. To achieve a high spatial resolution and stable signals the sample is moved and the laser focus and the plasma light collecting optics are fixed. The typical scanning time for a measuring area of 1×1 cm^2 with a step size of 20 μm, corresponding to 250 000 single measurements, amounts to 11 min including acceleration and deceleration times for the sample translation stages. The gas and plasma residuals of the interaction area are exhausted and purged through a filter to minimize disturbing influences of consecutive measurements.

Table 14.4. *Data of the high-speed scanning LIBS system SML 1*

Sample positioning accuracy	<1 μm
Minimal point to point distance	5 μm
Spatial resolution[a]	<20 μm
Maximum measuring frequency	1000 Hz
Number of photomultipliers	41
Number of different detected elements	24
Wavelength range	130–777 nm
Maximum size of scan field	45 mm × 110 mm
Measuring atmosphere	argon

[a] Refers to the spectroscopic analysis of a metallic sample.

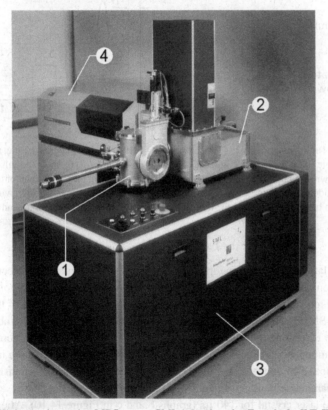

Figure 14.9. High-speed scanning LIBS system SML 1 developed at Fraunhofer ILT. The data of this system are given in Table 14.4. Numbers: 1 = pre-chamber; 2 = measuring chamber; 3 = housing comprising DPSSL, chiller, gas supply, integrator electronics, computer, power supply; and 4 = Paschen–Runge spectrometer.

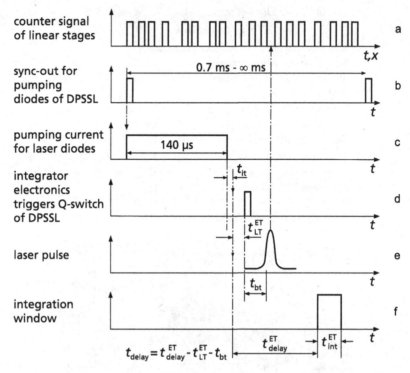

Figure 14.10. Timing diagram for high-speed scanning LIBS: t_{it} = internal transit time, t_{LT}^{ET} = time set in the event table (ET) for the laser trigger (LT), t_{bt} = build-up time (bt) for the laser pulse after activation of the Q-switch including transit times, t_{delay}^{ET} = delay time set in the event table for the start of the integration window, t_{int}^{ET} = time set in the event table for the width of the integration window, t_{delay} = true delay time between the start of the laser pulse and the start of the integration window.

14.4.2 Synchronization

Sample translation axes, laser, pumping unit, Pockels cell and integrator electronics have to be synchronized in order to achieve a correct measuring point location of better than 1 μm and a high signal stability. Figure 14.10 shows the timing diagram of the scanning LIBS system SML 1. The counter signal of the linear stages is counted by a synchronization electronics to determine the true position of the stages, see graph a in Figure 14.10 (the non-equidistant pulse sequence of the counter signal is a consequence of the interpolation procedure applied in connection with the measures integrated in both translation axes). After a programmable number of counts the pumping diodes of the DPSSL are triggered and pump the laser crystal for 140 μs (graphs b and c in Figure 14.10). After the falling edge of the pumping current and a fixed internal transit time t_{it} the integrator electronics is triggered and an event table – containing the temporal sequence for various trigger signals and the start and widths of the integration windows for the different signal channels – is started (graphs c and d). The Pockels cell is triggered by a signal initiated by the event table (graph d). This procedure ensures that the temporal jitter between the Pockels cell switching and the start of the integration windows is kept to less than 1 ns. After the laser pulse build-up time $t_{bt} = 400$ ns (including transit times), the laser pulse is emitted and the plasma is generated on the sample surface (graph e). The vertical broken arrow in Figure 14.10 running from the laser pulse peak in graph e to graph a illustrates that the laser ablation takes place at a defined position on the axes. The event table is processed step by step and at the end the measured values are digitized and transferred to a computer. During the whole time the synchronization electronics tracks the counter signal of the linear stages. When the next measuring position is reached the cycle will begin again. With this timing principle a spatial repeatability of the position of the laser-produced craters of better than 1 µm is achieved within a scanning area of up to 45 mm × 110 mm.

14.4.3 Maximum plasma generation frequency

The maximum plasma generation frequency for serial scanning with single measurement evaluation is limited by the plasma lifetime, the plasma residuals and the gas exchange. The gas exchange is limited by turbulences, which may cause an enlarged laser beam focus owing to local gradients of the refractive index, thus reducing the laser intensity in the focus and decreasing the pulse-to-pulse stability of the intensity distribution propagating to the target. Therefore a gas filtering system was integrated in the measuring chamber to remove ablated and recondensed particles from the argon atmosphere. The gas is exhausted via a nozzle positioned close to the interaction region. To study the efficiency of this approach a steel sample was scanned with 100 Hz and 1000 Hz laser repetition rates. The overall number of measuring points in each case amounts to 250 000. The ratios of the average intensities of the 1000 Hz and the 100 Hz measurements were calculated for each element channel i and plotted in a diagram versus the wavelength λ_i of the respective emission line; see Figure 14.11. From all detected element lines only the four argon lines – representing the ambient atmosphere – showed a significant change in the average signal intensity ratio. The upper energy level of the transitions of the atomic and ionic lines of the species originating from the steel sample lies in the range between 3.12 eV (Ca(II) at 396.85 nm) and 10.74 eV (O(I) at 777.19 nm). Figure 14.11 shows that for these lines no significant difference in the intensity is observed within the relative standard deviation of the intensity ratio. This is an indication that the conditions of the laser-induced plasma with respect to the excitation of the sample species do not significantly change while increasing the repetition rate from 100 Hz to 1000 Hz. Obviously the gas exhaust approach is sufficient to remove residues of ablated sample species in the vicinity of the interaction region in a time scale of less than 1 ms. In contrast to this the argon lines show an enhancement by about 15%–20%. The physical reason is not yet clear. All argon lines plotted in Figure 14.11 are argon ion lines with upper energy levels between 19.26 eV and 21.14 eV. A slight temperature rise in the argon atmosphere adjacent to the sample surface caused by the increased energy dissipation while irradiating a 10 times higher average laser power (from 100 Hz × 2 mJ = 200 mW to 1000 Hz × 2 mJ = 2 W) may cause a layer of reduced argon particle density

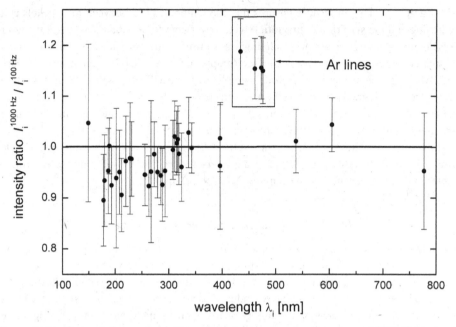

Figure 14.11. Ratio of element intensities acquired at a measuring frequency of 1000 Hz to those gained at 100 Hz at the same steel sample as a function of the wavelength λ_i of 37 atomic and ionic lines. The error bars are the standard deviation of the 250 000 measuring data for each plotted data point. The four data points between 430 nm and 480 nm are argon lines representing the ambient gas atmosphere in the measuring chamber; see Figure 14.8.

close to the surface. This density reduction favors a faster shock wave expansion after the irradiation of the laser pulse, which in turn excites the ambient argon more strongly, leading to an increase in the argon line intensities. Further experimental studies are necessary to clarify whether this interpretation can be validated.

The principal limitation of the plasma generation frequency for LIBS is defined by the temporal behavior of the line emission of the laser-induced plasma, see Figure 14.6. Assuming a laser pulse energy of 2 mJ as typical value for high-repetition-rate DPSSL systems, the element-specific emission extends over a period of about 15 µs at maximum. Hence the principal limit is estimated to be at about 70 kHz, which is nearly two orders of magnitude beyond the present achieved maximum frequency for LIBS of 1 kHz described here. However, exchange times for plasma residuals and ambient gas as well as processing times of the signal electronics will impose further practical limits.

Recent developments of DPSSL systems based on a slab geometry of the laser crystal have demonstrated the feasibility to generate Q-switch laser pulses with the following data: pulse energy 5 mJ, pulse duration 5 ns, beam quality $M^2 < 2$, repetition frequency 5 kHz [12]. Hence, there is still further potential to push high-speed LIBS to the range well above 1 kHz.

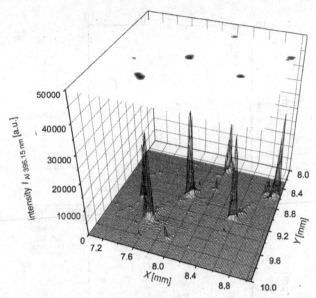

Figure 14.12. Three-dimensional plot of the intensity distribution of the aluminum line 396.15 nm of a part of a scanned surface of a steel sample with aluminum-containing inclusions. Measuring parameters as those given in Figure 14.16.

14.4.4 Results with high-speed scanning LIBS using DPSSL

In this section results gained with high-speed, high-resolution scanning LIBS are presented. Steel samples were investigated to detect inclusions and segregations influencing the material properties [13]. The homogeneity and cleanness of steel are a key issue for the production of spring steel, thin sheets, and wires. Inclusions in steel consist of, for example, Al_2O_3, SiO_2, CaO, TiN, CaS, or MnS. The typical dimensions of these inclusions are in the range 0.1–100 µm. For product development, process control, and quality assurance a simple and fast method is required to identify and to quantify inclusions [40].

At first the samples are ground using grinding paper with grain size of 80 and grains consisting of ZrO_2 and Al_2O_3. A further treatment of the sample surface is not necessary. Then the sample is mounted in a cassette holder, which is brought via the pre-chamber to the measuring chamber of SML 1, see Section 14.4.1. The measuring chamber is always kept under an argon atmosphere. The scanning field, the step size between subsequent laser pulses, the delay times and widths of integration windows can be programmed at the operational terminal of SML 1.

An example for the type of data generated is shown in Figure 14.12 as a three-dimensional plot of the intensity of the aluminum line 396.15 nm versus the spatial coordinates x and y of a part of a scanned surface of a steel sample with aluminum-containing inclusions. The different peaks correspond to localized enriched concentrations of aluminum in the iron matrix. The dynamic range of the intensity axis covers 15 bit. For the presentation of the

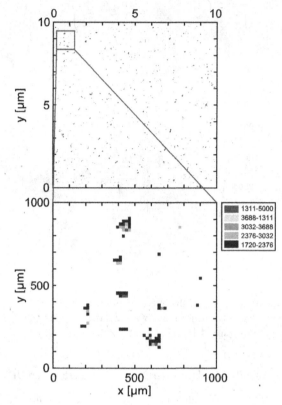

Figure 14.13. Manganese mapping with high-speed scanning LIBS. Absolute intensities of the Mn 263.82 nm line exceeding a defined threshold (see text) are shown with a gray scale in arbitrary units, step size 20 μm, measuring frequency 500 Hz, laser pulse energy 2 mJ, atmosphere Ar 4.8, 900 mbar. Top, scanned area of 1 cm^2; bottom, enlarged part of 1 mm^2. For simplicity the origin of the axes was set to (0, 0) in both maps.

data of larger scan fields we will use the presentation shown on top of Figure 14.12, where those points at a position x, y having an intensity beyond a given threshold are shown in black or in a gray scale.

Figure 14.13 shows measuring results for an area of 1 cm^2 of a low-alloyed steel sample. At a step size of the translation axis of 20 μm, the map comprises 250 000 measuring points each generated by a single laser pulse of 2 mJ corresponding to a total data volume of 48 MB. The map shows absolute intensities of the manganese line at 263.82 nm. The integration window was set to accumulate the whole photomultiplier signal induced by a single laser pulse. Only those intensities are plotted which are greater than the average intensity value plus five times the standard deviation of the intensity distribution of the whole data ensemble. The intensities exceeding this threshold are displayed in a gray scale.

Figure 14.14. Manganese maps of two steel samples of different quality. Absolute intensities of the Mn 263.82 nm line exceeding a defined threshold are shown with a gray scale in arbitrary units. Both samples were measured with the same parameter set: 500 Hz, 2 mJ, 500 × 500 measuring points, step size 20 μm, scanned area 10 × 10 mm^2.

The Mn mappings reveal an irregular distribution of localized Mn enrichments across the scanned sample surface. The enlarged part of Figure 14.13 (bottom) shows that, besides spot-like Mn peaks corresponding to a single laser pulse, clusters of enhanced Mn intensities occur with lateral dimensions in the range 50–100 μm. Within these clusters different intensities are observed indicating variations in the local Mn concentration.

In Figure 14.14 the manganese mappings of two different steel samples are shown. Again only those intensities are plotted which are greater than the average intensity plus five times the standard deviation. In the left-hand map 968 points fulfill this condition, whereas for the right-hand map the number is only 492. Hence the steel sample of the measurement shown in the left of Figure 14.14 is less clean with respect to the number of manganese enrichments than the other sample.

The maps for different elements are all measured at the same time, hence for one scanned area of a sample SML 1 yields up to 24 element maps. Figure 14.15 shows as example two maps of the elements manganese and sulfur measured at the same time at the same locations. The map at the bottom corresponds to manganese (263.82 nm line), the top to sulfur (180.73 nm line). The vertical arrow illustrates that the locations of enhanced manganese emission are spatially correlated to an enhanced sulfur emission. This type of inclusions can be attributed to MnS.

14.4.5 Detection of light elements

The detection of the light elements nitrogen and oxygen with LIBS is difficult, owing to the high excitation energies of 9.5 eV to 10.7 eV and their emission wavelengths in the vacuum

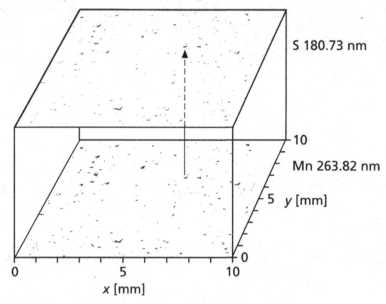

Figure 14.15. Bottom, manganese map; top, sulfur map. The measuring parameters are the same as for Figure 14.14.

ultraviolet (VUV) and infrared (IR) spectral regions. In the VUV the transparency of optical materials, such as entrance windows for spectrometers and of certain gases (oxygen in air), is low. In the infrared region detectors for high-speed LIBS, such as photomultipliers, have a low quantum efficiency of less than 5%.

The high-speed LIBS system SML 1 was tested with artificial samples provided by ThyssenKrupp Stahl to validate the ability to detect the elements oxygen and nitrogen with single laser shots on a microscopic scale. A sample area of 10×10 mm^2 of a pure iron sample doped with Al_2O_3 inclusions was scanned with a step size of 20 µm. Figure 14.16 shows mappings of two aluminum channels gained in the same scanning measurement. The data were evaluated in the following way. For each measuring point the signal intensity of the respective element channel is determined. These 250 000 values are sorted in a declining series starting with the highest signal intensity found. The first 1000 values are then taken and the corresponding locations are plotted as dark spots in Figure 14.16. A comparison of these two aluminum mappings shows that most of the points have identical positions. Hence it can be concluded that these locally enhanced measuring values belong to real aluminum inclusions and are not the result of any measuring artefacts. The same comparison can be made in Figure 14.17 with oxygen mappings measured with two different wavelengths at the same time. A comparison of these two mappings shows that most points have identical positions. Some smaller points do not match. This effect is attributed to the high excitation energies of oxygen (9.5 eV and 10.7 eV) influencing the sensitivity. Comparing the aluminum mappings of Figure 14.16 with the oxygen mappings of Figure 14.17 shows

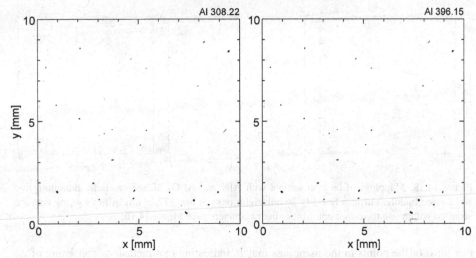

Figure 14.16. Aluminum mappings of an artificial pure iron sample doped with Al_2O_3 inclusions. Left: aluminum line 308.22 nm; right: aluminum line 396.15 nm. Points where increased signals have been measured coincide in both maps. Measuring frequency 1000 Hz, measuring time 11 min, atmosphere Ar 5.0, delay time $t_{delay} = 1.3$ μs, width of integration window $t_{int} = 8$ μs. All other measuring parameters as in Figure 14.14.

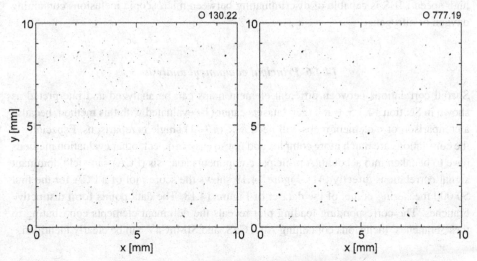

Figure 14.17. Oxygen mappings of an artificial pure iron sample doped with Al_2O_3 inclusions measured in the same measurement as the mappings of Figure 14.16. Left: oxygen line 130.22 nm; right: oxygen line 777.19 nm. Measuring parameters as for Figure 14.16.

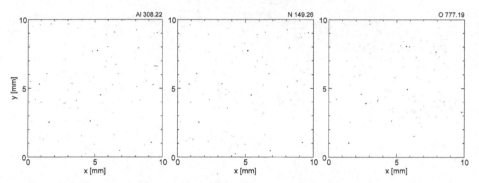

Figure 14.18. Mappings of an iron sample with AlN and Al_2O_3 inclusions. Left: aluminum line 308.22 nm; middle: nitrogen line 149.26 nm; right: oxygen line 777.19 nm. All measuring data are acquired within one measurement. Measuring parameters as in Figure 14.16.

that most of the points in the mappings match, indicating that inclusions consisting of Al and O have been detected and can be identified definitely.

To ensure that nitrogen is also detectable a sample with aluminum nitride and aluminum oxide inclusions was scanned. The results are shown in Figure 14.18. Correlations between the aluminum mapping and the nitrogen mapping can be seen, but some points in the aluminum mapping have no corresponding partner in the nitrogen mapping. These remaining aluminum intensity peaks have corresponding partners in the oxygen channel. Obviously high-speed LIBS is capable of discriminating between microscopic inclusions containing oxygen or nitrogen.

14.4.6 Principal component analysis

Spatial correlations between different element maps can be analyzed and interpreted as shown in Section 14.4.5. But larger datasets cannot be evaluated with this method, because a comparison of n element maps will need $n \times (n - 1)$ single comparisons. Especially if the correlations are much more complex and not so easy to detect, other evaluation methods have to be taken into account. A principal components analysis (PCA) can yield dominant signal correlations directly [41]. Figure 14.19 shows the score plot of a PCA for the first 50 000 measuring points of the dataset of Figure 14.18. The data points form distinctive branches. The corresponding loading plot reveals the dominant elements contributing to these branches. Inclusions containing AlN, CuS, and SiMnCaV can be clearly identified.

14.5 Laser-induced crater geometry and spatial resolution of high-speed, high-resolution scanning LIBS with DPSSL

The achievable spatial resolution with LIBS is determined by the size of the laser craters produced within the material. The crater diameter depends not only on the laser focus diameter, but also on the heat affected zone and therewith on the thermal diffusivity and

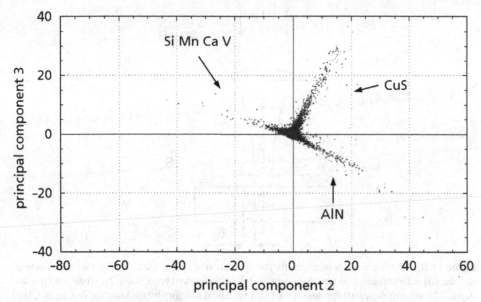

Figure 14.19. Score plot of the principal components analysis for the first 50 000 data points of the dataset of Figure 14.18. The principal component (PC) 2 is plotted in the *x*-direction, PC 3 is plotted in the *y*-direction. The elements contributing to the distinctive branches have been identified by the corresponding loading plot.

on the laser pulse duration. Figure 14.20 illustrates these characteristic dimensions in the interaction region of the focused laser beam with the specimen. The dependence of the crater radius produced by the laser on the radius of the beam waist and the heat penetration depth is described in a simplified approach for constant laser pulse energy and beam profile

$$r_c \propto w_b + \delta_h$$
$$\delta_h \propto \sqrt{\kappa_s \cdot \tau_l}$$

where r_c = radius of the produced laser crater, w_b = radius of the beam waist on the surface of the specimen, δ_h = heat penetration depth, κ_s = thermal diffusivity of the sample material, and τ_l = duration of the laser pulse. A shorter laser pulse duration will lead to a decreased heat penetration depth and hence to a smaller crater radius corresponding to an enhanced spatial resolution for scanning microanalysis.

Smaller laser craters can also be achieved by further reducing the applied laser pulse energy while keeping the beam waist radius constant. However, this will lead to smaller intensities within the focus and therefore elements with high excitation energies could not be excited sufficiently, with the consequence of higher detection limits.

Figure 14.21 shows two photographs of laser craters produced by single laser pulses of different pulse duration. The nanosecond pulse is produced by the DPSSL described in Section 14.2 (see Figures 14.1, 14.2, and Table 14.2). The picosecond pulse is generated also by a DPSSL. Although the intensity within the laser focus is approximately 21 times higher

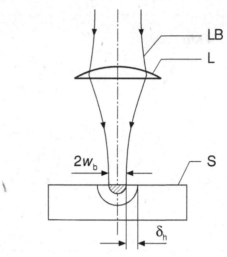

Figure 14.20. Characteristic dimensions for the interaction region of the laser beam with the sample surface. LB = laser beam, L = lens, S = sample, $2w_b$ = diameter of beam waist, δ_h = heat penetration depth. The achievable spatial resolution for LIBS is limited not only by the laser focus diameter, but also by the heat penetration depth, which depends on the thermal diffusivity of the analyzed material and the duration of the laser pulse.

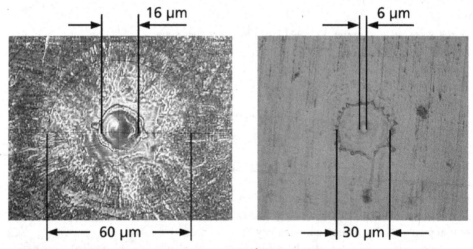

Figure 14.21. Microscopic photographs of craters produced in a steel sample by single laser pulses generated with DPSSL systems. Left: crater generated with a laser pulse of 5.5 ns duration, 2 mJ pulse energy, and a focus intensity of 1.2 TW cm^{-2}. Right: crater generated with a laser pulse of 40 ps duration, 0.4 mJ pulse energy, and a focus intensity of 25 TW cm^{-2}. The laser wavelength in both cases is 1064 nm. The crater diameter is reduced from 16 μm diameter for the nanosecond laser to 6 μm for the picosecond laser, the diameter of the debris area is reduced from 60 μm to 30 μm.

for the picosecond laser used in the right-hand picture, the crater and the debris diameter are smaller than those for the nanosecond laser in the left-hand picture of Figure 14.21. Hence picosecond lasers offer the potential to further improve the spatial resolution for scanning LIBS. A further increase of resolution would be gained with the use of femtosecond lasers. However, an application of such lasers has to cope with the fact that these lasers are, at present, twice as expensive as picosecond lasers.

Whereas subnanosecond laser pulses allow reduction of the crater diameters, their efficiency for plasma excitation for LIBS is expected to decrease. A compromise has to be found to optimize the ablation geometry and the plasma excitation. There is no reason to assume that the laser pulse durations applied so far for LIBS are already at this optimum. It is well known that the temporal pulse structure has a strong influence on the ablated mass and the efficiency of plasma excitation [42]. Different approaches were studied based on combinations of nanosecond pulses, which can easily be generated by conventional lasers [43]. In the near future further progress in the development of DPSSL systems will open up new perpectives to generate temporally modulated laser pulses within time scales from picoseconds to nanoseconds. This will give us the chance to find the optimum tailored pulse structures for the various LIBS applications.

The authors thank the European Coal and Steel Community (ECSC, funded under research contract no. 7210-PA/PB/PC/PD/168), the Fraunhofer Society, Germany, and national and international industrial partners, especially ThyssenKrupp Stahl AG, Duisburg, for the support of the presented work.

14.6 References

[1] A. Whitehouse, J. Young, I. Botheroyd *et al.*, Remote material anaysis of nuclear power station steam generator tubes by laser-induced breakdown spectroscopy. *Spectrochim. Acta B*, **56** (2001), 821–830.

[2] C. J. Lorenzen, C. Carlhoff, U. Hahn and M. Jogwich, Application of laser-induced breakdown emission spectral analysis for industrial process and quality control. *J. Anal. Atom. Spectrorom.*, **7** (1992), 1029–1035.

[3] I. Mönch, R. Noll, R. Buchholz and J. Worringer, Laser identifies steel grades. *Stainless Steel World*, **12** (4) (2000), 25–29.

[4] S. Rosenwasser, G. Asimellis, B. Bromley *et al.*, Development of a method for automated quantitative analysis of ores using LIBS. *Spectrochim. Acta B*, **56** (2001), 707–714.

[5] C. Geertsen, J. Lacour, P. Mauchien and L. Pierrard, Evaluation of laser ablation optical emission spectrometry for microanalysis in aluminium samples. *Spectrochim. Acta B*, **51** (1996), 1403–1416.

[6] J. Vadillo, S. Palanco, M. Romero and J. Laserna, Applications of laser-induced breakdown spectrometry (LIBS) in surface analysis. *Fresenius J. Anal. Chem.*, **355** (1996), 909–912.

[7] P. Lucena, J. Vadillo and J. Laserna, Mapping of platinum group metals in automotive exhaust three-way catalysts using laser-induced breakdown spectrometry. *Anal. Chem.*, **71** (1999), 4385–4391.

[8] M. Mateo, L. Cabalin, J. Baena and J. Laserna, Surface interaction and chemical imaging in plasma spectrometry induced with a line-focused laser beam. *Spectrochim. Acta B*, **57** (2002), 601–608.

[9] P. Peuser and N. Schmitt, *Diodengepumpte Festkörperlaser* (Berlin: Springer-Verlag, 1995).

[10] W. Koechner, *Solid-State Laser Engineering*, 5th edition (Berlin: Springer-Verlag, 1999).

[11] D. Hoffmann, G. Bonati, P. Kayser *et al.*, Modular, fiber coupled, diode-pumped solid state laser with up to 5 kW average output power. *Proceedings of Advanced Solid-State Lasers*, OSA, Seattle (2001), pp. 33–35.

[12] K. Du, D. Li, H. Zhang *et al.*, Electro-optically Q-switched Nd:YVO$_4$ slab laser with high repetition rate and short pulse width. *Opt. Lett.*, **28** (2003), 87–89.

[13] H. Bette, R. Noll, H.-W. Jansen *et al.*, Schnelle, ortsaufgelöste Materialanalyse mittels Laseremissionsspektrometrie (LIBS). *LaserOpto*, **33** (2001), 60–64.

[14] International Standard ISO 11146: 1999(E), Lasers and laser-related equipment – Test methods for laser beam paramters – Beam widths, divergence angle and beam propagation factor.

[15] E. Muller, A. Lauritzen and P. Eggimann, Laser spark optical emission spectrometry: a revolutionary method for the iron and steel industry. *Steel Times*, **6** (1998), 224–225.

[16] Product information of Spectro Analytical Instruments GmbH & Co KG, Kleve, Germany (April 2000).

[17] I. Whiteside and R. Jowitt, Laser liquid metal analysis, in R. Nauche (editor), *Progress of Analytical Chemistry in the Iron and Steel Industry* (Brussels: Commission of the European Communities, 1992), pp. 135–141.

[18] *Final report of the Brite-Euram project*: Study of emission spectroscopy on laser produced plasma for localised multielemental analysis in solids with surface imaging, Coordinator Saclay, France, Contract MAT 1-CT-93-0029 (1996).

[19] K. Yamamoto, D. Cremers, M. Ferris and L. Foster, Detection of metals in the environment using a portable laser-induced breakdown spectroscopy instrument. *Appl. Spectrosc.*, **50** (1996), 222–233.

[20] L. Paksy and B. Nemet, Production control of metal alloys by laser spectroscopy of the molten metals. Part 1. Preliminary investigations. *Spectrochim. Acta B*, **51** (1996), 279–290.

[21] M. Sabsabi and P. Cielo, Quantitative analysis of aluminium alloys by laser-induced breakdown spectroscopy and plasma characterization. *Appl. Spectrosc.*, **49** (1995), 499–507.

[22] L. Cabalin, D. Romero, J. Baena and J. Laserna, Effect of surface topography in the characterization of stainless steel using laser-induced breakdown spectrometry. *Surf. Interface Anal.*, **27** (1999), 805–810.

[23] C. Aragon, J. Aguilera and F. Penalba, Improvements in quantitative analysis of steel composition by laser-induced breakdown spectroscopy at atmospheric pressure using an infrared Nd:YAG laser. *Appl. Spectrosc.*, **53** (1999), 1259–1267.

[24] M. Hatcher, Breakdown spectroscopy finds nuclear application. *Opto Laser Europe*, April 2000, pp. 20–22.

[25] V. Sturm, L. Peter, R. Noll *et al.*, Elemental analysis of liquid steel by means of laser technology. In *International Meeting on Chemical Engineering, Environmental Protection and Biotechnology*, ACHEMA 2000, Materials Technology and Testing (2000), pp. 9–11.

[26] H. Sattler, Metallschrotte nach Analyse automatisch sortieren. *Materialprüfung*, **35** (1993), 312–315.

[27] R. Noll, V. Sturm, L. Peter and I. Whiteside, Analysis using lasers. *Proceeding of the 49th Chemists' Conference*, Scarborough (1997), pp. 22–27.

[28] D. A. Cremers, R. Wiens, M. Ferris, R. Brennetot and S. Maurice, Capabilities of LIBS for analysis of geological samples at stand-off distances in a Mars atmosphere. *Technical Digest, Conference: Laser-Induced Plasma Spectroscopy and Applications*, September 25–28, Orlando (2002), pp. 5–7.

[29] U. Panne, R. Neuhauser, M. Theisen, H. Fink and R. Niessner, Analysis of heavy metal aerosols on filters by laser-induced plasma spectroscopy. *Spectrochim. Acta B*, **56** (2001), 839–850.

[30] B. Charfi and M. A. Harith, Panoramic laser-induced breakdown spectrometry of water. *Spectrochim. Acta B*, **57** (2002), 1141–1153.

[31] O. Samek, D. Beddows, H. Telle *et al.*, Quantitative laser-induced breakdown spectroscopy analysis of calcified tissue samples. *Spectrochim. Acta B*, **56** (2001), 865–875.

[32] Thermo-ARL 4460 with SparkDat, http://www.thermoarl.com

[33] J. Carranza, B. B. Fisher, G. Yoder and D. Hahn, On-line analysis of ambient air aerosols using laser-induced breakdowm spectroscopy. *Spectrochim. Acta B*, **56** (2001), 851–864.

[34] D. Body and B. Chadwick, Optimization of the spectral data processing in a LIBS simultaneous elemental analysis system. *Spectrochim. Acta B*, **56** (2001), 725–736.

[35] A. I. Whitehouse, J. Young, I. Botheroyd *et al.*, Remote material analysis of nuclear power station steam generator tubes by laser-induced breakdown spectroscopy. *Spectrochim. Acta B*, **56** (2001), 821–830.

[36] K. Lo and N. Cheung, ArF Laser-induced plasma spectroscopy for part-per-billion analysis of metal ions in aqueous solutions. *Appl. Spectrosc.*, **56** (2002), 682–688.

[37] S. Palanco, M. Klassen, J. Skupin *et al.*, Spectroscopic diagnostics on CW-laser welding plasmas of aluminium alloys. *Spectrochim. Acta B*, **56** (2001), 651–659.

[38] HR 2000 from www.oceanoptics.com, DU 401 from www.andor-tech.com, Spec-10:100B from www.prinst.com

[39] V. Sturm, L. Peter and R. Noll, Steel analysis with laser-induced breakdown spectrometry in the vacuum ultraviolet. *Appl. Spectrosc.*, **54** (2000), 1275–1278.

[40] H. Bette, R. Noll, G. Müller *et al.*, High-speed scanning laser-induced breakdown spectroscopy at 1000 Hz with single pulse evaluation for the detection of inclusions in steel. *J. Las. Appl.*, **17** (2005), 183–190.

[41] H. Martens and T. Mæs, *Multivariate Calibration* (New York: John Wiley & Sons, 1998).

[42] R. Sattmann, V. Sturm and R. Noll, Laser-induced breakdown spectroscopy of steel samples using multiple Q-switch Nd:YAG laser pulses. *J. Phys. D: Appl. Phys.*, **28** (1995), 2181–2187.

[43] D. Stratis, K. Eland and M. Angel, Dual-pulse LIBS using a pre-ablation spark for enhanced ablation and emission. *Appl. Spectrosc.*, **54** (2000), 1270–1274.

15

Laser-induced breakdown spectroscopy using sequential laser pulses

Jack Pender, Bill Pearman, Jon Scaffidi, Scott R. Goode and S. Michael Angel

*Department of Chemistry and Biochemistry,
The University of South Carolina, USA*

15.1 Introduction

In laser-induced breakdown spectroscopy (LIBS), first reported by Brech and Cross in 1962 [1], a laser is used to ablate and atomize material from a sample and to form a plasma. Emission from the plasma is used to identify and quantify elements within the sample. The ability to form a plasma on unprepared samples makes LIBS an amazingly versatile analytical technique. It is one of the few techniques that can be used for non-contact elemental analysis, making LIBS uniquely suited to measurements of hazardous materials and materials in difficult-to-reach locations [2–16]. The sampling is virtually non-destructive, making LIBS useful for such unique applications as the analysis of price-less works of art and archeological relics [17–21]. Other applications that benefit from the unique advantages of LIBS include environmental [22–28], industrial [2–4, 23, 24, 29–36], geological [22, 25–27], planetary [22], art [28–42], medical [43, 44], and dental [45] measurements. Recently, many researchers have coupled LIBS with other techniques such as ICP-MS [12, 41, 42, 46–48].

Despite the increasing popularity of LIBS, the sensitivity and precision of the technique are relatively poor compared with other forms of atomic spectroscopy and there are significant matrix effects and relatively high background signals [49]. There are also many fundamental studies aimed at improving the sensitivity and precision of LIBS [34, 50–54]. These studies have led to investigations of multiple-pulse LIBS, which can give greatly enhanced emission signals and improved signal to background ratio [55–58]. Several investigators have used multiple pulses for LIBS and report enhanced emission in a variety of samples and sample matrices [55, 56, 58–61]. Recently we described a new dual-pulse LIBS technique based on the formation of an air spark prior to ablation [56–58, 62, 63]. In this technique one laser pulse is used to form an air plasma a few millimeters above the sample several microseconds prior to ablation (pre-ablation spark dual-pulse LIBS). Up to 100-fold emission enhancements have been observed for many samples by using this technique. In this chapter we describe dual-pulse LIBS experiments with emphasis on the

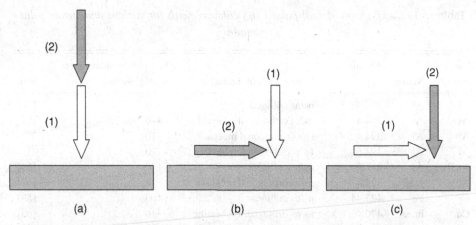

Figure 15.1. Commonly used beam geometries for dual-pulse LIBS. (a) Collinear dual-pulse arrangement, wherein both pulses are focused upon the same point on the sample. (b) Orthogonal reheating configuration, for which an air spark is formed after ablation. (c) Orthogonal pre-ablation spark dual-pulse configuration, in which an air plasma is formed above the sample surface prior to ablation.

pre-ablation air spark configuration. We also include updates of previously reported work, new results combining femtosecond and nanosecond laser pulses for dual-pulse LIBS, and bulk aqueous-phase measurements using orthogonal dual-pulse LIBS.

15.2 Dual-pulse LIBS

15.2.1 Dual-pulse LIBS using an ablation pulse followed by a re-excitation pulse

Most of the studies on dual-pulse LIBS have been performed using a collinear beam orientation (Figure 15.1a), with both pulses incident on the same spot on the sample surface [59–61, 64–67]. Enhancements for neutral atomic emission are generally 10-fold or less using this configuration, though there has been one report of 100-fold enhancement for an ionic species using this configuration [55]. Authors generally explain emission increases in collinear dual-pulse LIBS through a combination of increased material ablation, an increase in the plasma volume, and plasma reheating by the second pulse. An orthogonal beam orientation has also been described [68] (Figure 15.1b), wherein the second laser pulse is focused parallel to the sample surface and serves to reheat or re-excite material ablated by the first laser pulse. Table 15.1 presents a summary of dual-pulse LIBS emission enhancements that have been reported for a number of analytes. This table also includes data from our laboratory that have not been previously reported.

The first use of multiple pulses for enhanced emission in LIBS was reported by Cremers *et al.* [64]. In that study enhanced emission from dissolved species in aqueous solution was reported by the formation of a cavitation bubble by the first laser pulse into which the second laser pulse was focused to create the plasma. Both orthogonal and collinear laser geometries were investigated and the authors determined that the collinear geometry gave the largest emission enhancements.

Table 15.1. *Comparison of dual-pulse LIBS enhancements for various analytes in solid samples*

Analyte	Matrix	Atomic line (nm)	Dual-pulse configuration	Dual-pulse enhancement vs. single-pulse	Ref.
Al	Al	394.4	ns/ns collinear	3–4	[60]
Al	Al	394.4	ns/ns orthogonal prespark	4–6	[67]
Al	Al	394.4	fs/ns orthogonal prespark	10	**b
Al	Al	394.6	fs/ns orthogonal reheating	80	**b
Al(II)	Al	281.6	ns/ns collinear	100	[64]
Al	glass	394.4	ns/ns orthogonal prespark	22	[67]
Cr	Fe^a	425.44	ns/ns collinear	~100	[59]
Cu	brass	470.46	ns/ns orthogonal reheating	~10	[63]
Cu	brass	578.2	ns/ns orthogonal prespark	27	[68]
Cu	Cu	521.8	ns/ns orthogonal prespark	120	**b
Cu	Cu	510.3	ns/fs orthogonal reheating	30	**b
Cu	Cu	510.3	ns/fs orthogonal prespark	4	**b
Cu	Cu	521.8	ns/ns orthogonal prespark	~30	[69]
Cu	Fe^a	324.75	ns/ns collinear	~100	[59]
Fe	Al	281.33	ns/ns collinear	3	[60]
Fe	Fe	404.6	ns/ns orthogonal prespark	28	[68]
Fe	Fe	426.1	ns/ns orthogonal prespark	~30	[69]
Fe	glass	404.6	ns/ns orthogonal prespark	11	[67]
Mg	Al	285.21	ns/ns collinear	4	[60]
Mn	Al	279.83	ns/ns collinear	28	[60]
Mn	Fe^a	404.58	ns/ns collinear	~30	[59]
Pb	Pb	500.5	ns/ns orthogonal prespark	11	[65]
Si	Al	288.16	ns/ns collinear	30	[60]
Si	Al	288.16	ns/ns collinear	3	[60]
Si	Fe^a	288.16	ns/ns collinear	~100	[59]
Sn	Al	284.00	ns/ns collinear	3	[60]
Ti	glass	391.4	ns/ns orthogonal prespark	10	[67]
Ti	Ti	391.4	ns/ns orthogonal prespark	4–6	[67]
Zn	brass	468.01	ns/ns orthogonal reheating	~10	[63]
Zn	brass	636.2	ns/ns orthogonal prespark	16	[68]

^{a} Iron sample, underwater.
^{b} Angel, unpublished results.

Uebbing *et al.* [68] were among the first to use dual pulses in LIBS for solid samples, using an orthogonal geometry (Figure 15.1b) where the first pulse is incident on the sample surface and the second pulse is focused parallel to the surface to reheat material ablated by the first pulse. By using this geometry, an increase in Al and Mn emission and improved detection limits were observed in glass and steel samples.

Sattmann *et al.* [59] compared single, double and multiple sequential pulses in a collinear geometry (Figure 15.1a) from a Q-switched Nd:YAG laser on steel samples. The results showed increases in material ablation, greater electron densities, and higher electron temperatures for double-pulse and multiple-pulse bursts compared with single-pulse bursts. Line intensities increased by a factor of two using double pulses even with the second pulse only half the energy of a single pulse, and relative standard deviations were slightly smaller using double-pulse bursts compared with single-pulse experiments when performing multi-elemental analysis. Double-pulse bursts also improved the detection limits for some trace elements, but the improvement was element dependent.

Increases in atomic line intensity in aluminum alloys were reported by St.-Onge *et al.* [65] using double-pulse bursts, and enhancement factors of 3 to 4 were achieved for neutral Al lines. This enhancement was attributed to a larger plasma volume from the greater ablation mass and the presence of a pre-plasma into which the second laser pulse was absorbed. As in previous studies, they found an increase in electron temperature; however, the electron density was found to be the same with single and double pulses. It was also pointed out that intensity improvements are line dependent and also dependent on the time between the pulses. Other improvements include a decrease in relative standard deviation when comparing single with double-pulse bursts.

More recently, Sturm *et al.* [67] used triple-pulse excitation for multi-elemental analysis of steel samples in an effort to obtain lower detection limits than conventional LIBS for an on-line method. The authors examined the effect of pulse delay time on detection limits. These pulses were generated from a Q-switched Nd:YAG laser, with timing separations of 25 μs between the first and second pulse and 40 μs between the second and third pulse for best results. Improved signal-to-noise ratios were obtained, and limits of detection below 10 μg g^{-1} were achieved for phosphorus, sulfur, and carbon in steel for the first time using LIBS.

In very recent work St.-Onge *et al.* reported their investigation of collinear dual-pulse LIBS of aluminum alloys using a 1064 nm first pulse followed by a second pulse using a 266, 355, 532, or 1064 nm laser line [55]. Signal enhancements of approximately 30 for the Si(I) 288.16 nm line and 100 for the Al(II) 281.62 nm line were calculated in the case of a 1064 nm first pulse followed by a 266 nm second pulse. These researchers also reported a substantial increase in plasma temperature, with larger enhancements for lines with higher excitation energy.

Colao *et al.* recently compared single- and double-pulse LIBS of aluminum samples using a collinear geometry [69]. In this study the total laser energy per pulse was held constant. They found a lowering of the second pulse plasma threshold (e.g. the energy required to produce a plasma) and significantly improved analytical capabilities.

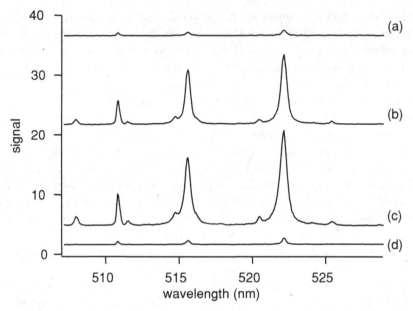

Figure 15.2. Copper LIBS spectra using (a) single-pulse excitation and collinear dual-pulse excitation with pulse delays of (b) 1 μs, (c) 20 μs, and (d) 300 μs. The spectra have been offset for clarity.

The initial dual-pulse LIBS studies in our group were carried out in a collinear geometry using the configuration shown in Figure 15.1a. These first studies produced 16- to 17-fold signal enhancements for copper using a 1–20 μs delay time between the laser pulses (Figure 15.2). For these measurements we used 100 mJ and 180 mJ for the first laser beam and the second laser beam, respectively, and each signal was optimized for beam and sample alignment at each delay time. As shown by the emission signal versus pulse delay curve in Figure 15.3 the enhancement becomes a maximum for a pulse delay of 1–2 μs, and little or no enhancement was observed for delay times greater than about 300 μs.

15.2.2 *Pre-ablation spark dual-pulse LIBS*

We have observed very large emission enhancements by using an orthogonal dual-pulse configuration (see Figure 15.1c), where the sample ablation pulse is preceded by a pulse that forms an air spark above the sample. In this technique (pre-ablation spark dual-pulse LIBS) the air above a solid sample is ionized using a laser pulse focused parallel to, and millimeters above, the sample. Several microseconds later, a second pulse ablates material from the sample surface and generates a laser-induced plasma in the region previously occupied by the air plasma. We previously reported the use of this technique for the analysis of metals and glasses [56–58, 62, 63], finding enhancements to be both matrix and element dependent. The enhancement is also dependent on the emission line being

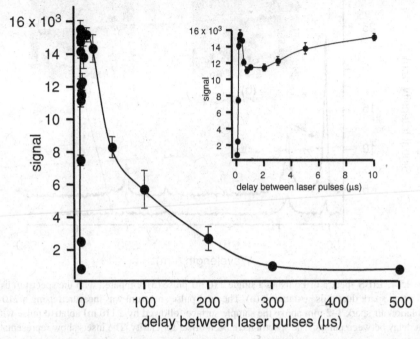

Figure 15.3. LIBS emission intensity versus pulse delay for collinear dual-pulse excitation for the neutral copper line at 521.8 nm using a 100 mJ, 7 ns pre-ablation pulse followed by a 180 mJ, 7 ns ablation pulse. Enhancements were calculated relative to a fully optimized 180 mJ, 7 ns single ablative pulse. Error bars represent one standard deviation, $n = 5$.

observed, as shown by the single-pulse (a) and dual-pulse (b) lead spectra in Figure 15.4. The 500.5 nm emission line shows about a 12-fold enhancement for the dual-pulse sepectrum (upper) compared with the single-pulse spectrum (lower) while the ∼505 nm emission line shows very little enhancement. The insets in Figure 15.4 show images of the abla-tion craters that were formed using single-pulse (upper right) and dual-pulse (upper left) excitation, indicating a substantial increase in the ablation volume for the latter.

Recently we have noted that the magnitude of the enhancement depends strongly upon pulse alignment and, with careful alignment, we have observed up to 120-fold enhancements for copper emission from a pure copper sample compared with optimized single-pulse LIBS (Figure 15.5, described below). Like other dual-pulse LIBS techniques, we also observe greater sample ablation and very well-defined craters (Figure 15.5 inset), and the increase in sample ablation is qualitatively consistent with the magnitude of the emission enhancement.

The delay time between laser pulses is important for large enhancements. A typical emission enhancement versus pulse delay curve is shown in Figure 15.6. The negative time values indicate that the air spark is created before the ablation pulse arrives at the sample. The enhancement increases rapidly at a pulse delay of 1–2 μs, followed by a slow decrease out to 100–150 μs. Prior to 1 μs, there is either no enhancement or a decrease

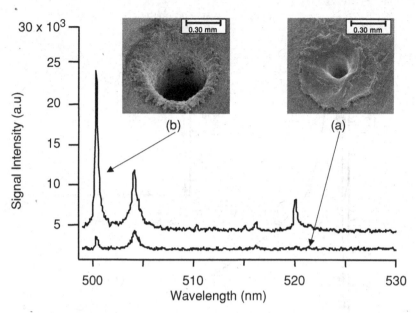

Figure 15.4. LIBS spectra of Pb using a single 110 mJ pulse (a) compared with the spectrum using pre-ablation spark dual pulse excitation (b). The dual-pulse spectrum was measured using a 210 mJ pre-ablation air spark 1–2 mm above the sample surface followed by a 110 mJ ablative pulse with a 1.0 μs delay between pulses. The spectra have been offset for clarity. The insets show representative ablation craters from previous single-pulse (a) and pre-ablation spark dual-pulse (b) experiments.

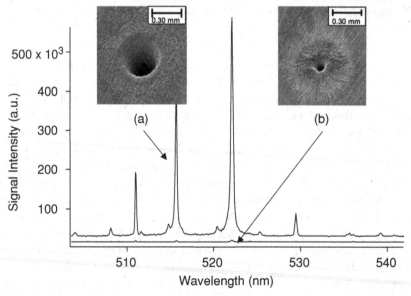

Figure 15.5. LIBS spectra of Cu using a single 100 mJ pulse (a) compared with the spectrum using pre-ablation spark dual-pulse excitation (b). The dual-pulse spectrum was measured using a 200 mJ pre-ablation air spark 1 mm above the sample surface followed by a 100 mJ ablative pulse with a 5 μs delay between pulses. The spectra have been offset for clarity. The insets show representative ablation craters from previous single-pulse (a) and pre-ablation spark dual-pulse (b) experiments.

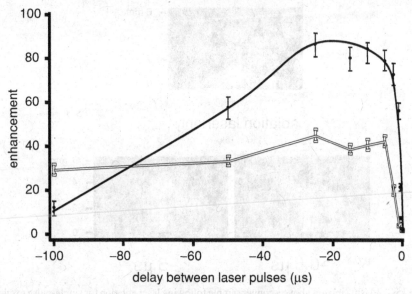

Figure 15.6. Orthogonal pre-ablation spark dual-pulse enhancement vs. inter-pulse delay for the neutral copper line at 521.8 nm using a 100 mJ, 7 ns ablative pulse after formation of a pre-ablation air spark with a 170 mJ, 7 ns pulse focused 0.25 mm above the surface of a brass sample. Enhancements were calculated relative to the fully optimized 100 mJ, 7 ns ablative pulse. Error bars represent one standard deviation, $n = 5$.

in the LIBS signal, probably because of plasma shielding, similar to the collinear dual-pulse results. The similarity in the enhancement versus pulse delay curves is one indication that the enhancement has a similar origin in both cases. We also observed that large LIBS enhancements are observed only when the pre-ablation spark is formed close to the sample surface. This result indicates that the first pulse might affect the sample surface directly. However, the effect is clearly not the result of a permanent change in the sample surface, because the magnitude of the enhancement decreases rapidly for pulse delays beyond a few hundred microseconds.

There are several notable features of a dual-pulse plasma. Time-resolved plasma images show very large changes in plasma shape and volume as functions of pulse delay for pre-ablation spark dual-pulse LIBS compared with single-pulse LIBS. Figure 15.7 shows copper LIBS plasma images for the plasma formed using the ablative pulse alone and for dual-pulse LIBS using an air spark 0.5 μs and 5 μs prior to ablation. The copper plasma height increases from approximately 0.5 mm with a single pulse to 4 mm when using a pre-ablation spark 5 μs before the ablative pulse. Although the air spark pulse usually requires relatively high pulse energy, the energy of the ablation pulse is usually the same as the single-pulse experiment. Similar increases in plasma volume have been observed in all pre-ablation spark dual-pulse LIBS experiments. It is interesting to note that these images look qualitatively similar to images of a LIBS plasma formed in a reduced-pressure atmosphere [23]. Increased ablation

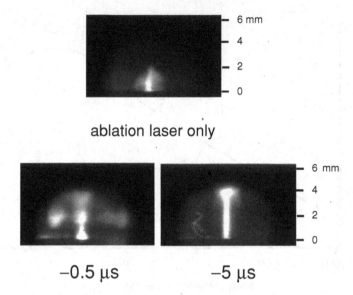

ablation laser only

−0.5 µs   −5 µs

Figure 15.7. Plasma images above a copper sample following laser ablation for single-pulse excitation (top) and dual-pulse excitation (lower two images) at pulse delay times of 0.5 µs and 5 µs. The negative time values indicate that the air spark pulse arrives before the ablation pulse.

is also reported for LIBS in low-pressure environments. In addition to the change in plasma size and shape, we also measure higher electronic temperatures using dual-pulse excitation for metal samples but there is no increase in the plasma temperature for glass samples.

In our early pre-ablation spark dual-pulse LIBS experiments we observed 11-, 28- and 40-fold signal enhancements for lead [56], iron [57], and copper [56] samples, respectively. We also found enhancements of 11- to 20-fold for Ti, Al, and Fe in glass samples, though enhancements were much lower for the same elements in the pure metals [57]. Since that time we have found that beam alignment is critical to obtaining large enhancements. Figure 15.5 shows the largest emission enhancement we have observed for a solid sample using dual-pulse LIBS. This figure shows the emission spectrum of a copper sample using the pre-ablation spark dual-pulse configuration, compared with a spectrum obtained using single-pulse excitation. In this case, a 120-fold signal enhancement was obtained for the 515 nm copper line using a pulse delay time of 2.5 µs.

Pre-ablation spark dual-pulse LIBS works well for a variety of sample types including non-conductive samples. Figure 15.8 shows SEM images of the ablation craters formed after 50 consecutive samplings of the same region of a glass sample using a single pulse (S-P) and using a pre-ablation spark at various times ranging from 0.2 µs to 100 µs prior to ablation [56]. As found in experiments with metal samples, the signal enhancement for analytes in glasses (11-, 11-, and 20-fold for Fe, Ti, and Al, respectively) seems qualitatively comparable to the increases in material ablation.

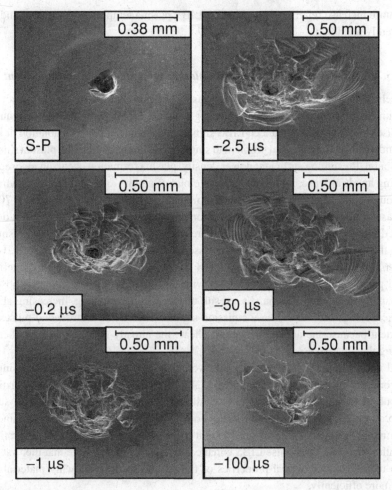

Figure 15.8. SEM images of pre-ablation spark dual-pulse ablation craters from glass after 50 consecutive measurements using various pulse delay times. The laser pulse timings that result in the largest signal enhancements (-50 μs to -0.2 μs) also result in the greatest amount of material ablation. The upper left image (S-P) is the result of 50 single-pulse ablations and is included for comparison.

Unlike the case of pre-ablation spark dual-pulse LIBS of metal samples, the plasma temperature for glass samples is similar for both the single and dual-pulse plasmas. For a steel sample, the plasma electronic temperature calculated using a previously described method [70] increased from ~11 000–12 000 K using a single laser pulse to ~16 000 K using a pre-ablation spark. For glass samples, the plasma temperature was found to be 9800 ± 1900 K for single-pulse excitation, and 9200 ± 900 K when using the pre-ablation spark dual-pulse configuration. The reasons for this large difference in dual-pulse plasma temperature increase relative to single-pulse LIBS plasma temperatures for steel and glass are currently unknown.

15.2.3 Possible mechanisms of pre-ablation spark dual-pulse enhancement

The mechanisms of emission enhancement using the pre-ablation air spark are not completely understood and are still under investigation. To explain how the use of sequential laser pulses might provide enhanced LIBS signals and enhanced material ablation, it is crucial to understand a fundamental problem in single-pulse laser ablation: the plasma shielding phenomenon. Presuming sufficient power density, plasma formation occurs in picoseconds, generating an ultra-dense, rapidly expanding cloud of atoms, ions, and electrons which, at high enough electron densities, can almost completely absorb or reflect light [49, 70–73]. In the case of ultra-fast laser pulses, absorption and reflection by the high electron density generated by the leading edge of the pulse are not an issue – by the time the plasma has formed, the entire laser pulse has already passed. Most lasers typically used for LIBS, on the other hand, have a relatively long (~7–8 ns) pulse width, so only a tiny fraction of the total pulse energy can directly interact with the sample – the remainder is absorbed by the rapidly expanding plasma. As a result of this plasma shielding, increasing ablative pulse energy does little to increase emission signals when using nanosecond pulses, to a large extent because the added energy only heats the already-formed plasma rather than increasing the ablative volume [74].

Based upon the greatly increased crater volumes seen in dual-pulse LIBS using two nanosecond pulses (see insets in Figures 15.4 and 15.5), it seems clear that, relative to single-pulse LIBS, much more energy from the ablative pulse reaches the sample surface before plasma electron densities rise to the point where they can absorb the remainder of the ablative pulse. Although they are thought to play only a minor role in generation of pre-ablation air spark dual-pulse LIBS enhancements, it is also possible that the first pulse changes the thermal or optical properties of the sample surface, allowing material to be ablated more efficiently.

15.2.4 The potential role of pre-ionization

At short delay times in pre-ablation spark dual-pulse LIBS, approximately 1 µs or less, emission is generally much weaker than that seen in single-pulse LIBS (see Figure 15.6 and inset of Figure 15.3), consistent with the hypothesis that high electron densities early in the evolution of the air plasma absorb a large fraction of the second, ablative pulse. At longer delay times, where large enhancements are observed, the air spark formed by the first pulse has had time to expand and cool before the ablative pulse is fired, so that it does not significantly shield the sample surface. At the same time, however, lingering electron density in the pre-ionization region may allow the ablative plasma formed by the second pulse to quickly expand above the sample surface, thereby reducing plasma shielding within the ablative spark itself, yielding a larger, more homogeneous plasma with a slightly lower average electron density than is seen in single-pulse LIBS. This result is exactly what is observed in a dual-pulse plasma. Thus, a combination of increased plasma volume, increased sample ablation due to decreased plasma shielding, and increased plasma temperature may well generate the majority of the observed dual-pulse LIBS enhancements. We have also seen evidence that surface heating may play a small role in the enhanced ablation. Specifically, we have measured about a 2.5-fold increase in emission for copper using a single laser pulse as the sample was heated from 38 °C to 250 °C. However, this effect seems to be sample dependent, as no enhancement was observed for a tin sample heated up to the melting point.

15.2.5 Dual-pulse experiments using a combination of nanosecond and femtosecond pulses

In previous work we found that the use of 140-femtosecond laser pulses for single-pulse LIBS produced better ablation precision and better reproducibility in the measured emission compared with nanosecond ablative pulses [70]. Thus, it seems logical to combine nanosecond and femtosecond pulses in the orthogonal dual-pulse configuration in an attempt to further increase the precision of this dual-pulse method. However, pre-ablation spark dual-pulse experiments might not be expected to enhance emission when using a femtosecond ablative pulse. For example, because of the high peak power of the femtosecond pulse (6 × 10^{10} W for a pulse energy of 6 mJ and a pulse width of 100 fs) and the multi-photon nature of plasma formation by femtosecond laser pulses [74, 75], it would be expected that any electronic or energetic effects (such as increased sample or atmospheric temperature) introduced by the use of a nanosecond pre-ablation spark would not be observed using a femtosecond ablation pulse. On the other hand, any environmental changes above the sample surface (reduced pressure, for example), that might be responsible for a part of the dual-pulse enhancement might be observed using either a nanosecond or a femtosecond ablative pulse. In the reverse configuration (a femtosecond pre-ablation spark followed by a relatively low-power nanosecond ablative pulse), dual-pulse enhancements might be expected to be observed regardless of the source of dual-pulse enhancements.

15.2.6 Pre-ablation spark dual-pulse results using combinations of nanosecond and femtosecond laser pulses

Copper and aluminum samples were measured by using combinations of femtosecond and nanosecond pulses in a pre-ablation spark dual-pulse configuration (Figure 15.1c). Although the results differ significantly for the two samples (perhaps owing to the non-optimized nature of the study, or perhaps because of the analytes themselves), significant enhancements were seen with either the femtosecond or the nanosecond laser as the ablation pulse. Nanosecond ablation with a femtosecond pre-ablation spark yielded four-fold

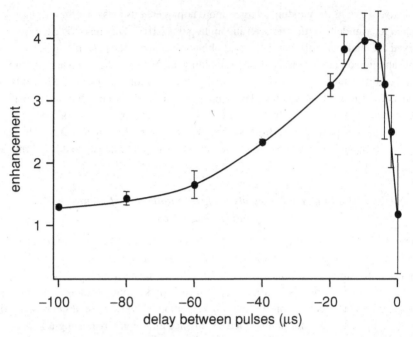

Figure 15.9. Pre-ablation spark dual-pulse enhancement versus inter-pulse delay time for copper using a 6 mJ, 100 fs air spark pulse focused 0.25 mm above the sample surface, followed by a 30 mJ, 7 ns ablative pulse. Enhancements were calculated relative to fully optimized single-pulse excitation using a 30 mJ, 7 ns pulse. Error bars represent one standard deviation, $n = 3$.

enhancements for copper compared with fully optimized nanosecond single-pulse LIBS. This result was obtained using a pulse delay of 6 μs and a 1 mm air spark height above the sample surface. These enhancements tailed off as pulse delay was increased (Figure 15.9), finally becoming statistically equal to nanosecond single-pulse LIBS signals at an inter-pulse delay of 100 μs. No dual-pulse enhancements were seen for aluminum in this pulse configuration.

For femtosecond ablation with a nanosecond pre-ablation air spark, on the other hand, large dual-pulse enhancements were observed for aluminum, but not for copper. Figure 15.10 shows emission spectra for aluminum using a single femtosecond pulse (line A) compared with dual-pulse using a femtosecond ablation pulse following a nanosecond air spark (line B). Figure 15.11 shows the pre-ablation spark dual-pulse LIBS enhancement, relative to femtosecond single-pulse excitation, versus pulse delay curve for aluminum at delay times ranging from zero to 100 μs at an air spark height of 0.5 mm. Ten-fold emission enhancements were observed at a pulse delay of 11 μs, and remain statistically separable from femtosecond single-pulse ablation signals up to pulse delays somewhere between 25 μs and 40 μs. The precipitous decrease in dual-pulse enhancements between a pulse delay of 11–15 μs may be significant regarding the causes of dual-pulse enhancement in this

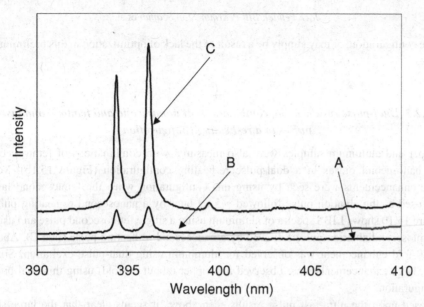

Figure 15.10. LIBS spectra of aluminum metal using a single 100 fs pulse (line A), pre-ablation spark dual-pulse excitation with a femtosecond ablation pulse following a nanosecond air spark (line B), and a femtosecond ablation pulse followed by a nanosecond re-heating pulse (line C).

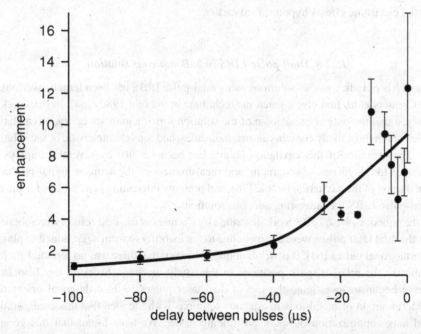

Figure 15.11. Dual-pulse enhancement versus inter-pulse delay time for aluminum when using a 6 mJ, 100-femtosecond ablative pulse preceded by formation of an air spark using a 30 mJ, 7-nanosecond pulse focused 0.5 mm above the sample surface. Enhancements were calculated relative to fully optimized single-pulse LIBS using a 6 mJ, 100-femtosecond pulse. Error bars represent one standard deviation, $n = 3$.

pulse configuration, or may simply be a result of the lack of optimization in this preliminary work.

15.2.7 Dual-pulse results using combinations of nanosecond and femtosecond laser pulses in a re-heating configuration

Copper and aluminum samples were also measured with combinations of femtosecond and nanosecond pulses in a dual-pulse re-heating configuration (Figure 15.1b). Very large enhancements were seen by using this configuration when the femtosecond laser was used as the ablation pulse followed $\sim$5 µs later by a nanosecond re-heating pulse. Figure 15.10 shows LIBS spectra of aluminum using a single femtosecond pulse and using a femtosecond ablation pulse followed by a nanosecond re-heating pulse (line C). About an 80-fold enhancement was observed for aluminum using dual-pulse excitation. Similarly, large enhancements were observed for copper (about 30-fold) using this dual-pulse configuration.

Based upon the ultra-fast pulse results seen above, it seems clear that the large signal, signal-to-background, and signal-to-noise improvements observed in orthogonal pre-ablation air spark dual-pulse LIBS must be the result of a combination of factors, perhaps including changes in pressure, increases in sample and atmospheric temperature, and the potential electronic effects hypothesized earlier.

15.2.8 Dual-pulse LIBS in bulk aqueous solution

The analysis of bulk aqueous solutions using dual-pulse LIBS has been largely overlooked since Cremers *et al.* first investigated the technique in the mid 1980s [64]. In that work, it was noted that the optical breakdown of the solution forms a gaseous cavity, or cavitation bubble, which most likely contains atoms, molecules, and ions characteristic of the solution itself. The presence of this cavitation bubble has been verified by several groups using techniques like Schlerein photography and measurement of the acoustic signal produced by oscillation of the cavitation bubble [76], and presents interesting possibilities for the use of dual-pulse LIBS for measuring aqueous solutions.

In the aqueous-phase LIBS work described by Cremers *et al.*, best results were obtained when the dual laser pulses were collinear due to the inability to form a reproducible plasma with orthogonal pulses [64]. It is worth noting, however, that there was no reported attempt to optimize the relative spark positions in this study. In recent work in our laboratory we have begun to investigate the use of dual-laser pulses in an orthogonal orientation for determination of metal ions in aqueous solutions with the idea that this configuration would make optimization of spark positioning easier. We have found that this geometry, combined with emission collection orthogonal to both pulses, allows better optimization of the two laser pulses and provides higher signal enhancements than the collinear configuration.

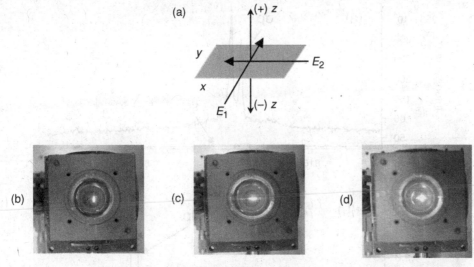

Figure 15.12. Detail of spark alignment in x, y plane with emission collected along the z axis (a). Also included are pictures of the plasma in the sample cell for pulse 1 alone (b), pulse 2 alone (c), and with both pulses using an inter-pulse delay time of 240 μs (d).

For solution measurements a Teflon cell with quartz windows was used and the emission was collected from beneath the cell to minimize scattering of the signal by bubbles that were produced by the laser. Figure 15.12 shows the orientation of the two laser pulses (E_1 and E_2) inside the sample cell and also the two possible directions for orthogonal collection of the emission ($+/- z$). Figure 15.12 also shows images of the plasma viewed from the top ($+z$ direction) for single-pulse (b and c) and for dual-pulse excitation (d). Because of the large number of experimental parameters in this measurement, such as inter-pulse timing, spark position and intersection, gate delay, gate width, and collection position and focus, we found it useful to use a simplex optimization scheme. The simplex technique allowed rapid optimization of the key experimental parameters to give large enhancements and reproducible results. The study also showed that beam alignment and collection focus were among the most critical parameters that could be adjusted for best accuracy and precision. This alignment was best accomplished by back-focusing a helium–neon laser through the sample cell.

A number of dissolved species were measured by using the orthogonal dual-pulse method including Na, Ca, Cr and Zn as well as oxygen that was produced from water breakdown. During set-up and optimization we found that the oxygen signal was an excellent indicator of alignment. Greater than 300-fold emission enhancements were observed for both sodium and oxygen (see Figures 15.13a and b, respectively), compared with the emission signal obtained using a single pulse. Laser energy per pulse was not investigated as part of this study. For the single-pulse measurements the sample was excited with ~175 mJ per pulse at 1064 nm. For the orthogonal dual-pulse measurements the energy was ~95 mJ and ~175 mJ

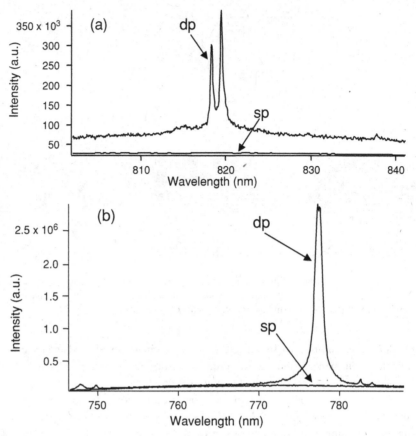

Figure 15.13. (a) LIBS spectrum of an aqueous sodium solution using single-pulse (sp) and orthogonal dual-pulse (dp) excitation. In this experiment the concentration of sodium was not measured. (b) LIBS spectrum of water showing an oxygen line using single-pulse (sp) and orthogonal dual-pulse (dp) excitation.

for pulse 1 (E_1) and pulse 2 (E_2), respectively. Enhancements were not determined for the other dissolved species; however, detection limits of 42 p.p.b, 1 p.p.m., and 17 p.p.m. were experimentally determined for Ca, Cr, and Zn, using the neutral emission lines at 422, 425, and 472 nm, respectively. These measurements represent an order-of-magnitude improvement over published detection limits for these elements in aqueous solution using LIBS (see Table 5.2 for comparison with several literature values) [7, 64, 76–80].

15.3 Summary

Dual-pulse LIBS using collinear, orthogonal or pre-ablation spark configurations provides significantly enhanced emission signals compared with single-pulsed LIBS. The

Table 15.2. *Comparison of optimization results and detection limits for LIBS of metal ions in aqueous solution*

Optimization results	Angel, unpublished	Published	Ref.
Δt value for maximum oxygen emission	range: 70–330 μs	18 μs	[56]
Enhancement of oxygen signal with dual-pulse	314-fold	50-fold	[56]
Max gate delay (td) for detection of oxygen emission	2400 ns	500 ns	[56]
Cavitation bubble "lifetime"	480 μs	650 μs	[56]
Detection limits			
Ca (aq)	41.7 p.p.b.	0.6 p.p.m.	[37]
		0.09 p.p.m.	[77]
		0.13 p.p.m.	[78]
Cr (aq)	1.04 p.p.m.	0.1 p.p.m.	[80]
		20 p.p.m.	[79]
		200 p.p.m.	[37]
Zn (aq)	17 p.p.m.	1 p.p.m.[a]	[76]

[a] Zn solution deposited on carbon planchet and evoporated prior to LIBS analysis.

pre-ablation spark method provides up to 120-fold enhancements for solids with enhanced sample ablation. The mechanism of enhancement is not completely understood, but might involve a combination of effects including increased plasma volume and temperature, increased laser/sample interactions, formation of a pre-ionized region above the sample, and direct surface effects such as heating. Orthogonal dual-pulse LIBS is also useful for aqueous solution measurements providing significantly lower detection limits than single-pulse measurements.

We thank the US Department of Energy for support of this work under grant number DEFG0796ER62305. We also express thanks to the Office of Naval Research for support of this work under grant number N0014-97-1-0806. Travel support was provided by the Office of Research at The University of South Carolina. The authors also gratefully acknowledge Dr. Chance Carter and Dr. Bill Colston of the Medical Technology Program, and Dr. Dwight Price of the Ultrashort Pulse Laser Facility at Lawrence Livermore National Laboratory, for providing laser facilities and technical assistance in support of the femtosecond dual-pulse LIBS experiments.

15.4 References

[1] F. Brech and L. Cross, Optical microemission stimulated by a ruby MASER. *Appl. Spectrosc.*, **16** (1962), 59.

[2] R. Noll, H. Bette, A. Brysch *et al.*, Laser-induced breakdown spectrometry – applications for production control and quality assurance in the steel industry. *Spectrochim. Acta B*, **56** (2001), 637.

[3] J. Gruber, J. Heitz, H. Strasser, D. Bäuerle and N. Ramaseder, Rapid in-situ analysis of liquid steel by laser-induced breakdown spectroscopy. *Spectrochim. Acta B*, **56** (2001), 685.

[4] L. Barrette and S. Turmel, On-line iron-ore slurry monitoring for real-time process control of pellet making processes using laser-induced breakdown spectroscopy: graphitic vs. total carbon detection. *Spectrochim. Acta B*, **56** (2001), 715.

[5] R. Nyga and W. Neu, Double-pulse technique for optical-emission spectroscopy of ablation plasmas of samples in liquids. *Opt. Lett.*, **18** (1993), 747.

[6] A. I. Whitehouse, J. Young, I. M. Botheroyd *et al.*, Remote material analysis of nuclear power station steam generator tubes by laser-induced breakdown spectroscopy. *Spectrochim. Acta B*, **56** (2001), 821.

[7] O. Samek, D. C. S. Beddows, J. Kaiser *et al.*, Application of laser-induced breakdown spectroscopy to in situ analysis of liquid samples. *Opt. Eng.*, **56** (2000), 2248.

[8] G. A. Theriault and S. H. Lieberman, Field deployment of a LIBS probe for rapid delineation of metals in soils. *SPIE- Int. Soc. Opt. Eng.*, **2835** (1996), 83.

[9] G. A. Theriault, S. Bodensteiner and S. H. Lieberman, A real-time fiber-optic LIBS probe for the in situ delineation of metals in soils. *Field Anal. Chem. Technol.*, **2** (1998), 117.

[10] B. J. Marquardt, S. R. Goode and S. M. Angel, In situ determination of lead in paint by laser-induced breakdown spectroscopy using a fiber-optic probe. *Anal. Chem.*, **68** (1996), 977.

[11] B. J. Marquardt, B. M. Cullum, T. J. Shaw and S. M. Angel, Fiber optic probe for determining heavy metals in solids based on laser-induced plasmas. *SPIE- Int. Soc. Opt. Eng.*, **3105** (1997), 203.

[12] B. J. Marquardt, D. N. Stratis, D. A. Cremers and S. M. Angel, Novel probe for laser-induced breakdown spectroscopy and Raman measurements using an imaging optical fiber. *Appl. Spectrosc.*, **52** (1998), 1148.

[13] C. M. Davies, H. H. Telle, D. J. Montgomery and R. E. Corbett, Quantitative analysis using remote laser-induced breakdown spectroscopy (LIBS). *Spectrochim. Acta B*, **50** (1995), 1059.

[14] C. M. Davies, H. H. Telle and A. W. Williams, Remote in situ analytical spectroscopy and its applications in the nuclear industry. *Fres. J. Anal. Chem.*, **355** (1996), 895.

[15] R. E. Neuhauser, U. Panne and R. Niessner, Laser-induced plasma spectroscopy (LIPS): a versatile tool for monitoring heavy metal aerosols. *Anal. Chim. Acta*, **392** (1999), 47.

[16] S. Palanco and J. J. Laserna, Full automation of a laser-induced breakdown spectrometer for quality assessment in the steel industry with sample handling, surface preparation and quantitative analysis capabilities. *J. Anal. Atom. Spectrom.*, **15** (2000), 1321.

[17] D. Anglos, S. Couris and C. Fotakis, Laser diagnostics of painted artworks: laser-induced breakdown spectroscopy in pigment identification. *Appl. Spectrosc.*, **51** (1997), 1025.

[18] D. Anglos, C. Balas and C. Fotakis, Laser spectroscopic and optical imaging techniques in chemical and structural diagnostics of painted artwork. *Amer. Lab.*, **31** (1999), 60.

[19] D. Anglos, C. Balas and C. Fotakis, Laser spectroscopic and optical imaging techniques in chemical and structural diagnostics of painted artwork. *Amer. Lab.*, **31** (1999), 62.

[20] D. Anglos, Laser-induced breakdown spectroscopy in art and archaeology. *Appl. Spectrosc.*, **55** (2001), 186A.

[21] K. Melessanaki, M. Mateo, S. C. Ferrence, P. P. Betancourt and D. Anglos, The application of LIBS for the analysis of archaeological ceramic and metal artifacts. *Appl. Surf. Sci.*, **197–198** (2002), 156.

[22] A. K. Knight, N. L. Scherbarth, D. A. Cremers and M. J. Ferris, Characterization of laser-induced breakdown spectroscopy (LIBS) for application to space exploration. *Appl. Spectrosc.*, **54** (2000), 331.

[23] M. Tran, Q. Sun, B. Smith and J. D. Winefordner, Direct determination of trace elements in terephthalic acid by laser induced breakdown spectroscopy. *Anal. Chim. Acta*, **419** (2000), 153.

[24] P. Fichet, P. Mauchien, J. F. Wagner and C. Moulin, Quantitative elemental determination in water and oil by laser induced breakdown spectroscopy. *Anal. Chim. Acta*, **429** (2001), 269.

[25] R. Barbini, F. Colao, R. Fantoni, A. Palucci and F. Capitelli, Application of laser-induced breakdown spectroscopy to the analysis of metals in soils. *Appl. Phys. A*, **69** (Suppl.) (1999), S175.

[26] V. Lazic, R. Barbini, F. Colao, R. Fantoni and A. Palucci, Self-absorption model in quantitative laser induced breakdown spectroscopy measurements on soils and sediments. *Spectrochim. Acta B*, **56** (2001), 807.

[27] R. T. Wainner, R. S. Harmon, A. W. Miziolek, K. L. McNesby and P. D. French, Analysis of environmental lead contamination: comparison of LIBS field and laboratory instruments. *Spectrochim. Acta B*, **56** (2001), 777.

[28] J. O. Cáceres, J. Tornero López, H. H. Telle and A. González Ureña, Quantitative analysis of trace metal ions in ice using laser-induced breakdown spectroscopy. *Spectrochim. Acta B*, **56** (2001), 831.

[29] M. Tran, S. Sun, B. W. Smith and J. D. Winefordner, Determination of C:H:O:N ratios in solid organic compounds by laser-induced plasma spectroscopy. *J. Anal. Atom. Spectrom.*, **16** (2001), 628.

[30] Q. Sun, M. Tran, B. W. Smith and J. D. Winefordner, Determination of Mn and Si in iron ore by laser-induced plasma spectroscopy. *Anal. Chim. Acta*, **413** (2000), 187.

[31] C. Aragón, J. A. Aguilera and F. Peñalba, Improvements in quantitative analysis of steel composition by laser-induced breakdown spectroscopy at atmospheric pressure using an infrared Nd:YAG laser. *Appl. Spectrosc.*, **53** (1999), 1259.

[32] L. M. Cabalín and J. J. Laserna, Surface stoichiometry of manganin coatings prepared by pulsed laser deposition as described by laser-induced breakdown spectrometry. *Anal. Chem.*, **73** (2001), 1120.

[33] P. Lucena and J. J. Laserna, Three-dimensional distribution analysis of platinum, palladium and rhodium in auto catalytic converters using imaging-mode laser-induced breakdown spectrometry. *Spectrochim. Acta B*, **56** (2001), 177.

[34] J. Amador-Hernández, J. M. Fernández-Romero and M. D. Luque de Castro, Three-dimensional analysis of screen-printed electrodes by laser induced breakdown spectrometry and pattern recognition. *Anal. Chim. Acta*, **435** (2001), 227.

[35] A. De Giacomo, V. A. Shakhatov and O. De Pascale, Optical emission spectroscopy and modeling of plasma produced by laser ablation of titanium oxides. *Spectrochim. Acta B*, **56** (2001), 753.

[36] V. Detalle, R. Héon, M. Sabsabi and L. St.-Onge, An evaluation of a commercial Echelle spectrometer with intensified charge-coupled device detector for materials analysis by laser-induced plasma spectroscopy. *Spectrochim. Acta B*, **56** (2001), 1011.

[37] Y. Yoon, T. Kim, M. Yang, K. Lee and G. Lee, Quantitative analysis of pottery glaze by laser induced breakdown spectroscopy. *Microchem. J.*, **68** (2001), 251.

[38] L. Burgio, R. J. H. Clark, T. Stratoudaki, M. Doulgeridis and D. Anglos, Pigment identification in painted artworks: a dual analytical approach employing laser-induced breakdown spectroscopy and Raman microscopy. *Appl. Spectrosc.*, **54** (2000), 463.

[39] M. Castillejo, M. Martin, D. Silva, T. Stratoudaki, D. Anglos, L. Burgio and R. J. H. Clark, Analysis of pigments in polychromes by use of laser induced breakdown spectroscopy and Raman microscopy. *J. Moloc. Struct.*, **550** (2000), 191.

[40] V. Tornari, V. Zafiropulos, A. Bonarou, N. A. Vainos and C. Fotakis, Modern technology in artwork conservation: a laser-based approach for process control and evaluation. *Opt. Las. Eng.*, **34** (2000), 309.

[41] L. Burgio, K. Melessanaki, M. Doulgeridis, R. J. H. Clark and D. Anglos, Pigment identification in paintings employing laser induced breakdown spectroscopy and Raman microscopy. *Spectrochim. Acta*, **56** (2001), 905.

[42] M. Bicchieri, M. Nardone, P. A. Russo *et al.*, Characterization of azurite and lazurite based pigments by laser induced breakdown spectroscopy and micro-Raman spectroscopy. *Spectrochim. Acta*, **56** (2001), 915.

[43] O. Samek, D. C. S. Beddows, H. H. Telle *et al.*, Quantitative laser-induced breakdown spectroscopy analysis of calcified tissue samples. *Spectrochim. Acta*, **56** (2001), 865.

[44] O. Samek, D. C. S. Beddows, H. H. Telle *et al.*, Quantitative analysis of trace metal accumulation in teeth using laser-induced breakdown spectroscopy. *Appl. Phys. A*, **69** (Suppl.) (1999), S179.

[45] O. Samek, M. Liska, J. Kaiser *et al.*, Clinical application of laser-induced breakdown spectroscopy to the analysis of teeth and dental materials. *J. Clin. Las. Med. Surg.*, **18** (2000), 281.

[46] C. C. Garcia, J. M. Vadillo, S. Palanco, J. Ruiz and J. J. Laserna, Comparative analysis of layered materials using laser-induced plasma spectrometry and laser-ionization time-of-flight mass spectrometry. *Spectrochim. Acta B*, **56** (2001), 923.

[47] F. Hilbk-Kortenbruck, R. Noll, P. Wintjens, H. Falk and C. Becker, Analysis of heavy metals in soils using laser-induced breakdown spectrometry combined with laser-induced fluorescence. *Spectrochim. Acta B*, **56** (2001), 933.

[48] H. H. Telle, D. C. S. Beddows, G. W. Morris and O. Samek, Sensitive and selective spectrochemical analysis of metallic samples: the combination of laser-induced breakdown spectroscopy and laser-induced fluorescence spectroscopy. *Spectrochim. Acta B*, **56** (2001), 947.

[49] L. J. Radziemski and D. A. Cremers (editors), *Laser-Induced Plasmas and Applications* (New York: Marcel Dekker, 1989).

[50] G. Colonna, A. Casavola and M. Capitelli, Modelling of LIBS plasma expansion. *Spectrochim. Acta B*, **56** (2001), 569.

[51] A. Ciucci, S. Palleschi, S. Rastelli *et al.*, CF-LIPS: a new approach to LIPS spectra analysis. *Las. Part. Beams*, **17** (1999), 793.

[52] A. Ciucci, M. Corsi, S. Palleschi *et al.*, New procedure for quantitative elemental analysis by laser-induced plasma spectroscopy. *Appl. Spectrosc.*, **53** (1999), 960.

[53] I. B. Gornushkin, A. Ruíz-Medina, J. M. Anzano, B. W. Smith and J. D. Winefordner, Identification of particulate materials by correlation analysis using a microscopic laser induced breakdown spectrometer. *J. Anal. Atom. Spectrom.*, **15** (2000), 581.

[54] G. Galbacs, I. B. Gornushkin, B. W. Smith and J. D. Winefordner, Semi-quantitative analysis of binary alloys using laser-induced breakdown spectroscopy and a new calibration approach based on linear correlation. *Spectrochim. Acta B*, **56** (2001), 1159.

[55] L. St.-Onge, V. Detalle and M. Sabsabi, Enhanced laser-induced breakdown spectroscopy using the combination of fourth-harmonic and fundamental Nd:YAG laser pulses. *Spectrochim. Acta B*, **57** (2002), 121.

[56] D. N. Stratis, K. L. Eland and S. M. Angel, Enhancement of aluminum, titanium, and iron in glass using pre-ablation spark dual-pulse LIBS. *Appl. Spectrosc.*, **54** (2000), 1719.

[57] S. M. Angel, D. N. Stratis, K. L. Eland *et al.*, LIBS using dual- and ultra-short laser pulses. *Fres. J. Anal. Chem.*, **369** (2001), 320.

[58] D. N. Stratis, K. L. Eland and S. M. Angel, Effect of pulse delay time on a pre-ablation dual-pulse LIBS plasma. *Appl. Spectrosc.*, **55** (2001), 1297.

[59] R. Sattmann, V. Sturm and R. Noll, Laser-induced breakdown spectroscopy of steel samples using multiple Q-switch Nd-YAG laser-pulses. *J. Phys. D: Appl. Phys.*, **28** (1995), 2181.

[60] S. Nakamura, Y. Ito, K. Sone, H. Hiraga and K. Kaneko, Determination of an iron suspension in water by laser-induced breakdown spectroscopy with two sequential laser pulses. *Anal. Chem.*, **68** (1996), 2981.

[61] A. E. Pichahchy, D. A. Cremers and M. J. Ferris, Elemental analysis of metals under water using laser-induced breakdown spectroscopy. *Spectrochim. Acta B*, **52** (1997), 25.

[62] D. N. Stratis, K. L. Eland and S. M. Angel, Dual-pulse LIBS: why are two lasers better than one? *SPIE-Int. Soc. Opt. Eng.*, **3853** (1999), 385.

[63] D. N. Stratis, K. L. Eland and S. M. Angel, Dual-pulse LIBS using a pre-ablation spark for enhanced ablation and emission. *Appl. Spectrosc.*, **54** (2000), 1270.

[64] D. A. Cremers, L. J. Radziemski and T. R. Loree, Spectrochemical analysis of liquids using the laser spark. *Appl. Spectrosc.*, **38** (1984), 721.

[65] L. St.-Onge, M. Sabsabi and P. Cielo, Analysis of solids using laser-induced plasma spectroscopy in double-pulse mode. *Spectrochim. Acta B*, **53** (1998), 407.

[66] D. N. Stratis, K. L. Eland and S. M. Angel, Characterization of laser-induced plasmas for fiber optic probes. *SPIE-Int. Soc. Opt. Eng.*, **3534** (1999), 592.

[67] V. Sturm, L. Peter and R. Noll, Steel analysis with laser-induced breakdown spectrometry in the vacuum ultraviolet. *Appl. Spectrosc.*, **54** (2000), 1275.

[68] J. Uebbing, J. Brust, W. Sdorra, F. Leis and K. Niemax, Reheating of a laser-produced plasma by a 2nd pulse laser. *Appl. Spectrosc.*, **45** (1991), 1419.

[69] F. Colao, V. Lazic, R. Fantoni and S. Pershin, A comparison of single and double pulse laser-induced breakdown spectroscopy of aluminum samples. *Spectrochim. Acta B*, **57** (2002), 1167.

[70] K. L. Eland, D. N. Stratis, D. M. Gold, S. R. Goode and S. M. Angel, Energy dependence of emission intensity and temperature in a LIBS plasma using femtosecond excitation. *Appl. Spectrosc.*, **55** (2001), 286.

[71] K. J. Grant and G. L. Paul, Electron temperature and density profiles of excimer laser-induced plasmas. *Appl. Spectrosc.*, **44** (1990), 1349.

[72] E. E. B. Campbell, D. Ashkenas and A. Rosenfeld, Ultra-short-pulse laser irradiation and ablation of dielectrics. *Mater. Sci. Forum, Las. Mater. Sci.*, **301** (1999), 123.

[73] R. Sattmann, V. Sturm and R. Noll, Laser-induced breakdown spectroscopy of steel samples using multiple Q-switch ND-YAG laser-pulses. *J. Phys. D.*, **28** (1995), 2181–2187.

[74] R. E. Russo, X. Mao and S. S. Mao, The physics of laser ablation in microchemical analysis. *Anal. Chem.*, **74** (2002), 71A.

[75] P. K. Kennedy, D. X. Hammer and B. A. Rockwell, Laser-induced breakdown in aqueous media. *Prog. Quant. Electr.*, **21** (1997), 155.

[76] R. L. Vander Wal, T. M. Ticich, J. R. West, Jr. and P. A. Householder, Trace metal detection by laser-induced breakdown spectroscopy. *Appl. Spectrosc.*, **53** (1999), 1226.

[77] T. Bundschuh, J. I. Yun and R. Knopp, Determination of size, concentration, and elemental composition of colloids with laser-induced breakdown detection/spectroscopy (LIBD/S). *Fres. J. Anal. Chem.*, **371** (2001), 1063.

[78] R. Knopp, F. J. Scherbaum and J. I. Kim, Laser induced breakdown spectroscopy (LIBS) as an analytical tool for the detection of metal ions in aqueous solutions. *Fres. J. Anal. Chem.*, **355** (1996), 16.

[79] P. Fichet, A. Toussaint and J. F. Wagner, Laser-induced breakdown spectroscopy: a tool for analysis of different types of liquids. *Appl. Phys. A*, **69**(Suppl.) (1999), S591.

[80] G. Arca, A. Ciucci, V. Palleschi, S. Rastelli and E. Tognoni, Trace metal analysis in water by the laser-induced breakdown spectroscopy technique. *Appl. Spectrosc.*, **51** (1997), 1102.

16

Micro LIBS technique

Pascal Fichet, Jean-Luc Lacour, Denis Menut, Patrick Mauchien,
Annie Rivoallan

CEA Saclay, France Cécile Fabre, Jean Dubessy, Marie-Christine

Boiron

Equipes Interactions entre Fluides et Minéraux, Université Henri Poincaré, France

16.1 Introduction

The LIBS technique (laser-induced breakdown spectroscopy) has been applied mainly for the bulk analysis of solids [1], liquids [2], or gases [3], but more sparsely for elemental microanalysis of solid surfaces. In this chapter we describe different results obtained with a micro LIBS device devoted to element distribution analysis on solid surfaces and to localized analysis. The crater diameter and its shape are two crucial parameters that have to be well controlled to obtain reliable results. After a description of different published results concerning micro plasmas and recent applications of surface analysis, a complete description of a laboratory micro LIBS device is reported. The smallest crater diameter achieved with the experimental set-up and that can be used for analytical purposes is 3 μm. An original device offering an attractive feature to obtain regular spaced craters is also presented. The characteristics of the system in terms of quantitative analysis are highlighted. Different element distributions on surfaces of ceramics and steel samples are shown to demonstrate the very high potential of micro LIBS for elemental microanalysis. Finally, the micro LIBS technique is presented as a powerful analytical method for geological samples.

The use of a microscope combined with a laser has been reviewed previously [4]. This present chapter provides information on technical details of manufactured microanalyzers combined with a spark-gap device, positioned above the sample surface, to make localized analysis. With the different manufactured systems, crater diameters from laser ablation could vary from 10 μm to 1 mm. Various experimental set-ups devoted to scanning the sample surfaces at different locations with a laser microanalysis system have been described. A special device provided an application of quantitative element distribution on the surface, where an analytical specimen was moved with one system spirally on a special stage under a microscope objective. Up to 100 steps could be recorded for one spiral and the length of the steps varied from 100 μm to 250 μm in four stages. The distance between the coils of the spiral was 250 μm. Another mechanical stage was tested to study different craters on the surface of the material. A rectangular area was scanned and the optimum size of the crater was between 60 μm and 100 μm. Moenke-Blankenburg [4] mentioned also a lot of fields of applications and the fact that the lack of sample preparation for a laser microanalyzer was found to be a great advantage compared with standard techniques for surface analysis. Some commercial devices were also presented such as the LMA 10 (Carl Zeiss), the Mark III (Jarrel Ash), and the JLM 200 (Jeol Ltd). In spite of the technical advances they represented, they never gained wide acceptance compared with standard techniques for microanalysis because of the difficulties of obtaining quantitative measurements with good repeatability and reliability. Moreover, another paper [5] describing results obtained with the laser microprobe system LMA 10 illustrated the ability of such a system to analyze quantitatively a lateral profile across a blemish (0.2 mm diameter) on a ceramic tube.

After these first set-ups, the aim of developing a micro LIBS system has been to use the laser not only as a source of solid vaporization but also as a source of excitation. Therefore, the surface composition has been characterized directly by plasma emission recording.

The LIBS technique devoted to element distributions on solid surfaces and not to bulk analysis has only recently been published. An optical set-up [6] was developed in order to produce small craters, which are of great interest in this paper. In this application, the aim of micro LIBS analysis [6] was to inspect the way the coating is distributed on a coated paper. In the experimental set-up the laser is focused on the sample surface by a 3.5 cm focal-length lens to obtain craters about 30 μm in diameter and 2 μm in depth. A computer controlled XY-translation stage is used to move the sample with respect to the laser beam. The possibility of performing cartography with a frequency of 20 Hz is described in detail. To perform the analysis, the authors decided to accumulate, in PMT detectors, the emission coming from eight contiguous craters. The accumulated signal along this profile was recorded continuously to investigate quantitatively macroscopic surface up to 50×50 mm^2. Each experiment was replicated many times (up to eight) on the same location to give a three-dimensional distribution. PMT detectors were used to improve the frequency, but only single element information could be obtained for a given surface, which was a drawback. Moreover, the plasma emissions were not recorded separately, which was a serious problem in investigating local heterogeneity ($\approx$10 μm).

A LIBS system for element distribution mapping, which is an important application of micro LIBS, was also proposed by Wiggenhauser et al. [7] and Romero et al. [8] but the crater diameters were, respectively, 1 mm and 125 μm, which are far larger than those obtained in the micro LIBS system (typically less than 40 μm).

Another micro LIBS arrangement for solid surface analysis has been reported [9–12]. An exact synchronization of the sample positioning with the laser pulse was tested to acquire element distribution on the surface. A minimum lateral resolution of up to 30 μm has been achieved. Increments of the lateral movement could be automatically modified but the system was optimized to avoid overlapping between adjacent craters. The system has been devoted to silicon, titanium, and carbon analysis on photovoltaic cell surfaces. The LIBS device is based on the use of a pulsed nitrogen laser ($\lambda = 337.1$ nm), and an iris diaphragm allows a portion of the laser beam to be isolated. Several diaphragm sizes were tested to reduce the crater size in order to improve the lateral resolution. For diaphragm apertures below 3 mm, no plasma emission was observed. In the experimental set-up described, the laser is focused on the sample by a planoconvex lens with $f = 45$ mm and an f-number of 1.5. The lens–target distance remains constant during all experiments at 40 mm from the lens and therefore above the focal point. No microscope device is described in this experimental set-up. The laser beam was precisely focused, inducing a reproducibility of the spectral intensity of around 5%. A three-dimensional analysis was also proposed, taking into account that five replicated laser pulses were applied on each crater. The capabilities of LIBS for depth profiling were discussed and evaluated. For the authors, micro LIBS was found to be interesting to fill the gap between standard surface analysis techniques (SIMS, SEM, XPS) and bulk analysis techniques. Considering the different advantages of the micro LIBS technique (almost no sample preparation, relatively inexpensive equipment) and the major interest of microanalysis in industrial applications, it can be said that a tremendous effort to develop such a LIBS set-up must be realized.

Even if depth profile studies using the LIBS technique are very interesting and promising [13], the crater depths obtained with different experimental set-ups are obviously dependent on the matrix. Thus, this is a drawback for this particular application. Nevertheless, some investigations on coating thickness for Zn/Ni and Sn on steel sample have already been published [14]. Moreover, a three-dimensional analysis by micro LIBS is difficult to achieve because a tightly focused laser beam is absolutely necessary to insure good results in terms of repeatability. However, some authors [15, 16] recently published very interesting results of two- and three-dimensional element mapping performed by LIBS in automobile converters; but the experiments were made with crater diameter of 1.75 mm, which is far larger than those researched for micro LIBS experiments. Obviously the precision of the laser focusing system with such crater diameter is not so tight.

Another development of a micro LIBS set-up was recently described by Yoon *et al.* and applied to the analysis of element distribution over a polished rock surface [17]. The experimental arrangement used a Nd:YAG laser emitting at 532 nm. The plasma was formed on the solid surface with a tilted pathway (60° relative to the normal incidence) using a planoconvex 3 cm focal-length lens. The crater sizes formed were about 10 μm in diameter. The analyzed sample was positioned on an XY-translation stage 10 mm above the focal point. Each measurement of the element distribution on the surface was performed by an average of three shots on the same location.

Gornushkin *et al.* [18] developed a compact LIBS system with microscopic sample imaging, aiming at providing an instant, reliable classification of different groups of solid materials (stainless steel and cast iron standards). In their experimental set-up, a very compact laser was attached to a modified microscope. A dichroic mirror was placed inside the microscope to guide the 1064 nm laser beam towards a 10× objective (magnification 500×). This objective was the only one used because it allowed a working distance of 5 mm between the objective and the sample surface. Other objectives had a working distance that was too short and that could induce damage on the objective tip. Their device has produced craters of 20 μm diameter and 30 μm depth. No cartography was produced because it was not the topic of this study, but a moving stage in only one axis was mounted on the microscope to allow sample moving. For this application, a very small system, a minispectrometer from Ocean

Optics, was chosen as a detector. Even if the temporal resolution of such a spectrometer is nearly impossible, results obtained seem to be very interesting and the detector price is far lower compared with standard ICCD.

Kossakovski and Beauchamp [19] also proposed an Ocean Optics minispectrometer as a detector for a micro LIBS development for the TOPOLIBS experiment. In this experimental set-up, an optical fiber that guides a pulse nitrogen laser is placed very close to the sample surface. LIBS experiments were performed to analyze the surface of a meteorite and of basalt samples. To obtain a spectrum the maximum laser power that could be coupled into the fiber was delivered to the sample. The typical crater sizes were around tens of micrometers. However, until then without any analytical signal, submicrometer resolution was reached with such system. Ablation crater sizes as small as 200 nm diameter have been reported with a 248 nm excimer laser coupled with a fiber probe [20]. Kossakovski and Beauchamp [19] have demonstrated the high potentiality of TOPOLIBS but the probe contamination by species ejections induced by plasma formation seems to still remain a problem for the analysis of different surface locations.

The very high potential of the micro LIBS experiment developed in the CEA Saclay laboratory has been described in a European project [21] and also published by Geertsen *et al.* [22]. The full description of the experimental set-up will be presented later in the chapter. But major parts of this first experimental set-up have been modified since these first papers [21, 22]. The lateral resolution obtained was close to 6 μm but taking into account some recent changes this characteristic has been improved to 3 μm.

Briefly, the micro LIBS device is composed of a quadrupled Nd:YAG laser used in connection with a home-made ablation head based on a modified microscope. The analytical performances were assessed for aluminum alloys, steel samples, ceramics, and minerals [21]. The use of an ultraviolet (UV) laser is, for a large part, responsible for the improvements in terms of repeatability and ablation efficiency [23]. Moreover, the high numerical aperture of the microscope objective has allowed major progress in the ablation process.

The development of a micro LIBS system has also been dedicated to the surface analysis of some nuclear fuels (Section 16.3.1.) The laser probe is a very powerful technique to study radioactive or more generally dangerous samples and it has been developed since the discovery of the laser [24]. The smallest craters obtained by Adams and Tong on a UO_2 surface had approximately 5 μm diameter and the largest had 200 μm diameter. But, in this study, a laser micro probe was used to volatilize a chosen area and not to make any analytical measurement by LIBS. Vaporized material was then condensed on a transparent cover plate held less than 1 mm above the specimen surface. The laser microprobe arrangement used a modified microscope.

The potentialities of LIBS experiment for Pu sample [1, 25] analysis have been studied. LIBS has also been used to measure uranium line emissions [1, 25–29], and is sometimes preferable to radiological measurements because nuclear detectors may not be able to differentiate between U, Pu and Np. To this end, Dr. Andrew Whitehouse established Applied Photonetics Ltd in 1998 [30] with the aim of providing a remote analysis service by LIBS to the nuclear industry. So LIBS research for nuclear applications continues because of the main advantages of the technique such as small sample preparation and the possibility of multielemental analysis.

Several papers have described the LIBS experimental set-up for the bulk analysis of steel samples [31, 32]. For the present chapter, chemical analyses of some inclusions at the surface of steel samples were studied and are presented in detail (Section 16.3.2.).

Researchers in geology also initiated a great interest in the plasma analysis coming from very small craters. Accurate analysis of small fluid inclusions is a prerequisite for reconstructing the paleo-chemical equilibria between the fluid phase and the rock-forming minerals they originate from [33]. During the European project, Fabre *et al.* [21] also built a micro LIBS system developed in partnership with CEA Saclay. Since the first experiments [33], many other localized analyses (crater diameter of about 6 μm) have been made by Fabre *et al.* [34–36] and are described in Sections 16.3.3 and 16.3.4.

16.2 Experimental set-up for the micro LIBS system

In this section, an original design of a micro LIBS system under patent [37] will be described. This system has been studied and improved for many years in the CEA laboratory. Either mapping the distribution of different elements on the surface of any type of sample or just performing localized analysis, at air pressure, without complex preparation, is the driving force for the development of a micro LIBS system. In the case of microanalysis, the most important features are the lateral resolution of the measurements and the analytical performance in terms of sensitivity and accuracy. Moreover the sample displacements with regard to the fixed position of the laser microprobe must be well controlled and as rapid as possible to obtain very fast analytical response.

16.2.1 Laser

A schematic representation of the experimental set-up is shown in Figure 16.1. As previously mentioned [22], a Nd:YAG Minilite-20 laser (Continuum, USA) was chosen operating at 20 Hz with an output energy of 25 mJ per pulse in the fundamental (6–8 ns duration, FWHM) and up to 3.0 mJ in the quadrupled harmonic 266 nm. This small laser has been used because it delivers enough energy for the micro LIBS process and because the fourth harmonic crystal is able very quickly to reach its optimum temperature. And thus the laser energy remains almost constant during the experiment duration (pulse to pulse fluctuations of about 5%). During the European project [21] different wavelengths of the Nd:YAG laser were tested on the same sample and with the same experimental conditions. The plasma images obtained with a pulsed CCD camera with a constant delay after the laser pulse show undoubtedly that plasmas obtained with the infrared (IR) beam show important changes (four images at the bottom of Figure 16.2) but plasmas produced by UV always remain stable shot by shot (two images at the top of Figure 16.2). Thus, this result has induced the choice of a UV laser in the micro LIBS set-up to improve the plasma emission stability.

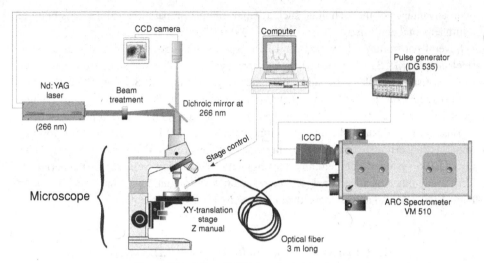

Figure 16.1. Micro LIBS experimental set-up.

The laser beam is first guided up to an optical system specially designed for the purpose (Figure 16.1). A diaphragm is used to isolate a portion of the laser. Different apertures were tested and can be used in order to obtain different crater sizes. Energy of 7 µJ can be measured after the smallest diaphragm and therefore a very small but sufficient fraction of the laser energy is effectively used for ablation process.

16.2.2 Microscope

At the top of a modified microscope (Olympus BX-40, Japan) a dichroic mirror (100% reflectance at 266 nm and at 45°) reflects the laser on a 15× refractory microscope objective (OFR, Santa Clara, USA) having a working distance of 8 mm. The numerical aperture of the objective is 0.32. Images of plasmas induced by the laser were also recorded with different numerical apertures. It was proved that using large numerical aperture induces a better stability of the plasma emission.

With the objective chosen, the working distance is sufficient to avoid any damage to the optical part from particulates ejected from the plasma. For the different applications, the same objective is always maintained but the diaphragm is sometimes changed in order to obtain crater diameters from 3 µm up to 10 µm. The corresponding diaphragm diameter is, respectively, 80 µm and 400 µm. The objective chosen was optimized for the ablation process. A CCD camera is placed above the microscope in order to control and adjust the sample position on the translation stage. The CCD is designed to cover a visible image of the sample surface of about 200 µm × 200 µm.

Samples are positioned and fixed with screws on a powerful XY-driver (ETEL DSA2, Switzerland) that can move very quickly and precisely. Taking into account all the different parameters, the driver is used at 20 Hz with a movement of 3 µm (repeatability <10%)

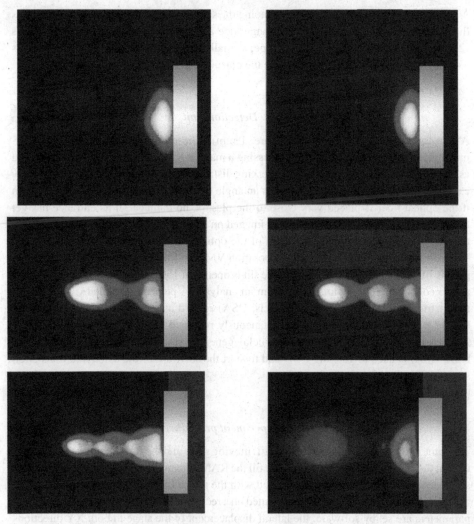

Figure 16.2. Images of plasmas obtained on the same sample with a Nd: YAG emitting at 266 nm (two images at the top) and at 1064 nm (four images at the bottom).

between consecutive laser shots. After the experiment starting from the detector described below, the ETEL translation stage automatically governs the laser shooting.

Movements in the XY-plane are automatically performed with servo drivers, and the Z-plane motion remains manual. The Z-axis is used for laser focusing on the sample surface. A He–Ne laser (although this is not represented in Figure 16.1) is used as a pointer to obtain an exact and reproducible focusing point. This system allows very precise positioning, evaluated at about 1 μm of repeatability.

It was experimentally proved that sweeping the sample surface with an Ar flow increases the plasma emission by two orders of magnitude and reduces the self-absorption process [21]. So, on a fixed part of the microscope, a small tube is locked to add the Ar buffer gas (3 l min^{-1} Ar flow) near the point where the plasma is generated.

16.2.3   Detection unit

A 3 m long optical fiber (high OH$^-$ core, 1 mm), whose entrance, fixed also on a non-moving part of the microscope and possessing a manual XYZ adjustment system, is placed as close as possible to the plasma. The working distance of the objective induces a relatively easy adjustment. The entrance is tilted at an angle of about 45° to the objective axis. Even if the optical fiber is placed very close to the plasma, no damage on its entrance has yet been observed. Then, the plasma light is imaged on a slit of a spectrometer with two lenses respecting the numerical apertures both of the optical fiber and of the spectrometer. The spectrometer is an Acton Research Corporation VM510, 1 m focal length equipped with a grating of 2400 lines mm^{-1}. The entrance slit is opened at 150 μm to have a good resolution and to collect enough photons to perform an analysis. A pulsed intensified CCD (Princeton Instruments I-MAX, 256*1024 pixels, USA) is used as a detector with its controller (ST 133). The wavelength range simultaneously recorded is 9 nm and the resolution is $\lambda/\Delta\lambda = 10\,000$. A programmable pulse-delay generator (DG 535, Stanford Research Systems, USA) is used to gate the CCD and to start the software of the automatic translation stage (ETEL DSA2).

16.2.4   Experiment procedure

The same computer controls both the ETEL moving stage and the ICCD camera. Software is used at the beginning of an experiment to fill the RAM (memory) of the ETEL moving stage servo drivers. Each experiment performed with the micro LIBS system corresponds to the analysis of many craters regularly distributed on a rectangular area. So, at the beginning, two parameters are set by software: the lateral displacement of the stage in both XY directions and the number of points recorded for the mapping.

Then the CCD data acquisition software (Winspec/32 2.2.3.1, Princeton Instruments) is started. The number of spectra recorded in the data file exactly corresponds to the number of displacements of the translation stage.

Once the data acquisition software is started, the experiment continues automatically. The signal coming from the NOT SCAN pulse coming from the DG 535 pulser, driving the ICCD, starts the servo drivers. Then, the time procedure used to record the mapping can be described as follows. The servo drivers send to the XY-axis of the stage an electric pulse to move. Once the stage is fixed the servo drivers deliver another pulse to a DG 535, which commands the laser shot. Finally, the trigger output of the laser starts the ICCD exposure time. And all these signals start again up to the end of the cartography.

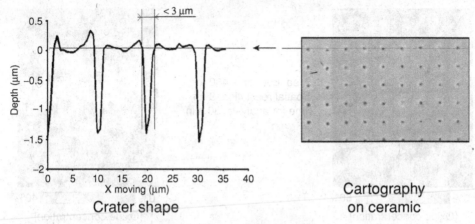

Figure 16.3. Crater shapes obtained with the micro LIBS system (left) and surface mapping with craters regularly separated by 10 μm (right).

16.3 Results and discussion

16.3.1 Results on ceramics

The micro LIBS device described in Section 16.2 has been first applied to observe very small aggregates on ceramics surfaces (up to 10 μm wide). With a spatial resolution of 3 μm combined with a well-controlled displacement of 3 μm described in the previous section, it is obvious that at least one entire crater can describe qualitatively and quantitatively the composition of a 10 μm wide aggregate. Smaller inclusions of impurities in a matrix can be observed but certainly the quantification is less accurate because the scanned surface does not represent the actual aggregate size.

With the micro LIBS experimental set-up described above, each crater formed and analyzed has been obtained with a single laser shot. The shape of a typical crater is depicted in Figure 16.3. On the left-hand part of the picture different craters obtained on a ceramic surface show a well-controlled 3 μm diameter and a depth of about 1.5 μm. Diameter measurements correspond to the top of the crater induced by a single laser shot. The crater sizes were estimated with a surface-mapping microscope (Phase Shift Technology Inc., USA). In this picture, each crater is separated by a 10 μm distance to be clearly depicted but when surface mapping is carried out the moving always remains at a value of 3 μm to obtain adjacent craters. A controlled morphology of the crater is a very important factor, which affects the spatial resolution of the micro LIBS system. On the right-hand side of Figure 16.3, the top of an elemental cartography is shown.

For the analysis, every plasma emission is recorded by the ICCD. The experimental set-up allows the surface analysis at a frequency of 20 Hz. As an example, it takes about half an hour to record the emission of 32 000 separated craters. The corresponding rectangular area is 160 μm × 200 μm with a lateral resolution of 3 μm.

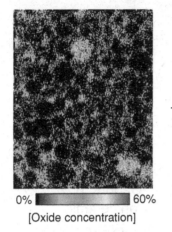

Area: 600 µm × 480 µm
Spatial resolution: 3 µm
Time for analysis: 30 min

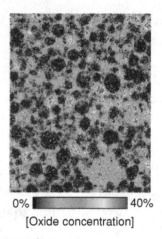

0% ▬▬▬▬▬ 60%   0% ▬▬▬▬▬ 40%

[Oxide concentration]   [Oxide concentration]

Figure 16.4. Element distribution in two ceramics surfaces.

Two different distributions of one oxide in another obtained on two different samples are shown in Figure 16.4. The different points in the surface mapping correspond to the intensity in arbitrary units of an emission line of an element composing one oxide, recorded by the ICCD camera. A home-made program in Visual Basic (Microsoft, USA) is used to extract automatically the interesting pixel values coming from the data acquisition software of the ICCD. No image smoothing was used in these cartographies and gray grading represents a quantitative oxide distribution on the ceramic surface. A good deal of information can be extracted from these different mappings in terms of aggregate repartitions, mean sizes and so on. For simplicity, no color was used in the cartography picture described in Figure 16.4, but of course colored grading can more easily allow small aggregate observation.

With the micro LIBS system, because sample preparation and evacuation chamber are not necessary and because the spectral record is instantly digitized and stored, very rapid surface mapping can be performed. Even if the micro LIBS technique does not need sample preparation (cutting, pulverizing), it is obvious that performing analyses with samples having an important roughness remains difficult. If a laser probe is designed to investigate a crater of 3 µm wide, the laser must be tightly focused (with an accuracy of about 1 µm). Thus, samples with relatively flat surfaces are preferred.

A comparison of a qualitative element distribution on a ceramic surface with our micro LIBS system and a standard technique SEM/WDS (scanning electron microprobe/wavelength dispersion spectroscopy) for surface sample analysis is illustrated in Figure 16.5. The resolution of the electron microprobe is around 1 µm, which is better than our system, but of the same order of magnitude. A lateral resolution of 10 µm was chosen for both systems to realize the comparison of Figure 16.5. When comparing these two results, one can easily conclude that micro LIBS is a very powerful system for obtaining elemental cartography on sample surface, because the same aggregate shapes can be clearly observed. Moreover, two main advantages of the micro LIBS may be noted: the possibility

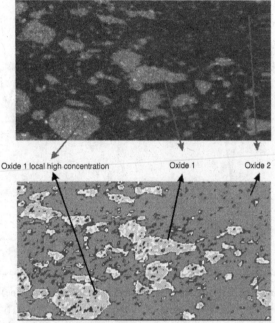

Figure 16.5. Comparisons of SEM/WDS and micro LIBS for ceramic surface analysis.

of analyzing very quickly all types of materials without any preparation except a polishing, and the fact that the experiment can be performed at atmospheric pressure. Note that a crater with 1 μm diameter corresponding to the exact spatial resolution of SEM/WDS can be obtained by using the micro LIBS device, but the emission signal is too low to allow reproducible analysis.

To highlight the potential of the micro LIBS system for quantitative analysis, a comparison of the results obtained with a SEM/WDS system and the micro LIBS system can be seen in Figure 16.6 for a straight line on the surface of a ceramic. The concentrations of an oxide (in mass %) along this line are very similar, which confirms the high potential of micro LIBS compared with the standard technique. However, LIBS remains a destructive analytical technique even if mass ablated was evaluated at about 1 ng per laser shot.

Moreover, to illustrate the quantitative results obtained, a calibration curve is shown in Figure 16.7 representing the ratio of the emission line intensity of oxide 1 and oxide 2 versus the concentration of the corresponding mass ratio. Each data point of the calibration curve corresponds to five replicate accumulations of line emissions of 100 different plasmas

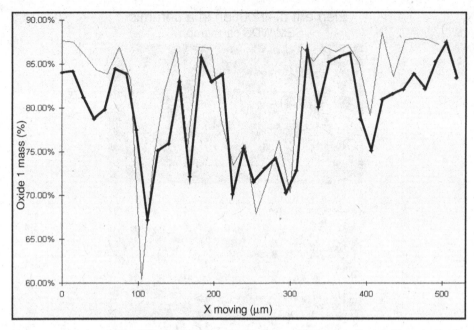

Figure 16.6. Comparisons of SEM/WDS (thin line) and micro LIBS (heavy line) for the quantitative measurement of oxide 1 in oxide 2 along a straight line.

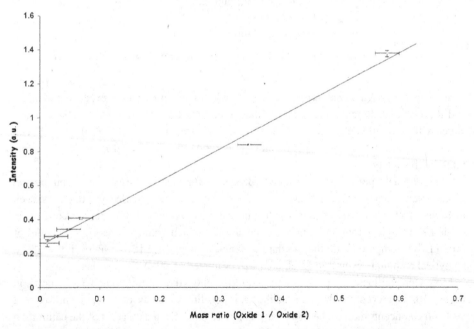

Figure 16.7. Calibration curve obtained with the micro LIBS set-up for oxide 1 in oxide 2.

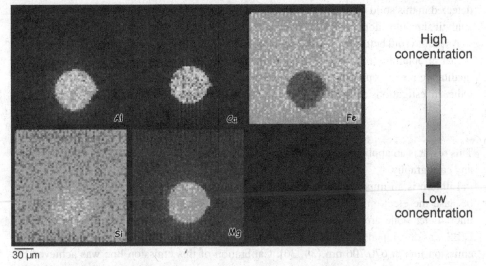

Lateral resolution : 3 μm

Figure 16.8. Observation of an inclusion containing Al, Ca, Si, Mg on a steel sample surface.

obtained on the surface of ceramic standards. The repeatability of the quantitative results is of the order of 10%. However, this repeatability is influenced by two factors: one is inherent in the technique and the other is caused by the reliability of the standards. More precisely, it is an actual problem to prepare standards that are homogeneous at the micrometer scale. To obtain quantitative results, the calibration curve is then applied for all single plasma emissions obtained on an unknown sample.

The three-dimensional experiments are difficult to perform with the micro LIBS system described above, because it assumes that for each layer examined the craters induced by laser have the same sizes. Until now, different tests have been performed on different matrices. Naturally, the laser was focused tightly for the recording of the different layers. In our opinion the first crater disturbs results coming from a second layer (second plasma inside the first crater). That induces an actual problem for three-dimensional mapping.

16.3.2 Results on steel

The analytical applications of the micro LIBS system have been mainly applied on ceramics and on steel samples. Micro LIBS results concerning studies of aluminum alloys are presented in more detail in the paper by Geertsen *et al.* [22].

The quality of steel can be influenced by the presence of some impurities inclusions. The micro LIBS system has been tested to investigate some solid inclusions present on steel sample surfaces and an example is shown in Figure 16.8. An inclusion of 50 μm diameter can be clearly seen. This inclusion was studied qualitatively. Ca lines were found to be very sensitive in this inclusion. Moreover, other elements such as Al, Si and Mg were also detected in this solid inclusion. With a spatial resolution of 3 μm, about 17 craters provided analytical results along a profile realized across this inclusion.

As pointed out before, one of the main problems in qualifying a microanalytical technique is that of obtaining reliable samples on the micrometer scale. Thus, for the moment, only qualitative results on steel samples have been investigated with this micro LIBS system. Other investigations are going on with this kind of sample.

16.3.3 Lithium analysis in mineral

This result is an application of the micro LIBS technique for localized analysis but without any cartography.

Lithium is an important geochemical tracer for geological fluids and solids. However, because the electron microprobe cannot detect Li, variations of its concentration at the micrometer scale are most often estimated from bulk analyses of other elements. Micro LIBS was developed to perfect the analysis of Li at the micrometer scale using the intense emission line at 670.706 nm [34, 36]. Calibration of this emission line was achieved by using synthetic glasses and natural minerals and the detection limit for Li was found to be about 5 p.p.m.

Then, lithium-rich minerals were analyzed with micro LIBS. The Li_2O concentrations in spodumene ($LiAlSi_2O_6$) and petalite ($LiAlSi_4O_{10}$) from granite pegmatite dikes (Portugal) are respectively 7.6 ± 1.6 wt.% and 6.3 ± 1.3 wt.%. These values agree well with ion microprobe and/or bulk analyses (7.65 ± 0.3 wt.% and 4.5 ± 0.2 wt.%, respectively) [35].

Lithium was also analyzed as a trace element in hydrothermal quartz veins from the Spanish Central System (Sierra de Guadarrama) (Figure 16.9). The highest Li concentrations (250–370 p.p.m) were found in specific growth bands in conjunction with the observed variation in optical cathodoluminescence intensity [35]. It is worth noting that lithium was analyzed in fluid inclusions (see Section 16.3.4) of this quartz sample (Na/Li between 3 and 14). The source of fluid responsible for the Li enrichment in quartz is probably high-salinity fluids derived from sedimentary basins.

16.3.4 Analysis of fluid inclusions present in natural or synthetic minerals

Fluid inclusions are intracrystalline cavities inside minerals that contain relicts of fluid or silicate melts that circulated during the growth of the mineral or later when it was fractured. As a crystal may record a long geological history, only the analysis of individual fluid inclusions is relevant in most of the geological samples. Their dimensions are typically between 5 μm and 20 μm. The amount of aqueous solution is generally around 10^{-9} g, and analysis of the ion content is geochemically relevant for concentrations down to 0.01 molal (number of moles per kilogram of water). Therefore the challenge of LIBS analysis is the analysis of amounts of matter between 10^{-13} g (0.01 molal) to 10^{-11} g (1 molal).

Micro LIBS with a spatial resolution around 3–10 μm is convenient for the size of such objects. A typical emission spectrum obtained on fluid inclusion shows the wide potential of this method with the identification of Li, Na, K, Ca, and Mg (Figure 16.10) [36].

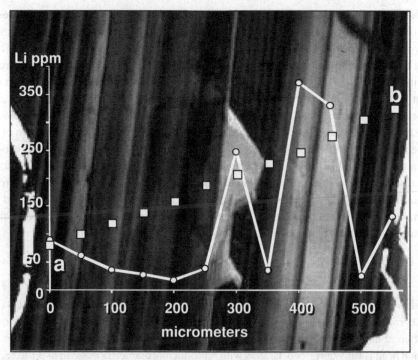

Figure 16.9. Cathodoluminescence photomicrograph of hydrothermal quartz showing zonation of luminescence intensities. The a–b profile is shown by the squares. Each square corresponds to one laser shot and to one LIBS analysis. The Li estimates (in p.p.m.) are also shown.

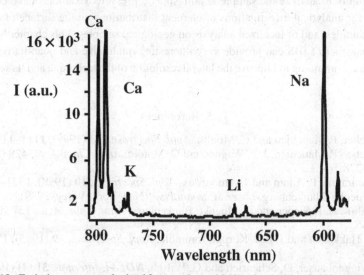

Figure 16.10. Emission spectrum obtained for a one laser ablation shot inside a fluid inclusion showing Li, Na, K, Ca, and Mg.

Matrix effects produce variations of the amount of ablated matter and temperatures of the plasma. However, only determination of elemental ratios is sufficient for fluid inclusion analysis. Emission intensity is well known to be a function of plasma temperature. Thus, using the copper emission lines at 515.32 nm and 510.55 nm for the respective electronic transitions ($3d^{10}4p^1 \rightarrow 3d^{10}4d^1$) and ($3d^94s^2 \rightarrow 3d^{10}4p^1$), the plasma temperatures in different matrixes can be determined. Fortunately, temperatures of the plasmas formed from copper-bearing aqueous solution and silicate glasses are identical [34]. This feature justifies the establishment of calibration curves using synthetic glasses for the determination of the elemental ratios Na/Li, Na/K, Na/Ca, and Ca/Mg. Limits of detection are typically around 10 p.p.m for Na and Li, 20 p.p.m for Ca, and 750 p.p.m for K. The method was validated on natural samples with a single fluid generation using the bulk analytical method and technique. These analyses were used for the reconstruction of palaeofluid chemistry and the determination of the temperature of circulation using chemical equilibrium between fluid and minerals on the example of quartz crystal formation in the Alps [36].

16.4 Conclusion

A review of the main results already published in the literature showed that micro LIBS (use of LIBS with craters ≤ 40 μm) is a very powerful new technique that has been developed in different laboratories. An experimental set-up applying a modified microscope for the ablation head is fully described. Craters with reproducible diameters as small as 3 μm can be obtained. The full design has taken into account the main advantages of the more common LIBS technique such as the lack of sample preparation (except perhaps a polishing) and the opportunity to analyze the sample at atmospheric pressure. Examples have been fully described for analytical investigations of element distributions on the surface of ceramics and steel samples, and of localized analysis on geological samples. This chapter highlights the fact that micro LIBS can provide very interesting qualitative and quantitative results. Now, work is continuing to improve the lateral resolution of the technique and its sensitivity.

16.5 References

[1] P. Fichet, P. Mauchien and C. Moulin, *Appl. Spectrosc.*, **53** (1999), 1111–1117.
[2] P. Fichet, P. Mauchien, J. F. Wagner and C. Moulin, *Anal. Chim. Acta*, **429** (2001), 269–278.
[3] L. Dudragne, P. Adam and J. Amouroux, *Appl. Spectrosc.*, **10** (1998), 1321–1327.
[4] L. Moenke-Blankenburg, *Laser Microanalysis* (New York: Wiley, 1989).
[5] Y. Talmi, H. P. Sieper and L. Moenke-Bankenburg, *Anal. Chim. Acta*, **127** (1981), 71–85.
[6] H. J. Häkkänen and J. E. I. Korppi-Tommola, *Appl. Spectrosc.*, **49** (1995), 1721–1728.
[7] H. Wiggenhauser, D. Schaurich and G. Wilsch, *NDT&E Internat.*, **31** (4) (1998), 307–313.

[8] D. Romero, J. M. Fernandez-Romero and J. J. Laserna, *J. Anal. Atom. Spectrom.*, **14** (1999), 199–204.

[9] J. M. Vadillo, S. Palanco, M. D. Romero and J. J. Laserna, *Fresenius J. Anal. Chem.*, **355** (1996), 909–912.

[10] M. Hidalgo, F. Martin and J. J. Laserna, *Anal. Chem.*, **68** (1996), 1095–1100.

[11] D. Romero and J. J. Laserna, *Anal. Chem.*, **15** (1997), 2871–2876.

[12] D. Romero and J. J. Laserna, *J. Anal. Atom. Spectrom.*, **13** (1998), 557–560.

[13] J. M. Vadillo, C. C. Garcia, C. Palanco and J. J. Laserna, *J. Anal. Atom. Spectrom.*, **13** (1998), 793–797.

[14] D. A. Anderson, W. M. Cameron, T. English and A. T. Smith, *Appl. Spectrosc.*, **49** (1995), 691–701.

[15] P. Lucena and J. J. Laserna, *Spectrochim. Acta*, **56**B (2001), 177–185.

[16] P. Lucena, J. M. Vadillo and J. J. Laserna, *Appl. Spectrosc.*, **55** (2001), 267–272.

[17] Y. Y. Yoon, T. S. Kim, K. S. Chung, K. Y. Lee and G. H. Lee, *Analyst*, **122** (1997), 1223–1227.

[18] I. B. Gornushkin, B. W. Smith, H. Nasajpour and J. D. Winefordner, *Anal. Chem.*, **71** (1999), 5157–5164.

[19] D. Kossakovski and J. L. Beauchamp, *Anal. Chem.*, **72** (2000), 4731–4737.

[20] C. C. Davies, W. A. Atia, A. Gungor. *et al.*, *Las. Phys.*, **7** (1997), 243–248.

[21] Contract MAT1 (Measurement and testing), CT-93-0029 ((1993–1996).

[22] C. Geertsen, J. L. Lacour, P. Mauchien and L. Pierrard, *Spectrochim. Acta*, **51**B (1996), 1403–1416.

[23] N. André, C. Geertsen, J. L. Lacour, P. Mauchien and S. Sjöström, *Spectrochim. Acta*, **49**B (1994), 12–14.

[24] M. D. Adams and S. C. Tong, *Anal. Chem.*, **12** (1968), 1762–1765.

[25] H. N. Barton, *Appl. Spectrosc.*, **23** (1969), 519–520.

[26] S. W. Allison, PhD thesis, University of Virginia (1979).

[27] J. R. Wachter and D. Cremers, *Appl. Spectrosc.*, **41** (1987), 1042–1048.

[28] E. Stoffels, P. Van de P. Weijer and J. Van der Mullen, *Spectrochim. Acta*, **46**B (1991), 1459–1470.

[29] W. Pietsch, A. Petit and A. Briand, *Spectrochim. Acta*, **53**B (1998), 751–761.

[30] A. I. Whitehouse, J. Young, I. M. Botheroyd *et al.*, *Spectrochim. Acta*, **56**B (2001), 821–830.

[31] C. Aragon, J. A. Aguilera and F. Penalba, *Appl. Spectrosc.*, **53** (1999), 1259–1267.

[32] V. Sturm, L. Peter and R. Noll, *Appl. Spectrosc.*, **9** (2000), 1275–1278.

[33] M. C. Boiron, J. Dubessy, N. André *et al.*, *Geochim. Cosmochim. Acta*, **55** (1991), 917–923.

[34] C. Fabre, M. C. Boiron, J. Dubessy and A. Moissette, *J. Anal. Atom. Spectrom.*, **14** (1999), 913–922.

[35] C. Fabre, M. C. Boiron, J. Dubessy *et al.*, *Geochim. Cosmochim. Acta* (2001), in press.

[36] C. Fabre, M. C. Boiron, J. Dubessy, M. Cathelineau and D. A. Banks, *Chem. Geol.* (2001), in press.

[37] European Patent Office no. 00974635.5-2204 (10/8/2001).

17

New spectral detectors for LIBS

Mohamad Sabsabi and Vincent Detalle

Industrial Materials Institute
National Research Council of Canada
Boucherville, Québec,
Canada

17.1 Chapter organization

The purpose of this chapter is to provide the reader with an evaluation of new commercial echelle spectrometers equipped with charge transfer device (CTD) detectors for laser-induced breakdown spectroscopy (LIBS). Because it is difficult to separate the detector from the dispersion optics, a brief review of detectors, spectrometers and multichannel detection used in LIBS is given in Section 17.3. An evaluation of the echelle spectrometer/intensified charge coupled device (ICCD) performances is carried out in Section 17.4 in terms of plasma diagnostics (temperature, electron density) and spectrochemical analysis (wavelength coverage, spectral response, and detection limit). The advantages and limitations of such a system are discussed in Section 17.5. In order to provide a critical evaluation of the echelle spectrometer for LIBS applications, the performance of an echelle spectrometer coupled with an ICCD detector is compared with one equipped with an interline CCD camera in Section 17.6. Finally, a summary is presented in Section 17.7.

17.2 Introduction

Laser-induced plasma spectroscopy (LIPS), also known as laser-induced breakdown spectroscopy (LIBS), is a form of atomic emission spectroscopy (AES). LIBS is being used as an analytical method by a growing number of research groups. The growing interest in LIBS, particularly in the past decade, has led to an increasing number of publications on its applications, both in the laboratory and in industry. Despite all these activities, the LIBS technique is still not widely accepted in analytical chemistry. Nevertheless, LIP spectrochemical analysis has become an important tool for answering chemical questions and diagnosing industrial problems.

Recent developments in technology and research in spectroscopic detectors have suggested a promising future and an improvement of measurements in plasma spectroscopy. Undoubtedly, the advent of high-quality solid-state detectors is revolutionizing the field of atomic spectroscopy [1–3]. New optical technologies, when coupled with new generations

Laser-Induced Breakdown Spectroscopy: Fundamentals and Applications, ed. Andrzej W. Miziolek, Vincenzo Palleschi and Israel Schechter. Published by Cambridge University Press. © A. W. Miziolek, V. Palleschi and I. Schechter 2006. A. W. Miziolek's contributions are a work of the United States Government and are not protected by copyright in the United States.

of optical detectors, provide powerful tools for plasma diagnostics and spectrochemical analysis.

Unlike inductively coupled plasma atomic emission spectroscopy (ICP-AES) or other techniques based on optical emission spectroscopy (OES), the LIBS technique requires a time-resolved detection owing to its transient nature. LIBS is concerned with the radiation emitted by the microplasma induced by focusing a powerful laser on the sample. LIP emission is space and time dependent. In the initial moments, the plasma emission consists of an intense radiation continuum superimposed with very broadened lines. These lines are strongly broadened by the Stark effect, owing to the high electron density that exists in the plasma at this time. Thus, temporally gating off the earlier part of the plasma is essential for spectrochemical analysis. This dictates an important difference based on time-gated detection of the atomic emission between the detector requirements for LIBS and ICP-OES or other techniques based on OES.

The combination of spectrometer and detector in LIBS requires a compromise between wavelength coverage, spectral resolution, read time, dynamic range, and detection limit. For LIBS applications as well as for other forms of OES techniques, the wavelength region of interest starts from approximately 165 nm and reaches up to 800–940 nm. Over this region there are tens of thousands of emission lines with width ranging from approximately 0.002 nm to 0.01 nm and intensities ranging over many orders of magnitude. Of course not all these lines are of equal analytical interest, and there are large regions between the lines for which there is no analytical interest. Furthermore, the transient nature of the LIBS signal and the repetition rate of the laser place time constraints on the data acquisition scheme.

In the 1960s and 1970s, most of the LIBS work published in the literature was carried out using a conventional Czerny–Turner spectrometer with photomultipliers as detectors. Intensified photodiode arrays (IPDA) coupled with Czerny–Turner spectrometers were used in the 1980s. In the 1990s the advent of the intensified CCD gradually replaced the photomultipliers and photodiode array detectors and they are presently widespread in the LIBS community. On the other hand, Paschen–Runge spectrometers equipped with photomultipliers are also still used [4–6]. Such a combination of spectrometer/ICCD or IPDA can yield either a wide spectral coverage or a high spectral resolution, which results in limited applicability for multielement analysis. Linear photodiode arrays or ICCDs, when they are attached to a conventional Czerny–Turner spectrometer, allow only a limited spectral coverage (typically a few nanometers wide). Effective use in analytical spectroscopy of the two-dimensional CTDs, which include the CCD and coupled-injection device (CID), requires the use of optical configurations different from the Czerny–Turner system. A CTD can be efficiently exploited for high-resolution spectroscopy when the instruments are coupled with an echelle spectrometer.

Since the beginning of the 1990s, echelle-based systems coupled with two-dimensional CTD for simultaneous measurements of analyte lines at different wavelengths have been developed and commercialized in ICP [7–12] and used in microwave-induced plasma (MIP) analysis [13]. In the past five years few were used in LIBS [14–21], and it is presently commercialized, to our knowledge, by three companies [22].

This chapter is not intended to be a comprehensive survey of detectors or spectrometers. A detailed description (principles, characteristics, etc.) of the solid-state detectors can be found elsewhere [1, 23], and especially in publications from the Bonner Denton group [24, 25]. Also, information on spectrometers can be found in most books on spectroscopy [26, 27]. The instrument chosen as an example in this chapter was selected to illustrate the impact of solid-state array detectors on LIBS. Our purpose is to provide the reader with information about the use of the echelle spectrometer with an intensified CCD detector in LIBS and to evaluate how this combination can increase the analytical capabilities of LIBS by summarizing the advantages and limitations of the system.

17.3 Multidetection in LIBS

17.3.1 Desirable requirements for atomic emission detectors

The combination of spectrometer and detector is an important factor to consider in optical emission spectroscopy for any plasma characterization or analytical spectrochemistry. The experimentalist must choose the type of measurements to be made. Based on that, the appropriate system can be designed. For example, using non-resonance lines of non-metals such as F, Cl, Br, I, S, N, and O requires working in the near infrared (NIR) region (<940 nm), while the resonance lines of these elements are located in the vacuum ultraviolet (VUV) (below 185 nm). The requirements for the ideal spectrometer and detector to provide for simultaneous determination of any combination of elements in the spectrum include a high resolution of 0.01–0.003 nm to resolve the lines of interest and avoid interferences. Second, wide wavelength coverage is needed, typically from 165 nm to 800–950 nm to be able to detect simultaneously several elements. Third, a large dynamic range is necessary to provide the optimum signal-to-noise ratio (SNR) for a large range of elemental intensities; the detector has to have a wide dynamic range, typically 6–7 orders of magnitude. The spectrometer/detector combination should have a high sensitivity and a linear response to radiation. The detector has to have high quantum efficiency, particularly in the near infrared and UV, and low noise characteristics. Furthermore, for rapid analysis the readout and data acquisition time should be shorter, at least less than the time lap between the laser pulses. Before proceeding further, it is necessary briefly to review the characteristics of detectors and spectrometers used in LIBS applications so as to appreciate the impact the echelle spectrometer/ICCD has had on LIBS.

17.3.2 Photomultiplier detectors

The photomultiplier tube (PMT) has been the mainstay of low-light detectors for well over 50 years. Photomultipliers are extremely sensitive light detectors that provide a current output proportional to light intensity. They have a large area of light detection, wide range of spectral sensitivity (UV to visible), wide dynamic range (over six orders of magnitude), high quantum efficiency or QE (10%–25% at the optimum response, see Figure 17.1), fast temporal response (typically 10 ns to a few picoseconds), low detector cost and ease of

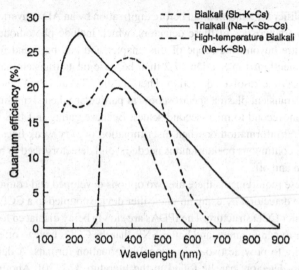

Figure 17.1. Quantum efficiency for photomultiplier tubes.

use. The photomultiplier starts at the photocathode, followed by an assembly of dynodes (9–16 dynodes), each at a potential of about 100 V with respect to the last, and each emitting several secondary electrons for every one that hits it. The total gain of the multiplier depends on the number of stages and the total applied voltage. A typical PMT will have a gain of 10^6–10^7 for each incident photon. The electrical current measured at the anode is proportional to the amount of light that strikes the photocathode. This current is then converted into voltage signal, which is transferred via an analog-to-digital (AD) converter to a computer for processing purposes. The noise from a PMT comes almost entirely from the photocathode dark current and it can be reduced by cooling. The response of the photomultiplier to a delta pulse of light is governed by the electron trajectories within the tube and their transit time (typically 10 ns to a few picoseconds). For more details about PMTs we refer the reader to the following references [26]–[29]. The major drawbacks of PMTs are that they are relatively bulky and are single-channel, i.e. one PMT per analyte line. As explained above, to separate the analyte signal from the earlier part of the strong background LIP emission, some form of time gating is carried out with boxcar integration to acquire information about analytical lines. Transient events in LIP are difficult to monitor using a single-channel mode of detection. The spectral bandwidth of a PMT (typically 6–100 pm) in the range 170–800 nm corresponds to a small part and represents only 10 p.p.m. (10^{-5}) of the information available. As a consequence, single-channel methods of detection are not optimal for many plasma diagnostic and analytical spectrochemistry tools.

17.3.3 Solid-state detectors

Solid-state array detectors consist of series of radiation-sensitive semiconductor picture elements or pixels that convert photons to a quantifiable charge. This charge is transferred to a readout amplifier for amplification before digitization by an AD converter. Description, operation and characteristics of these detectors, which include photodiode array (PDA), CCD and CID, are beyond the scope of this chapter and can be found in the literature [1, 23–25]. We mentioned in Section 17.2 that, because of its transient nature, the LIBS technique requires time-resolved detection. Therfore the CTD and PDA detectors are often not adequate for transient plasma spectroscopy, in particular when signal-integration time is ultra-short (nanosecond to microsecond scale). Because gating in LIBS is essential for acquiring temporal information or when discrimination of very weak lines superimposed on a very intense continuous background is needed, solid detectors need to be gated, that is rapidly turned on and off.

To address these requirements there are two options developed and commercially available to gate these detectors: by using an intensifier device coupled to a CCD or PDA or by an interline-transfer CCD structure. The IPDAs are slowly being displaced for the increased sensitivity, wider dynamic range, and better SNR that the CCDs have to offer, in particular the ICCDs' ability to provide two-dimensional information formats. A detailed comparison of these two detectors can be found in the literature [23, 30]. Another subclass of the CTD detectors are the CIDs. The CID differs from the CCD primary by the mode of readout of the photogenerated charge [23–25, 30]. To our knowledge, these detectors have not been used in LIBS applications and are not available commercially with intensifiers. As a consequence, they will not be studied here. This section focuses on highlighting some characteristics of the ICCD and the interline CCD without giving details, to allow the reader to realize the advantages of the ICCD detectors or of the interline CCD for their specific application. In particular, these detectors are generally two-dimensional and can be coupled with appropriate dispersing systems such as the echelle spectrometer to take advantage of this characteristics.

The interline CCD detectors

A CCD detects light through the creation of electron–hole pairs in silicon. The electrons are trapped by an imposed electric field and read out as charge by the camera electronics. CCDs measure the photogenerated charge accumulated in pixels by sequentially shifting the charge to an on-chip preamplifier located at the periphery of the device, and the signal is converted to computer-readable form by an analog-to-digital (AD) converter (also known as an ADC). Depending on the architecture of the device (*full frame, frame transfer* or *interline transfer*), the specifics of how this is done vary. Interline charged coupled device architecture is designed to compensate for many of the shortcomings of frame-transfer CCDs. The interline CCD is a hybrid sensor with photosensitive diodes on one part of the pixel that are electrically coupled to a CCD type storage region which resides under a mask structure. The masks are long structures running along the vertical axis of the CCD alternating with the open regions, hence the name interline CCD. In this approach, the sensor is subdivided into interdigitized lines of photoactive regions and masked registers. The photoactive regions integrate light and then pass the charge to the adjacent masked registers.

Table 17.1. *CCD characteristics and their limitations*

Characteristics	Typical range	Limitations
Full-well capacity	10 000–500 000 electrons	Defines dynamic range
Pixel dimensions	6–30 μm	Dictate spectral or spatial resolution
Array format	Related to pixel size and number	Dictates the active area
Number of pixels	58×512 to 2048×2048	Dictates number of resolution elements and acquisition time
Quantum efficiency	0%–80%	Defines ultimate sensitivity limit
Blooming	Present with materials having strong lines in the range 200–900 nm	Cannot observe weak lines near strong lines, ghost lines
Read noise	Few electrons	Excess noise limits weak light detection

Like the full-frame and frame-transfer architectures, interline-transfer CCDs undergo readout by shifting rows of image information in a parallel fashion, one row at a time, to the serial shift register. The serial register then sequentially shifts each row of image information to an output amplifier as a serial data stream. The entire process is repeated until all rows of image data are transferred to the output amplifier and off the chip to an analog-to-digital signal converter integrated circuit.

During the period in which the parallel storage array is being read, the image array is busy integrating charge for the next image frame, similar to the operation of the frame-transfer CCD. A major advantage of this architecture is the ability of the interline-transfer device to operate without a shutter or synchronized strobe, allowing for an increase in device speed and faster frame rates. Image "smear," a common problem with frame-transfer CCDs, is also reduced with interline CCD architecture because of the rapid speed (only one or a few microseconds) in which image transfer occurs. Drawbacks include a higher unit cost to produce chips with the more complex architecture, and a lower sensitivity due to a decrease in photosensitive area present at each pixel site since only one-half of the array is photoactive. Also, this affects the spatial resolution on one axis owing to the gaps between photoactive regions. The advantage of the interline-transfer approach is that there is virtually no optical smearing during transfer and possibility of gating. Table 17.1 lists the main factors to be taken into consideration when selecting a CCD for LIBS applications.

The image intensifier

The image intensifier in front of the CCD chip acts as a superfast shutter capable of operating on nanosecond timescales. An image intensifier is a cylindrical vacuum device typically 18 or 25 mm in diameter. It consists of three major components: photocathode, microchannel plate (MCP) and the phosphor screen [23, 31]. The photocathode transduces the photon image to an electron image. The MCP amplifies the electrons and the phosphor screen converts the amplified electron image to a corresponding photon image. The MCP is a disk of

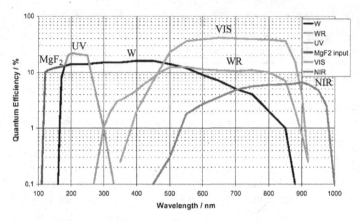

Figure 17.2. Quantum efficiency for photocathodes.

special glass with many individual hollow microchannels, each acting as an electron multi-plier. When the photocathode is struck with incident photons, this causes the release of free electrons by photoelectric effects. A gate voltage (typically 200 V) accelerates the emitted electrons to the MCP. After leaving the MCP the resulting electrons are further accelerated into a phosphor screen by high voltage (kilovolts) then converted to photons and detected by the CCD. The photons detected by the CCD are always monochromatic, independent of the incident photon wavelength received by the photocathode of the intensifier. This means the ICCD sensitivity is dependent on the photocathode characteristics.

Figure 17.2 shows a typical quantum efficiency of the ICCD photocathodes. The ability rapidly to gate the potential applied to the photocathode allows electronic shuttering of the camera. The result is the ability to image and resolve short-lived events. The shortest gate times are limited by the speed of switching the gate potential. Intensifier manufacturers have addressed this issue by lowering the impedance of the photocathode by using a con-ductive nickel undercoating on one of the cathode surfaces. However, this approach lowers the effective QE in the UV because of absorption of the light by nickel [23, 31]. Presently the shortest gate is in the range 1–3 ns, while it is 20–40 ns for the slower gate.

17.3.4 Spectrometers

A spectrometer separates light into its component wavelengths. Separation of light is achieved in all modern instruments by the use of a diffraction grating. A grating consists of a series of closely spaced lines, which are ruled or etched onto the surface of a mirror. When light strikes the grating it is diffracted at a certain angle, according to the following

$$\sin\alpha \pm \sin\beta = kn\lambda \tag{17.1}$$

where α and β are respectively the incidence and diffracted angles, n is the grating density (number of grooves per millimeter), λ is the wavelength, and k is diffraction order. Modern gratings are usually blazed, which means that the groove is shaped to concentrate a large

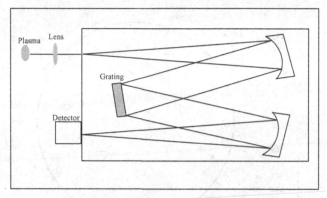

Figure 17.3. Czerny–Turner spectrometer system.

fraction of the incident intensity into diffraction at a particular angle. Thus it is possible to have a blaze diffraction grating that is more efficient for a specific wavelength region. In this case, the grooves are ruled at a specified angle (known as the blaze angle).

Czerny–Turner optical configuration

Figure 17.3 shows an optical layout incorporating the Czerny–Turner mounting, which is the most commonly used. This spectrometer typically consists of entrance, exit optics, and grating. The grating groove density used is between 100 and 4800 grooves per millimeter. The Czerny–Turner optical configuration operates from the first order up to the fourth. The resolution or the ability of the spectrometer to separate wavelengths or lines is specified as the spectral bandpass. This is defined as one-half of the wavelength distribution passing through the exit slit. According to the Rayleigh criteria of resolution for diffraction line profiles, two lines are considered to be just resolved if the central maximum of the first line profile coincides with the first minimum of the following line profile. In other words, to completely resolve two lines or wavelengths, the spectral bandpass must be no greater than one-half their difference in wavelength. For example, if two lines are separated by 0.2 nm, the spectral bandpass must be 0.1 nm or less. In this configuration the selection of the desired wavelength is achieved by rotation of the grating within its mounting, which brings individual spectral lines into focus at a fixed position at the exit of the spectrometer. This approach allows any wavelength to be observed sequentially, one wavelength at a time if the spectrometer is equipped with a PMT at the exit. It also allows the measurement of a small section of the spectrum when it is attached to a linear diode array or a CCD detector. The spectral window obtained is only a few nanometers wide depending on the reciprocal linear dispersion of the spectrometer and the detector size.

Paschen–Runge mounting

Polychromators generally use a Paschen–Runge optical configuration. The grating, entrance slit and the multiple exit slits are fixed around the periphery of a circle, called the

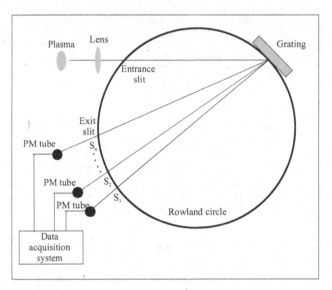

Figure 17.4. Paschen–Runge spectrometer system.

Rowland circle (Figure 17.4). In this configuration, the grating is concave and does not rotate. This approach allows simultaneously the acquisition of multiple wavelengths by using several PMT detectors. However, to cover the wavelength range available (for example from 160 nm to 770 nm) with 6 pm as spectral bandpass, 100 000 detectors would be necessary. The total length of the detector must be 200 cm with 14 μm pixel size. Thus, 40 ICCD detectors (25 mm size) would be necessary to cover the whole spectrum, and this would be very expensive. Another alternative simultaneously to collect several thousand independent channels of spectral information is the use of an echelle grating with a cross-disperser to form a two-dimensional spectrum. This spectrometer cuts a vertical spectrum into several segments horizontally. The format of the spectrum can therefore be made to match the rectangular format of the two-dimensional CCD detectors.

Echelle spectrometer

Echelle spectrometers are orthogonal dual-dispersion devices usually employing a refractive element (prism) as an order sorter and a high-dispersion element (echelle grating) for wavelength resolution. The major characteristic of the echelle spectrometer in terms of components is the grating. In contrast to the blazed grating used in the Czerny–Turner mounting the design of the echelle grating utilizes the spectral order for maximum wavelength coverage [32]. The resolution R of a diffraction grating is related to the groove density n and the spectral order k ($R = kn$). In this case, instead of using a grating with a large number of grooves, resolution is improved by increasing both the blaze angle and the spectral order. For example, if we take $\sin\alpha + \sin\beta = 1$ and $\lambda = 333$ nm, the product kn will be 3000: for $k = 1$, $n = 3000$ grooves per millimeter, while, for $k = 50$, $n = 60$

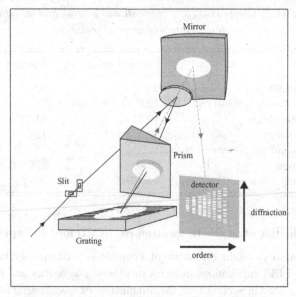

Figure 17.5. Echelle monochromator configuration.

grooves per millimeter. The problem with an echelle grating is the overlapping of spectra. To prevent the overlapping of spectral orders a secondary dispersion stage or order sorter is required. This is accomplished by using a prism or another grating. If the prism is placed so that light separation occurs perpendicular to the diffraction grating (Figure 17.5), a two-dimensional spectral map is produced where the data are sorted into spectral order vertically and wavelength horizontally (the orthogonal dispersive element, either a prism or grating, provides the extended wavelength coverage by controlling the position at which wavelengths are imaged in space before being dispersed by the echelle grating). This gives the echelle spectrometer its two-dimensional wavelength dispersion character. Wavelength is dispersed horizontally within a given order and it decreases vertically as order number increases. The dimensional display of spectra is ideally suited for the use of solid-state detectors, in particular the ICCD or interline CCD which are of interest in this chapter. However, the intensifier cylindrical shape (18 mm or 25 mm diameter) adds some restrictions on the active pixels. For example, the pixels outside the intensifier tube will not be sensitive; this induces some gaps in the spectrum, as discussed later in Section 17.6. The echelle grating can give excellent resolution and dispersion in a relatively compact spectrometer. One commercial echelle spectrometer provides an average reciprocal linear dispersion of 0.12 nm mm^{-1} and an average resolution of 0.003 nm in an instrument with 0.75 m focal length. For comparison, a conventional 0.75 m grating spectrometer with 2400 grooves per millimeter gives a reciprocal linear dispersion of 0.54 nm mm^{-1} and a resolution of 0.03 nm in the first order. Table 17.2 gives a comparison of the spectral features of a conventional grating and an echelle grating.

Table 17.2. *Comparison of a conventional grating and an echelle grating in terms of spectral features*

Feature	Conventional grating	Echelle grating
Focal length (m)	0.5	0.5
Grating density (grooves mm^{-1})	1200	79
Diffraction angle	10°22'	63°26'
Width (mm)	52	128
Spectral order at 300 nm	1	75
Resolution	62 400	758 400
Resolving power at 300 nm (nm)	4.807E-3	396

17.4 Evaluation of an echelle spectrometer/ICCD for LIBS applications

In this section, we focus on the evaluation of a combination of commercial ICCD/echelle spectrometer for LIBS applications in terms of plasma diagnostics and plasma spectrochemistry. As described in Section 17.2, the combination of spectrometer and detector is an important choice to be considered in any plasma characterization. The experimentalist must choose the type of measurement to be made, and then the appropriate detection scheme can be designed. The two-dimensional character of the ICCD can be fully exploited with the echelle spectrometer, and an ICCD/echelle system offers substantial experimental flexibility for plasma diagnostics and plasma spectrochemistry. Firstly, it allows simultaneously the measurement of not only the Boltzmann excitation temperature but also gas-kinetic-temperature studies (from molecular Boltzmann studies), ionic temperature (from the Sahas' equilibrium), and electron density measurements (from Stark broadening). Secondly, for spectrochemistry it provides a choice of several analytical lines for the same element and simultaneously for several elements present in the sample.

17.4.1 Plasma diagnostics

In order to evaluate the echelle spectrometer/ICCD for plasma diagnostics, a brief discussion of the LIP conditions and description of the system used in these experiments appear to be necessary and may be helpful to the reader. Because only a few applications of echelle spectrometers involving diagnostics of LIP have been reported [14, 21], most of the experimental results presented here were performed with LLA instrumentation ESA 3000 [21, 22].

The spectrometer has a focal length of 25 cm with a numerical aperture of 1:10. A quartz prism positioned in front of the grating separates the different orders and produces a two-dimensional pattern. The operating diffraction orders range from 30 to 120. The flat image plane is 24.85 × 24.5 mm^2. This system is a compromise that offers maximum resolution in the wavelength range between 200 nm and 780 nm. The reciprocal linear dispersion per pixel ranges from 0.005 nm (at 200 nm) to 0.019 nm (780 nm), according to $\lambda/\Delta\lambda =$ 40 000. The detector is an ICCD camera, comprising a Kodak KAF-1001 CCD array of 1024×1024 pixels (24×24 μm^2) and a microchannel plate type BV2562 of 25 mm diameter from Proxitronix coupled with a UV-enhanced photocathode. The fitted spectral length is approximately 1204 mm, with 50172 channels of a width of 24 μm. The CCD temperature is controlled by a closed water-cooling system. A fast pulse generator delivers a pulse of 5 ns width to the intensifier to ensure synchronization of the measurements with the laser pulse.

The energy source used in our experiments was a Nd:YAG laser (Surelite I 10, Continuum). This laser can supply an energy of 450 mJ, with a maximum repetition rate of 10 Hz. The pulse duration is 6 ns FWHM (full width at half maximum), the wavelength is 1064 nm, the beam is quasi-Gaussian, and its divergence is 0.6 mrad. In order to have better control and reproducibility of the laser material interaction, the laser beam was passed through a 1000 μm diameter diaphragm. The pinhole allows selection of the central, approximately homogenous, part of the Gaussian beam. Plasma is formed on the target by focusing the laser beam using two lenses having 50 cm and 25 cm focal length respectively, allowing an imaging of 2:1 (500 μm diameter spot on target). The energy of the laser beam in the focal point can be adjusted by using an optical group composed of a half-wave plate and a polarizer. The energy used on target was 65 mJ, corresponding to 33 J cm^{-2}.

The image collection system uses an unconventional approach, the plasma being directly collected by optical fiber (800 μm of core) without a focusing lens. This mode allows the collection of plasma emission while preventing problems related to the dependence on wavelength that arise when focusing a very large spectral range. The optical fiber is located a few millimeters from the plasma and the collection is carried out in the axis of expansion of the plasma, shifted in order not to block the laser beam. The optical fiber is then directly coupled to the echelle spectrometer.

A typical linear echelle spectrum is presented in Figure 17.6. The spectrum arises from an aluminum alloy (sample 4104 AC) containing Mg, Cu, and Mn. The capability for multi-elemental applications presented by the echelle spectrometer is illustrated by this example. In order to show the very high resolution and good dynamic range of the system, we have enhanced a spectral zone of 2 nm in the same spectrum. It is shown that 13 p.p.m. of Be can be distinguished in our experimental acquisition conditions.

Because the echelle system is composed of a wavelength dispersion device (grating and prism) and a detection device (ICCD) the spectral response will depend on both components. Moreover, the response of the detection device will depend both on the cathode (intensifier) and on the CCD. In the case of the echelle spectrometer, the spectral response in one order of diffraction, as observed for example using the blackbody radiation from a deuterium lamp, is nonlinear. As can be seen in Figure 17.7, the response has a maximum in the center of a given order, which is centered on the CCD. Each diffraction order presents the same response shape but not necessarily the same overall sensitivity [14, 20, 21]. This implies that the spectrum must be corrected when performing physical measurements using lines from different regions of the response curve (e.g. determination of the excitation

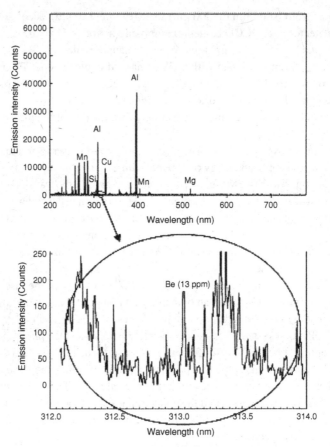

Figure 17.6. Typical echelle spectrum.

temperature from relative line intensities), especially if one line of interest is located at the edge of one order. A blackbody radiation calibration spectrum has to be recorded to obtain the intrinsic response of the echelle/ICCD system, which is then used to normalize the acquired spectrum. The continuum radiation source used in our experiments is from a combination of calibrated halogen and deuterium lamps. Thus, a spectrum can be recorded independently of the system response, and the physical measurements can be realized after this data treatment.

For the physical study, spectra from individual laser shots were analyzed. Indeed, the variability of the laser–material interaction and of the laser-induced plasma essentially makes each measurement independent. Therefore, no physical measurement results from the accumulation of shots. The reproducibility of measurements is evaluated once physical calculations are carried out.

The main difficulty in evaluating the quality of new equipment stems from the dependence of the result on the laser–material interaction. This is why part of the results presented in

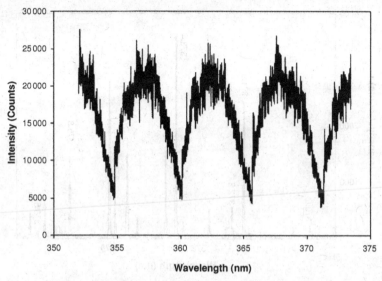

Figure 17.7. Spectral response in successive echelle orders.

this study were realized on the same setup but with a conventional spectrometer. In the following, the values of the plasma parameters such as temperature and electron density are spacially averaged; no Abel inversion was used.

Temperature

In order to verify the performance of the system Detalle *et al.* [21] carried out measurements of the excitation temperature by the Boltzmann plot method. An aluminum sample with 0.29% of Fe was chosen for our experiments. Using lines from a minor element (e.g. Fe) to construct the plots avoids possible problems of self-absorption. Moreover, Fe atoms have a rich spectrum of optical transitions from excited levels in a wide energy range (from $27\,395$ cm^{-1} to $54\,655$ cm^{-1}), and the Einstein coefficients for Fe(I) transitions are known with high accuracy. One of the major advantages of the echelle system is the possibility it offers of working with all lines of the dispersed spectrum available from one measurement. Moreover, its high resolution enables the use of weak lines because their corresponding signal-to-noise ratio is enhanced. They used 11 lines already reported in the literature [33, 34] for temperature measurements. These lines are present in the spectrum of Figure 17.8. The excitation temperature was determined by using Boltzmann plots, assuming that the plasma is in local thermodynamic equilibrium (LTE). The slope of the linear regression of the Boltzmann plot gives the temperature value. Details about the validity of LTE conditions in plasma are out of the scope of this chapter and for more details the reader is referred to the literature [26, 35–42].

Fifty temperature measurements were performed to assess the reproducibility of the echelle/ICCD system (see Figure 17.9). The average temperature for a delay of 3 µs and 1 µs gate width is approximately 8000 K and the shot-to-shot relative standard deviation, caused

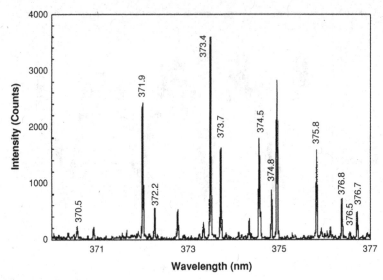

Figure 17.8. Emission spectrum from an aluminum sample showing the iron lines used for the Boltzmann plot.

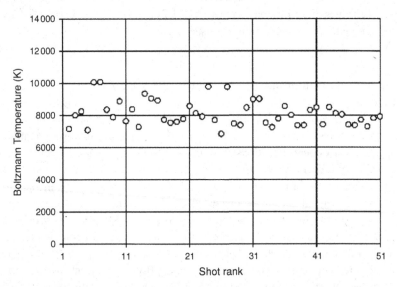

Figure 17.9. Evaluation of the reproducibility of Boltzmann plot temperature measurements.

by variations in the laser–matter interaction, is less than 10%. Two other measurements performed at delays of 900 ns (with 100 ns width) and 10 μs (with 1 μs width) give temperatures of 9800 K and 7500 K respectively.

Bauer *et al.* [14] showed that the determination of excitation temperatures from line intensities of different elements by the Boltzmann plot method could be easily realized with

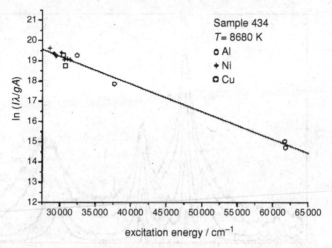

Figure 17.10. Boltzmann plot with intensities *I* of Al (◯), Cu (▢) and Ni (+) lines measured with sample 434 (λ, wavelength; *g*, statistical weight; *A*, transition probability). The intensities of the Cu and Ni lines are normalized to the intensities of the Al lines taking into account the concentration ratios c_{Cu}/c_{Al} and c_{Ni}/c_{Al}, respectively.

an echelle instrumentation. The Boltzmann plots are obtained in aluminum samples using Al, Cu, and Ni lines with very different excitation energies. The intensities of the Cu and Ni lines were normalized to the Al lines taking into account the concentration ratios c_{Cu}/c_{Al} and c_{Ni}/c_{Al}, respectively. Figure 17.10 shows the Boltzmann plot and the temperature obtained in their conditions with argon atmosphere, which was 8680 K ($\pm$ 500 K) with 2 μs of delay time and 3 μs width.

The results obtained [14, 21] show that measurements carried out with the echelle system have a good reproducibility, similar to that achievable with classical instrumentation. The measured temperatures are comparable to those obtained with a Czerny–Turner spectrometer [33, 43] using similar laser irradiance conditions.

Electron density

Spectral emission lines (atomic or molecular) can be broadened by several factors. Line broadening in plasma can arise from thermal motion, collisions, and electric field interactions. Broadening due to both collisional and electrostatic plasma environment is called Stark broadening [26, 27, 36, 39, 42]. Stark broadening measurement of electron density is the most common lineshape-based plasma diagnostic technique. The Stark broadening is more prevalent at high electron density and has only a weak temperature dependence. The Stark broadening for the determination of electron density is commonly used in LIP diagnostics, generally by using a conventional spectrometer [44], and to our knwoledge only in one case with an echelle spectrometer [21]. Figure 17.11 shows the contribution of the Stark effect on the broadening and shift of the Al(II) 281.6 nm line, for five different

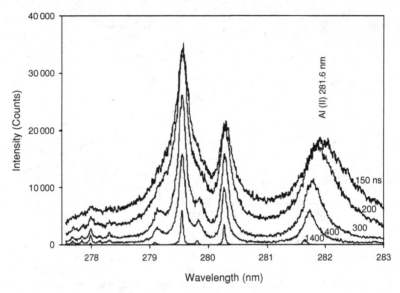

Figure 17.11. Profile of the 281.6 nm line of Al(II) obtained at different delay times (150, 200, 300, 400, and 1400 ns).

delays (150, 200, 300, 400, and 1400 ns) by using the echelle spectrometer with ICCD detector.

Because of the wide spectral coverage offered by the echelle spectrometer, it is possible to determine the electron density on the same laser shot, therefore the same plasma, using Stark-broadened lines of different elements present in the sample. We can thus follow at the same time different lines having different lifetimes, in order to have a more complete description of the plasma evolution. Moreover, this enables a comparison of various measurements, and thus an improvement in the precision.

Detalle *et al.* [21] used the evolution of the Al(II) 281.62 nm, Si(I) 288.16 nm, and Mg(II) 280.27 nm line profiles to obtain the electron density. They assumed in their conditions that the main contributions to line broadening were the instrumental function and Stark effect. Assuming a Lorentzian line shape, the FWHM attributable to the Stark effect is the difference between the measured FWHM and the instrumental FWHM. This latter, instrumental parameter was approximated using a Fe line located in the same spectral zone at a long delay (20 μs). At the end of the plasma lifetime the broadening becomes negligible and the line width results mainly from the instrumental characteristics. The value of the instrumental FWHM is approximately 0.01 nm. Using the deduced Stark width $\Delta\lambda_{\text{Stark}}$ (FWHM) [42], the electron density is obtained from the relation

$$\Delta\lambda_{\text{Stark}} \approx 2w_{\text{ref}}\left(\frac{N_e}{N_e^{\text{ref}}}\right), \tag{17.2}$$

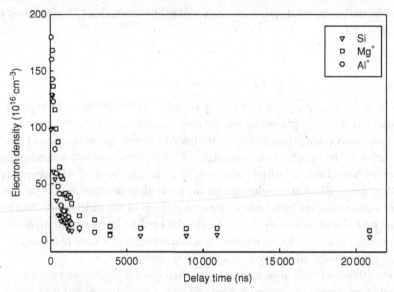

Figure 17.12. Temporal characteristics of the electron density deduced from Stark broadening of the Si (288.16 nm), Al(II) (281.62 nm) or Mg(II) (280.27 nm) lines in an aluminum alloy (7075 AB).

w_{ref} being the half width at half maximum Stark parameter for a reference electron density N_e^{ref}. The w_{ref} parameters determined experimentally by different authors were used: $w_{ref}(Si) = 0.095$ nm (obtained at $N_e^{ref} = 5.6 \times 10^{17}$ cm^{-3}) [45], $w_{ref}(Mg^+) = 0.0044$ nm ($N_e^{ref} = 10^{17}$ cm^{-3}) [46], and $w_{ref}(Al^+) = 0.00212$ nm ($N_e^{ref} = 10^{16}$ cm^{-3}) [47]. The determination of electron density using this method is independent of any assumption about LTE. The evolution of the electron density is shown in Figure 17.12. The electron density data deduced from the ionized aluminum line stop after 3 µs because the emission intensity for this transition becomes too weak. However, measurement of the electron density from 50 ns to 21 µs is possible with the two other lines. The same temporal evolution is found in each case. In fact, the results obtained with the silicon and aluminum ion lines are quasi-superimposable. The electron density found with the magnesium ion line is always slightly shifted by the same relative value compared with the other values, possibly due to an error in the available Stark parameters. All three values are thus comparable.

These results show that the quantity of information contained in the spectrum is enormous and the possibility of achieving physical measurements using various lines increases the reliability of the results, especially when they are correlated. In conclusion, the results obtained with the echelle spectrometer confirm those obtained with classical instrumentation [33, 43], but provide a more complete description of the temporal evolution of the laser-induced plasma. Furthermore, the determination of the ionic Saha temperature and the excitation Boltzmann temperature simultaneously enables the determination of LTE electron density values that can be compared with the electron density values (Stark boadening), which are independent of LTE conditions. The comparison of the electron density values obtained by the two methods provides a new tool to check whether LTE conditions prevail in the LIP or not.

17.4.2 Spectrochemical analysis

Generally, a classical analytical LIBS study begins with the choice of the spectral region where the lines of interest presenting the best response are located. Several factors intervene in this choice such as the dynamic range of the line (depending on the variation of the analyte concentration in the samples), the possibility of interference and self-absorption. In fact, it is not rare with classical instrumentation to be compelled to work with several spectral windows to cover different wavelengths for the same element. Moreover, if one wishes to carry out standardization with a line of a major element of the matrix, this often prevents using the best analytical line available. Lastly, the need for carrying out multi-elemental measurements often calls for an increase in spectral range, generally obtained by using a less dispersing grating, which involves a loss of resolution.

The availability of the entire emission spectrum from a plasma generated by a single laser shot in an echelle spectrometer equipped with a two-dimensional detector as ICCD or interline CCD makes it possible to eliminate most of these problems. The choice of the lines can be made subsequently to the measurement, and it is possible to use several lines to follow the same element depending on the variations of concentration encountered in the samples.

In the following, an evaluation of the echelle spectrometer coupled to an ICCD for spectrochemical analysis and exploitation of its characteristics is presented. The echelle spectrometer/ICCD was used for LIBS analysis of colloidal material [16], liquid [19] and solid samples [14], an aluminum sample [21], and identification of metal alloys [18].

Calibration by single shot

Quantitative spectral analysis involves relating the spectral line intensity from an element in the transient plasma to the concentration of that element in the target. Detalle *et al.* [21] investigated a set of eight samples of aluminum alloy. They analyzed beryllium, magnesium, manganese, copper, and silicon in these samples. In order to gate off the early part of the plasma emission to avoid the intense continuum emission and to improve the resolution, a delay is required before the measurement. In their experimental conditions, they have seen that after 3 μs the electron density has fallen (reducing the Stark broadening) and the temperature has stabilized. Therefore, they chose a delay time of 3 μs with a short gate width of 1 μs to avoid saturating the detector by emission lines of the major element in the matrix.

Figure 17.13 shows the calibration curves obtained with a conventional Czerny–Turner spectrometer coupled with intensified diode array [21], and with an echelle spectrometer/ICCD. They chose two elements for which the range of concentration is very different. The manganese content varies from 25 p.p.m. to 1.09% and the beryllium content varies

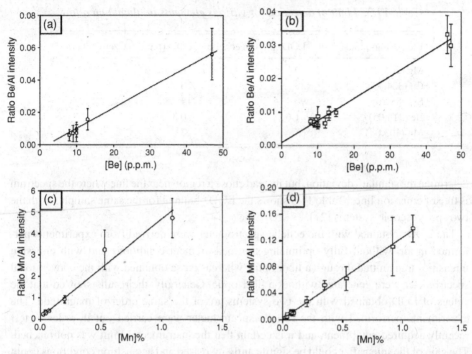

Figure 17.13. Calibration curves vs. concentration in aluminum alloys, obtained with an echelle spectrometer (a, c) and with Czerny–Turner spectrometer (b, d): (a) Be (313.0 nm)/Al (308.2 nm); (b) Be (313.0 nm)/Al(308.2 nm); (C) Mn (403.1 nm)/Al(305.0 nm); and (d) Mn (403.1 nm)/Al (394.4 nm).

from 8 p.p.m. to 47 p.p.m. In the case of beryllium, the calibration for both systems uses the Be (313.0 nm) normalized to the Al (308.2 nm) line. It is found that both curves are approximately linear and are comparable in terms of sensitivity (slope) and precision. In the case of manganese, they followed the evolution of the 403.1 nm line. With the classical detection system, one is forced to use a nearby aluminum line (at 394.4 nm) for normalization. With the echelle system, one can look for such a line even very far in the spectrum. The aluminum line used (at 305.0 nm) is located 100 nm from the Mn(I) 403.1 nm line; it is chosen because it has the same intensity as the manganese line and is less prone to self-absorption. A useful calibration curve is obtained with the two setups. Both systems are thus equally suitable for carrying out LIBS analysis. The same conclusion was reached by other groups [14, 16, 19].

Limit of detection

Detalle *et al.* [21] have calculated the limit of detection (LOD) for various elements present in the aluminum alloy sample. They chose the strongest line for each element and based the calculation on the 3σ-IUPAC definition. However, they did not use an analytical blank to

Table 17.3. *Limit of detection for various elements in aluminum alloys*

Element and line (nm)	LOD (p.p.m.) echelle	LOD (p.p.m.) Czerny–Turner
Mg (285.21)	9	0.5
Cu (324.75)	33	10
Si (288.16)	90	14
Be (313.05)	1.6	0.3
Mn (403.08)	65	7

determine the standard deviation, but instead chose a region near the line where the spectrum is free of emission lines. Table 17.3 shows the LOD obtained for the same samples with the two spectroscopic systems [21].

The LODs obtained with the echelle spectrometer were derived from experiments performed in air, without fully optimizing the measurement conditions and with emission intensities transmitted through a fiber optic, whereas those obtained with the conventional spectrometer were generated without a fiber optic. Generally, the results showed that the values of LOD obtained with the two systems are of the same order of magnitude. The experiments performed with the echelle spectrometer were carried out in order to test recently acquired equipment, and it is certain that the instrumental limit was not reached. The gain of the intensifier could be significantly increased and acquisition conditions could be improved. One can suspect that the echelle spectrometer was less luminous because of its mode of dispersion, leading to higher limits of detection. Nevertheless, the results obtained at the time of this study showed that the level of performance of the echelle system in terms of detectability is quite comparable with that of classical instrumentation. Haisch *et al.* [16] conducted measurements with both systems in imaging the same plasma, and noted no significant differences, in terms of LODs, between the echelle spectrometer/ICCD and the conventional Czerny–Turner spectrometer with intensified diode array. Additionally, the LODs obtained for several elements on aluminum alloy samples by Bauer *et al.* [14] with a customized echelle spectrometer with ICCD are similar to those known in the LIBS literature obtained by using a conventional spectrometer coupled to IPDA or ICCD detectors.

17.5 Advantages and limitations

All of the results presented by Detalle *et al.* [21] present an assessment of only two experiments carried out over a one day period. This again shows that the echelle system allows for the acquisition of a significant quantity of data in a very reduced time, from which one could extract relevant physical or chemical information. Calibrations of various elements were obtained, limits of detection were determined, and the excitation and ionic temperatures as well as the electron density were measured. An evaluation of the reproducibility of the measurements was also achieved. These experiments revealed that the dynamic range of the detector (coding 16 bits) makes it possible to follow the intensity of the major elements in the sample as well as those of the minor elements or traces (e.g. 8 p.p.m. of Be). The good detectability for elements present at low concentrations was facilitated by the excellent resolution of this system, which serves to principally improve the signal to noise ratio. These results thus made it possible to highlight the intrinsic performance of the equipment. The analytical results were comparable with those obtained by associating a monochromator of the Czerny–Turner type with an intensified photodiode array. One should then ask the natural question of what then are the advantages and disadvantages presented by the echelle spectrometer coupled with an ICCD.

17.5.1 Advantages

The first significant advantage presented by the echelle system is its excellent resolution and its compactness. For example, the reciprocal linear dispersion of a 25 cm focal length echelle spectrometer (ESA 3000 system of LLA) is comparable with that of a Czerny–Turner spectrometer of one meter focal length including a grating of 3600 grooves per millimeter. This resolution makes it possible to limit spectral interference.

The second feature of this system is the access it gives to a very broad spectral range: 200–780 nm with LLA [22], or 200–1000 nm and less resolution with the Mechelle 7500 or Echelette SE 200 [22]. That makes it possible to monitor several lines of the same element. Thus, a large dynamic range of concentrations can be measured simultaneously, since a saturated line can be replaced by another one of the same element in the spectrum. Moreover, it is possible to increase the accuracy of the analysis by retaining several analytical lines for a given element. The system is also appropriate for the analysis of complex matrices such as alloys containing a great number of elements [18]. The possibility of covering a broad spectral range, combined with the good resolution of the system, always permits selection of a suitable line. For the same reason, the capacity of LIBS for multi-elemental analysis can be fully exploited. For example, in the work of Detalle *et al.* [21], the concentration of five elements was measured from each single shot. Of course, this does not constitute a limit to the number of elements that can be analyzed simultaneously. Multi-elemental analysis was also demonstrated by several LIBS groups [14, 16, 18, 19]. The possibility of reaching the wavelengths located beyond 500 nm also makes it possible to measure the atomic and molecular emission from atmospheric species.

An echelle system becomes, with the addition of a spectral data bank, a tool for instantaneous recognition of the elements present in an unknown sample, as long as the element is present in concentrations within the range usually detected by LIBS (starting from the p.p.m. level for most elements).

Another important feature is the possibility of normalization. One of the main difficulties presented by the LIBS technique is the variability of the recorded measurements, mainly caused by the great number of parameters that influence the laser–matter interaction. The solution often used for this problem is internal standardization. Generally, one calculates the ratio of the line intensity of the measured element by the line intensity of the major element of the matrix or the ratio of the line intensity to the continumm of the plasma. Other research groups showed the possibility of normalizing the analyte signals using the variation of the excitation temperature of the plasma [48]. This is difficult to realize with a classical system, but becomes straightforward with an echelle system. In the latter case, because the total emission spectrum is accessible, another solution is to follow the evolution in emission of all the elements present in the matrix and assess the sample knowing that the sum of concentrations must always be 100%. Finally, it should be possible to more fully exploit the information contained in the whole spectrum in order to correct for matrix effects or to obtain the plasma composition by using the LTE equations and the plasma temperature in a scheme called calibration-free LIBS by Palleschi's group [49] and described in Chapter 3 of this book. In all cases, the system becomes a tool that allows a new approach to the physics of the laser–material interaction and of the laser-induced plasma.

17.5.2 Limitations

Several limitations of the present equipment must, however, be mentioned. Firstly, despite the access to a broad spectrum offered by the echelle spectrometer, the detected spectrum depends on the sensitivity of the detector. For example, the response of the intensifier is not equal throughout the spectral range; it presents a maximum around 400 nm (see Figure 17.2 for the UV-enhanced intensifier). The sensitivity becomes low for wavelengths above 600 nm. The consequence is a loss of luminosity and resolution of the system for the highest wavelengths. It should also be noted that this mode of dispersion involves the loss of zones of the spectrum located between the different orders, called dead zones. Although this has little consequence for the UV region, the dimension of these dead zones increases with the wavelength to reach several nm beyond 700 nm. Thus, some spectral lines are no more accessible. Moreover, as we saw, the dispersive system involves a nonlinear response on the same order of dispersion. The total response of the spectrometer is thus the convolution of these two contributions (response in one order, and intensification system). Thus, for any physical measurement or absolute measurement of emission lines, it is necessary to correct the signal. This consists in recording, under the conditions of the experiment, a reference spectrum corresponding to the radiation of a black body, whose emission is linear in the spectral range covered by the spectrometer. Furthermore, in contrast to the Czerny–Turner configuration, which presents an almost constant linear dispersion, the mode of dispersion by an echelle spectrometer involves a change in resolution with the wavelength. This mode is problematic when a high resolution is needed for lines in the visible.

The second limitation is related to the full-well capacity of the pixel. The CCD detector can receive a limited number of electrons on each one of its pixels. When the gain or the gate width of the intensifier is increased, the strong emission lines will saturate some pixels of the CCD. The electrons accumulated on the saturated pixels will spill over into the adjacent pixels. This is the phenomenon of blooming, illustrated in Figure 17.14. This figure presents the image recorded by the CCD. One sees that blooming clearly appears for the

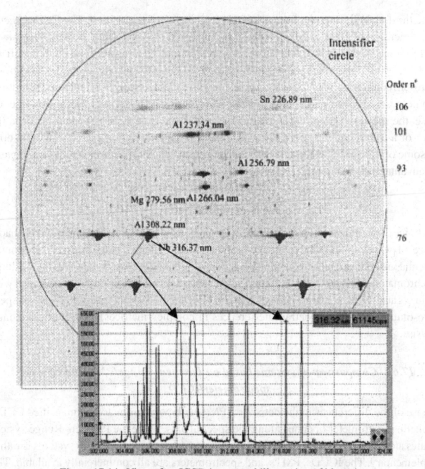

Figure 17.14. View of an ICCD image and illustration of blooming.

Al (308.22 nm) emission line. The pixels located below, which correspond to another order and other wavelengths, are affected by the signal of this emission line. The direct consequence is the appearance on the rebuilt linear spectrum (shown at the bottom of Figure 17.14) of one line corresponding to the theoretical position Nb (316.37 nm). Indeed, the sample is completely free of this element; the line does not exist, it is a "ghost line." It is thus always necessary to make sure that one works under conditions of non-saturation, or to check that the saturation does not affect the measurement of the elements of which one follows the emission. Moreover, if measurements are carried out under accumulation mode, there exists a risk of deterioration of the intensifier, if a too strong current is induced locally. The dynamic range of the system thus presents a limit, not because of a lack of gain of the intensifier, but because of the impossibility of eliminating the signal from the major lines, which was possible with a conventional system by selecting a spectral window that does not contain strong lines.

The third main limitation of the system is its low rate of acquisition. The CCD detector has a dimension of 1024×1024 pixels (it is composed of 1.05 million pixels). The rate of transfer of the PC board is 500 kHz for 16 bit resolution. The transfer time of all data from the CCD is thus higher than 2 s. One acquisition is really possible only every 3 s. Therefore, the present configuration of the system does not make it possible to carry out rapid sampling, often necessary for physico-chemical analysis or during analysis on a production line. Of course, the speed of analysis is a relative concept, but nowadays it is a main factor in the choice of a technique or a method over another. In the end, although LIBS analysis does lose some of its rapidity when it is associated with an echelle spectrometer system, it gains new advantageous features.

17.6 Choice of an optical setup for LIBS

The choice of an optical setup for LIBS will depend on the type of measurements to be made. Choice of wavelengths, analyte concentration range, dynamic range, matrix, and multi-elemental analysis are all of concern. We have seen that both systems, the Paschen–Runge polychromator equipped with photomultipliers, and the echelle spectrometer equipped with interline gated CCD or with intensified CCD, offer the necessary spectral coverage, spectral resolution (avoiding spectral interferences in complex matrices), and multi-elemental analysis.

17.6.1 Comparison of systems based on Paschen–Runge/PMT and echelle spectrometer/ICCD

Because of the differences in the detectors (PMT with polychromator or intensified CCD), an echelle spectrometer operates along the spectral axis, while the Paschen–Runge system operates along the temporal axis. Hence, the information gained from both systems is rather complementary. The ICCDs, PMTs, and spectrometers are all commercially available. The high-performance gated integrators needed for the PMT are commercially available but are expensive. However, low-performance gated integrators can be built in house from available components at lower cost ($200 per channel)[50]. Based on the price of spectrometers and different detectors in the market, it is believed that for the detection of a small number of elements (<10) the combination of a Paschen–Runge spectrometer with boxcar integrators can be a less expensive alternative to conventional systems or echelle spectrometer systems coupled with an ICCD [50]. For spectrochemical measurements with both systems, the Paschen–Runge/PMT [4–6] and echelle spectrometer/ICCD [14, 16, 19, 21, 50] show similar performances in terms of LOD and multi-elemental analysis.

17.6.2 Comparison of systems based on ICCD and interline CCD

Recently, we have seen the arrival of a new interline CCD detector system coupled to an echelle spectrometer [17, 22], which enables simultaneous measurements of spectral lines in a wide range with enhanced resolution. To our knowledge, no evaluation for LIBS applications in terms of plasma diagnostics or spectrochemistry of such a commercial system has been published. The interline CCD detectors have been commercialized by several companies (Roper, Hamamatsu, Dicom, etc.) and also integrated into the LIBS system using a conventional Czerny–Turner spectrometer ([51]; see also Chapter 9 of this book). More details about ICCD and CCD detectors can be found in the excellent book by Sweedler *et al.* [23]. We summarize here the relative advantages and drawbacks related to LIBS applications of such detectors whether coupled to a conventional or echelle spectrometer.

The interline CCD, which is a particular case of CCD, has wider simultaneous wavelength coverage at low cost, extremely low readout noise, excellent visible-to-NIR (<1 μm) spectral response, and, recently, good UV response [12]. Its drawbacks are mainly low dynamic range (with small pixels at least for those that are commercially available), slow gating capabilities (0.2–1 μs), and lower sensitivity due to a decrease in photosensitive area present at each pixel site. In particular this last drawback affects the LIBS sensitivity.

In contrast, the ICCD has fast gating capabilities (1–3 ns), allows detection of weak transient signals superimposed on an intense background, is usable in the vacuum UV, and has good dynamic range (related to the larger pixel size of the CCD available in the market). Its drawbacks include the high cost when compared with LIBS alternative technologies, fragility and relative complexity.

In its basic form for LIBS applications, the system based on the interline CCD can be used to achieve moderate detection limits (∼0.01%) at an affordable price. However, with the system using an incorporated image intensifier in the front of the CCD (or ICCD) the detection limit can be improved with the corresponding penalty of high cost.

Owing to the opacity and absorption of photons by the atmosphere or through the optical fiber, there are some difficulties involved in observing emission in the vacuum UV region below 185 nm. This requires the use of non-resonant lines for some elements like F, N, O, Cl, Br, I, and S in the near-infrared (NIR) region to enable the determination of these elements. The interline CCD with its high quantum efficiency seems to be an appropriate choice for the detection in the NIR. Until recently, the upper wavelength range of the ICCD was restricted by the sensitivity of the photocathode to about 850 nm. Since the beginning of 2001, this upper limit has been increased to 960 nm owing to the availability a new intensifier in the infrared (see Figure 17.2). However, the detection limits for these NIR lines are less than satisfactory for applications demanding high sensitivity.

17.7 Conclusions

For effective exploitation of the modern CTDs in spectrometry, the optical mounting of the spectrometer must be carefully adapted to the detector dimensions. When an ICCD or interline gated CCD is coupled with echelle spectrometers, a powerful system for measuring emission lines in LIP is created. This system has many desirable properties, including high resolution, high sensitivity, wide dynamic range, and large wavelength coverage. Furthermore it achieves simultaneous multi-channel detection of signal and background. Further, it has the ability to detect of part or all of the entire spectrum from 165 nm to

1000 nm and to store it digitally. The echelle spectrometer permits complete elemental analysis in a single shot, as spectral lines of major, minor and trace constituents, as well as plasma parameters, are measured simultaneously. This enables a real-time identification of unknown matrices and an improvement in the analytical precision by selecting several lines for the same element. As well, it improves the analytical figures of merit (precision, accuracy, and detection limit) and permits information on inhomogeneities of solid samples to be obtained.

Unfortunately, improvements in conventional CCD technology are required to accommodate the wide range of intensities found in LIBS and to address the blooming problem. In atomic emission spectroscopy, and particularly in LIBS, where strong and weak lines occur in close spatial proximity, blooming from strong analyte lines will typically create ghost lines, cause interference, and decrease the sensitivity and the dynamic range of the device. Moreover, the minimum ICCD readout time does not allow a sufficiently fast data acquisition rate, which may be critical for some applications.

Nevertheless, coupling the improvements in array detector and new software as well as new optical technologies promises to provide powerful tools for plasma diagnostic and spectrochemical analysis by LIBS.

The authors would like to thank Dr. Ralph Sturgeon, Dr. L. St-Onge, and Dr. Jean-Pierre Monchalin for their comments in reading this chapter.

17.8 References

[1] J. M Harnly and R. E. Fields, *Appl. Spectrosc.*, **51**(9) (1997), 334A–351A.

[2] F. M. Pennbaker, D. A. Jones, C. A. Gresham *et al.*, *J. Anal. Atom. Spectrom.*, **13**(9) (1998), 821–827.

[3] Q. S. Hanley, C. W. Earle, F. M. Pennebaker, S. P. Madden and M. B. Denton, *Anal. Chem.*, **68**(12) (1996), 661A–667A.

[4] R. E. Neuhauser, B. Ferstl, C. Haisch, U. Panne and R. Niessner, *Rev. Sci. Instrum.*, **70**(9) (1999), 3519–3522.

[5] M. Hemmerlin, R. Meilland, H. Falk, P. Wintjens and L. Paulard, *Spectrochim. Acta B*, **56** (2001), 661–669.

[6] R. Noll, H. Bette, A. Brysch *et al.*, *Spectrochim. Acta B*, **56** (2001), 637–649.

[7] M. J. Pillon, M. B. Denton, R. G. Scleicher, P. M Moran and S. B. Smith, *Appl. Spectrosc.*, **44**(10) (1990), 1613–1620.

[8] T. W. Barnard, M. J. Crockett, J. C Ivaldi and P. L Lundberg, *Anal. Chem.*, **65** (1993), 1225–1230.

[9] G. M Hieftje, *J. Anal. Atom. Spectrom.*, **11**, (1996), 613–622.

[10] A. T. Zander, *J. Anal. Atom. Spectrom.*, **13** (1998), 459–461.

[11] A. T. Zander, R.-L. Chien, C. B. Cooper and P. V. Wilson, *Anal. Chem.*, **71** (1999), 3332–3340.

[12] S. Luan, R. G. Schleicher, M. J. Pilon, F. D. Bulman and G. N. Coleman, *Spectrochim. Acta B*, **57** (2001), 1143–1157.

[13] L. Hiddemann, J. Uebbing, A. Ciocan, O. Desenne and K. Niemax, *Anal. Chim. Acta*, **283** (1994), 152–159.

[14] H. E. Bauer, F. Leis and K. Niemax, *Spectrochim. Acta B*, **53** (1998), 1815–1825.

[15] H. Becker-Ross and S. V. Florek, *Spectrochim. Acta B*, **52** (1997), 1367–1375.

[16] C. Haisch, U. Panne and R. Niessner, *Spectrochim. Acta B*, **53**, (1998), 1657–1667.

[17] P. Lindblom, *Anal. Chim. Acta*, **380** (1999), 353–361.

[18] S. R. Goode, S. L. Morgan, R. Hoskins and A. Oxsher, *J. Anal. Atom. Spectrom.*,
15(9) (2000), 1133–1138.

[19] P. Fichet, P. Mauchien, J.-F. Wagner and C. Moulin, *Anal. Chim. Acta*, **429** (2001),
269–278.

[20] S. Florek, C. Haisch, M. Okruss and H. Becker-Ross, *Spectrochim. Acta B*, **56**
(2001), 1027–1034.

[21] V. Detalle, R. Héon, M. Sabsabi and L. St-Onge, *Spectrochim. Acta B*, **56** (2001),
1011–1025.

[22] www.multichannel.se, 1999; www.lla.de, 1999; www.catalinasci.com, 2000.

[23] J. V. Sweedler, K. Ratzlaff and M. B. Denton, *Charge Transfer Device in
Spectroscopy* (New York: VCH Publishers, 1994).

[24] R. B. Bilhorn, J. V. Sweedler, P. M. Epperson and M. B. Denton, *Appl. Spectrosc.*,
41(7) (1987), 1114–1124.

[25] F. M. Pennebaker, R. H. Williams, J. A. Norris and M. B. Denton, Developments in
detectors in atomic spectroscopy. In *Advances in Atomic Spectroscopy*, ed.
J. Sneddon, volume 5, chapter 3 (Stamford, CT: JAI Press, 1999).

[26] A. P. Thorne, *Spectrophysics* (New York: Chapman and Hall, 1988).

[27] W. Demtroder, *Laser Spectroscopy: Basic Concepts and Instrumentation*, second
edition (Berlin: Springer-Verlag, 1981).

[28] K. W. Busch and A. M. Busch, *Multielement Detection Systems for Spectrochemical
Analysis* (New York: Wiley, 1990).

[29] J. Mika and T. Torok, *Analytical Emision Spectroscopy: Fundamentals*, translated by
A. P. Floyd (New York: Crane, Russak & Company, 1974), pp. 473–496.

[30] M. J. Pelletier, *Appl. Spectrosc.*, **44** (1987), 1699–1705.

[31] C. Earle, *Laser Focus World*, **10** (1999), 69–76.

[32] J. Olesik, *Spectrosc.*, **14** (1999), 36–42.

[33] M. Sabsabi and P. Cielo, *Appl. Spectrosc.*, **49** (1995), 499–510.

[34] V. Detalle, J. L. Lacour, P. Mauchien and A. Semerok, *Appl. Surf. Sci.*, **138–139**
(1999), 299–301.

[35] R. W. P. McWhirter, in *Plasma Diagnostic Techniques*, ed. R. H. Huddlestone and
S. L. Leonard (New York: Academic Press, 1965), pp. 201–264.

[36] G. V. Marr, *Plasma Spectroscopy* (New York: Elsevier, 1968).

[37] J. Z. Richter, *Astrophysik*, **61** (1965), 57–70.

[38] P. M. J. M. Boumans, *Theory of Spectrochemical Excitation* (London: Adam Hilger,
1966).

[39] H. R., Griem, *Plasma Spectroscopy* (New York: McGraw-Hill, 1964).

[40] H. W. Drawin, Validity conditions for local thermodynamic equilibrium. In
Progress in Plasmas and Gas Electronics, vol. 1, ed. R. Rompe and M. Steenbeck
(Berlin: Akademie Verlag, 1975), pp. 591–660.

[41] S. Vacquié, *Arc électrique et ses applications*, Tome 1 (Paris: CNRS, 1984).

[42] G. Bekefi, *Principles of Laser Plasmas* (New York: Wiley-Interscience, 1976).

[43] L. St-Onge, M. Sabsabi and P. Cielo, *J. Anal. Atom. Spectrom.*, **12** (1997), 997–1004.

[44] L. Radziemski and D. A. Cremers, *Laser-Induced Plasmas and Applications* (New
York: Marcel Dekker, 1989).

[45] A. Sreckovic, S. Burvi and S. Djenize, *Phys. Scripta*, **57** (1998, 225–227.

[46] W. W. Jones, A. Sanchez, J. R. Greig and H. R. Griem, *Phys. Rev. A*, **5** (1972), 2318–2328.

[47] C. Colón, G. Hatem, E. Verdugo, P. Ruiz and J. Campos, *J. Appl. Phys*, **73** (1993), 4752–4758.

[48] U. Panne, C. Haisch, M. Clara and R. Niessner, *Spectrochim. Acta B*, **53** (1998), 1957–1968

[49] A. Ciucci, M. Corsi, V. Palleschi *et al.*, *Appl. Spectrosc.*, **53**(8) (1999), 960–964.

[50] U. Panne, R. E. Neuhauser, M. Theisen, H. Fink and R. Niessner, *Spectrochim. Acta B*, 56 (2001) 839–850.

[51] D. Body and B. L. Chadwick, *Rev. Sci. Instrum.*, **72** (3) (2001), 1625–1629.

18

Spark-induced breakdown spectroscopy: a description of an electrically generated LIBS-like process for elemental analysis of airborne particulates and solid samples

Amy J. R. Hunter and Lawrence G. Piper

Physical Sciences Inc., USA

18.1 Introduction

Spark-induced breakdown spectroscopy (SIBS) is a plasma-based atomic emission analytical technique that draws from both traditional spark spectroscopy and laser-induced breakdown spectroscopy (LIBS). Like traditional spark spectroscopy, the plasma is formed electrically. Like LIBS, the sparks are generally made in ambient air and detection is timed to eliminate the initial very bright breakdown. SIBS has been applied to a variety of interesting and challenging analytical problems. This chapter describes the basic hardware required to perform SIBS analyses. Here we also present the application of SIBS to real-time monitoring of airborne microparticulate lead in workplace hygiene scenarios, and the use of SIBS as a field-screening analyzer for metal contamination in soils.

SIBS can be regarded as a "marriage" of two other pulsed plasma techniques for elemental analysis: laser-induced breakdown spectroscopy and traditional spark spectroscopy. The latter is generally performed upon conductive samples (metals) and currently is predominantly used in alloy analysis. The sample itself acts as the cathode, and an anode is brought very near to the surface. An electrical discharge is formed between the electrode and the sample and ablates some material while simultaneously creating a plasma. The ablated material from the sample is vaporized, atomized, and electronically excited in the plasma. Elemental components of the sample are identified by the presence of their persistent emission lines. This detection is performed either immediately after the plasma is created or after a delay to enable plasma cooling, and spectral simplification, to take place. The history of spark spectroscopy has been ongoing for 150 years, and innumerable journal articles, review articles and books have been devoted to the topic. A very small fraction of this literature is referred to in references [1–5].

Laser-induced breakdown spectroscopy (LIBS) is the topic of this book and requires no elaboration here; it suffices to say that it is a technique that is very similar to spark spectroscopy, except the spark is generated with an optically induced breakdown. The

Laser-Induced Breakdown Spectroscopy: Fundamentals and Applications, ed. Andrzej W. Miziolek, Vincenzo Palleschi and Israel Schechter. Published by Cambridge University Press. © A. W. Miziolek, V. Palleschi and I. Schechter 2006. A. W. Miziolek's contributions are a work of the United States Government and are not protected by copyright in the United States.

energy for spark initiation can therefore be delivered remotely to the surface or volume to be analyzed. LIBS can be used to analyze both conducting and insulating materials.

We are developing SIBS as a simple, low-cost alternative to LIBS. One of the advantages of LIBS is the ability to easily process and analyze non-conductive material, and SIBS has a similar capability because the spark is generated between two rod electrodes. Material residing between the active electrodes is incorporated into the spark-produced plasma, and can therefore be analyzed. As with LIBS, the plasma is pulsed, and detection is delayed until the plasma has expanded and cooled and a significant amount of recombination has taken place. The efficiency with which the material to be sampled is taken into the spark is a function of the characteristics of the material. Various strategies are being used to make the SIBS output quantitative for different types of samples.

SIBS development at Physical Sciences Inc. (PSI) originally focused on measuring lead in airborne particulate material, specifically microparticulates in indoor rifle ranges. Inhaling this material is known to be harmful, and the current analytical methods cannot protect personnel in real-time. With a simple power supply and a detection system composed of interference filters and miniaturized PMTs, we were able to measure lead in real-time with a detection limit of $10 \ \mu g \ m^{-3}$. Since that time, we have expanded the application of SIBS to measuring other airborne metals in real-time, developing a prototype field-screening monitor for trace metals in soil (with detection limits of about $20 \ mg \ kg^{-1}$ for Pb, Cr, Ba, and Hg) and monitoring major components in cement. This wide range of applications demonstrates the utility and versatility of SIBS. Future developments will involve process control, both of particles entrained in gas flow and of solid materials, and other environmental applications, especially water analysis.

In the following three sections we present details of our SIBS development. Section 18.2 discusses the basic components common to all SIBS systems. Section 18.3 goes into more detail about the implementation specifics for two different applications, aerosols monitoring and soils monitoring. Finally Section 18.4 presents results from our investigations related to each of the aforementioned applications.

18.2 Basic description of SIBS processes and hardware

This section focuses on basic hardware used to apply SIBS to a variety of measurement scenarios. The first subsection focuses on the production of the SIBS spark; the second discusses detection strategies. The following section addresses details for implementation-specific elemental analysis of airborne particulate and of soil.

18.2.1 Spark characteristics

The heart of the SIBS apparatus is the spark-generating system, which consists of a pulsed high-voltage power supply connected to a closely spaced pair of electrodes. A high-voltage pulse applied to the electrodes causes the air between them to break down with the resultant production of a high-temperature plasma (we have calculated Boltzmann temperatures from iron lines as high as 20 000 K). Sample material in this plasma is vaporized and ionized. As

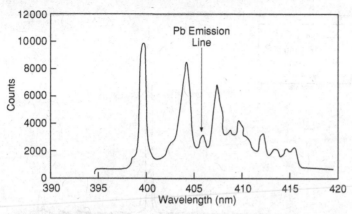

Figure 18.1. SIBS spectrum of Pb aerosol with no delay between spark and detection.

is the case with LIBS, the radiation arising from the SIBS plasma is analyzed to identify the species present.

The spark power supply integrates a standard pulsed, high-voltage supply (<2000 V, 100 J s^{-1}) with a capacitor bank. After the capacitors are charged, a trigger pulse initiates the creation of an ion channel between the electrodes through a high-voltage (15–40 kV), low-current pulse provided by a transformer. The threshold for this event is about 2.5 GW cm^{-1}. The ion channel creates a low-resistance path for the prompt discharge of the capacitor bank. The total discharge energy can be varied from 1 J to 5 J with <0.5 J arising from the initiator pulse. The spark repetition rate varies from a single pulse to about 10 Hz, the upper limit being determined by the power supply. Because of the large amount of current involved with the discharge of the capacitor bank (~ 500 A), the SIBS process can be considered a high-voltage breakdown followed by a classic arc discharge.

The spark occurs between two separated electrodes (typical gap is 5–6 mm) placed in or near the sample matrix. The electrode gap currently employed is the largest that enables reproducible spark formation. The electrodes are composed of proprietary materials, which have been chosen for high corrosion resistance, high melting temperature and low ablation propensity. Additionally, the electrode materials are unlikely analytes. In most cases, the electrode material does not add spectral complexity to the analyte signals.

The spark formed in SIBS has a visible volume on the order of 0.07 cm^3. This is considerably larger than a typical LIBS plasma (Radziemski and Cremers [6] report the spark volume from a 100 mJ Nd:YAG laser to be 0.003 cm^3). The greater sampling volume has some advantages, one of which is a greater likelihood of processing particles during each plasma event when measuring metals in airborne particulate material. Another advantage is that the amount of averaging necessary to achieve acceptable reproducibility often can be reduced, thus shortening analysis time.

The character of the light emitted in the breakdown region depends upon the delay after spark ignition. At times immediately after ignition (a few microseconds), the radiation is a spectral continuum characteristic of the emission of hot free electrons in the presence of

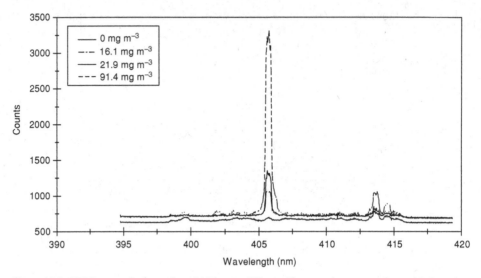

Figure 18.2. SIBS spectra in the region of 406 nm at different Pb-aerosol concentrations with detection delayed 75 µs after the excitation spark.

positive ions (Bremsstrahlung radiation), with little additional spectral structure, as shown in Figure 18.1. As the plasma cools, structured emission, characteristic of the species with high internal energy, principally ionized atoms, becomes more prominent. Continued cooling results in recombination of the ions with electrons and produces species of lower energy, which then emit light at their own characteristic wavelengths. As a result, the spectra become less complex at long times (>20 µs) after ignition, as shown in Figure 18.2. Typical delay times between 20 and 200 µs reject the bright continuum radiation due to plasma recombination and allow ready detection of the persistent atomic emission from the analyte metals. As in LIBS, the optimal detection delay is a function of power input: the higher the input power, the longer the optimal delay. Additionally, the optimal delay depends upon the element itself. For example, we have observed that sodium allows a longer delay for optimal detection sensitivity than does iron.

18.2.2 Detection systems

Analyzing the SIBS-produced plasma involves monitoring light emission intensities at a few selected wavelengths. Depending on the nature of the sample, the detection system is based on one or the other of two distinct strategies. The first, appropriate for spectral surveys to identify bright lines, as well as to analyze metals in complicated matrices, uses wavelength-dispersive spectroscopy. Here we use a grating spectrograph for the dispersion element. This approach is necessary in the analysis of soil, where iron is relatively abundant. Iron has a large number of lower-lying excited states that emit in the region where the features of the analytes of interest are normally found (300–450 nm). The ability to resolve spectrally the analyte features from iron lines is critical, since it is a dominant element in soils. Sometimes other elements also interfere with the analysis, but iron is the most common interferent.

In less complicated matrices, such as air, a combination of narrow-band interference filters and miniaturized photomultiplier tubes (PMTs) is used to detect the optical signals. The conditions under which this approach is used are very carefully screened beforehand with the wavelength-dispersive technique described above. In these studies, all likely interfering elements are added to the sample stream in quantities about $1000\times$ their normal abundance. If, in this situation, there are no observed spectroscopic interferences, the PMT approach is deemed acceptable.

The wavelength-dispersive detection system consists of a lens to collect light generated in the spark region, an optical fiber to transmit the light to the entrance slit of a spectrometer, and an intensified linear photodiode array at the exit plane of the spectrometer (the spectrometer/photodiode array configuration is known as an optical multichannel analyzer or OMA). For most of these studies, we have used a spectrometer having a 0.32 m focal length and a $2400\,\text{g mm}^{-1}$ grating. This system has a spectral dispersion varying from about $0.03\,\text{nm pixel}^{-1}$ at 250 nm to about $0.02\,\text{nm pixel}^{-1}$ at 600 nm. The overall spectral resolution, defined as the full-width at half-maximum of an isolated atomic line, varied from about 0.15 nm in the ultraviolet to about 0.10 nm in the visible. For wavelengths longer than 450 nm, a 420 nm cut-on glass filter is placed in front of the monochromator slit to remove second-order spectral features.

As mentioned above, the fluorescence detection relies upon a timing system that delays the detection of the atomic emission a suitable time after spark initiation to eliminate interference from the prompt spark-plasma emission. Our spark-generating power supply is driven by a delay generator that allows the sparks to occur at set intervals. Spark events trigger the detection system by activating a second delay generator. This generator delays the turn-on time of the OMA, so that the OMA will not be saturated by the very bright early plasma of the spark. The OMA is then kept activated for a set time (gate time) and detects emission from the cooling plasma.

By carefully choosing appropriate delay and gate times, we can optimize detection conditions for a given element. Generally, longer delay times are beneficial because, even though total intensity is reduced, the signal-to-noise ratio is better. As mentioned, typical delay times vary between 20 and 200 μs. Typical gate times are 10–50 μs. The OMA can be set to accumulate signal from a series of sparks or, as is often the case, the light from a single spark provides sufficient signal for a complete spectrum.

In the case of a relatively simple matrix, optical signals are collected radiometrically with a pair of filtered, miniature photomultiplier tubes. The delay time is typically 20–50 μs with an equivalent integration (gate) time. The advantages of the filter and PMT approach include simplicity and low cost. The strategy behind the radiometer is to accumulate the signal associated with the atomic line (on-line) with one filter and to subtract any background signal with a nearby filter (off-line) that has no analyte or electrode atomic features inside its bandwidth. For lead, the Pb(I) line at 405.8 nm is used with 400 nm for background. For chromium the Cr(I) line at 427.5 nm is used with 420 nm for background. The filters have pass-bands of 1 nm FWHM (full-width, half-maximum). Figure 18.3 shows the transmission curves of the on-line and off-line filters for lead analysis superimposed over a wavelength-dispersive spectrum taken of lead aerosol.

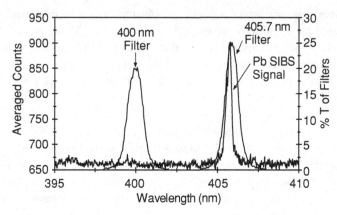

Figure 18.3. Transmission curves of interference filters superimposed on aerosol lead spectrum.

18.3 Application-specific considerations

18.3.1 Aerosol metals monitoring

Metals in airborne particulate are a recognized respiratory health hazard in many venues, especially in factory workplaces. Most monitoring of these materials is now performed with a relatively long-term filter collection period followed by laboratory analysis of species captured on the filter. Currently, there is no widespread method to monitor the hazardous components of airborne particles, and therefore no way to protect human respiratory health in real-time.

Aerosol monitoring hardware and analytical methods

Aerosol monitoring is accomplished by pumping contaminant-laden air through the spark gap at a known rate with a pump. Fiber optic cables transmit the optical emission from the spark-induced plasma to the detection system. Figure 18.4 shows a diagram of this apparatus.

Airborne particulate usually provides a relatively simple matrix and the PMT/interference filter approach is sufficient. The PMTs supply output voltages that are proportional to the amount of light in the pass-band of the interference filter. These analog signals are input into a 1.2 MHz A/D data acquisition board incorporated into a PC (100 MHz Pentium). The temporal traces of the PMT signals are analyzed using custom routines developed with Lab Windows CVI software [7]. The on-line and off-line temporal traces are subtracted and the difference is then integrated. This difference signal is then compared with a previously acquired calibration curve (see below) and presented in strip-chart fashion as concentration versus time. In most cases, the data are averaged and reported at 2 min intervals.

Recently, we have packaged the components of this breadboard into a single unit, as shown in Figure 18.5. The air to be sampled can be drawn though an optional cyclone, to eliminate non-respirable particles if desired, prior to flowing through the spark chamber.

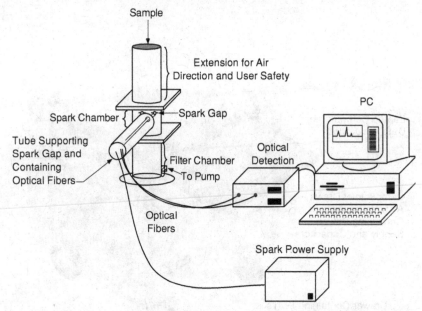

Figure 18.4. Diagram of laboratory breadboard SIBS monitor for lead in airborne particulate.

The air then passes through a filter, which is used to collect a sample for laboratory analysis and correlation, and exhausted out the rear of the unit. The sub-unit containing the sampling cell can be separated from the data acquisition sub-unit for monitoring in hazardous environments. In this situation, the sampling unit can be placed in the environment being monitored, and the user and data acquisition unit can safely operate from a remote location.

Dry aerosol calibration system

To calibrate the instrument for dry metal-laden aerosols we devised a system to produce well-characterized flows of dry aerosols. The method involves injecting droplets of known size and formation rate made from a solution of known metal concentration into the top of a drying column. As the droplets traverse the column, all water is evaporated, leaving solid particles of known size and concentration. We have used two different droplet generators in our work. The first of these was based on the design demonstrated by Berglund and Liu [7]. Similar systems have been employed by other researchers; see, for example, the work of Zynger and Crouch [8]. The operating principle involves the imposition of a periodic instability on a thin stream of fluid flowing through a pinhole orifice. The instability is induced by coupling a high-frequency vibration into the fluid using a piezoelectric transducer driven by a square wave function generator. Under certain conditions of fluid flow rate and instability frequency, monodisperse aerosols are generated. For our usual operating conditions, this system produced droplets 48 μm in diameter.

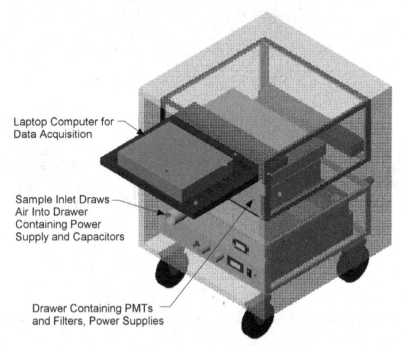

Laptop Computer for
Data Acquisition

Sample Inlet Draws
Air Into Drawer
Containing Power
Supply and Capacitors

Drawer Containing PMTs
and Filters, Power Supplies

Figure 18.5. Diagram of alpha-prototype SIBS monitor for lead in airborne particulate. This instrument was designed for application to workplace hygiene measurements.

The second droplet generator is a commercial atomizer nozzle (Sonotek). The Sonotek unit generates polydisperse droplets ultrasonically with a narrow distribution centered at 18 μm in diameter. It is much the easier of the two to operate, and the non-monodisperse character of the output does not affect the SIBS signals. This is primarily because of the large number of particles produced by the generator. It produces an output stream of 2.4×10^5 particles per second, so many particles (40–50) are processed by each spark.

The droplet generator is situated at the top of a Plexiglas tube in a downward-directed orientation toward the drying column (Figure 18.6). Air is drawn at a fixed rate through a porous, stainless steel inner sleeve in order to minimize loss of particles on the drying column walls. The flow rate ($0.028 \text{ m}^3 \text{ min}^{-1}$) exceeds the particle terminal velocities. The air is heated to 110 °C to evaporate solvent (water) from the droplets and is drawn out the bottom of the column by a filtered pumping system. The dry aerosol (0.1–3 μm diameter) is trapped on a filter placed at the bottom of a Plexiglas extension under the drying column that houses the spark chamber. This filter prevents particles from exiting the generator and contaminating the air pump. We have verified gravimetrically that 80%–95% of the dry mass exiting the droplet generator passes through the spark region to be collected by the filter.

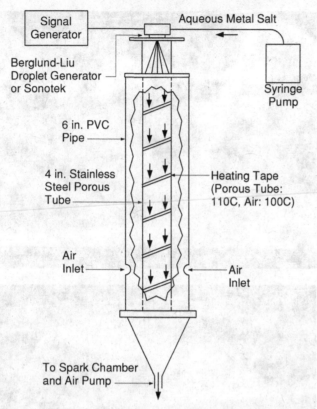

Figure 18.6. Dry aerosol generator with Berglund–Liu droplet generator.

Speciation, particle size and realistic samples in aerosol measurements

We have studied the possible effects of speciation on the airborne particulate signal in a variety of ways. Possible changes in signal level between various metal salts were easily studied with the previously described aerosol generator using dry aerosols of identical concentrations (440 μg m^{-3}) of lead chloride, nitrate and acetate. These aerosol concentrations were sampled one after another, allowing the signal to return to near baseline between additions. The resulting data showed no detectable distinction between the three types of lead species. Similar assays were performed with a variety of water-soluble chromium compounds.

Actual airborne analysis will, however, not likely involve water-soluble metal salt particles. Instead, these measurements will involve elemental and oxide particulate. It is more challenging to look for this type of speciation effect, since these compounds are not water-soluble and test particles cannot be generated with the dry aerosol generator previously described. For these tests, other methods of particle generation have had to be used. The SIBS signals obtained for these aerosols were time averaged and compared with correlative filter samples analyzed for total metal content by an independent laboratory.

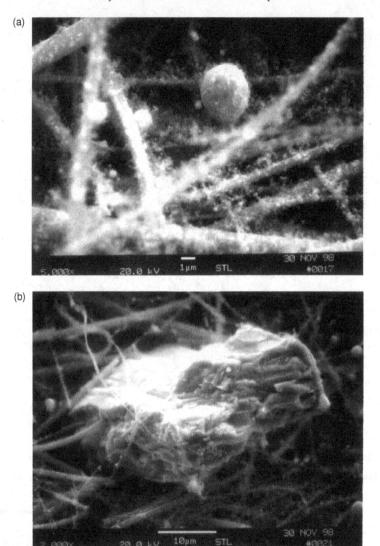

Figure 18.7. Gun plume particle elemental analysis – SEM photomicrographs showing (a) small and medium size particles and (b) a large particle. The larger particles are probably spalled bullet material remaining in the muzzle, the smaller ones created by combustion of the lead styphnate within the primer material.

The first of these test aerosols was generated by firing blanks from a handgun (Smith and Wesson 357 Magnum) in the laboratory. We drew samples of the plume from several shots through the SIBS system and subsequently trapped the particles on an internal filter. The particles trapped on the internal filter were submitted to an independent analytical laboratory for scanning electron microscopy (SEM) and energy-dispersive X-ray analysis

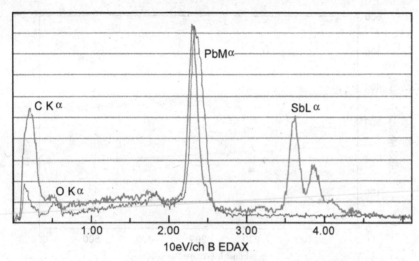

Figure 18.8. EDAX analysis of particles obtained by firing blanks from a handgun.

(EDAX) analysis to verify the presence of lead and lead oxide (Figures 18.7 and 18.8) and to quantify the total amount of lead collected. The SIBS lead mass was obtained by temporally integrating the data and convolving the air-sampling rate (1.3 cfm). The correlative data were obtained by acid digestion of the internal filters and subsequent ICP analysis of the digestate.

The integrated signal from the SIBS monitor was consistent with the total lead mass measured from the internal filters. The results showed 140 μg of lead collected on the filter compared with 144 μg determined by the SIBS monitor. A second comparison produced 69 μg of lead on the filter versus 100 μg predicted. This shows that the SIBS monitor responds to lead oxide similarly to other lead compounds. More thorough verification of the ability of the SIBS monitor to measure lead in elemental and oxide forms is currently under way.

The SEM analysis shown in Figures 18.7a and 18.7b also indicates at least three classes of particle size fractions, one at 50–100 nm in diameter, another at 0.5–3 μm and a third at 15–45 μm diameter. The largest are likely the result of residual spalled lead material in the gun barrel. From the SEM photographs, we estimate that the first two mass fractions account for the majority of the mass. Our initial analysis indicates that SIBS is accurately responding to these size lead/lead oxide particles (up to at least a few micrometers). The EDAX analysis (Figure 18.8) shows the presence not only of lead, but of oxygen, indicating the presence of at least an oxide layer on these particles. Also observed were antimony features (from the lead styphnate primer) and carbon (from filter material).

To further test the ability of SIBS to analyze realistic samples, we prepared an aqueous solution of lead nitrate and a small mass of coal ash. The ash used in this test had a median particle diameter of 0.5 μm and was composed mostly of iron, aluminum, and silicon oxide. This mixture readily passed through the ultrasonic aerosol generator to produce individual

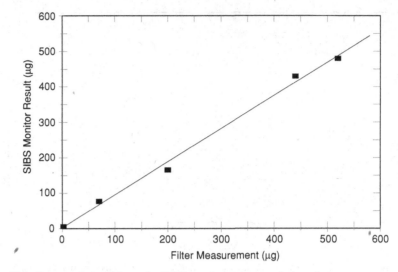

Figure 18.9. Integrated signal from the SIBS monitor versus total chromium mass measured from the internal filters. Good correlation between the two implies that SIBS monitoring of chromium is free from speciation effects, and also that real-time SIBS measurements could replace the filter-based ones.

and agglomerated ash particles containing condensed lead nitrate such that the final predicted lead concentration in air was 120 μg m^{-3}. The SIBS monitor produced the lead response predicted from our calibration curve (made with dilute solutions of lead nitrate) indicating that these particles, which are principally composed of aluminum, silicon, and iron oxides, are being vaporized and excited in the SIBS spark.

We also addressed the problem of chromium speciation and have determined that SIBS exhibits equivalent response to two chromium valences (Cr(III) and Cr(VI)). We used SIBS to measure chromium above a hard chromium plating tank and compared our results with those from filter samples of mixed Cr(III) and Cr(VI) oxides. Figure 18.9 shows the excellent correlation between the SIBS and filter results. The filter analysis is based on OSHA draft method ID-215, a filter digestion followed by ICP analysis [9]. The detection limit associated with this analysis is about 5 μg m^{-3}. This detection limit is similar to those usually reported in LIBS analyses. There have been optimized LIBS systems that are capable of monitoring airborne chromium down to 0.4 μg m^{-3}, but the hardware needed is more complex than that described here. Note also that the SIBS results are available in real-time, unlike those from the filter analysis.

Interference measurements

The primary interference in SIBS is spectral overlap of analyte emission lines with emission from other species present in the sample. This is an especially important issue in the case of the airborne metals monitors, because for the most part, signal acquisition is performed by narrow bandwidth interference filters and miniature photomultipliers. The 405.7 nm Pb atomic line and 400 nm off-line wavelengths were initially chosen because they were suitably free from electrode lines. To verify that these lines are free from interference by other species likely to be present in a real-world sample, we prepared solutions of iron, aluminum, silicon, calcium, and magnesium such that >1000 $\mu g\, m^{-3}$ concentrations would result in the aerosol generator.

Spectrally resolved data were then taken with a Princeton Instruments optical multi-channel analyzer/monochromator combination. The results indicated there to be no spectral interferences from these elements in these wavelength regions at the bandwidth (1 nm FWHM) of the radiometer filters. Similar tests have been conducted with lead paint samples where the following atomic lines in the 392–418 nm region are observed: Al(I) – 394.4 and 396.15 nm; Mn(I) – 403.31, 403.45, 404.51, and 406.35 nm; Fe(I) – 404.58 and 406.36 nm; Pb(I) – 405.78 nm. None of these features falls within the spectral bandwidth (1 nm FWHM) of the lead feature and so they are not expected to provide a spectral interference in the SIBS sampling. This is particularly important for airborne monitoring of lead paint abatement. We have also taken similar data for lead paint that show prompt Ti lines in this region from this paint component. Proper choice of delay time readily discriminates against these features. Additionally, they do not spectrally overlap the Pb line.

18.3.2 *Metals monitoring in soil*

In general, disposition and/or treatment of contaminated soil is controlled under TCLP (toxicity characteristic leaching protocol, found in 40CFR 266 Appendix VII) limits [10]. TCLP is a process where the soil is gathered (either from the surface or from core samples) and submitted to a laboratory leaching procedure. This procedure, performed with weak acids, has been developed to imitate long-term leaching conditions in a landfill, and regulations written around the possibility of ground-water contamination. Because the material is diluted $20 \times$ in solution during this test, if the amount of total metal in the sample is less than $20 \times$ the TCLP limit it cannot fail to meet the regulations. For this reason, $20 \times$ TCLP is our sensitivity goal for each metal in soil.

Despite the desire of the remediation community, there are no widely used field-screening methods for the on-site determination of metals in soil. There are two EPA-approved field-screening methods for metals in soil, a general method for portable X-ray fluorescence analyzer use [11] and an antibody-based method for mercury, but these methods are of limited value to most remediation efforts [12]. While the XRF analyzer is sensitive enough for some elements (Pb), it is inadequate for detecting others (Cr, Hg, etc.) [13]. We have started to develop a SIBS-based soil monitoring method which should work for a larger number of contaminant species.

Soil monitoring hardware and analytical methods

With airborne particle analysis, samples are positioned and refreshed by drawing air through the electrode gap with an air pump. Solid samples require a different approach. In our initial investigations, small quantities of the loose samples are placed on a Plexiglas sheet. The electrode assembly (see Figure 18.10) is then placed over the sample. In most cases, one of the electrodes is in direct contact with the soil sample. The spark process creates a

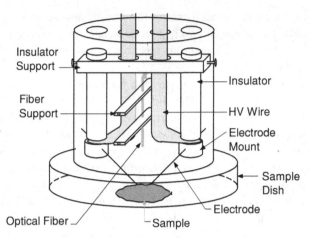

Figure 18.10. Diagram of SIBS soil analysis sampling interface.

shock wave that lifts some of the sample into the spark region where it is subsequently vaporized and excited to emitting electronic states. The shock waves can also disperse some of the sample, so the electrodes are repositioned, as needed, between sparks to ensure they remained in close proximity to the sample. The basic procedure has been to take 10 successive spectra, each one consisting of light collected from two successive sparks, and then to average the 10 spectra together. We generally acquire three such sets of spectra for each sample. The whole measurement process takes 2–3 min.

The electrodes are cleaned between samples by first swabbing them with a dilute solution of nitric acid, rinsing with deionized water, and finishing by swabbing with methanol. Then the system is allowed to spark in air for a minute or so before a new sample is poured out onto the Plexiglas sheet. This procedure guards against results being biased by material adhering to the electrodes from previous samples. The efficacy of this approach is proven; there is no effect of analysis order on signal level.

SIBS can be used to quantify metals in soils using standard addition analysis. This is a well-developed and accepted method in analytical chemistry in which known quantities of the analyte of interest are added to weighed sub-samples. These sub-samples are then analyzed. Extrapolation of the signal versus "added concentration" line to the y-axis yields the signal associated with the unknown. Dividing the y-axis intercept by the slope of the line yields the concentration of the unknown.

This method is important in quantitation because it allows the effects of the matrix to be removed from the analysis. Because soils are a widely varying and inhomogeneous matrix, standard addition analysis allows SIBS to be applied to different types of soil in a way that a single calibration curve cannot. Practically, this will not be onerous in fieldwork, because at each site there will usually be only a few types of soil present. In these cases, then, it will be possible to construct a standard addition plot for each soil type, and then use that plot as a calibration curve for other samples having the same physical and chemical characteristics. In our research, size of the sub-samples is usually about 1 g, but this can be varied with the textural scale of the soil.

Because no effort is currently made to control the amount of material processed with each discharge, there is considerable variation in intensity between spectra taken during a sample run and even between the 10-spectrum averages from each run at a given concentration. As a result, a meaningful standard addition curve cannot, in general, be made by plotting just the absolute intensity of the analyte lines for each sample at each concentration. There is, however, good linear correlation between the ratio of the intensity of the analyte features to one of several nearby iron lines, which are also excited in the spark plasma, and the concentration of metal in the sample. The reason for the normalization procedure is that differences in the line intensities between spark events are indicative of differences in the strength of the spark excitation itself and also in the amount of material present in the spark region during the analysis. The iron content of the soil sample does not vary with the amount of analyte in it as standard addition is applied (although, of course, it would be different for soils taken from different geographical locations), so provides a useful marker for both the strength of excitation of the sample and the amount of material in the spark region.

Spectral surveys to find optimum lines for analysis

There are a number of potential lines one could use for analysis of soil for metals content. Three important criteria in choosing an appropriate line for further investigation include the intensity of the line itself, whether it be free from interference by lines from other elements in the soil, and whether there be nearby lines from iron, or some other relatively abundant and constant component of soil. These lines can be used to normalize the contaminant spectral intensities so as to compensate for variations in spark strength and in amount of material in the spark region. The order of these priorities changes with characteristics of the analyte atom. The more toxic and tightly regulated the target analyte, the more sensitive must be the measurement. Therefore, in the case of mercury, finding a sensitive line is important (for example in order to pass TCLP, total mercury in soil must be <4 mg kg^{-1}). In other cases, emission lines of the analyte are bright, but the material is regulated at a much higher level. An example of this situation is the measurement of barium (which has a $20 \times$ TCLP limit of 2000 mg kg^{-1}).

The line intensities in the averaged spectra are calculated as peak heights from a local baseline. The local baselines for each spectral feature are calculated as an average of the local minima on either side of the line. This procedure is more essential for cases where there is some partial overlap with bright adjacent lines or when the background continuum radiation is non-uniform. This more detailed spectral analysis improves the overall data quality considerably compared with using a peak height above a common baseline, because of reduced scatter between replicate sets of data.

18.4 Applications and results

In order to demonstrate the versatility of SIBS, we present two applications: (a) the measurement of lead in airborne particulate, and (b) the analysis of soil for trace metals.

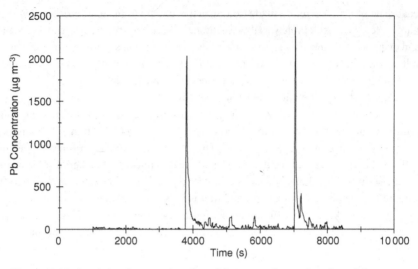

Figure 18.11. Lead signal versus time from firing range data, at location of shooters.

18.4.1 *Lead in airborne particulate application: firing range tests*

We performed two field tests of the airborne lead monitor at a nearby firing range. The range is a sophisticated facility with a filtered recirculating ventilation system. The airflow entrains the lead and carries it away from the shooters downrange through the ventilation system and into a set of HEPA filters. For the first test the lead monitor was positioned immediately in front of one of the 20 firing booths. Active handgun firing was occurring at a booth 10 ft to the right of the test position. During these tests 15 booths (shooter locations) were active.

Quantitative data acquired during this test are shown in Figure 18.11. The vast majority of the data show the measured lead concentration to be at or near our detection limit of 10 μg m^{-3}, well below the OSHA regulated level of 50 μg m^{-3}. A few events recorded lead concentrations exceeding 100 μg m^{-3}. The events showing lead concentrations above the detection limit involved a shooter discharging firearms directly above the sample inlet (vertical distance < 1 ft). The first of these occurred near 3900 s when four revolver rounds were fired in quick succession. The second event near 5900 s occurred when two full magazines from a semiautomatic pistol were fired, again directly over the sampling inlet of the instrument. Finally, in the time 7000–7300 s two different weapons were fired. It may be that in these cases the instrument is measuring larger particles as they settle into the sample inlet. Smaller particles (i.e., respirable ones) may still be pulled downrange by the air handling system.

In the time period 6000–7000 s we tested lead-free ammunition. No significant lead readings were obtained when this ammunition was discharged directly over our sample inlet. Additionally, at 5100 s approximately 12 individuals simultaneously fired two or

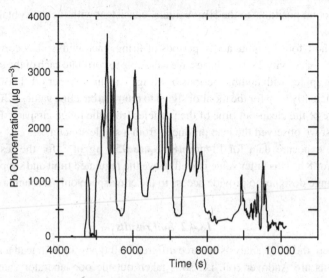

Figure 18.12. Lead signal versus time from firing range data, downrange directly before air filtration.

more clips consecutively. The signals appeared simultaneously with the smell of smoke at the sampling location.

This well-designed range is clean and free of lead contamination. Many other firing ranges, particularly those of older construction and in private use, often have transient airborne lead levels approaching 2000 μg m^{-3}. The time-weighted average (TWA) of lead data for the test period was 54.3 μg m^{-3}, which is above the regulated 50 μg m^{-3} limit. However, this is an extreme exposure upper limit as the majority of the data were deliberately introduced by firing above the instrument inlet and were not representative of the shooter's airspace. Even with this caveat, however, an 8 h TWA of the data works out to 13.5 μg m^{-3}, well below the action level because the shooting during the test period was the only range activity on that day.

A second test was also performed at this range. For this entry, the lead monitor was placed downrange in the ventilation crawlspace immediately before the contaminated air passed through the filters. The data are shown in Figure 18.12. We have excluded all of the data before 4000 s because this was before any firing occurred and there were no measurable lead signals.

Slightly before 5000 s the ventilation system was activated. The turbulent airflow agitated lead particulate in the ventilation room, producing the first visible lead signal in this data set.

When active firing took place the concentration reached as high as 3000 μg m^{-3} in the ventilation room. This is consistent with published values in firing ranges without active, filtered ventilation systems. This sophisticated range reaches this level only in the ventilation crawlspace just before the filters. The airborne lead and lead oxide are carried by the airflow from the shooting positions to the HEPA filters, consistent with the data obtained with our SIBS monitor.

The data show four separate active periods of firing, each with spikes corresponding to different levels of activity. During the test we were able to correlate each of the enhanced lead concentration spikes with audible periods of firing. The inactive periods between shooting were not sufficiently long for the monitor signal to return to baseline values. This is probably representative of the clean-out time of the particles from the range airspace. Finally, at the end of the test we observed the lead produced from a single shooter.

The TWA exposure data for Figure 18.12 are 950 μg m^{-3} for the test period and 178 μg m^{-3} for 8 h. This latter value is well above the regulated limit and supports the decision of the range designer to provide access to the crawlspace only through a sealed panel.

18.4.2 Soil results

We present here the SIBS analysis of three different soil types for both lead and chromium. The three soils are Andover soil, i.e. soils taken outside our laboratory, and two NIST Standard Reference Material (SRM) soils, SRM 2709, or San Joaquin soil, and SRM 2710, or Montana soil. The SRM soils were certified to have specific levels of lead and chromium. We sent samples of the Andover soil to an outside laboratory (Chelmsford, MA) for analysis for independent verification of our results. We also did a limited number of tests using fine SiO$_2$ particles as a simulant for sand.

Initially, the Andover soil samples were tested in three forms: loose, compressed into pellets, and compressed with a binder compound. These forms were chosen because of their similarity to those used in the XRF analysis protocol. Loose samples were prepared by drying to remove residual water and manually separating out large stones. Drying is necessary to allow comparison between our method and other laboratory methods, which report metals content in dried samples. To prepare compressed pellets, small aliquots ($\sim$0.5 g) of soils were pressed in a hydraulic press to between 8 and 9 tons. The pellets so prepared were 13 mm in diameter and about 2 mm thick.

Compressing the soil with added binder (17% cellulose) improved the integrity of the pellets, but at a cost of greatly reduced signal levels. These pellets were quite hard, and the sparks ablated only very small amounts of material from them. Compared with loose soil sample signals, pelletizing the Andover soil resulted in reductions in signal levels by factors of 3 to 5 when no binder was used and by factors of 10 to 20 when binder was added. The reduced signal levels had the effect of increasing the uncertainty in our measurements substantially.

A typical spectrum of loose Andover soil is shown in Figure 18.13. The most prominent features in the spectrum are iron lines at 404.58, 406.36, and 407.17 nm, and a line at 407.77 nm which is a strontium line. The lead line appears at 405.78 nm.

As previously detailed, data analysis involved measuring the peak height of the feature of interest, subtracting a local baseline and normalizing the intensity to that of a nearby iron line. The relationships between intensity ratios and concentration of lead in the sample are

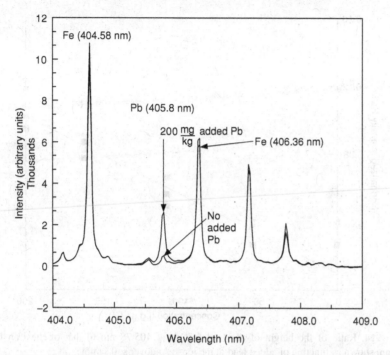

Figure 18.13. Spectra of Andover soil between 404 and 409 nm with and without added lead.

shown for the three samples in Figures 18.14–18.16 for loose soil, pelletized soil, and sand respectively. In Figure 18.14 the 404.58 nm iron line was used for normalization. In Figures 18.15 and 18.16 it was the 406.36 nm line. Derived sensitivities were the same regardless of which of the three iron lines was used for normalization (normalizing to the 405.17 nm line also gave results similar to those shown here). All plots were well behaved and linear.

When pelletized, the Andover soil had (Pb/Fe) ratios similar to the loose soils, but overall signal intensities were much lower. The net effect of this overall decrease in intensity was an increase in variability in the measurement and therefore an increase in the uncertainty in quantitation. The SIBS spark is largely a thermal event that happens near to the sample surface, as opposed to the LIBS process where energy is directly focused onto a surface for analysis. In the case of SIBS, shock waves can raise loose material into the thermal plasma, but have little effect in ablating material from the surfaces of compressed pellets. Thus, in the case of pellets, only a relatively small amount of material is removed into the plasma via thermal ablation induced when the outer edges of the plasma impinge on the pellet's surface. Despite the drop in overall intensity in the case of the pellets, it is important to note that the means of the signal from both the pelletized and loose samples overlay each other. The best-fit lines of both sample types coincide, but almost all of the raw data from the pelletized sample lie outside the ±1σ range of the loose soil data. This increased variability is reflected in the higher detection limits associated with pelletized samples shown in Table 18.1.

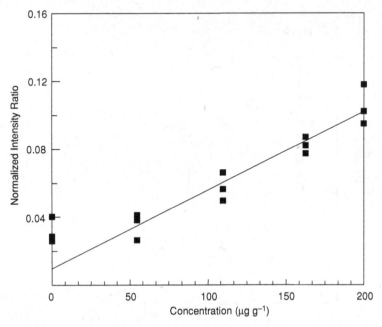

Figure 18.14. Ratio of the height of the lead feature at 405.78 nm to that of the iron line at 404.58 nm shown as function of added lead in the loose Andover soil sample.

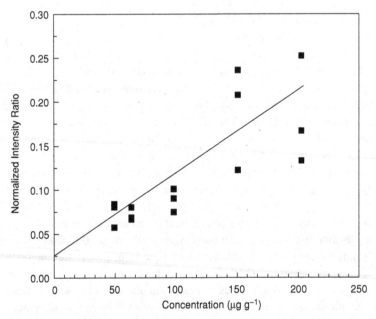

Figure 18.15. Ratio of the height of the lead feature at 405.78 nm to that of the iron line at 406.36 nm as a function of added lead in pelletized Andover soil.

Table 18.1. *Loose versus pelletized versus pelletized with binder SA results*

Preparation technique	Detection limits ($1\,\sigma$)
Loose soil (dried)	<20 mg kg^{-1}
Pelletized (dried)	~30 mg kg^{-1}
Pelletized with binders (dried)	~70 mg kg^{-1}

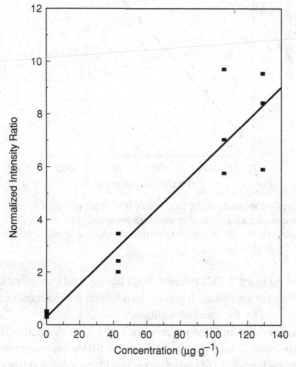

Figure 18.16. Ratio of the height of the lead feature at 405.78 nm to that of the iron line at 406.36 nm as a function of added lead in loose sand.

Samples were also prepared with added levels of 500 mg kg^{-1} and 1000 mg kg^{-1} of lead to examine calibration curve linearity at high levels of contamination. Figure 18.17 shows a standard addition curve combining data taken at lower levels with the results of soils spiked at higher levels. The curve is linear up to 500 mg kg^{-1} of added lead, but falls off at the highest concentration plotted. The intercept from this plot (fit through 500 mg of added Pb per kilogram of soil) gives results consistent with our data shown in Figure 18.14 as well as with out-of-house laboratory analysis, namely a natural level of lead in the soil of 33 ± 18 mg Pb per kilogram of soil. This rollover will probably not affect actual use of a field-screening monitor since it occurs at high lead concentrations ($\gg100$ mg kg^{-1},

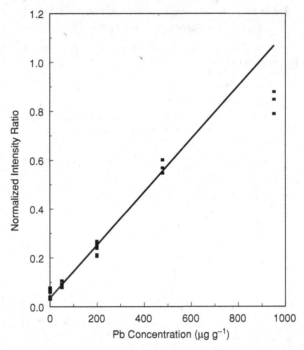

Figure 18.17. Ratio of the intensity of the lead line at 405.78 nm to that of the iron line at 404.58 nm as a function of the concentration of lead in loose Andover soil. The results show good linearity up to 500 mg Pb per kilogram of soil, but some fall-off in response at higher levels.

which is the 20×5 mg kg^{-1} TCLP limit). Soils having lead concentrations so high that remediation is necessary are easily identified, even though the exact concentration may not be accurately measured by the standard addition.

Analysis of the San Joaquin SRM material for lead was performed in the same manner as the Andover soil: that is each data point from the SRMs represents the average of 10 successive spectra, each made up of two spark events. Using loose soil (the pelletized SRMs exhibited an even greater signal drop than the pelletized Andover soil) and the standard addition technique, we took three sets of spectra for each sample concentration. Figure 18.18 shows a graph of the ratio of the intensity of the lead line at 405.78 nm to the iron line at 406.4 nm as a function of lead added to the San Joaquin soil sample. The data exhibit good linearity and low scatter. The original concentration of lead in the soil can be inferred from the ratio at the intercept at zero added lead to the slope of the standard addition curve. This value is 41 ± 18 mg Pb per kilogram of soil. This value is about a factor of 2 larger than the certified value of 19 mg Pb per kilogram of soil but does agree within our measurement uncertainty.

Comparing the San Joaquin reference material with the Andover soil illustrates the effect of iron content on instrument response. The slopes of the standard addition curves of the

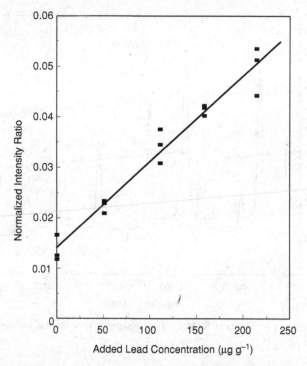

Figure 18.18. Ratio of the intensity of the lead line at 405.78 nm to that of the iron line at 406.36 nm as a function of lead concentration in San Joaquin soil (SRM 2709).

Andover soil and of the San Joaquin SRM, that is the ratio of the intensity of the lead line at 405.78 nm to the iron line at 406.4 nm, are 3.2×10^{-4} (intensity ratio/(milligrams of Pb per kilogram of soil)) and 9.8×10^{-4} (intensity ratio/(milligrams of Pb per kilogram of soil)), respectively. This difference is caused by the different iron concentrations in these samples (0.8% versus 3.3%).

The Montana SRM has a very high certified value for lead (5%). It is quite clear from the spectrum (Figure 18.19) that the lead feature is dominant. The intensity of the lead line at 405.78 nm is comparable to those of the iron lines in this sample.

Because of the very similar iron content of the two SRMs, we could estimate the lead content of the Montana soil using the San Joaquin soil standard addition curve. This procedure results in an estimate 4700 mg Pb per kilogram of soil for the Montana soil (certified value is 5532 mg kg^{-1}). Our estimate is within 15% of the NIST value, thus demonstrating the potential of the SIBS technology as a quick screening tool.

We also used SIBS to test soil for chromium. The spectra of samples with and without added chromium are shown in Figure 18.20. Figures 18.21–18.23 show standard addition curves for chromium in Andover soil, San Joaquin and Montana SRMs, respectively. The graphs indicate original chromium concentration in each soil of 18 ± 27 mg Cr per kilogram

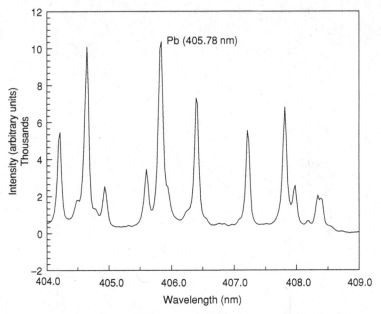

Figure 18.19. Spectrum of Montana SRM (2710) showing a very intense emission from lead at 405.78 nm.

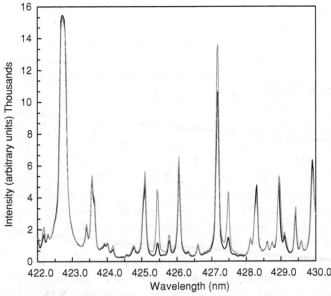

Figure 18.20. Spectra of Montana soil between 422 and 430 nm with (solid black line) and without (gray line) added chromium.

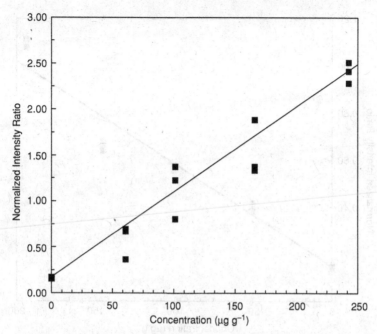

Figure 18.21. Ratio of the height of the chromium feature at 425.43 nm to that of the iron line at 426.06 nm as a function of added chromium in loose Andover soil.

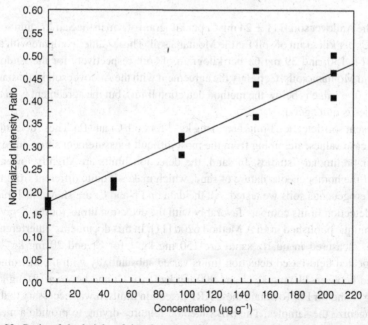

Figure 18.22. Ratio of the height of the chromium feature at 425.43 nm to that of the iron line at 427.15 nm as a function of added chromium in loose San Joaquin soil.

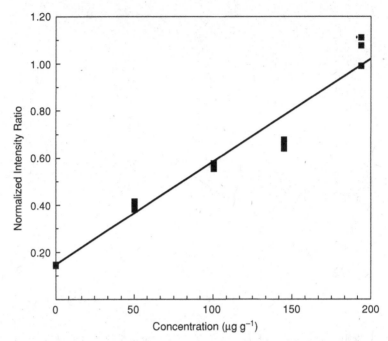

Figure 18.23. Ratio of the height of the chromium feature at 425.43 nm to that of the iron line at 427.15 nm as a function of added chromium in loose Montana soil.

of soil in the Andover soil, 111 ± 24 mg Cr per kilogram of soil in the San Joaquin soil and 33 ± 17 mg Cr per kilogram of soil for the Montana soil. These values compare with analyzed values of 14, 130, and 39 mg Cr per kilogram of soil, respectively for the Andover, San Joaquin and Montana soils. Certainly the agreement with the Andover soil is not statistically significant (the value is below the method detection limit), but the agreement with the other two samples is quite good.

Our present soil detection limits are 20 mg kg^{-1} for both Cr and Pb. These detection limits are worst-case values, stemming from the most difficult measurement situations examined during our preliminary studies. In sand, the detection limits are slightly lower. This is because of the homogeneous nature of sand, which makes it quite different from the other highly heterogeneous soils we tested. All the data on Pb and Cr are shown in Table 18.2.

These detection limits compare favorably with the detection limits for the X-ray fluorescence technique published in EPA Method 6200 [11]. In this document, "interference-free" limits (i.e. measured in quartz sand) are 150 mg kg^{-1} for Cr and 20 mg kg^{-1} for Pb. Actual repeated field-based detection limits varied substantially with the instrument used and ranged between 110 mg kg^{-1} and 900 mg kg^{-1} for Cr and between 40 mg kg^{-1} and 100 mg kg^{-1} and for Pb. To obtain these XRF detection limits, it was necessary to dry, grind and homogenize the samples. The SIBS method requires drying to provide a meaningful dry weight, but no other processing besides the application of the standard addition.

Table 18.2. *Certified values/lab measurements and SIBS values for Pb and Cr*

Soil	SIBS analysis (mg kg^{-1})		Independent lab analysis (mg kg^{-1})		NIST certification (mg kg^{-1})	
	Pb	Cr	Pb	Cr	Pb	Cr
Andover soil	32 ± 3^a	18 ± 27	43 ± 4	14 ± 4	NA	NA
NIST SRM 2709 San Joaquin soil	41 ± 18	111 ± 24	< 4	73	18.9 ± 0.5	130 ± 4
NIST SRM 2710 Montana soil	4700	33 ± 17	5600	20	5532 ± 80	29^b

^{a} Lower error associated with increased number of averages.
^{b} NIST analyzed, but not certified.

Compound sensitivity

During our development of SIBS for applications to metal-containing airborne particulate material, we have carefully evaluated the sensitivity of the method to different chemical formulations of the same analyte. We did this both for various forms of lead-bearing particulate (Pb, PbO, Pb(NO$_3$)$_2$) as well as for chromium aerosols of differing chemical identifies. In all cases, our SIBS-based measurements were in good agreement with the filter samples, which demonstrates that the chemical form of the metal does not influence the system response.

We have revisited this issue in developing SIBS for soils analysis because of a recent publication documenting such behavior in LIBS measurement of Ba, Cr and Pb in soil. In this publication, Cremers and co-workers doped soils to high levels (0.1%) with various lead and barium compounds [14, 15]. They reported a large amount of variability in signal intensity in the Pb(I) line at 405.8 nm (normalized to a Cr(I) line) in the compound series PbO, PbCO$_3$, PbCl$_2$, PbSO$_4$ and Pb(NO$_3$)$_2$. In their analysis, for example, lead chloride had three times the intensity of lead nitrate. This subsection documents preliminary efforts to uncover any changes in signal levels associated with chemical form. To date we have investigated this issue for soils containing two chemical forms of lead.

For the lead studies, we doped two samples of Andover soil at the 200 mg kg^{-1} level using two different lead compounds. ICP solutions of lead nitrate Pb(NO$_3$)$_2$ and lead chloride (PbCl$_2$) were used for this determination. These solutions were diluted with distilled water, quantitative aliquots were added to the soils, followed by drying, all in the manner described above. Following the sample preparation, the lead content of each sample was measured. The soils doped with lead nitrate and lead chloride had identical intensity ratios (Pb/Fe). We will address this important issue in more detail in future work.

Potential interference from organic contamination

In order to test the effects of organic contamination on the measurements, we divided a sample of soil which had been spiked previously with lead to a level of 225 mg Pb per

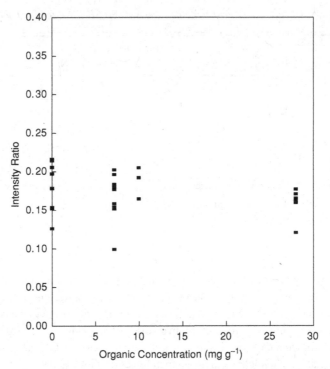

Figure 18.24. Ratios of the height of the lead feature at 405.78 nm to that of the iron feature at 404.58 nm in loose Andover soil with various quantities of organic contamination.

kilogram of soil into three portions. One we left alone, and the other two we spiked with organic material. After several measurements on one of the samples, we contaminated it further and made some additional determinations. The organic spiking procedure involved dissolving a measured amount of vacuum pump oil in trichloroethylene. Then, measured quantities of this solution were added to the previously weighed soil samples, the samples were thoroughly mixed and allowed to stand overnight at room temperature.

The results are shown in Figure 18.24, which displays repeated measurements of the ratio of the lead signal at 405.78 nm to that of the iron line at 406.36 nm as a function of organic contamination in the soil. Because of its high volatility and high surface area of the samples, we expect that most of the trichloroethylene evaporated from the samples prior to testing. Thus, the concentrations of organic contaminant shown in Figure 18.24 refer to the amount of pump oil added to the soil samples. Although the data are a bit more scattered than in some of our previous runs, they show that quite high levels of organic contamination do not interfere with the determination of lead in the sample. That is, within statistical uncertainties, ratios at all levels of organic contamination are the same. The total signal levels were reduced somewhat at higher levels of organic contamination, but the intensity ratios were not. Thus our technique is equally well suited to soils containing high levels of organic contamination as well as to those that are relatively clean.

Table 18.3. *Comparison of SIBS and LIBS*

	SIBS	LIBS
Excitation system	Electrical	Optical
Excitation power	1–6 J	75–350 mJ
Detection systems applied	Wavelength dispersive and filter based	Wavelength dispersive only
Development of fieldable system?	On-going	Accomplished
Detection limits (using Pb as an example)	In air 5 μg m^{-3} In soil 25 mg kg^{-1}	In air 190 μg m^{-3} [16] In soil 57 mg kg^{-1} [15]
Standoff detection possible?	No	Yes

18.5 Discussion and future directions

The set of applications described here shows the wide range of utility of SIBS. SIBS can be applied both to the elemental analyses of airborne particulate material and to solids, both soils and powders. An advantage of SIBS in both of these arenas is the high-energy spark. The large amount of energy deposited allows a large plasma to be generated, which helps to ensure good sampling statistics and sampling in the aerosol monitoring case. For airborne particle analysis, there are definite advantages to using an electrically generated 5 J spark over an optically generated 100 mJ spark (these sources having similar cost). In the case where the possibility of spectral interference has been evaluated and is low, this energetic excitation source can be coupled with the simple and inexpensive interference filter and PMT detection. This results in a simple and easy-to-use, real-time air monitoring system that can be used to protect workers or optimize processes.

SIBS solids applications also benefit from the large and energetic excitation spark. In this case, it allows processing of a larger sample without changing spark characteristics. In soil sampling, we are able to achieve detection limits in 10 or 15 min that are nearly as sensitive and reproducible as an analytical lab using a process taking several hours per sample, not including sampling and transport time. This type of quick analytical instrumentation is highly desired as field-screening for toxic metals in soil. In order for the device to be sufficiently portable, we are currently working on hardware miniaturization of the soil system. Table 18.3 briefly compares the SIBS results discussed here with LIBS characteristics from literature. SIBS and LIBS are very similar in both results and application. The most important differences are the fact that LIBS can be used in a stand-off fashion and SIBS has a simpler and more energetic plasma source.

The authors acknowledge the contributions of the many people at PSI who have contributed to the development of SIBS. Most particularly, we would like to acknowledge Dr. Karl Holtzclaw and Dr. Mark Fraser, who made significant efforts early in the development of SIBS. Additionally, we acknowledge the newer members of the SIBS team, Dr. Rick Wainner and Dr. Brian Decker, who have been involved in furthering the soil applications, and in the airborne lead monitor development, respectively. Dr. Steven Davis has been our constant supporter, both personally and professionally, and has our deep gratitude. This work has been funded by the US Department of Energy, US Environmental Protection Agency, US Army, and the National Institute of Occupational Safety and Health.

18.6 References

[1] R. M. Barnes (editor), *Emission Spectroscopy* (Stroudsburg, PA: Dowden, Hutchison & Ross, 1973).

[2] H. R. Griem, *Principles of Plasma Spectroscopy* (Cambridge: Cambridge University Press, 1997).

[3] M. N. Hirsch and H. J. Oskam (editors), *Gaseous Electronics, Volume 1: Electrical Discharges* (New York: Academic Press, 1978).

[4] J. M. Meek and J. D. Craggs (editors), *Electrical Breakdown of Gases* (New York: John Wiley and Sons, 1978).

[5] R. Payling, D. G. Jones and A. Bengtson (editors), *Glow Discharge Optical Emission Spectroscopy* (Chichester: John Wiley and Sons, 1997).

[6] L. J. Radziemski and D. A. Cremers, Spectrochemical analysis using laser plasma excitation. In *Laser Induced Plasmas and Applications*, L. J. Radziemski and D. A. Cremers (editors), chapter 7 (New York: Marcel Dekker, 1989), p. 295.

[7] R. N. Berglund and B. Y. Liu, *Anal. Chem.*, **7** (1973), 147.

[8] J. Zynger and S. R. Crouch, *Appl. Spectrosc.*, **29** (1975), 244.

[9] *Hexavalent Chromium in Workplace Atmospheres* (OSHA Method No. ID-215) (Salt Lake City, UT: OSHA, 1998).

[10] M. Martin and M.-D. Cheng, *Appl. Spectrosc.*, **54** (2000), 1279.

[11] Code of Federal Regulations, Title 40, Volume 17, Parts 266–299, 40CFR266, pp. 5–119, *Standards for the Management of Specific Hazardous Wastes and Specific Types of Hazardous Waste Management Facilities*, Revised as of July 1, 1997.

[12] US Environmental Protection Agency, Method 6200, *Field Portable X-Ray Fluorescence Spectrometry for the Determination of Elemental Concentrations in Soil and Sediment*, Revision 0, January 1998.

[13] US Environmental Protection Agency, Method 4500, *Mercury in Soil by Immunoassay*, Revision 0, January 1998.

[14] M. E. Fraser, T. Panagiotou, A. J. R. Hunter *et al.*, *Plating and Surface Finishing*, **87** (2000), 82.

[15] A. S. Eppler, D. A. Cremers, D. D. Hickmott, M. J. Ferris and A. C. Koskelo, *Appl. Spectrosc.*, **50** (1996), 1175.

[16] B. T. Fisher, H. A. Johnsen, S. G. Buckley and D. W. Hahn, *Appl. Spectrosc.*, **55** (2001), 1312.

Index

Page numbers in italic refer to figures. Page numbers in bold signify entries in tables.

ablation of sample surface 15–16, *16*
  effect of pulse duration 478–9
  stochastic 125–7
absorption/shadow photographic plasma imaging
    54–5, *55*
  time-integrated morphology 104–7
acousto-optic tunable filter (AOTF) 31–2
  scanning of laser-enhanced ionization 58–9
  time-resolved plasma imaging 46–8, *48*, *49*
aerosols
  chemical analysis 203–9, *206*
    toxic metal limits for industrial plants **204**
  fundamental principles 194–203
    EMEP source categories **201**
    sources and sinks of atmospheric particles *197*
    typical composition *195*
    worldwide anthropogenic emissions **200**
    worldwide atmospheric emissions **198**
  LIBS analysis 217–18, 222
    applications 242–5
    conditional analysis 227–31
    ensemble averaging 222–7, *223*, *230*
    future directions 245
    indirect analysis 236–8, *237*, *238*
    individual particle analysis 231–6, *234*
    principal methods *218*
    sampling 218–21, *221*
    SEM images *239–40*
ambient atmosphere, effect on plasma emission 63,
    76–7
analytical applications of LIBS
  advantages 17
    in situ analysis 20
    sample preparation 19
    speed of analysis 20
    variety of measurement scenarios 17–19, *18*
  combination techniques 26–7
    laser-induced fluorescence (LIBS-LIF) 26
  matrix effects 21–2

    performance 23–6
      detection limits **24–5**, **166–7**
    safety considerations 23
    sample homogeneity 20–1
      bulk non-uniformity 21
      non-representative surface composition 21
    sampling geometry 22–3
angle of incidence 91, *93*
atomic emission spectroscopy (AES) 3–4

bacteria, LIBS analysis 308–9
Balmer series atomic spectra 176–7, *177*
Bessel function 174–5
biomedical applications 282–3, 309
  bio-fluids, investigations of 301–4, *304*
  calcified tissue materials 283–4
    bones 291–2, *292*
    calibration standards 294, *295*
    methodology 284–5
    other bio-samples 292–4, *293*
    teeth 285–91, *286*
  microscopic bio-samples 304–9, *306*, *307*
  soft tissue materials
    human and animal tissue 295–7
    plant tissue 297–301, *299*, *300*
    soft-tissue organisms 301
blooming 578–9, *579*
Boltzmann distribution 128
Boltzmann plot 130, *131*, *138*
bones, LIBS analysis 291–2, *292*

calibration curves for quantitative analysis 150–1
  correction methods for shot-to-shot fluctuations
    151–2
  single shot analysis 152–3
calibration function 136–7
  analytical 137–46, *141*, *142*
CFD-ACE computer program for solving fluid
    equations 187–9

615

characteristics of LIBS 123–5
  efficiency function 147–8
  excitation/ionization function
    calibration function 136–7
    calibration function, analytical 137–46, *141, 142*
    electron density measurement 132–3
    plasma temperature measurement 129–32, *131*
    thermodynamic equilibrium 127–9
    thermodynamic equilibrium, evaluation 133–6
  method 5–9
    images of plasma formation *6*
    physics and chemistry of laser plasma 9–13, *10,
      11, 12,* **13**
    plasma formation in gases, liquids and solids
      13–16, *15, 16*
    spectrum evolution during cooling *8*
    time phases *7*
    typical apparatus *6*
  stochastic ablation 125–7
charge-coupled-devices (CCDs) 35
  direct plasma imaging 45–6, *46, 47*
  time-integrated morphology 83–9, *84, 85, 86, 87,
    88, 89*
  plasma imaging 43–4
  spectrometer slit plasma imaging *44*
charge-injection devices (CIDs) 35
chemical imaging of surfaces 254–5, 277–8
  depth profiling 264–6, *265, 266*
  mapping 266–77, *268, 270, 273, 276, 267*
  operational modes 255–8, *257*
  scan analysis 262–4, *263*
  spatial resolution 258
    depth resolution 261–2, *262*
    lateral resolution 258–9
    surface sensitivity 259–61, *260*
chemical maps 65, 266–77, *267, 268, 270, 273, 276*
civilian and military environmental contamination
      studies 368–9, 396
  calibration 375–8
    lead (Pb) *376, 377, 378*
  containment in liquid 387–8
    double laser pulse excitement *394*
    limit of detection **393**
    liquid configuration 388–90, *389,* **390**
    signal enhancement 390–6, *391, 392, 395*
  continuous emission monitor 382–7, *384, 386, 387*
  DIAL LIBS system 381–2, *382*
  field surveys 372
    Sierra Army Depot (SIAD) 372–5, *373, 374, 375*
  field-portable LIBS system 370–2, *371*
  geological applications
    desert varnish 378–80, *379, 380*
    hydrothermal alteration 380–1, *381*
cloud condensation nuclei (CCNs) 196
compact probe analysis LIBS *18,* 19
cultural heritage applications 332–3, 363
  analytical parameters and methodology 338–9
    irradiation–detection parameters 339–41, *340*
    material parameters 339
    qualitative analysis 341–3, *342*
    quantitative analysis 343–4

art and analytical chemistry 333
    techniques employed **334**
art and archaeology 344
    biological samples 356
    laser cleaning 356–7, *358*
    marble, stone, glass, and geological samples
      353–4
    metals 354–6
    metals *355*
    pigments 344–52, **346**, *348, 349, 350, 351, 352*
    pottery 352–3, *353*
  combination techniques 357
    fluorescence spectroscopy and LIBS 359–61,
      360, 361
    mass spectrometry and LIBS 362, 357–9, *359*
  instrumentation 336–8, *336, 338*
  physical principles 335–6
  rationale for LIBS 333–5
Curve of Growth (COG) method 137–46, *141, 142*
Czerny–Turner spectrometers 563, *563*

Debye sphere 132
diatomic molecular emission spectra 177–9, *178*
diode-pumped solid-state lasers (DPSSL) 491–4, **491**,
      492, 493, 494
direct analysis LIBS 17

echelle spectrometers *33,* 564–5, *565,* **566**
  evaluation 566
    detection limit 575–6
    electron density 571–4
    plasma diagnostics 566–74
    plasma temperature 569–71
    single shot calibration 574–5
    spectral response *569, 570*
    spectrochemical analysis 574–6
    typical spectrum *568*
efficiency function 147–8
electron density within plasmas 78, *79*
  measuring 132–3
elemental maps 72, *73*
elements, major emission wavelengths **347**
end-of-life waste electric and electronic equipment
      (EOL-WEEE) 411–12, *412, 413*
enhancement factor 454
ensemble averaging 222–7, *223, 230*
European Monitoring and Evaluating Program
      (EMAP) 201
    aerosol source categories **201**
excitation/ionization function
  electron density measurement 132–3
  plasma temperature measurement 129–32, *131*
  thermodynamic equilibrium 127–9
  thermodynamic equilibrium, evaluation 133–6

fiber optic delivery LIBS 17–19, *18*
fiber techniques
  single fiber scanning 48, *49*
    time-integrated morphology 94–5
    time-integrated morphology, long laser pulses
      110–11

time-resolved fiber-assisted simultaneous
   spectroscopy 48–50, *50*
   time-integrated morphology 95–102, *97, 103*
finite difference equations (FDE) 188–9
fluid dynamic simulations 186–9, *187*
fluorescence spectroscopy and LIBS, applications
   359–61, *360, 361*
focal volume irradiance distribution 173–6, *174*
Fourier transform plasma imaging spectroscopy 55–6,
   56
   time-integrated morphology *69*, 70–3, *71, 73*
Franck–Condon factor 179
Fresnel–Kirchoff integral 174–5
full-width at half-maximum (FWHM) 176

gases, LIBS experiments and simulations 171–2,
   189–90, 209–17
   computational fluid dynamic simulations 186–9,
     187
   dependence upon laser pulse width *214*
   diatomic molecular emission spectra 177–9, *178*
   focal volume irradiance distribution 173–6, *174*
   hydrogen Balmer series atomic spectra 176–7,
     177
   laser-induced ignition 172–3
   NEQAIR simulations 179–82, *181*
     experimental details 182
     results and simulations 182–6, *183*
geological materials, LIBS applications 353–4
   desert varnish 378–80, *379, 380*
   hydrothermal alteration 380–1, *381*
giant pulse laser 5
glass identification, LIBS applications 415–17, *415,
   416*, **417**

high temperature combustion sources (HTCS) 202
high-pressure liquid chromatography (HPLC),
   pharmaceutical applications 317–18
   advantages and disadvantages **318**
high-speed, high-resolution LIBS 490–1
   diode-pumped solid-state lasers (DPSSL) 491–4,
     491, *492, 493, 494*
   laser-induced crater geometry 510–13, *512*
   scanning LIBS 498
     detection of light elements 507–10, *509, 510*
     maximum plasma generation frequency 503–4,
       504, 505
     principle component analysis 510, *511*
     results 505–7, *506, 507, 508*
     set-up 498–500, *499, 500, 501, 502*
     synchronization 502–3
     state of the art 494–8, *495, 496*
history of LIBS 1, 36
   discovery 4–5
   timeline 5
   publications over past 11 years *2*
holography, double-pulse 56–7
   time-integrated morphology 108
Hönl–London factor 179
human and animal tissue, LIBS analysis 295–7
hydrogen Balmer series atomic spectra 176–7, *177*

industrial applications 400
   metals and alloys processing 400–1
     liquid steel analysis 406–9, *409, 410*, **410**
     pipe fittings 401–4, *402, 403*
     slag analysis 404–6, *405*, **406**, *407*
   miscellaneous applications
     analysis of steel pipes at high temperature 435–6
   nuclear power generation and spent fuel
     reprocessing 417–18
     fiber optic probe instruments 418–27
     telescope LIBS instruments 427–35
   scrap material sorting and recycling 409–10
     glass identification 415–17, *415, 416*, **417**
     polymer identification 411–15, *412, 413*
instrumentation for LIBS
   components 30
     detectors 34–6
     laser systems 30–1
     spectral resolution 31–2, *33, 34*
   early instruments 27
     field-portable instruments *28, 29*
     industrial instruments 29–30
     transportable instruments 27–8, *28*
interaction function 125–7
inverse bremsstrahlung (IB) 211
ionic temperature within plasmas 78–9, *79*
irradiation wavelength, effect on plasma emission 63

laser indentification of fittings and tubes (LIFT) 401,
   403
laser-induced ignition 172–3
laser-supported combustion (LSC) 10, 209
laser-supported detonation (LSD) 10, 209
lens effect 70
lens-to-sample distance (LTSD) 22–3, 90–1
line broadening 63–4
liquid crystal (LC) tunable filters 47
   time-integrated morphology 113–14, *115, 116,
     117*
   time-integrated morphology, medium laser pulses
     89–94, *94*

mass spectrometry and LIBS, applications 362
matrix effects on LIBS 21–2, *34*, 89, *89*, 97–9, *101*
   chemical matrix 22
   physical matrix 22
   quantitative analysis 155–7
     calibration-free technique 160–4
     correction methods 157–60
McWhirter criterion 128
metals and alloys, LIBS applications 354–6, 400–1
   liquid steel analysis 406–9, *409, 410*, **410**
   pipe fittings 401–4, *402, 403*
   slag analysis 404–6, *405*, **406**, *407*
metals, LIBS applications *355*
micro LIBS technique 539–43, 554
   experimental set-up 543, *544*
     detection unit 546
     laser 543–4
     microscope 544–6
     procedure 546

micro LIBS technique (*cont.*)
  results
    ceramics 547–51, *548*, *549*
    fluid inclusions in minerals 552–4, *553*
    lithium in minerals 552, *553*
    steel 551–2, *551*
multiphoton absorption and ionization (MPI) 210

Nd:YAG lasers 29–31
near-infrared spectroscopy (NIR), pharmaceutical
    applications 318–19, **319**
NEQAIR (non-equilibrium air radiation) computer
    program 177–8, 179–82, *181*
  experimental details 182
  results and simulations 182–6, *183*
nonresonance laser-induced fluorescence 57–8
  time-integrated morphology 111–12
non-uniform magnetic fields 86
nuclear power generation and spent fuel reprocessing,
    LIBS applications 417–18
  fiber optic probe instruments 418
    AGR pressure vessel 424–7, *426*
    AGR superheater tubes 420–4, *422*, *423*, *424*,
      425
    batch identification of Magnox reactor control
      rods 418–19, *419*
    trace metal analysis of uranium metal fuel 420
  telescope LIBS instruments 427
    contamination analysis of high-level waste
      containers 432–5, *434*, *435*
    cooling water tubes 428–30, *430*, *431*
    non-invasive contamination analysis 430–2, *432*
    oxide thickness inside Magnox reactors 427–8,
      429

one-dimensional imaging 65

Paschen–Runge spectrometers 563–4, *564*
pharmaceutical analysis applications 314–16, 330–1
  blend and tablet uniformity 323–7, *325*, *326*, *327*
  dose strength identification 328, *329*
  drug component mapping 328–30, *329*
  excipients used in pharmaceutical manufacture **320**,
    321
  film coating uniformity 328
  industry needs 316–17
  instrumentation 319–22, *322*
  LIBS compared to current technologies 317–19
    advantages and disadvantages **318**
  signal characteristics 323, *324*
photodiode arrays (PDAs) 34–5
  direct plasma imaging 45–6, *46*
photography, high-speed 42
  time-integrated morphology 110
photomultiplier tubes (PMTs) 34–5
pigments, LIBS applications **346**, *348*, *350*, *351*, *352*
pipe fittings 401–4, *402*, *403*
plant tissue, LIBS analysis 297–301, *299*, *300*
plasma, primary region 74
plasma, secondary region 74

plasma formation 6–7, *6*
  physics and chemistry of laser plasma 9–13, *10*
    air plasma electron density *12*, **13**
    air plasma temperature *11*
  sample state 13
    gases 14
    liquids 14
    particles 14
    solids 14–16, *15*, *16*
  spectrum evolution during cooling *8*
  time phases *7*
plasma morphology 40–1, 118
  experimental imaging techniques 41–2
    acousto-optical scanning of laser-enhanced
      ionization 58–9
    direct CCD imaging 45–6, *46*, *47*
    double-pulse holography 56–7
    Fourier transform imaging spectroscopy 55–6,
      56
    high-speed gated X-ray cameras 53–4, *54*
    high-speed photography 42
    monochromatic imaging spectroscopy 43
    near resonant photographic absorption/shadow
      imaging 54–5, *55*
    nonresonance laser-induced fluorescence 57–8
    resonance laser-induced fluorescence 57, *58*
    single fiber scanning 48, *49*
    slitless spectroscopy 42–3
    spectrometer slit imaging 43–4, *44*
    streak cameras 50–2, *51*, *52*, *53*
    time-resolved fiber-assisted simultaneous
      spectroscopy 48–50, *50*
    time-resolved imaging through tunable filters
      46–8, *48*, *49*
  time-integrated morphology 59
  time-integrated morphology, double laser pulses
    113, *114*
    time-resolved imaging through tunable filters
      113–14, *115*, *116*, *117*
  time-integrated morphology, long laser pulses 110
    high-speed photography 110
    nonresonance laser-induced fluorescence 111–12
    single fiber scanning 110–11
    spectrometer slit imaging 110
  time-integrated morphology, medium laser pulses
    73–4
    direct CCD imaging 83–9, *84*, *85*, *86*, *87*, *88*, *89*
    double-pulse holography 108
    high-speed gated X-ray cameras 104
    near resonant photographic absorption/shadow
      imaging 104–7
    resonance laser-induced fluorescence 108–9, *109*
    single-fiber scanning 94–5
    spectrometer slit imaging 74–81, *75*, *76*, *77*, *78*,
      82
    streak cameras 102–4
    streak cameras *104*
    time-resolved fiber-assisted simultaneous
      spectroscopy 95–102, *97*, *103*
    tunable filters 89–94, *94*

time-integrated morphology, short laser pulses
112–13
Fourier transform imaging spectroscopy *69,
70–3, 71, 73*
monochromatic imaging spectrometry 66–8, *67,
68*
spectrometer slit imaging 59–65, *61, 62, 63, 64*
plasma temperature 63, 78, *78*
measuring 129–32, *131*
pollen, LIBS analysis 304–5, *306, 307*
polymer waste identification, LIBS applications
411–15, *412, 413*
pottery, LIBS applications 352–3, *353*
principles of LIBS 1–3
atomic emission spectroscopy (AES) 3–4
prisms for spectral resolution 32

Q-switched laser 5
quantitative analysis 122–3, 148, 164–5
calibration curves 150–1
correction methods for shot-to-shot fluctuations
151–2
single shot analysis 152–3
conditional analysis 155
detection limits 153–4
internal standard methods 149
matrix effects 155–7
calibration-free technique 160–4
correction methods 157–60
multivariate analysis 154–5

radial emission distribution 84
Raman microscopy and LIBS, applications 357–9, *359*
red blood cells 470–2, *471, 472*
repetitive single spark (RSS) technique 9
repetitive spark pair (RSP) technique 9
resonance laser-induced fluorescence 57, *58*
time-integrated morphology 108–9, *109*
resonance-enhanced LIBS 440–1, 473–4
gaseous samples 473
liquid samples 463
analytical performance 468–70, **469**, *470*
ArF laser ablation of aqueous samples 465–8,
466, **467**
photoresonant pumping of water molecules
463–5, *464*
sodium and potassium in single red blood cells
470–2, *471, 472*
summary 472–3
RELIPS analysis of solids 451
experiments 451–3, *452, 453*
important experimental parameters 454–6, *455*
multielement analysis 456–8, *458, 459, 460*
resonance enhancements 453–4
summary 461–3
time-resolved characterization of plasma plume
458–61, *462*
spectrochemical excitation 441
conceptual summary of RELIPS 450–1
continuum emissions 441–2

laser-induced cool plasmas 447–9
line emissions 442–3
line-to-continuum ratio 444–5
maximum line-to-continuum ratios 441–5
suitability 446–7, *448*
sustaining non-thermal plasmas 449–50

Saha equation 158, 442, *443*
Saha–Boltzmann method 133
Saha–Boltzmann relation 131–2
sampling geometry 90
Sedov equation 99
self-absorption 59, 64–5, 124
Semi-Implicit Method of Pressure-Linked Equations
Consistent (SIMPLEC) algorithm 188–9
sequential pulse LIBS 516–17, 532–3
dual-pulse LIBS
ablation followed by re-excitation 517–20, *517*
aqueous solution 530–2, *531, 532*
enhancement vs. single-pulse performance **518**,
520, 524
nanosecond and femtosecond pulses 527–30,
528, 529
possible mechanisms for enhancement 526
pre-ablation spark dual-pulse LIBS 520–6, *521,
522, 523, 525*
pre-ionization 526–7
results with reheating configuration 530
optimization **533**
short-pulse LIBS 477, 483–7, 487–8
effect of pulse duration on ablation 478–9
effect of pulse duration on plasma 479–80, *480*
femtosecond-induced electron plasma 482–3, *484*
picosecond-induced electron plasma 480–2, *481,
482, 483*
shot-to-shot fluctuations, correction methods 151–2
silicon plasma 80, *80*
similarity maps 70–2
singal-to-noise maps 72
spark-induced breakdown spectroscopy (SIBS)
585–6
applications 599
aerosol metals monitoring 590–7
aerosol monitoring hardware and methods
590–1, *591, 592*
compound sensitivity 611
dry aerosol calibration system 591–2, *593*
interference measurements 596–7
lead in airborne particulates 600–2, *600, 601*
metals monitoring in soil 597–9
potential interference from organic
contamination 611–12
soil monitoring hardware and methods 597–9,
598
soils 602–10, *603, 604, 605, 606, 607, 608, 609,
610*
speciation, particle size and realistic samples
593–6, *594, 595*
spectral surveys 599
future directions 613–14

spark-induced breakdown spectroscopy (SIBS) (*cont.*)
  processes and hardware 586, *587, 588*
    detection systems 588–9
    spark characteristics 586–8
spectral detectors for LIBS 556–8, 581–2
  assessment 576–7
    advantages 577–8
    limitations 578–80, *579*
  evaluation of echelle spectrometers 566
    detection limit 575–6
    electron density 571–4
    plasma diagnostics 566–74
    plasma temperature 569–71
    single shot calibration 574–5
    spectral response *569, 570*
    spectrochemical analysis 574–6
    typical spectrum *568*
  multidetection
    desirable requirements 558
    image intensifier 561–2, *562*
    interline CCD detectors 560–1, **561**
    photomultiplier detectors 558–9, *559*
    solid-state detectors 559–60
    spectrometers 562–5
  optical set-up 580
    comparison of ICCD and interline CCD 580–1
    comparison of Paschen–Runge/PMT and
      eschelle spectrometer/ICCD 580
spectral maps 72
spectrometry
  monochromatic imaging spectrometry 66–8, *67, 68*
  slit imaging 43–4, *44*

time-integrated morphology, long laser pulses
  110
time-integrated morphology, medium laser
  pulses 74–81, *75, 76, 77, 78, 82*
time-integrated plasma morphology 59–65, *61,
  62, 63, 64*
spectroscopy
  Fourier transform plasma imaging 55–6, *56*
  monochromatic imaging 43
  slitless plasma imaging 42–3
  time-resolved fiber-assisted simultaneous
    spectroscopy 48–50, *50*
stand-off analysis LIBS *18*, 19
Stark effect 132
Stefan–Boltzmann law of black-body radiation 100
stone, LIBS applications 353–4
streak cameras 50–2, *51, 52, 53*
  time-integrated morphology 102–4, *104*

teeth, LIBS analysis 285–6, *286*
  determination of carious and healthy tissue 288–90,
    289
  spatial mapping of elemental content 286–8, *287,
    288*
  tracing environmental influences 290–1
    tooth pastes *291*
transport equations **188**
two-dimensional imaging 65

X-ray cameras 53–4, *54*
  time-integrated morphology 104
X-ray fluorescence (XRF) spectrometry 404